# ATOMIC AND MOLECULAR RADIATION PHYSICS

L. G. Christophorou, *Head of Atomic and Molecular Radiation Physics Group, Oak Ridge National Laboratory, and Professor of Physics, University of Tennessee*

## CONTENTS

*Atomic*
*and Molecular*
*Radiation Physics*

# Atomic
# and Molecular
# Radiation Physics

**L. G. Christophorou**

*Head of Atomic and Molecular Radiation Physics Group,*
*Oak Ridge National Laboratory, and Professor of Physics,*
*University of Tennessee*

## WILEY–INTERSCIENCE

a division of John Wiley & Sons Ltd

London   New York   Sydney   Toronto

Library of Congress catalog card number
72–129159
ISBN 0 471 15629 9

Printed in Belgium by Ceuterick
Brusselse straat 153   3000-Louvain
Adm.-dir. L. Pitsi
Bertemse baan 25   3008-Veltem-Beisem

TO MY WIFE ERATOULA

# Preface

Many physical processes take place when ionizing radiation interacts with matter. The understanding of the various dissipative processes requires studies on the atomic and molecular level. The discussions in this book are on matter rather than on radiation and concentrate mainly on aspects concerned with the interaction of radiation with single atoms and single molecules. Phenomena associated with radiation interaction under multiple collision conditions in gases are discussed also. These form the transition from single particle behaviour to that of the liquid and the solid. Interaction of radiation with liquids and solids is discussed very briefly.

Special emphasis is accorded to photophysical and low-energy electron-atom (molecule) interaction processes. Such studies have realized an impressive progress recently mainly as a result of new sophisticated instrumentation. This progress has enriched our understanding of atoms and molecules and has had an impressive impact on other fields of pure and applied science.

Certain of the processes discussed are by no means completely understood. On the contrary, there is need for further and deeper understanding. Our knowledge of polyatomic molecules, in particular, is less accurate than for atoms and simple molecules, and of a more qualitative and descriptive character. However, a wealth of information on polyatomic molecules has appeared during the last decade or so. This, in part, is the result of the growing interest of the physicist in polyatomic systems and also of the prevailing and overwhelming opinion that answers to basic problems in Radiation and Life Sciences are to be searched for both on the atomic and molecular level.

The book is aimed at the postgraduate student and the researcher in the field. It is divided into nine chapters. The number of chapters is kept purposely small as we are keen in giving a coherent picture of many seemingly diverse physical phenomena. In Chapter 1 the slowing-down of ionizing radiation in matter and in Chapter 2 the total ionization produced in gaseous media by the complete absorption of charged particles are discussed. Both of these chapters help to show the need for a detailed study of the many and distinct processes which accompany the interaction of ionizing radiation with matter. This is attempted in the subsequent six chapters: Photophysical Processes (Chapter 3); Elastic Scattering of Slow Electrons (Chapter 4); Inelastic Electron Scattering (Chapter 5); Negative Ions (Chapter 6);

Electron Detachment from Negative Ions and Electron Affinity of Atoms and Molecules (Chapter 7); Heavy Neutral and Charged Particle Interactions (Chapter 8). In the last chapter (Chapter 9) two specific areas, biophotophysics and bioelectronics, are discussed in an effort to show the need for a better physical understanding of biological problems and a wider interdisciplinary interaction between the Physical and Life Sciences.

I wish to express my sincere thanks to many of my colleagues and reseach students at the Oak Ridge National Laboratory and the University of Tennessee for their help and advice and for reading parts of the manuscript:

| | |
|---|---|
| V. E. Anderson | J. Kirby |
| R. D. Birkhoff | C. E. Klots |
| R. P. Blaunstein | D. L. McCorkle |
| J. G. Carter | P. McGill |
| E. L. Chaney | P. Pisanias |
| A. A. Christodoulides | D. Pittman |
| P. M. Collins | R. H. Ritchie |
| C. E. Easterly | H. C. Schweinler |
| K. Gant | J. E. Turner |
| J. A. Harter | |

I am also indebted to Dr. J. B. Birks of the University of Manchester and to Dean A. H. Nielsen of the University of Tennessee for their encouragement and moral support.

Many thanks are due to Mrs. Norma Brashier and Mrs. Beverly Varnadore who typed the entire manuscript and greatly helped in organizing the tables and the references. I also acknowledge the excellent work of the Graphic Arts Department of the Oak Ridge National Laboratory.

Finally, it is with a deep debt of gratitude that I thank my wife for her understanding and invaluable help.

L. G. CHRISTOPHOROU

*Oak Ridge, Tennessee*
*November 1970*

# Contents

# 1 Slowing-down in matter of ionizing radiation

## 1.1 Introduction

### 1.1.1 *A Definition*

Radiation Physics deals with the physical action of ionizing radiation and the phenomena that accompany this action. It treats the energy exchanges between radiation and matter.

### 1.1.2. *Principal Types of Radiations*

Two broad groups of radiations can be distinguished with respect to origin: (i) atomic and (ii) nuclear. Atomic radiations originate from the electronic system of the atom (and molecule) and include photons (visible and ultraviolet light and x-rays) and electrons. Nuclear radiations include neutrinos, electrons, mesons, nucleons, hyperons, light and heavy ions, and electromagnetic radiations such as x-rays and $\gamma$-rays. Nuclear radiations from radioactive materials are $\alpha$ particles, $\beta$ particles (accompanied by neutrinos) and $\gamma$-rays. Nuclear radiations are sometimes accompanied by atomic radiations (e.g. x-rays following $K$-electron capture by the nucleus, and internal conversion electrons associated with $\gamma$-ray emission).

We shall divide the atomic and nuclear radiations into three broad types: charged particles, electromagnetic radiation, and neutrons. The interaction of charged particles and electromagnetic radiation is predominantly with the atomic electrons. Neutrons differ in this respect in that their primary interactions are with the nucleus rather than with the electronic system of the atom. No discussion will be made of neutron interactions. It might be mentioned, however, that neutron scattering (producing a recoil nucleus) and neutron absorption (producing an excited nucleus which decays by $\gamma$-ray or charged-particle emission) are the two main types of neutron interaction with matter.

Atomic and nuclear radiations constitute not only essential tools for the exploration of matter, but they have become of increasing importance in life sciences. They undoubtedly help to unravel the secrets of life in the same way they have helped unravel the secrets of the atom and its nucleus.

### 1.1.3 *Primary and Secondary Processes in the Action of Ionizing Radiation with Matter*

As stated in the previous section, in the action of all kinds of ionizing radiation with matter (with the exception of fast neutrons) the transfer of energy to the electronic system of individual atoms and molecules constitutes the most important primary process. The primary products, therefore, are mainly excited and ionized atoms or molecules and electrons liberated in the ionization process. In condensed media collective excitations become important also. The spectrum of the primary activations is called the *excitation spectrum.*

In the slowing-down process a large number of secondary particles are quickly generated by the primary (incident) radiation. These are mostly electrons of all energies (for an incident monoenergetic electron beam from the primary energy down to zero), x-ray photons, ions, radicals, etc. It has become increasingly clear recently that the kinetic energy of the secondary electrons which are released upon ionization, constitutes the largest portion of the energy which is deposited in a medium. Further, the major damage to the medium results from the secondary electrons which enter, subsequent to their production, into exciting and ionizing collisions. The energy spectrum of the charged particles actually present in the irradiated medium is called the *degradation or slowing-down spectrum* (Section 1.2.3).

The primary products are the precursors of all observed consequences of radiation absorption. Usually, no distinction is made between the effects of the incident particle and those due to the secondary particles. However, virtually all final effects (especially excitation) are produced by the secondary electrons.

Platzman (1962, 1967) distinguished four temporal stages in the response of a medium to irradiation: (i) *the physical stage* in which the absorbed energy is degraded to the atomic level producing a number of primary products; (ii) *the physicochemical stage* in which the primary products undergo rapidly a large number of high- and low-energy secondary reactions (the sequence of events which leads from the primary products to the formation of typical chemical species is the least understood stage in the interaction of ionizing radiation with matter); (iii) *the chemical stage* in which reactive species such as free radicals and ions attained thermal equilibrium and undergo various reactions; and (iv), for a biological system, *the biological stage* during which the organism responds to the products of irradiation.

In subsequent chapters we shall be concerned mainly with the processes which occur during the first and second temporal stages. In Table 1.1 we list the main sequence of the predominant processes taking place when a swiftly moving charged particle slows-down in matter. It should be noted that the energy ranges indicated in Table 1.1 are not confined, but overlap considerably in both energy directions. It is seen that the slowing-down

**Table 1.1** Main sequence of predominant processes taking place when a swiftly moving charged particle slows-down in matter.

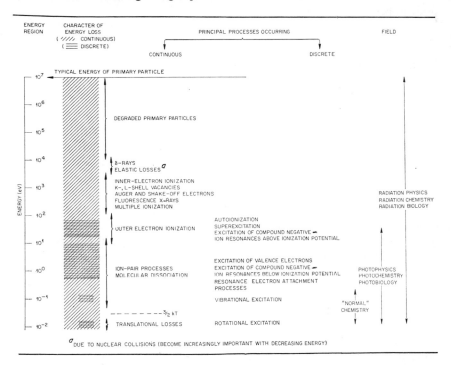

(absorption) in matter of ionizing radiation produces a large number of complex and diverse sequences of events at the atomic and molecular levels. Although the major work in the field has dealt with the gross effects and not directly with the details of energy deposition, a great deal of new information has been obtained in recent years.

Special emphasis will be accorded to the interactions of *subionization* and *subexcitation* electrons with atomic and molecular gases. Subionization electrons have insufficient kinetic energy to cause ionization, but sufficient to induce electronic transitions. Electronic excitation by electrons having energies very close to the excitation thresholds is of great importance in revealing new features of atomic and molecular structures. Subexcitation electrons attain kinetic energies which are below the energy of the first excited state of the medium. Thus, these electrons do not lose energy to electronic excitation. Their moderation takes place by elastic collisions (Chapter 4), excitation of rotational, vibrational and compound negative ion states (Chapters 4 to 6) and also through the various processes of electron attachment (Chapter 6). In the absence of strong energy losses to excitation of compound negative ion states and electron attachment, subexcitation electrons may have a prolonged existence as transient species, prior to being thermalized.

## 1.2 Slowing-Down in Matter of Charged Particles

### 1.2.1 *Energy Loss*

Heavy charged particles passing through matter have essentially straight paths—apart from the rare event of a nuclear collision where a large-angle scattering occurs—and slow-down in an almost continuous manner. Their behaviour is complicated in that they may be excited and ionized, and in that they may capture electrons from the atoms or the molecules of the medium. Bremsstrahlung is not an important energy loss mechanism for heavy charged particles. For electrons, except at high energies where bremsstrahlung is the dominant energy loss process, moderation occurs almost entirely through energy losses to the electronic system of the atoms and the molecules making up the medium.

#### 1.2.1.1 Semi-classical treatment

The characteristic structure and features of the theory of the slowing-down in matter of charged particles appeared in the semi-classical treatment of energy loss of Bohr (1913, 1915).

Consider the time dependence of the electric field vector $\mathbf{E}$ and its two components, the transverse ($\mathbf{E}_\perp$) and the longitudinal ($\mathbf{E}_\parallel$), due to the motion of a structureless charged particle of mass $M$, charge $ze$, and velocity $v$ ($= \beta c$; $c$ = speed of light in vacuum) passing at a distance $b$ (impact parameter) from an atom (or a molecule) at a point $P$. Figure 1.1 (A) illustrates the process. Let us now choose the time scale so that the particle passes closest to $P$ at time $t = 0$. If it is assumed that the particle's velocity is non-relativistic and that the impact parameter is large compared to the atom's (or molecule's) dimensions, the collision is a 'glancing' one, and the electric field at any time is essentially uniform throughout the atom (or molecule) at $P$.

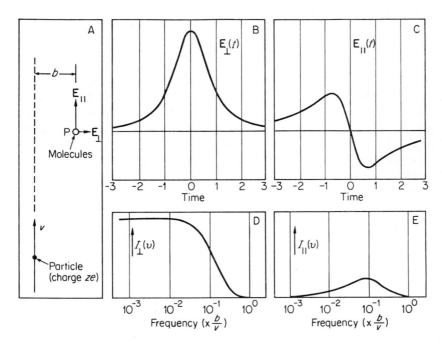

**Figure 1.1** A distant collision between a swiftly-moving charged particle and an atom or molecule at point $P$: A particle trajectory; B and C variation of the two $[E_\perp(t), E_\parallel(t)]$ electric-field intensity components during the collision; D and E spectrum of virtual photons; intensity $I(\nu)$, versus frequency $\nu$. Note that the unit of time is $b/v$ and the frequency unit $v/b$ (Platzman, 1962).

The electric field experienced by the atom at $P$ is independent of the particle's mass $M$ and is appreciable for only a short time interval. The two components of this 'impulsive' electric field, perpendicular and along the direction of the charged particle, are

$$E_\perp = \frac{zeb}{(v^2 t^2 + b^2)^{3/2}} \tag{1.1}$$

and

$$E_\parallel = \frac{zevt}{(v^2 t^2 + b^2)^{3/2}} \tag{1.2}$$

Since $\mathbf{E}_{\parallel}(t) = -\mathbf{E}_{\parallel}(-t)$ the only momentum change is that due to the $\mathbf{E}_{\perp}$ component. The time dependence of $\mathbf{E}_{\perp}$ and $\mathbf{E}_{\parallel}$ is shown in Figure 1.1 (B and C). If, for simplicity, we assume an electron (rather than an atom) located at $P$ the energy lost by the charged particle (and imparted to the electron) as a result of the collision is

$$\Delta E = \frac{(\Delta p)^2}{2m} = \frac{(\bar{F}\Delta t)^2}{2m} \simeq \frac{2z^2 e^4}{mv^2 b^2} \tag{1.3}$$

where $\bar{F}$ is the average force which acts on the electron of mass $m$ at right angles to the particle's path, $\Delta t$ (taken equal to $b/v$) is the duration of the collision and $\Delta p$ is the momentum transferred to the electron. The cross section per electron may appropriately be defined in terms of the electron impact parameter $b$.

Now, either field component ($\mathbf{E}_{\parallel}$ or $\mathbf{E}_{\perp}$) can be represented mathematically as a unique sum (integral) of contributions, each of which is a purely harmonic function of time. Each of the individual contributions $E_{\omega}(t)$ is equivalent to the electric field experienced by the molecule when it is traversed by a light wave of frequency $\nu = \omega/2\pi$. Thus, the effect of the collision of a charged particle on the molecule is equivalent to that sustained by the molecule if it were traversed by a beam of photons, often referred to as 'virtual photons'. To find the frequencies of these 'virtual photons' we take the Fourier components of $E_{\perp}(t)$ viz (Jackson, 1963)

$$E_{\omega\perp} = \frac{1}{2\pi} \int_{-\infty}^{+\infty} E_{\perp}(t)\, e^{i\omega t}\, dt \tag{1.4}$$

from which we have

$$E_{\omega\perp} = \frac{ze}{\pi bv} \left[ \frac{\omega b}{v} K_1\left( \frac{\omega b}{v} \right) \right] \tag{1.5}$$

where $K_1(x)$ is a modified Bessel function of the second kind; $K_1(x) \to 1/x$ for $x \to 0$. Hence,

$$E_{\omega\perp} \simeq \frac{ze}{\pi bv} \tag{1.6}$$

for $\omega \ll v/b$; $E_{\omega\perp} = 0$ for $\omega \gtrsim v/b$. Now, the total energy, $U$, associated with the electromagnetic waves is

$$U = \int_0^{\infty} U_{\omega}\, d\omega \tag{1.7}$$

where $U_\omega$ is the energy of the 'virtual photons' per unit angular frequency $\omega$ given by (for $\omega < v/b$)

$$U_\omega = \frac{2z^2 e^2}{\pi v} \ln\left(\frac{v}{\omega b_{min}}\right) \tag{1.8}$$

where $b_{min}$ is the minimum impact parameter.

Since $U_\omega d\omega = \hbar\omega N_\omega d\omega$, we obtain for the number of photons with energy between $\hbar\omega$ and $\hbar(\omega + d\omega)$ the expression

$$N_\omega = \frac{U_\omega}{\hbar\omega} = \frac{2z^2 e^2}{\pi \hbar v} \ln\left(\frac{v}{\omega b_{min}}\right) \frac{1}{\omega} \tag{1.9}$$

or

$$N_\omega \approx \text{const.} \frac{1}{\hbar\omega} \tag{1.10}$$

In most of the collisions involving ionizing radiation

$$\omega \times \frac{b}{v} \ll 1 \tag{1.11}$$

i.e. the frequency $v/b$ exceeds greatly the characteristic frequencies of the absorption spectrum. Therefore, the intensity of the 'virtual photons' (proportional to $N_\omega \omega$) is essentially constant over the range of frequencies normally involved in the excitation of atoms and molecules.

The intensities $I(v)$ of the 'virtual photons' corresponding to the field components $\mathbf{E}_\perp$ and $\mathbf{E}_\parallel$ are plotted in Figs. 1.1(D) and 1.1(E) (Platzman, 1962), respectively, as a function of the frequency multiplied by $b/v$. It is seen that for $I_\perp(v)$ the energy of the photons per unit frequency-interval is a constant from zero frequency up to about the 'cut-off' frequency $v_{max} = v/b$ and it declines sharply to zero as $v$ approaches $v/b$. On the other hand, $I_\parallel(v)$ is much less intense and has a distribution concentrated just below $v_{max}$. Hence for our purpose $I(v) \simeq I_\perp(v)$ is essentially constant, and the number of 'virtual photons' per unit interval, $N_\omega$, has the distribution given by expression (1.10). The features of (1.10) are essentially retained when the collisions are averaged over all effective values of $b$.

From the preceding discussion it is apparent that the effect of a swiftly-moving charged particle on an atom or molecule located at $P$ may be approximated by that of a beam of white light having the spectrum given by Equation (1.10). Each of these photons induces electronic transitions $n$, in proportion to the optical oscillator strength $f_n$ of the transition (Chapter 3). Thus, to a first approximation the number, $N_n$, of primary products in the excited state $n$, formed by the action of the swiftly-moving charged particle is

$$N_n \simeq \text{const.} \times \frac{f_n}{E_n} \qquad (1.12)$$

where $E_n$ is the energy of the particular primary product. Equation (1.12) is the *optical approximation* to the number of primary products produced by the ionizing particle.

Since in this approximation electronic transitions occur in proportion to $f_n/E_n$, optically-allowed and optically-forbidden transitions are equally allowed or forbidden, i.e. the excitation products are formed with probabilities corresponding to those of the optical absorption spectrum (Chapter 3). Restricting ourselves to high particle velocities, the optical approximation can be used to calculate the number of primary products with reasonable accuracy. Such calculations, however, are limited to transitions having large oscillator strengths, otherwise no information on low $f_n$ transitions can be obtained. A number of calculations in this approximation have been made for simple systems such as H, He, $H_2$ (e.g., Platzman, 1962).

As will be discussed in Chapter 3 the total oscillator strength for transitions originating in a given atomic shell of electrons is roughly the same as the number of electrons in that shell. Now, since in the optical approximation the primary products are formed in proportion to $f_n/E_n$ and $E_n$ is much greater for inner-shell electrons, the excitation and ionization of valence electrons predominates over that of the inner-shell ones. It might be noted that those excited states of the valence electrons which have the largest oscillator strengths lie at comparatively high energies. Our knowledge on these high-lying transitions is very little. This is because at the high excitation energies involved, processes such as dissociation, predissociation, preionization, and internal conversion complicate the experiments which are, in the first place, difficult to perform in this vacuum ultraviolet region (Chapter 3).

### 1.2.1.2 Bethe's quantum mechanical treatment

The non-relativistic expression for the differential cross section for scattering of an incident ion from a state of initial momentum $\hbar k_0$ to one of a final momentum $\hbar k$ with simultaneous excitation of the atom from its ground state 0 to an excited state $n$ in the first Born approximation is (see also Chapter 5)

$$d\sigma_n = \frac{k^2}{4\pi^2 \hbar^2 v^2} |\langle e^{ik \cdot r} \psi_n | V | e^{ik_0 \cdot r} \psi_0 \rangle|^2 d\Omega \qquad (1.13)$$

In expression (1.13) $V$ is the Coulomb interaction potential between the incident ion and the atomic electrons which can be written as

$$V = - \sum_{j=1}^{Z} \frac{ze^2}{|\mathbf{r}-\mathbf{r}_j|} \qquad (1.14)$$

where $\mathbf{r}$ is the position vector of the ion, $\mathbf{r}_j$ that of the $j$th electron, and the sum extends over all $Z$ atomic electrons. The exponential terms in Equation (1.13) are the wave functions for the free-particle eigenstates of the ion, $\psi_0$ and $\psi_n$ are the ground- and excited-state atomic wave functions, and $d\Omega$ is the solid angle in the direction of the final momentum of the scattered ion. The momentum transferred is $\mathbf{q} = \hbar\mathbf{k}_0 - \hbar\mathbf{k}$ and for an ion with mass $M \gg m$, $k \simeq k_0$. Defining $Q = q^2/2m$ the final expression for $d\sigma_n$ becomes (Turner, 1967)

$$d\sigma_n = \frac{2\pi z^2 e^4 Z}{mv^2} |F_n(\mathbf{q})|^2 \frac{dQ}{Q^2} \qquad (1.15)$$

where the inelastic form factor $F_n$ is given by

$$F_n(\mathbf{q}) = \frac{1}{\sqrt{Z}} \sum_{j=1}^{Z} \left\langle \psi_n \left| \exp\left( i \frac{\mathbf{q} \cdot \mathbf{r}_j}{\hbar} \right) \right| \psi_0 \right\rangle$$

$|F_n(\mathbf{q})|^2$ is the probability that the ion will be scattered with a momentum change $\mathbf{q} = \mathbf{K}\hbar$ and concomitant excitation of the atom (or molecule) from its ground state to the excited state $n$ (discrete or continuum). Expression (1.15) is usually written in the form (e.g. Bethe, 1930; Inokuti, Kim, and Platzman, 1967)

$$d\sigma_n = \frac{4\pi a_0^2 z^2}{E/R} \frac{f_n(K)}{E_n/R} d\ln(Ka_0)^2 \qquad (1.16)$$

where $a_0$ is the Bohr radius (0.529 Å), $R$ is the Rydberg energy (13.60 eV), $E$ is the kinetic energy of the ion, and $f_n(K)$ is the generalized oscillator strength for the transition from the ground state to the state $n$ and is averaged over degenerate substates (see Chapters 3, 4 and 5). From Equation (1.16) it is seen that if the logarithmic term is ignored, the probability of exciting a given state is proportional to $f_n/E_n$. This conclusion is identical with that reached earlier in the simple analysis of the optical approximation (Equation (1.12)).

### 1.2.2 Stopping Power of a Medium

#### 1.2.2.1 Heavy charged particles

A. *Semi-classical expression.* Consider an ion of charge $ze$ moving through a homogeneous medium of $NZ$ electrons per unit volume, where $N$

is the number density and $Z$ is the atomic number. The number of collisions per unit length at an impact parameter lying between $b$ and $b+db$ (Figure 1.2) is $2\pi NZbdb$. Thus from Equation (1.3) we find that the linear energy loss $dE(b)$ in the volume element defined above is

$$dE(b) = \frac{4\pi z^2 e^4 NZ}{mv^2} \cdot \frac{db}{b} \tag{1.17}$$

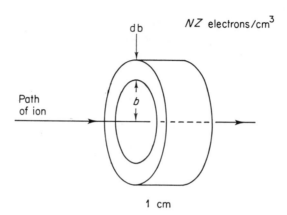

**Figure 1.2** $dE(b) = \dfrac{4\pi z^2 e^4 NZ}{mv^2} \dfrac{db}{b}$

Energy lost by the ion to the electrons, $dE(b)$, per unit distance travelled as a function of the impact parameter $b$

Integrating over all possible values of the impact parameter we have for the stopping power* $-dE/dx$, the expression

$$-\frac{dE}{dx} = \frac{4\pi z^2 e^4 NZ}{mv^2} \int \frac{db}{b} = \frac{4\pi z^2 e^4 NZ}{mv^2} \ln \frac{b_{max}}{b_{min}} \tag{1.18}$$

The (non-relativistic) limits of the integration are set as follows: denote by $b/v$ the duration of the field pulse due to the moving charged particle and by $1/\bar{v}$ the mean period of oscillation of the atomic electrons. Then, the probability of a transition to a higher state is small if $b/v$ is much longer than $1/\bar{v}$. Hence, $b_{max} \simeq v/\bar{v}$. Now, quantum mechanically a closer approach between the electron and the charged particle is meaningless if it is less than

*The stopping power of a medium for a swiftly moving charged particle is the mean rate of linear energy loss.

the de Broglie wavelength $\lambda = \hbar/mv$ of the electron (we view the collision from a system in which the charged particle is at rest). This imposes the condition

$$b > \lambda = \frac{\hbar}{mv} \quad \text{or} \quad b_{min} \sim \frac{\hbar}{mv}$$

Equation (1.18) can then be written as

$$-\frac{dE}{dx} = \frac{4\pi z^2 e^4 NZ}{mv^2} \ln \frac{mv^2}{h\bar{v}} \tag{1.19}$$

It is seen from the preceding semi-classical treatment that the charge $(ze)$ and the velocity $(v)$ of the incident particle as well as the electron density $(NZ)$ and the mean energy $(h\bar{v})$ of the stationary states of the atoms making up the medium are the only parameters which enter the stopping power formula. The last parameter $(h\bar{v})$ is defined precisely in the quantum mechanical treatment for the stopping power which follows.

B. *Quantum mechanical expression.* Using Equation (1.15) for $d\sigma_n$, the stopping power for an elemental medium is (Turner, 1967)

$$-\frac{dE}{dx} = N \sum_n (E_n - E_0) \int_{Q_{min}}^{Q_{max}} d\sigma_n = \frac{2\pi z^2 e^4 NZ}{mv^2} \sum_n (E_n - E_0) \times$$

$$\times \int_{Q_{min}}^{Q_{max}} |F_n(\mathbf{q})|^2 \frac{dQ}{Q^2} \tag{1.20}$$

In Equation (1.20) $Q$ is the energy that a free electron at rest would absorb as a result of the momentum lost by the ion. The total cross section $\int d\sigma_n$ for excitation of the atom to a state $n$, is weighted by the excitation energy $E_n - E_0$ and summed over all excited states including the continuum. With the aid of the sum-rule (Chapter 3) and the generalized oscillator strength (see Chapters 3 and 5; also Bethe, 1930, 1933) the sum over the excited states can be evaluated and the integration can be performed exactly yielding Bethe's stopping power formula

$$-\frac{dE}{dx} = \frac{4\pi z^2 e^4 NZ}{mv^2} \ln \frac{2mv^2}{I} \tag{1.21}$$

The mean excitation energy $I$ of the medium is defined through

$$\ln I = \sum_n \frac{2m}{\hbar^2} \frac{|\langle \psi_n | \Sigma_{jj} \mathbf{x} | \psi_0 \rangle|^2}{Z} \times (E_n - E_0) \ln (E_n - E_0) = \sum_n f_n \ln (E_n - E_0)$$

where $\langle \psi_n | \sum_j \mathbf{x}_j | \psi_0 \rangle$ is the dipole matrix element and $f_n$ are the optical oscillator strengths for electric dipole transitions (see Chapter 3). The term $Z \ln (2mv^2/I)$ in Equation (1.21) is a dimensionless quantity and is often called the stopping number. It also follows from Equation (1.21) and is supported by experiment (e.g., Phillips, 1953) that the stopping power should depend on the velocity of the incident particle but not on its mass. Further, the quantity $-1/(z^2)\, dE/dx$ should be independent of the incident particle's charge. This is a consequence of the Born approximation and is not expected to hold at low energies.

The main parameter, with respect to the medium, in the stopping power formula is the mean excitation energy $I$. This parameter has been calculated directly for a few simple atoms such as H and He (Table 1.2), but its estimation for other systems relies heavily on experiment. The difficulty in determining $I$ arises from the fact that the most important (intense) transitions correspond to excitation energies that lie high in the region between 10 to 1000 eV where the oscillator strengths are poorly known. However, $I$ represents an average quantity and moderately accurate values of this parameter are satisfactory for stopping power calculations. Bloch (1933) and Lindhard and Scharff (1953) using statistical models of the atom have shown that $I = \text{const.} \times Z$. From current estimates of $I$ and $I/Z$ for a number of elements shown in Table 1.2 (Fano, 1964), it is seen that $I/Z$ ranges from $\sim 20$ eV for He and molecular hydrogen, to $\sim 10$ eV for the heavy elements.

C. *Corrections to the Bethe formula.*   Three major corrections have been made to the Bethe formula for the stopping power (Equation (1.21)) to account for relativistic, inner-shell, and density effects. Inclusion of these corrections in Equation (1.21) yields for the stopping power the form

$$-\frac{dE}{dx} = \frac{4\pi z^2 e^4 NZ}{mv^2} \left( \ln \frac{2mv^2}{I} - \ln (1 - \beta^2) - \beta^2 - \frac{C}{Z} - \tfrac{1}{2}\delta \right) \quad (1.22)$$

The relativistic corrections involve the terms with $\beta$ and are introduced to account for the increase in energy loss with increasing energy. The inner-shell corrections (term $C/Z$) are introduced to account for the fact that the approximations made in the derivation of Equation (1.21) are not satisfied when the speed of the charged particle is not appreciably greater than the relativistic speeds of the inner-shell electrons. This is especially important for heavy elements. Since the ionization potentials of the $K$-shell electrons are several hundred eV even for light elements, Born's approximation ($ze^2/\hbar v \ll 1$) and especially Bethe's approximation ($E \gg M/(m)\, E_i$; $E$ is the particle's energy and $E_i$ is the ionization potential at the ($i$th) atomic electron) may not be fulfilled, and appropriate corrections need to be

**Table 1.2**    Estimates of $I$ for some elements.[a]

| Substance | $Z$ | $I^b$(eV) | $I/Z^c$ |
|---|---|---|---|
| H atomic | 1 | 15.0 (theor.) | 15 |
| H molecular | | $\begin{cases}19.0 \text{ (theor.)} \\ 18.3 \pm 2.6\end{cases}$ | 19 |
| He | 2 | $42 \pm 3$ | 21 |
| | | 41.8 (theor.) | |
| Li | 3 | 40; 38 | 13 |
| | | 45; 38.8 (theor.) | |
| Be | 4 | 64 | |
| | | 60; 66 (theor.) | 16 |
| C graphite | 6 | 81 | 13.5 |
| N molecular | 7 | 88 | 12.6 |
| O molecular | 8 | 101 | 12.6 |
| Al | 13 | 163 | 12.6 |
| Ar | 18 | 190 | 10.6 |
| Fe | 26 | 273 | 10.5 |
| Cu | 29 | 315 | 10.9 |
| Kr | 36 | 360 | 10.0 |
| Ag | 47 | $471^d$ | 10 |
| Au | 79 | $761^d$ | 9.6 |
| Pb | 82 | $788^d$ | 9.6 |
| U | 92 | $872^d$ | 9.5 |

[a]  U. Fano, Appendix A in *Studies of Penetration of Charged Particles in Matter* U. Fano (editor). Publication 1133 Nat. Acad. of Sciences — Nat. Research Council, Washington, D.C. (1964), p. 311. Recently, Turner, Roecklein and Vora (1970) reported new values of $I$ for chemical elements which do not differ significantly from those listed here.
[b]  Some of these values have been renormalized to $I_{Al} = 163$ eV or $I_{Cu} = 315$ eV.
[c]  Notice that $I/Z$ decreases fairly regularly with increasing $Z$, except for low $Z$. The high values of $I/Z$ at low $Z$ can be attributed to the 'extra-stiffness' that results from the tighter binding of valence electrons as compared to the prediction of the Thomas–Fermi model (Fano, 1964).
[d]  Values corrected for shell effects (Fano, 1964).

introduced. The last term ($\frac{1}{2} \delta$) in Equation (1.22) is introduced to account for the so-called density effect which results from the fact that at high ion velocities the atoms of the stopping medium cannot be regarded as isolated. Owing to the polarization of the medium the effective interaction between the incident ion and the material is decreased. The reduction in the field of the ion is greater the denser the medium and is negligible for gases (see further discussion in Sternheimer (1959) and Fano (1964)).

The above three corrections improve the agreement between the theory (Equation (1.22)) and experiment to within $\sim 0.5$ to 1%. We may note, however, that Equation (1.22) is not valid near the end of the ion's range since at low velocities the ion's charge $z$ is altered by electron capture and loss.* At the end of the ion's range $I$ changes also since the more tightly-bound electrons ($K$-shell electrons, for example) cease to participate in the absorption process. A summary of various formulae used to correct the stopping power was given by Fano (1964).[†]

In Figure 1.3 the stopping power (with various corrections) of water for protons is shown as a function of proton kinetic energy. In the region below $\sim 0.1$ MeV the curve was fitted to experimental data. The stopping power

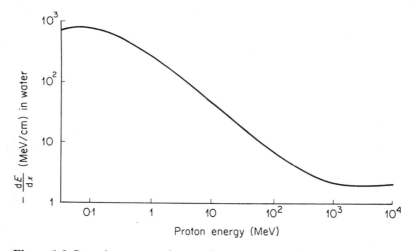

**Figure 1.3** Stopping power of water for protons as a function of proton energy (Turner, 1967)

reaches a maximum (Bragg peak) and then decreases with decreasing energy because the proton travels part of the time with no charge. The decrease in $-\mathrm{d}E/\mathrm{d}x$ at the right of the peak is mainly due to the factor $1/v^2$ in the stopping power formula. At relativistic energies $-\mathrm{d}E/\mathrm{d}x$ rises due mainly to the term $\ln(1-\beta^2)$. Shell corrections for the binding of inner-shell electrons are relatively unimportant for the atoms H and O. Their inclusion reduces the stopping power of only a few per cent in the region of the peak (Figure 1.3). Inclusion of the density effect causes the curve to rise somewhat less steeply to the right of the minimum.

---

*An average, effective, charge $z_{\mathrm{eff}}$ may be associated with the ion instead of $z$ and the energy loss may be obtained by replacing $z$ with $z_{\mathrm{eff}}$ (see, for example, Bell, 1953).
[†]See this reference for an enumeration of still unsolved problems connected with the passage of charged particles through matter.

## 1.2.2.2 Electrons

In their interactions with matter electrons are scattered in collisions with electrons and nuclei suffering appreciable deflection, so that their path is tortuous except at high energies when it is straight. They dissipate their energy, as stated earlier, in excitation and ionization of atomic electrons. At high energies radiation losses (bremsstrahlung) become an important energy loss mechanism also. Radiation losses increase as $Z^2$. Bethe and Heitler (1934) estimated the relative importance of the radiation and collision losses. An approximate expression for the radiative loss to collision loss is (Bethe and Heitler, 1934)

$$\frac{(\mathrm{d}E/\mathrm{d}x)\,\mathrm{rad}}{(\mathrm{d}E/\mathrm{d}x)\,\mathrm{coll}} \sim \frac{EZ}{800} \tag{1.23}$$

where $E$ is the total (kinetic plus rest mass) energy, and is measured in MeV.

Ignoring radiation losses, the Bethe theory discussed in Section 1.2.2.1 for heavy charged particles can be applied with some slight modifications to the case of fast electrons. Thus Equation (1.21) is now replaced by

$$-\frac{\mathrm{d}E}{\mathrm{d}x} = N\frac{4\pi e^4}{mv^2}\,Z\ln\frac{mv^2}{I} \tag{1.24}$$

which differs from Equation (1.21) by a factor of two in the logarithm. A further modification is necessary because of the indistinguishability of the two electrons in an ionizing collision. Defining the electron with the higher energy as the primary (Bethe, 1930), the maximum energy loss in any collision is $1/4\ mv^2$ (rather than $1/2\ mv^2$). Bethe (1930) obtained

$$-\frac{\mathrm{d}E}{\mathrm{d}x} = N\frac{4\pi e^4}{mv^2}\,Z\ln\frac{mv^2}{2I}\sqrt{\frac{e}{2}} \tag{1.25}$$

where $e$ under the square root symbol is the basis of the natural logarithms.

The difference between Equations (1.21) and (1.25) consists of a small factor in the logarithm. Protons and electrons having the same non-relativistic velocity are seen to be losing energy (Equations (1.21) and (1.25)) at approximately the same rate. The relativistic stopping power formula for electrons is (Bethe, 1933)

$$-\frac{\mathrm{d}E}{\mathrm{d}x} = N\frac{2\pi e^4}{mv^2}\,Z\left[\ln\frac{mv^2 E}{2I^2(1-\beta^2)} - (2\sqrt{1-\beta^2}-1+\beta^2)\ln 2\right.$$
$$\left. + 1-\beta^2+\tfrac{1}{8}(1-\sqrt{1-\beta^2})^2\right] \tag{1.26}$$

where $E$ is the total electron energy minus the rest energy.

The density effect discussed in Section 1.2.2.1 for ions is more easily observed for electrons since it depends on the particle velocity. Thus Equation (1.26) holds for electron energies larger than the binding energies of the atomic electrons but low enough that the density effect and radiation energy losses are small. The relativistic increase in the stopping power has been observed in many experiments (e.g., Birkhoff, 1958). In Figure 1.4 (Dalgarno, 1962) the stopping power for molecular hydrogen at 1-atm pressure is shown as a function of electron energy. The density, as well as the relativistic effects are clearly seen.

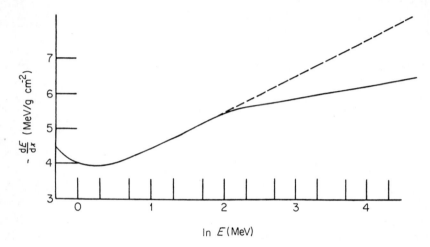

**Figure 1.4** Energy loss of electrons in molecular hydrogen at 15°C and 760 torr. The broken curve is obtained when the density effect is neglected (Dalgarno, 1962)

For positrons the magnitude of the stopping power differs slightly from Equation (1.26) due mainly to differences in the electron–electron and electron–positron collision cross sections, and also because the two particles $(e^+, e^-)$ are distinguishable after collision. Expressing the total energy, $\gamma$, in units of $mc^2$ (where $m$ is the rest mass of the electron) the stopping power for positrons is (Rohrlich and Carlson, 1954):

$$-\frac{dE^{(+)}}{dx} = \frac{2\pi Ne^4 Z}{mv^2}\left[\ln\left(\frac{E^2}{I^2}\frac{\gamma+1}{2}\right)+f^+(\gamma)\right] \qquad (1.27)$$

where

$$f^+(\gamma) = 2\ln 2 - \frac{\beta^2}{12}\left[23 + \frac{14}{\gamma+1} + \frac{10}{(\gamma+1)^2} + \frac{4}{(\gamma+1)^3}\right]$$

### 1.2.3. *Degradation Spectra*

When the incident radiation is made up of monoenergetic electrons the degradation spectrum is of one kind; it consists of electrons only. In this case the energy spectrum of the charged particles actually present in the irradiated medium has been referred to also, as the *slowing-down spectrum*. With other types of ionizing radiations two or more kinds of degradation spectra are present simultaneously. With $\gamma$-rays, for example, a spectrum of photons and a spectrum of electrons are present; with neutrons, a spectrum of neutrons, of electrons, and of recoil ions are present. The rest of the discussion will be confined, for simplicity, to the case of incident electrons.

The general theory of the electron degradation spectrum has been developed by Spencer and Fano (1954). They represented the spectral intensity by a function $y(E)$ defined by $y(E)dE$ = total distance traversed by all electrons (primary, secondary, etc.) present in the medium while in the energy range $E+dE$ and $E$. In this definition $y(E)$ has dimensions of length per unit energy (e.g. cm/eV). In the 'continuous slowing-down approximation' this is, for one particle, the inverse of its stopping power, i.e. $(dE/dx)^{-1}$. The function $y(E)$ depends on the initial energy spectrum. Its calculation requires knowledge of the differential ionization cross sections $\sigma_i(E, E')$ over a broad range of both $E$ and $E'$, i.e. knowledge of the energy distribution of secondary electrons of energy $E'$ ejected in collisions by electrons of energy $E$.

Spencer and Fano (1954) modified the 'continuous slowing-down approximation' to include secondary electrons produced by the violent collisions experienced by the primary electrons and also to include the effects of bremsstrahlung. Spencer and Attix (1955) gave a simplified treatment of the Spencer–Fano theory, while McGinnies (1959) evaluated it numerically. A modification of the Spencer–Attix treatment allowing calculation by successive generations of secondaries has been given by Hamm and coworkers (1965) (see also, McConnell and coworkers, 1968).

It has become apparent, from the preceding discussion, that the construction of a degradation spectrum appropriate to a given radiation and medium necessitates stopping powers and cross sections for secondary electron ejection. The former may be calculated with sufficient accuracy, but the latter appear to be more difficult to obtain. A bibliography on and calculations of electron flux spectra have been given by Danzker, Kessaris, and Laughlin (1959). See, also, a discussion by Birkhoff (1967).

Direct measurement of the slowing-down spectrum has been accomplished for a number of metals (McConnell and coworkers, 1965, 1966, 1968). In these experiments electrons were produced in a metal disc by a continuous $\beta$-ray source uniformly distributed within it and upon emerging from the metal they were energy-selected and collected in a Faraday cup. Figure 1.5 shows the slowing-down spectrum for $^{198}$Au $\beta$ particles in aluminium. The experimental data are compared with the predictions of the Spencer–Fano theory on an absolute basis. Above $\sim 100$ keV the shape of the electron flux is determined by the integral primary beta-spectrum. Below this energy the primary spectrum declines due to the increase in the stopping power, but the decline is matched by an increase in the number of secondary electrons in such a way that the flux is essentially constant down to $\sim 5$ keV. McConnell and coworkers (1968) reported that the flux rises approximately as $E^{-1}$ between 5 KeV and 100 eV and approximately as $E^{-3}$ below 100 eV.

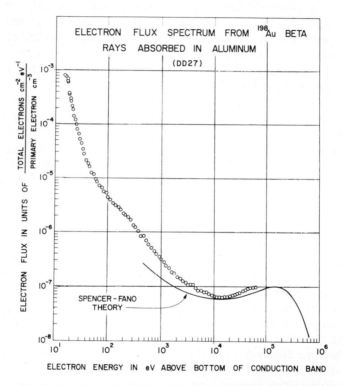

**Figure 1.5** Electron flux spectrum from [198]Au beta-rays absorbed in aluminium (McConnell and coworkers, 1968)

The experimental results of McConnell and coworkers (1968) agree very well with the Spencer–Fano theory for energies above $\sim 10$ keV but the theory fails to predict the large increase of low-energy secondary electrons. This may be due to the fact the theory neglects large energy losses resulting from inner-shell ionizations, and the associated Auger cascades.

The rapid rise in the electron flux at low energies clearly reflects the rise in the number of secondary electrons which contribute to $y(E)$. This skewness of the degradation spectrum at low energies depends but little on the nature and energy of the primary particle. It is then clear that the majority of the secondary electrons have rather low energy (electrons whose energy is less than $\sim 200$ eV can transfer energy only to the external electrons of a heavy atom). It is this abundance of low-energy electrons which makes the understanding of low-energy electron interactions with matter so demanding in radiation physics.

### 1.2.4 Yields of Primary Products

Once the degradation spectrum is calculated the total number, $N_n$, of a specific kind, $n$, of a primary product, formed in the complete absorption of a charged particle of initial kinetic energy, $E_0$, can be determined from

$$N_n(E_0) = N \int_{E_n}^{E_0} y(E)\, \sigma_n(E)\, \mathrm{d}E \qquad (1.28)$$

where $N$ is the number of molecules per unit volume and $E_n$ is the pertinent threshold energy. A theoretical determination of the 'energy spectrum of primary activations' (Platzman, 1967) $N_n(E_n)$ can be obtained from Equation (1.28) if $y(E)$ and $\sigma_n(E)$ are known. Also, a theoretical yield $G_{th}$ (mean number of events per 100 eV absorbed) can be computed for the primary product $n$ in terms of the energy absorbed ($E_0$ in eV) from

$$G_{th}(E_0) = \frac{100\, N_n}{E_0} \qquad (1.29)$$

Platzman (1961) studied the total ionization in helium at several energies by obtaining the degradation spectrum from the Spencer–Fano theory for $E > 300$ eV and constructing it semi-theoretically below this energy. Also Durup and Platzman (1961) calculated absolute yields for inner-electron ejection from atoms in a variety of media using Equation (1.28). It is worth mentioning also that McConnell and coworkers (1968) made use of the electron fluxes they measured as a function of electron energy in the slowing-down of $^{198}$Au $\beta$ particles in aluminium and obtained the number of interactions per cm$^3$ per primary electron per cm$^3$ for $K$-shell ionization, L-shell ionization, volume plasmon excitation, and electron–electron collisions. Figure 1.6 shows their results on $K$- and $L$-shells, and volume plasmon excitation. Here $Ey\lambda^{-1}$ ($\lambda^{-1}$ is the inverse mean free path) is plotted as a function of $\ln E$ so that the area under the curve between any two energies $E_1$ and $E_2$ is equal to the number of events occurring within the range $E_2 - E_1$. It is seen from the data shown in Figure 1.6 that when an electron slows-down in aluminium nearly all of its energy goes ultimately into volume plasmon excitation.

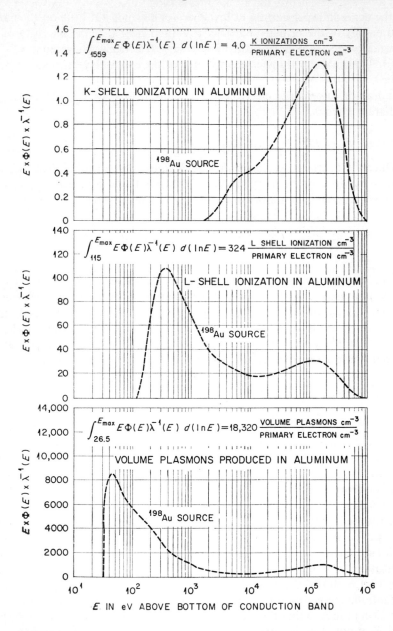

**Figure 1.6** Interactions per cm³ per primary electron per cm³ for
K-shell ionization, L-shell ionization, and volume plasmon excitation
for ¹⁹⁸Au beta-rays slowed-down in aluminium (McConnell and
coworkers, 1968)

From the preceding discussion a swiftly-moving charged particle loses its energy to excitation and ionization of outer-shell (predominantly) and to inner-shell (to a lesser extent) electrons. In addition (Table 1.1) a number of less frequent processes such as nuclear collisions, multiple ionizations, and

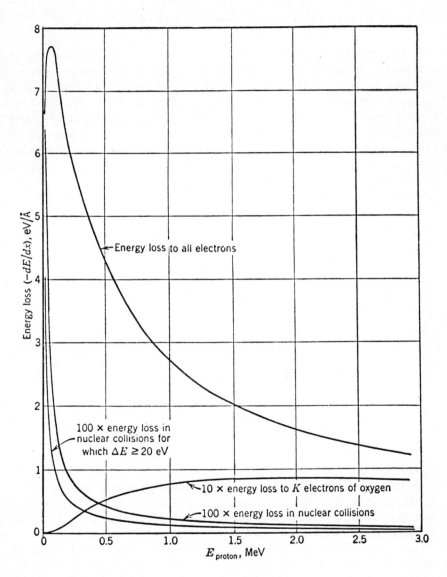

**Figure 1.7** Various modes of energy loss (stopping power) of protons in water (Platzman, 1952)

Auger disruptions take place. Figure 1.7 reproduced from Platzman (1952) helps demonstrate the relative importance of some of these dissipative processes.

## 1.3 Interaction of Electromagnetic Radiation with Matter

Electromagnetic radiations (photons of energy $E_p = h\nu$) are customarily classified according to their mode of origin and not their energy. Electromagnetic radiation in the visible and ultraviolet regions originating from electronic transitions of loosely bound electrons is generally known as luminescence (see Chapter 3), that emitted in atomic transitions of bound electrons between $K$-, $L$- and $M$-shells in atoms is known as characteristic x-rays, while that resulting from the acceleration of free electrons or other charged particles is known as bremsstrahlung or continuous x-rays. Electromagnetic radiations accompanying nuclear transitions, and electron–positron annihilation are known, respectively, as $\gamma$-rays and annihilation radiation.

Photons are absorbed or scattered in single events in contrast to charged particles which lose their energy continuously.

In Chapter 3 we shall focus attention on the visible and ultraviolet regions ($\sim 2 \leqslant E_p \lesssim 10$ eV) of the electromagnetic spectrum. This is the field of photophysics and photochemistry, characterized by strong, selective absorption by specific discrete atomic and molecular structures. Taking the photon energy to be below the ionization potential of the medium, electronic excitation takes place to discrete states, and only the direct action of the primary quantum is important. The primary spectrum so obtained is called the *optical or absorption* spectrum and is proportional to $f_n(E_n)$ where $f_n$ is the oscillator strength for the transition with energy $E_n$ (see Chapter 3). Many techniques have come into use recently in the field of atomic and molecular photophysics. These enabled detailed selective absorption and emission studies, which aided tremendously our understanding of the excitation process of the same systems exposed to ionizing radiation. The energy range extending from the far ultraviolet to the soft x-ray region ($10$ eV $\lesssim E_p \lesssim 60$ keV) has been less fully investigated although a growing interest has developed in recent years in this energy region. (The extremely interesting processes of superexcitation, autoionization, and photoionization will be discussed in Chapters 2 and 3.)

Quite generally, the secondary product of the absorption of a high-energy photon is an electron which upon release, enters into exciting and ionizing collisions. For a typical high-energy photon ($\sim 1$ MeV) the ionizing events due to these secondary electrons are far more numerous than the atoms which directly interact with radiation. The effect of radiation here, then, is that of randomly originating ionizing recoil electrons. A brief description of most of the processes by which γ-rays interact with matter is given in Table 1.3 (Davisson, 1965). In the energy range from $\sim 1$ KeV to $\sim 50$ MeV most of the interactions are due to the following three major processes: photoelectric effect, Compton effect, and pair production.

### 1.3.1 Photoelectric Effect

For the soft and medium x-ray region the principal type of interaction is the absorption of the quantum by an inner-shell atomic electron. The photoelectron emerges with kinetic energy $E$ equal to

$$E = E_{p} - B \qquad (1.30)$$

where $B$ is the binding energy at the electron prior to ejection. The energy $B$, subsequently, appears as characteristic x-rays and Auger electrons from the filling of the vacancy created by the ejected electron. The directional distribution of photoelectrons for different incident energies is shown in Figure 1.8.

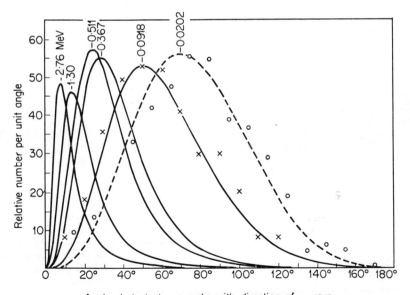

Angle photoelectrons make with direction of γ-rays

**Figure 1.8** Directional distribution of photoelectrons for different energies. Solid and dashed curves are calculations, crosses are measured values of E. Lutz at 0.0918 MeV (*Ann. Physik*, **9**, 853, 1931). Circles are measured values of E. J. Williams, J. M. Nuttall and H. S. Barlow at 0.0202 MeV (*Proc. Roy. Soc.* (*London*), **121A**, 611, 1928). (Evans (1955)

It is seen that with increasing photon energy the directional distribution of photoelectrons peaks closer to the direction of the incident beam.

The probability of photoelectric absorption increases with increasing binding energy of the electron, i.e. it is greater for the $K$-than the $L$-, than the $M$-shell. The photoelectric absorption cross section varies in a complex manner with $E_p$ and the $Z$ of the absorbing material (see Table 1.3 and also Evans (1955), and Davisson (1965)). In the energy region below $\sim 100\,keV$ the absorption cross section exhibits characteristic absorption edges due to the fact that the $K$-, $L$-, and $M$-shells cease to contribute to photon absorption as the photon energy $E_p$ becomes less than the binding energy of the corresponding shells.

**Table 1.3** Processes by which $\gamma$-rays interact with matter[a]

| Process | Kind of interaction | Approximate variation with atomic number, $Z$ | Approximate energy range of maximum importance |
|---|---|---|---|
| Photoelectric effect | With bound atomic electron (all energy given to electron) | $Z^5$ | Dominates at low $E$ (1 to 500 keV). Decreases as $E$ increases |
| Scattering from electrons<br>Coherent | With bound atomic electrons | $Z^2$ (small angles) $Z^3$ (large angles | $<1$ MeV and greatest at small scattering angles |
| | With free electrons | $Z$ | Independent of energy |
| Incoherent | With bound atomic electrons | $Z$ | $<1$ MeV; least at small scattering angles |
| | With free electrons | $Z$ | Dominates in region of 1 MeV. Decreases as $E$ increases |
| Photonuclear absorption<br>Nuclear photoeffect | With nucleus as a whole (emits $\gamma$ or particles) | | Above threshold has broad maximum in range of 10–30 MeV |

[a] From Davisson (1965).

Table. 1.3 (*continued*)

| Process | Kind of interaction | Approximate variation with atomic number, $Z$ | Approximate energy range of maximum importance |
|---|---|---|---|
| Nuclear scattering Coherent | With material as a whole (dependent on nuclear energy levels) | | Important only in very narrow resonance range |
| | With nucleus as a whole (dependent on nuclear energy levels) | $Z^2/A^2$ $A$ = atomic weight | Narrow resonance maxima at low energies. Broad maximum in range of 10–30 MeV |
| | With nucleus as a whole (independent of nuclear energy levels) | $Z^4/A^2$ $A$ = atomic weight | |
| Incoherent | With individual nucleons | | |
| Interaction with a Coulomb field Pair production | In Coulomb field of nucleus | $Z^2$ | Threshold about 1 MeV. Dominates at high $E$ (i.e. $E > 5$ or 10 MeV). Increases as $E$ increases |
| | In Coulomb field of electron | $Z$ | Threshold at 2 MeV Increases as $E$ increases |
| Delbrück scattering | In Coulomb field of nucleus | $Z^4$ | |

### 1.3.2 *Compton effect*

At photon energies much greater than the binding energy of the atomic electrons, the photons are scattered as if the electrons were free and at rest. The scattering of a photon from its original direction by a free electron, and the associated ejection of a free electron, is the well-known Compton effect. The process is illustrated in Figure 1.9. Compton scattering is the dominant mode of interaction at photon energies around 1 MeV. In Table 1.4 the various basic relations between the parameters shown in Figure 1.9 are

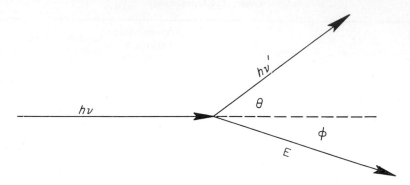

**Figure 1.9** Compton effect. Angular and energy notations

listed. The differential cross section per unit solid angle for the number of photons scattered at an angle $\theta$ (Equation (6) in Table 1.4) is shown in Figure 1.10. The angular distribution becomes increasingly peaked in the forward direction as the photon energy is increased.

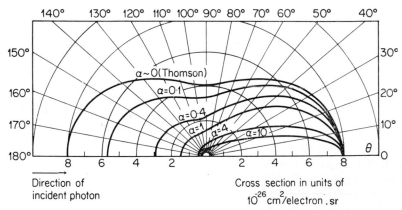

**Figure 1.10** Differential cross section per unit solid angle for the number of photons scattered at an angle $\theta$ (Equation (6) in Table 1.4) (Davisson, 1965).

The differential cross section for giving a free electron a kinetic energy between $E$ and $E + dE$ (Equation (7) in Table 1.4) is shown in Figure 1.11 (see, also, Nelms, 1953).

As far as the medium is concerned, in a Compton collision only a fraction of the photon energy is transferred to the electron, and, thus, is absorbed by the medium; the rest is carried away by the scattered photon. To deter-

**Table 1.4** Relations between the various parameters involved in the Compton scattering process.

| Explanation | Relation | Comments |
|---|---|---|
| Compton shift | $$\frac{c}{\nu'} - \frac{c}{\nu} = \lambda' - \lambda$$ $$= \frac{h}{mc}(1 - \cos\theta) \quad (1)$$ | $\dfrac{h}{mc}$ is usually called the Compton wavelength; $mc^2 = 0.51$ MeV |
| Energy of the scattered photon | $$h\nu' = \frac{h\nu}{1 + \alpha(1 - \cos\theta)} \quad (2)$$ | $\alpha = \dfrac{h\nu}{mc^2}$ |
| Energy of Compton electron | $$E = h\nu\,\frac{\alpha(1 - \cos\theta)}{1 + \alpha(1 - \cos\theta)} \quad (3)$$ | |
| Maximum energy of Compton electron | $$E_{max} = \frac{h\nu}{1 + (\frac{1}{2}\alpha)} \quad (4)$$ | Energy of recoil electron varies from zero ($\theta = 0°$, $\phi = 90°$) to a maximum, $E_{max}$, ($\theta = 180°$, $\phi = 0°$) |
| Relations between scattering angles $\theta$ and $\phi$ | $$\cot\phi = (1 + \alpha)\tan\tfrac{1}{2}\theta \quad (5)$$ | |
| Differential cross section $d_e\sigma$ per electron for the number of photons (unpolarized) scattered into the solid angle $d\Omega$ in the direction $\theta$ | $$\frac{d_e\sigma}{d\Omega} = \tfrac{1}{2}r_0^2 \left\{\frac{1}{[1 + \alpha(1 - \cos\theta)]^2}\right.$$ $$\left. \cdot\left[1 + \cos^2\theta + \frac{\alpha^2(1 - \cos\theta)^2}{[1 + \alpha(1 - \cos\theta)]}\right]\right\}$$ $$(6)$$ | This is the well-known Klein–Nishina formula (Klein and Nishina, 1929). See discussion and data in Davisson and Evans (1952), Evans (1955) and Davisson (1965). $$r_0 = \frac{e^2}{mc^2}$$ |
| Differential cross section for giving the free electron a recoil energy between $E$ and $E + dE$. | $$\frac{d_e\sigma}{dE} = \frac{\pi r_0^2}{\alpha^2 mc^2}\left\{2 + \left(\frac{E}{h\nu - E}\right)^2\right. \cdot$$ $$\left. \cdot\left[\frac{1}{\alpha^2} + \frac{h\nu - E}{h\nu} - \frac{2}{\alpha}\left(\frac{h\nu - E}{E}\right)\right]\right\} \quad (7)$$ | Equations (6) and (7) of this table are plotted in Figures 1.10 and 1.11, respectively |

mine the amount of energy transferred to a medium through Compton scattering the total cross section per electron $_e\sigma$ is separated into two components: $_e\sigma_s$, the cross section for the amount of energy in the scattered radiation and $_e\sigma_a$ the cross section for the amount of energy transferred to the electron (absorbed) viz

$$_e\sigma = {}_e\sigma_s + {}_e\sigma_a$$

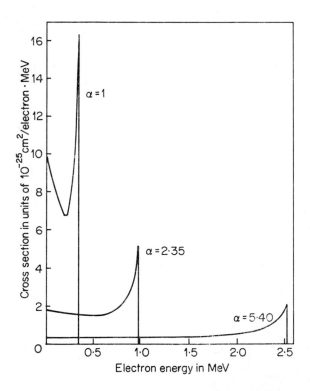

**Figure 1.11** Differential cross section for giving a free electron a recoil energy in the interval between $E$ and $E + dE$ (Eq. (7) in Table 1.4) (Evans, 1955)

### 1.3.3 *Pair Production*

When the energy $E_p$ of the incident photon exceeds 1.02 MeV ($= 2mc^2$) the production of an electron–positron pair $(e^-, e^+)$ can take place in the field of the nucleus or an electron. When in the field of the nucleus the photon is completely absorbed, its energy being converted into the rest-mass-energy

$(2mc^2)$ and kinetic energy $E$ of $e^-$ and $e^+$. The total kinetic energy of the pair is

$$E = E_p - 2mc^2 \qquad (1.31)$$

This kind of pair production is sometimes called coherent or elastic, in contrast to the incoherent or inelastic pair production which occurs when some of the photon energy is taken away by an atomic electron for either excitation or ionization. In the latter case, if the atom is ionized the interaction is called triplet production.

Pair production is related to bremsstrahlung and the two processes are often treated together theoretically. For a theoretical treatment of the pair production process see Bethe and Heitler (1934) (see also Davisson and Evans (1952), Evans (1955), and Davisson (1965)). For very high photon energies the angular distribution of the electron and positron is mainly forward. The average angle between the incident photon and the created electrons is $\sim mc^2/E(E \gg mc^2)$, but for photon energies of the order of $2mc^2$ the angular distribution is much more complicated. The cross section for the process varies approximately as the square of the atomic number. At the end of its range the positron interacts with an electron and two annihilation quanta each of energy 0.51 MeV are produced.

Each of the three basic interaction processes discussed is considered to act independently. In each process secondary electrons are produced whose kinetic energies $E$ are related in a different way to $E_p$: photoelectrons ($E = E_p - B$), Compton electrons ($E$ from zero to $E_{max}$; see, Equations (3) and (4) in Table 1.4), and pair-produced electrons ($E = E_p - 2mc^2$). The relative numbers of electrons produced in each of the three processes depends on the relative magnitudes of the associated probabilities for the three processes. Figure 1.12 shows the importance of the three processes as a function of the atomic number of the absorber and the primary photon energy (Evans, 1955).

A beam of well-collimated photons incident on an absorber undergoes true exponential attenuation. The fraction of the incident photons undergoing interactions, in passing an absorber of thickness $\Delta x$ is

$$f = 1 - e^{-\mu \Delta x} \qquad (1.32)$$

where $\mu^*$ is the linear attenuation coefficient in cm$^{-1}$. The coefficient $\mu$ can be separated into three parts which usually are designated by $\tau$ for the photoelectric effect, $\sigma$ for the Compton scattering, and $k$ for the pair-production process, i.e.

$$\mu = \tau + \sigma + k \qquad (1.33)$$

---

*The generally accepted symbol, $\mu$, is used for the linear attenuation coefficient in spite of the fact that the same symbol will be used for the electron mobility and permanent molecular electric dipole moment in subsequent chapters.

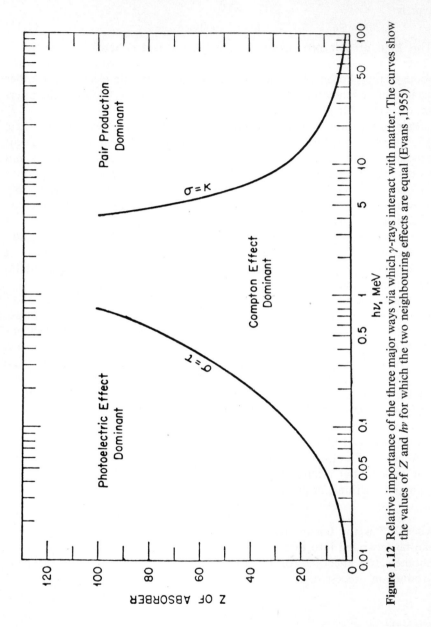

**Figure 1.12** Relative importance of the three major ways via which $\gamma$-rays interact with matter. The curves show the values of $Z$ and $h\nu$ for which the two neighbouring effects are equal (Evans ,1955)

Each of these quantities depends on the energy of the quantum and the nature of the absorber. Typical data are shown in Figure 1.13 for water (Evans, 1955).

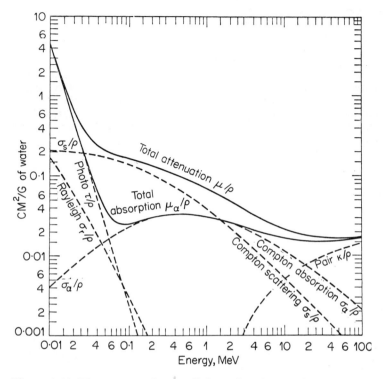

**Figure 1.13** Mass attenuation coefficients for photons in water. The curve marked 'total absorption' is $(\mu_\alpha/\rho) = (\sigma_\alpha/\rho)+(\tau/\rho)+(k/\rho)$, where $\sigma_\alpha$, $\tau$ and $\varkappa$ are the corresponding linear coefficients for Compton absorption, photoelectric absorption, and pair production, and $\rho$ is the density of the material. When the Compton scattering coefficient $\sigma_\mathrm{s}$ is added to $\mu_\alpha$, we obtain the curve marked "total attenuation," which is $(\mu/\rho) = (\mu_\alpha/\rho)+(\sigma_\mathrm{s}/\rho)$. The total Rayleigh scattering cross section $(\sigma_\mathrm{r}/\rho)$ is shown separately. Because the Rayleigh scattering is elastic and is confined to small angles, it has not been included in $\mu/\rho$
(Evans, 1955)

## 1.4 References

G. I. Bell (1953). *Phys. Rev.*, **90**, 548.
H. A. Bethe (1930). *Ann. Phys.*, **5**, 325.
H. A. Bethe (1933). In *Handbuch der Physik* **24** (1), Springer-Verlag, Berlin, 273.
H. A. Bethe and W. Heitler (1934). *Proc. Roy. Soc. (London)*, **A146**, 83.

R. D. Birkhoff (1958). In S. Flugge (Ed.), *Encyclopedia of Physics*, Vol. 34, Springer-Verlag, Berlin, p. 53.
R. D. Birkhoff (1967). In K. Z. Morgan and J. E. Turner (Eds.), *Principles of Radiation Protection*, John Wiley and Sons, New York, Chap. 8.
F. Bloch (1933). *Z. Physik*, **81**, 363.
N. Bohr (1913). *Phil. Mag.*, **25**, 10.
N. Bohr (1915). *Phil. Mag.*, **30**, 581.
A. Dalgarno (1962). In R. D. Bates (Ed.), *Atomic and Molecular Processes*, Academic Press, New York, Chap. 15, pp. 622–642.
M. Danzker, N. D. Kessaris and J. S. Laughlin (1959). *Radiology*, **72**, 51.
C. M. Davisson (1965). In K. Siegbahn (Ed.), *Alpha-, Beta-, and Gamma-Ray Spectroscopy*, Vol. 1, North-Holland Publishing Company, Amsterdam, pp. 37–78.
C. M. Davisson and R. D. Evans (1952). *Rev. Mod. Phys.*, **24**, 79.
J. Durup and R. L. Platzman (1961). *Discussions Faraday Soc.*, **31**, 156.
R. D. Evans (1955). *The Atomic Nucleus*, McGraw-Hill Book Company, New York.
U. Fano (Ed.) (1964). *Studies in Penetration of Charged Particles in Matter*, National Academy of Sciences-National Research Council, Washington, D.C. Publ. 1133.
R. N. Hamm, W. J. McConnell, R. D. Birkhoff and M. Berger (1965). *Bull. Am. Phys. Soc.*, **10**, 477.
M. Inokuti, Y-K. Kim and R. L. Platzman (1967). *Phys. Rev.*, **164**, 55.
J. D. Jackson (1963). *Classical Electrodynamics*, 3rd. ed. John Wiley and Sons., New York, Chaps. 11–15.
O. Klein and Y. Nishina (1929). *Z. Physik*, **52**, 853.
J. Lindhard, and M. Scharff (1953). *Kgl. Danske Videnskab. Selskab., Mat.-Fys. Medd.*, 2i7, No. 15.
R. T. McGnnies (1959). National Bureau of Standards (U.S., Circular No. 597).
W. J. McConnell, H. H. Hubbell, Jr., R. N. Hamm, R. H. Ritchie and R. D. Birkhoff (1965). *Phys. Rev.*, **138**, A1377.
W. J. McConnell, H. H. Hubbell, Jr., and R. D. Birkhoff (1966). *Health Phys.*, **12**, 693.
W. J. McConnell, R. D. Birkhoff, R. N. Hamm and R. H. Ritchie (1968). *Rad. Res.* 33, 216.
A. T. Nelms (1953). National Bureau of Standards Circular 542.
J. A. Phillips (1953). *Phys. Rev.*, **90**, 532.
R. L. Platzman (1962). *Vortex*, **23**, 372.
R. L. Platzman (1952). In J. J. Nickson (Ed.), *Symposium on Radiobiology*, National Research Council-National Academy of Sciences, John Wiley and Sons, New York, p. 97.
R. L. Platzman (1961). *Intern. J. Appl. Radiation and Isotopes*, **10**, 116.
R. L. Platzman (1967). In G. Silini (Ed.), *Proc. Third Intern. Congress of Radiation Research*, North-Holland Publishing Company, Amsterdam, pp. 20–42.
F. Rohrlich and B. C. Carlson (1954). *Phys. Rev.*, **93**, 38.
L. V. Spencer and F. H. Attix (1955). *Rad. Res.*, **3**, 239.
L. V. Spencer and U. Fano (1954). *Phys. Rev.*, **93**, 1172.
R. M. Sternheimer (1959). *Phys. Rev.*, **115**, 137.
J. E. Turner (1967). *Health Phys.*, **13**, 1255.
J.E. Turner, R.D. Roecklien and R.B. Vora (1970). *Health Phys.*, **18**, 159.

# 2 Total ionization produced in gases by high-energy charged particles

## 2.1 Energy per Ion Pair

The average energy spent by a charged particle to create an ion pair in a given gaseous medium is known as $W$. The quantity $W$ is thus defined as

$$W = \frac{E_0}{N_i} \tag{2.1}$$

where $E_0$ is the energy of the charged particle (totally absorbed) and $N_i$ is the total number of ion pairs produced. Theoretically, $W$ is the simplest and most direct measure of an aggregate effect of high-energy radiation (Platzman, 1961). However, since a large number of varying primary elementary processes (Table 1.1) occur in the slowing-down of a single charged particle in a gaseous medium, $W$ is a gross, macroscopic, physical parameter which necessitates detailed microscopic information for its theoretical interpretation. The entire degradation spectrum has to be considered and the total excitation, total, and differential ionization cross sections for the charged particles present have to be known. For high-energy monoenergetic incident electrons we may express $N_i(E)$ as (Platzman, 1961)

$$N_i(E) = p_i(E) + \sum_n p_n(E) N_i(E - E_n') +$$

$$+ \int_I^{\frac{E+I}{2}} p_i(E, E') \{N_i(E - E') + N_i(E' - I)\} \, dE' \tag{2.2}$$

where $p_i(E)$ (ratio of the total ionization cross section $\sigma_i(E)$ to the total cross section for all inelastic collisions, $\sigma_{in}(E)$) is the probability of ionization, $p_n(E)$ is the analogous probability for excitation to an energy level $E'_n$, and $p_i(E, E')\,dE'$ is the analogous probability for ionization with production of a secondary electron of kinetic energy $(E' - I)$; $I$ is the ionization potential. The direct determination of $N_i(E)$ through Equation (2.2) necessitates knowledge of the various cross sections involved. If, on the other hand, the degradation spectrum $y(E)$ and the total ionization cross section $\sigma_i(E)$ are known, then from an expression similar to Equation (1.28), i.e.

$$N_i(E_0) = N \int_I^{E_0} y(E)\sigma_i(E)\,dE \tag{2.3}$$

the number $N_i$ of the ions produced in the complete absorption of the charged particle can be calculated. Calculations of total ionization have been made using Equation (2.2) for He (Fano, 1946; Erskine, 1954; Miller, 1956) and atomic hydrogen (Dalgarno and Griffing, 1958), and using Equation (2.3) for He (Platzman, 1961). Lack of accurate cross sections seems to be the major obstacle in these types of calculations which have been restricted to only a few of the simplest atoms.

## 2.2  W for Pure Gases

The magnitude of $W$ is of practical importance in a number of areas such as nuclear radiation detection, radiation dose measurements, dose partition and radiation-induced chemical reactions. Abundant measurements of $W$ have been made since the early days of radioactivity. Experimental methods and results have been reviewed by a number of workers (see, for example, Gray (1936, 1944); Binks (1954); Weiss and Bernstein (1957); Valentine and Curran (1958) and Whyte (1963)). Usually, instead of an absolute measurement, values of $W$ have been obtained by comparison with standards, i.e. well-established $W$ values such as those for $N_2$ (36.4 eV/ip) and Ar (26.4 eV/ip) (Table 2.1). In such experiments it is essential to avoid ion recombination and ensure complete electron collection and gas purity. Special attention should be given also to complexities arising from highly electron attaching gases (Chapter 6) and densely ionizing particles. Electron attachment increases the measured $W$. The effect of impurities is mainly due to the well-known Jesse effect (Platzman, 1961) which is an 'excited-state effect', i.e. ionization of the impurity atom or molecule by electronic energy transfer from excited gas atoms or molecules under study (to be referred to as 'the parent' gas) (Section 2.4.3). Such an effect obviously decreases $W$

**Table 2.1**  *W* values for alpha and beta particles slowed-down in rare gases
and molecular vapours

| Atom or molecule | $W$ (eV/ip) | Ref. to $W^d$ | $I$ (eV) | Ref. to $I^d$ | $W^a$ used | $I^a$ used | $\dfrac{W_\alpha}{W_\beta}$ | $\dfrac{W_\alpha}{I}$ |
|---|---|---|---|---|---|---|---|---|
| 1. He | 41.3 | 1 | 24.581 | 26 | 46.0 | 24.581 | 1 | 1.87 |
|  | 29.6 | 2 |  |  |  |  |  |  |
|  | 31.7 | 3 |  |  |  |  |  |  |
|  | 42.7 | 4 |  |  |  |  |  |  |
|  | 46.0 | 5 |  |  |  |  |  |  |
|  | [42.3]$^b$ | 6, 33 |  |  | [42.3] |  |  |  |
| 2. Ne | 36.3 | 1 | 21.559 | 26 | 36.55 | 21.559 | 1 | 1.7 |
|  | 36.8 | 4 |  |  |  |  |  |  |
|  | [36.6] | 6, 33 |  |  | [36.4] |  |  |  |
|  | [36.2] | 34 |  |  |  |  |  |  |
| 3. Ar | 25.9 | 3 | 15.755 | 26 | 26.4 | 15.755 | 1 | 1.68 |
|  | 26.22 | 24 |  |  |  |  |  |  |
|  | 26.25 | 2 |  |  |  |  |  |  |
|  | 26.3 | 7 |  |  |  |  |  |  |
|  | 26.31$^c$ | 8 |  |  |  |  |  |  |
|  | 26.37 | 9 |  |  |  |  |  |  |
|  | 26.38 | 30 |  |  |  |  |  |  |
|  | 26.40 | 1, 4, 5, 12 |  |  |  |  |  |  |
|  | 26.5 | 10 |  |  |  |  |  |  |
|  | [26.4] | 6, 33 |  |  | [26.3] |  |  |  |
|  | [26.20] | 34 |  |  |  |  |  |  |
| 4. Kr | 23.90 | 24 | 13.996 | 26 | 24.00 | 13.996 | 1 | 1.72 |
|  | 23.99 | 11 |  |  |  |  |  |  |
|  | 24.04 | 12 |  |  |  |  |  |  |
|  | 24.1 | 4 |  |  |  |  |  |  |
|  | [24.2] | 6, 33 |  |  | [24.05] |  |  |  |
|  | [23.9] | 34 |  |  |  |  |  |  |
| 5. Xe | 21.32 | 24 | 12.127 | 26 | 21.7 | 12.127 | 1 | 1.79 |
|  | 21.39 | 14 |  |  |  |  |  |  |
|  | 21.5 | 12 |  |  |  |  |  |  |
|  | 21.9 | 4 |  |  |  |  |  |  |
|  | [22.0] | 6, 33 |  |  | [21.9] |  |  |  |
|  | [21.8] | 34 |  |  |  |  |  |  |
| 6. $H_2$ | 35.8 | 15 | 15.427 | 26 | 36.4 | 15.43 | $\sim$1 | 2.34 |
|  | 36.0 | 16 | 15.44 |  |  |  |  |  |
|  | 36.01 | 35 |  |  |  |  |  |  |

$^a$  These values have been used in all comparisons and discussions throughout this chapter.
See a complete table of *I* in Appendix II.
$^b$  Values in brackets are for beta particles.
$^c$  Average of six values for alpha particle energies ranging from 1.08 to 4.19 MeV.
$^d$  References to Table 2.1 (See numbered list p. 40).

Table 2.1 (*continued*)

| Atom or molecule | $W$ (eV/ip) | Ref. to $W^d$ | $I$ (eV) | Ref. to $I^d$ | $W^a$ used | $I^a$ used | $\dfrac{W_\alpha}{W_\beta}$ | $\dfrac{W_\alpha}{I}$ |
|---|---|---|---|---|---|---|---|---|
| | 36.3 | 4 | | | | | | |
| | 37.0 | 3, 5 | | | | | | |
| | [36.3] | 6, 33 | | | [36.30] | | | |
| 7. $D_2$ | [35.6] | 36 | | | [35.6] | | | |
| 8. HCl | | | 12.84 | 26 | | 12.71 | — | |
| | [25.3] | 31 | 12.74 | | [25.05] | | | |
| | [24.8] | 32 | 12.56 | | | | | |
| 9. $N_2$ | 36.0 | 3 | 15.6 | 26 | 36.39 | 15.59 | 1.05 | 2.33 |
| | 36.3 | 2, 5, 17 | 15.58 | | | | | |
| | 36.33 | 9 | | | | | | |
| | 36.38 | 18, 19 | | | | | | |
| | 36.4 | 7 | | | | | | |
| | 36.5 | 16 | | | | | | |
| | 36.6 | 4, 10 | | | | | | |
| | 36.65$^c$ | 8 | | | | | | |
| | [34.9] | 6 | | | [34.65] | | | |
| | [34.4] | 34 | | | | | | |
| | [34.1] | 37 | | | | | | |
| 10. NO | 28.85 | 35 | 9.25 | 26 | 28.85 | 9.252 | 1.049 | 3.12 |
| | [27.5] | 37 | 9.258 | | [27.5] | | | |
| 11. CO | 34.77 | 21 | 14.009 | 26 | 34.65 | 14.04 | 1.058 | 2.47 |
| | 34.53 | 35 | 14.01 | | | | | |
| | [32.2] | 37 | 14.1 | | [32.75] | | | |
| | [33.3] | 31 | | | | | | |
| 12. $O_2$ | 32.1 | 15 | 12.1 | 26 | 32.23 | 12.15 | 1.045 | 2.65 |
| | 32.2 | 2, 3, 5 | 12.2 | | | | | |
| | 32.23 | 35 | | | | | | |
| | 32.5 | 4 | | | | | | |
| | 32.9 | 7 | | | | | | |
| | [30.9] | 6 | | | [30.83] | | | |
| | [30.6] | 33 | | | | | | |
| | [31.0] | 37 | | | | | | |
| 13. $H_2O$ | 37.70 | 21 | 12.61 | 26 | 30.5 | 12.60 | 1.027 | 2.44 |
| | 30.5 | 38 | 12.60 | | | | | |
| | [29.9] | 39 | 12.59 | | [29.9] | | | |
| 14. $N_2O$ | 34.34 | 11 | 12.72 | 26 | 34.39 | 12.84 | 1.057 | 2.68 |
| | 34.43 | 12 | 12.90 | | | | | |
| | [32.9] | 37 | 12.9 | | [32.55] | | | |
| | [32.2] | 31 | | | | | | |

**Table 2.1** (*continued*)

| Atom or molecule | $W$ (eV/ip) | Ref. to $W^d$ | $I$ (eV) | Ref. to $I^d$ | $W^a$ used | $I^a$ used | $\dfrac{W_\alpha}{W_\beta}$ | $\dfrac{W_\alpha}{I}$ |
|---|---|---|---|---|---|---|---|---|
| 15. $CO_2$ | 34.04 | 18 | 13.79 | 26 | 34.26 | 13.81 | 1.045 | 2.38 |
| | 34.1 | 15 | 13.85 | | | | | |
| | 34.16 | 35 | 13.79 | | | | | |
| | 34.2 | 7 | | | | | | |
| | 34.3 | 5, 16 | | | | | | |
| | 34.45 | 12 | | | | | | |
| | 34.5 | 4 | | | | | | |
| | [32.9] | 6, 33 | | | [32.80] | | | |
| | [32.1] | 31 | | | | | | |
| | [33.5] | 37 | | | | | | |
| 16. $SO_2$ | | | 12.34 | 26 | | 12.6 | | |
| | | | 12.05 | | | | | |
| | [30.4] | 37 | 13.4 | | [30.4] | | | |
| 17. $C_2H_2$ | 27.35 | 35 | 11.40 | 26 | 27.5 | 11.40 | 1.068 | 2.41 |
| | 27.4 | 15 | 11.41 | | | | | |
| | 27.5 | 4, 11 | 11.40 | | | | | |
| | 27.64 | 12, 13 | | | | | | |
| | [25.6] | 20 | | | [25.75] | | | |
| | [26.1] | 33 | | | | | | |
| | [26.3] | 31 | | | | | | |
| | [25.3] | 37 | | | | | | |
| 18. $BF_3$ | 35.3 | 16 | 15.95 | 28 | 35.63 | 15.83 | — | 2.25 |
| | 36 | 5 | 15.5 | | | | | |
| | 35.6 | 34 | 16.05 | 29 | | | | |
| 19. $NH_3$ | | | 10.34 | 26 | | 10.3 | — | |
| | | | 10.40 | | | | | |
| | [26.5] | 37, 31 | 10.15 | | [26.5] | | | |
| 20. $CH_4$ | 28.9 | 15 | 12.99 | 26 | 29.1 | 12.99 | 1.073 | 2.24 |
| | 29.0 | 3, 16 | | | | | | |
| | 29.01 | 35 | | | | | | |
| | 29.1 | 7 | | | | | | |
| | 29.2 | 4 | | | | | | |
| | 29.26 | 12 | | | | | | |
| | 29.4 | 5 | | | | | | |
| | [27.3] | 6, 20 | | | [27.10] | | | |
| | [27.0] | 33 | | | | | | |
| | [27.6] | 31 | | | | | | |
| | [26.7] | 37 | | | | | | |
| 21. $CH_3I$ | 24.8 | 13 | 9.49 | 26 | 24.8 | 9.51 | — | 2.61 |
| | | | 9.51 | | | | | |
| | | | 9.54 | | | | | |

Table 2.1 (*continued*)

| Atom or molecule | $W$ (eV/ip) | Ref. to $W^d$ | $I$ (eV) | Ref. to $I^d$ | $W^a$ used | $I^a$ used | $\dfrac{W_\alpha}{W_\beta}$ | $\dfrac{W_\alpha}{I}$ |
|---|---|---|---|---|---|---|---|---|
| 22. $C_2H_4$ | 27.85<br>27.87<br>28.00<br>28.02<br>[25.5]<br>[25.9]<br>[26.2] | 15<br>35<br>4, 5, 12<br>14, 19<br>31<br>37<br>33 | 10.51<br>10.516<br>10.6 | 26 | 27.96<br><br><br><br>[25.8] | 10.54 | 1.084 | 2.65 |
| 23. $CCl_4$ | 25.79<br>[23.2] | 40<br>31 | 11.1<br>11.47 | 26 | 25.79<br>[23.20] | 11.29 | 1.11 | 2.28 |
| 24. $CCl_2F_2$ (Freon-12) | 29.5<br>29.60 | 5<br>13 | 11.8 | 26 | 29.55 | 11.8 | — | 2.5 |
| 25. $CH_3OH$ | [23.6]<br>[25.5] | 31<br>39 | 10.95<br>10.85 | 26 | [24.55] | 10.9 | — | |
| 26. $C_2H_6$ (Ethane) | 26.40<br>26.47<br>26.60<br>26.70<br>[23.6]<br>[24.5]<br>[24.6]<br>[24.8] | 15<br>35<br>4<br>12, 13<br>37<br>20<br>33, 31<br>6 | 11.65<br>11.65 | 26 | 26.57<br><br><br><br>[24.38] | 11.65 | 1.1 | 2.28 |
| 27. $CH_3CHO$ | [26.4] | 37 | 10.18<br>10.25<br>10.28<br>10.20 | 26 | [26.4] | 10.23 | — | |
| 28. $CH_3NH_2$ (Methylamine) | 25.89<br>[23.6] | 14<br>31 | 9.41<br>9.8<br>8.97 | 27 | 25.89<br>[23.6] | 9.39 | 1.097 | 2.76 |
| 29. c-$C_3H_6$ (Cyclopropane) | 25.8<br>25.82<br>26.11<br>[23.7] | 15<br>35<br>12<br>31 | 10.06<br>10.23<br>10.53 | 26 | 25.9<br><br><br>[23.7] | 10.27 | 1.093 | 2.52 |
| 30. $C_3H_6$ (Propylene) | 27.01<br>27.0<br>27.15<br>27.28<br>[24.8] | 35<br>15<br>14<br>12<br>37, 31 | 9.73<br>9.65<br>9.84 | 26 | 27.1<br><br><br><br>[24.8] | 9.73 | 1.09 | 2.78 |
| 31. $(CH_3)_2NH$ (Dimethylamine) | 25.40<br>[22.9] | 14<br>31 | 8.93<br>9.21 | 26 | 25.40<br>[22.9] | 9.05 | 1.109 | 2.81 |

Table 2.1 (*continued*)

| Atom or molecule | $W$ (eV/ip) | Ref. to $W^d$ | $I$ (eV) | Ref. to $I^d$ | $W^a$ used | $I^a$ used | $\dfrac{W_\alpha}{W_\beta}$ | $\dfrac{W_\alpha}{I}$ |
|---|---|---|---|---|---|---|---|---|
| 32. $C_3H_6O$ (Acetone) | 28.5 | 13 | 9.89 9.69 | 27 | 28.5 | 9.79 | — | 2.91 |
| 33. $C_3H_8$ | 26.1 26.25 26.3 26.15 [23.4] [23.5] | 15 11 12, 13 35 37 31 | 11.21 11.08 | 26 | 26.20 [23.45] | 11.15 | 1.11 | 2.37 |
| 34. 1-$C_4H_8$ (1-Butene) | 26.48 26.5 27.09 [24.4] | 35 11, 15 12 37 | 9.72 9.76 9.58 | 27 | 26.69 [24.4] | 9.69 | 1.094 | 2.75 |
| 35. i-$C_4H_8$ (Isobutylene) | 26.5 26.91 [24.4] [23.8] | 35 11 37 31 | 9.35 | 26 | 26.7 [24.1] | 9.35 | 1.108 | 2.86 |
| 36. 2-$C_4H_8$ (*Cis*- and *Trans*- Butene II) | 26.1 26.18 26.42 26.44 [23.9] [23.6] | 15 35 11 14 37 31 | 9.34 9.24 9.13 | 27 | 26.29 [23.75] | 9.24 | 1.107 | 2.85 |
| 37. n-$C_4H_{10}$ (n-Butane) | 25.72 26.18 26.4 [22.9] | 35 12 5 37, 31 | 10.63 10.80 | 27 | 26.10 [22.9] | 10.72 | 1.141 | 2.44 |
| 38. $(CH_3)_3N$ (Trimethyla-mine) | 25.29 [22.8] | 14 31 | 9.02 9.4 7.82 | 27 | 25.29 [22.8] | 8.75 | 1.109 | 2.89 |
| 39. i-$C_4H_{10}$ (Isobutane) | 26.17 26.24 26.52 [23.0] [23.4] | 35 11 12 37 31 | 10.55 10.79 | 26 | 26.31 [23.2] | 10.67 | 1.134 | 2.47 |
| 40. i-$C_5H_{12}$ | [23.9] | 37 | 10.6 10.3 | 26 | 10.45 [23.9] | | — | |
| 41. *cis*-$C_6H_{12}$ | 25.05 [22.7] | 40 39 | 9.88 10.30 | 26 | 25.05 [22.7] | 10.09 | 1.103 | 2.48 |

Table 2.1 (*continued*)

| Atom or molecule | $W$ (eV/ip) | Ref. to $W^d$ | $I$ (eV) | Ref. to $I^d$ | $W^a$ used | $I^a$ used | $\dfrac{W_\alpha}{W_\beta}$ | $\dfrac{W_\alpha}{I}$ |
|---|---|---|---|---|---|---|---|---|
| 42. $(C_2H_5)_2O$ | [23.8] [23.1] | 37 31 | 9.65 9.53 | 26 | [23.45] | 9.59 | — | |
| 43. $C_6H_6$ | 26.93 [23.3] | 40 31 | 9.245 9.210 9.247 | 26 | 26.93 [23.3] | 9.23 | 1.156 | 2.92 |
| 44. $CH_3Cl$ | [25.2] | 31 | 11.28 11.25 11.42 | 26 | [25.2] | 11.32 | — | |
| 45. $CH_2Cl_2$ | [24.5] | 31 | 11.40 11.35 | 26 | [24.5] | 11.37 | — | |
| 46. $CHCl_3$ | [24.5] | 31 | 11.42 | 26 | [24.5] | 11.42 | — | |
| 47. $CH_3CHCl_2$ | [26.4] | 31 | | | [26.4] | | — | |
| 48. $H_3CCOCH_3$ | [25.8] | 31 | 9.705 9.69 9.75 9.89 | 26 | [25.8] | 9.76 | — | |
| 49. $(CH_3)_2O$ | [23.9] | 31 | 10.00 10.15 | 26 | [23.9] | 10.07 | — | |
| 50. $CH_3CH_2NH_2$ | [25.1] | 31 | 9.19 9.5 | 26 | [25.1] | 9.35 | — | |
| 51. Air | 34.95 34.96 35.00 35.2 35.5 35.52[c] 35.6 [33.6] [33.7] [33.9± 0.5] [33.88± 0.06] | 16 19 5 3 4, 10 8 7 22 23 25 30 | | | 35.0 [33.80] | | | |

1. W. P. Jesse and J. Sadauskis, *Phys. Rev.*, **88**, 417 (1952).
2. W. Haeberli, P. Huber and E. Baldinger, *Helv. Phys. Acta*, **26**, 145 (1953).
3. J. M. Valentine and S. C. Curran, *Phil. Mag.*, **43**, 964 (1952).

4. W. P. Jesse and J. Sadauskis, *Phys. Rev.*, **90**, 1120 (1953).
5. T. E. Bortner and G. S. Hurst, *Phys. Rev.*, **93**, 1236 (1954).
6. W. P. Jesse and J. Sadauskis, *Phys. Rev.*, **107**, 766 (1957).
7. J. Sharpe, *Proc. Phys. Soc. (London)*, *A*65, 859 (1952).
8. S. E. Chappell and J. H. Sparrow, *Rad. Res.*, **32**, 383 (1967).
9. Z. Bay and F. D. McLernon, *Rad. Res.*, **24**, 1 (1965).
10. R. Genin, *J. Phys. Radium*, **17**, 571 (1956).
11. G. L. Gels, M.Sc. Thesis, University of Tennessee, 1965.
12. T. E. Bortner, G. S. Hurst, M. Edmundson and J. E. Parks, ORNL-3422 (1963).
13. C. E. Melton, G. S. Hurst and T. E. Bortner, *Phys. Rev.*, **96**, 643 (1954).
14. B. R. Fellers, M.Sc. Thesis, University of Tennessee, 1966.
15. T. D. Strickler, ORNL-3080 (1961).
16. C. Biber, P. Huber and A. Muller, *Helv. Phys. Acta*, **28**, 503 (1955).
17. F. Adler, P. Huber and F. Metzger, *Helv. Phys. Acta*, **20**, 234 (1947).
18. Z. Bay, P. A. Newman and H. H. Seliger, *Rad. Res.*, **14**, 551 (1961).
19. W. P. Jesse, *Rad. Res.*, **13**, 1 (1960).
20. W. P. Jesse, *Phys. Rev.*, **109**, 2002 (1958).
21. Quoted by G. S. Hurst, T. E. Bortner and R. E. Glick, *J. Chem. Phys.*, **42**, 713 (1965).
22. W. Gross, C. Wingate and G. Failla, *Rad. Res.*, **7**, 570 (1957). These authors used $^{35}S$ $\beta$ rays.
23. Z. Bay, W. B. Mann, H. H. Seliger and H. O. Wyckoff, *Rad. Res.*, **7**, 558 (1957).
24. C. E. Klots, *J. Chem. Phys.*, **46**, 3468 (1967).
25. This value is for $^{137}Cs$ and $^{60}Co$ gamma rays (see G. N. Whyte, *Rad. Res.*, **18**, 265 (1963).
26. V. I. Vedeneyev, L. V. Gurvich, V. N. Kondrat'yev, V. A. Medvedev and Ye. L. Frankevich, *Bond Energies, Ionization Potentials and Electron Affinities*, St. Martin's Press, New York (1966).
27. R. I. Reed, *Ion Production by Electron Impact*, Academic Press, New York, 1962.
28. R. W. Law and J. L. Margrave, *J. Chem. Phys.*, **25**, 1086 (1956).
29. R. J. Boyd and D. C. Frost, *Chem. Phys. Letters*, **1**, 649 (1968).
30. W. P. Jesse, *Rad. Res.*, **33**, 229 (1968).
31. R. M. Leblanc and J. A. Herman, *J. Chim. Phys.*, **63**, 1055 (1966); see this reference for additional data on $W_\beta$.
32. R. S. Davidov and D. A. Armstrong, *Rad. Res.*, **28**, 143 (1966).
33. W. P. Jesse and J. Sadauskis, *Phys. Rev.*, **97**, 1668 (1955).
34. G. N. Whyte, *Rad. Res.*, **18**, 265 (1963).
35. C. E. Klots, *J. Chem. Phys.*, **44**, 2715 (1966).
36. W. P. Jesse, *J. Chem. Phys.*, **38**, 2774 (1963).
37. G. G. Meisels, *J. Chem. Phys.*, **41**, 51 (1964).
38. R. K. Appleyard, *Nature*, **164**, 838 (1949).
39. P. Alder and H. K. Bothe, *Z. Naturforschg.*, **20a**, 1700 (1965), see this reference for additional data.
40. L. M. Hunter and R. H. Johnsen (quoted by C. E. Klots, in *Fundamental Processes in Radiation Chemistry*, P. Ausloos, Ed., John Wiley and Sons, New York, 1968, pp. 1–57).

and is important when the ionization potential (*I*) of the impurity is lower than the excitation energy levels of the parent gas. Another impurity effect (which also decreases *W*) arises from the relative magnitude of the first excitation potential of the parent gas and the ionization potential of the impurity. This is due to subexcitation electrons. Such electrons have insufficient energy to cause electronic excitation of the parent gas but may have sufficient energy to ionize the impurity. Clearly then, *W* for the rare gases (especially He) is critically dependent on gas purity. An impurity content of about one part per $10^4$ parts of He reduces the $W$ for He from 46 eV/ip to $\sim 31$ eV/ip (Bortner and Hurst, 1954).

An additional process which is believed to affect the value of $W$ for the rare gases and which may involve a short-lived intermediate is

$$R + R^* \rightarrow R_2^* \rightarrow R_2^+ + e \qquad (2.4)$$

where R is an unexcited and $R^*$ an excited rare gas atom, respectively, $R_2^*$ is an excited short-lived intermediate which may be an excimer†, and $R_2^+$ is a rare gas positive molecular ion. Evidence for process (2.4) comes from optical studies on the continuum emission from rare gases (see, for example, Tanaka (1955)), which is characteristic of Franck-Condon transitions in molecules and also from mass spectroscopic studies on the production of $R_2^+$ in rare gases (see, for example, Hornbeck and Molnar (1951) and Dahler and coworkers (1962)).

In Table 2.1, $W$ values for the rare gases and a number of molecular vapours irradiated by alpha or beta particles are summarized. Generally the $W$ for alpha particles, $W_\alpha$, obtained by different workers show much smaller discrepancies than $W$ values for beta particles, $W_\beta$. It is seen from Table 2.1 that $W$ values vary from $\sim 21$ to 46 eV/ip. For sufficiently large $E_0$, $W$ varies only slowly with $E_0$ and depends slightly on the particle type. Experiments with beta particles for mean energies from 3 to 50 keV established to within $\sim \pm 2\%$ the lack of energy dependence (Jesse and Sadauskis, 1957; Valentine, 1952). Miller (1956) predicted an energy dependence for monoenergetic electrons in the range 1 to $\sim 3$ keV, but for energies greater than $\sim 3$ keV he predicted a constant $W$. Excluding the extreme energy ends, the only slight dependence of $W$ on $E_0$ is understandable since most of the measured ionization is due to the low-energy portion of the degradation spectrum which is comparatively insensitive to the initial electron energy. For the case of alpha particles the end of the track, where electron capture and loss and elastic collisions become important, should be distinguished from the high-energy region where such processes have small probability of occurrence. $W_\alpha$ showed an increase with decreasing alpha particle energy (1–9 MeV) for $N_2$ and $C_2H_4$ (Jesse, 1961). Similar behaviour has been observed for air (Jesse, Forstat and Sadauskis, 1950; Ishiwari and coworkers, 1956).

Jesse and Sadauskis (1955a), and Jesse (1961) have found that the ratio $W_\alpha/W_\beta$ is equal to unity, within experimental error, for the rare gases and molecular hydrogen, but it exceeds unity for all molecules except $H_2$. From the data collected in Table 2.1 it is seen that for certain molecules this ratio is as much as 15% greater than unity. The deviation of the ratio $W_\alpha/W_\beta$ from

---

† Excited dimeric species with a repulsive ground state are known as excimers (Chapter 3).

unity is probably due to the energy dependence of $W_\alpha$ and it may have different explanations for different gases. Track-end effects most probably affect this ratio also. For further discussion on the dependence of $W$ on the energy and type of the ionizing particle, see Binks (1954), Valentine and Curran (1958), Platzman (1961), Whyte (1963), and Section 2.3.

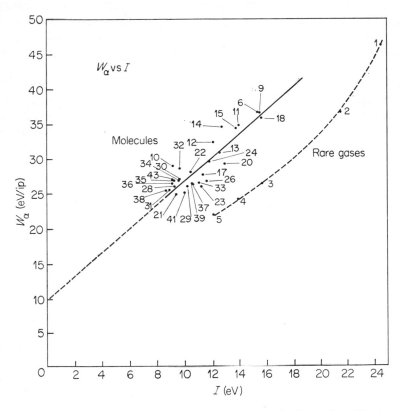

**Figure 2.1** $W_\alpha$ versus $I$. The numbers shown on the figure identify the species as shown in Table 2.1

In Figure 2.1, $W_\alpha$ (see Table 2.1) is plotted as a function of $I$, the atomic or molecular ionization potential. A straight-line least-squares fitting to the data for molecules gave

$$(W_\alpha)_{\text{molecule}} = 9.8 + 1.67\ I \tag{2.5}$$

Similarly,

$$(W_\beta)_{\text{molecule}} = 5.9 + 1.82\ I \tag{2.6}$$

Although the straight-line fitting is somewhat arbitrary (for Equation (2.5) the standard deviation for the intercept is 1.7 and for the slope 0.14), Equations (2.5) and (2.6) describe reasonably well the variation of $W$ with $I$. The difference in the intercept and the slope between (2.5) and (2.6) may be attributed to the effect of elastic collisions, which are known to be important at low ion energies (see Section 2.3). It might be noted that although there is evidence for a slight curvature in the $W_\alpha$ (or $W_\beta$) versus $I$ plot for the rare gases, the ratio $W/I$ is almost constant for the rare gases and equal to $\sim 1.75$. Platzman (1961) discussed the constancy of $W/I$ for the rare gases in terms of the balancing of the energy absorbed, $E_0$, through the equation

$$E_0 = N_i \bar{E}_i + N_{ex} \bar{E}_{ex} + N_i \bar{\varepsilon}_s \qquad (2.7)$$

where $N_i$ is the number of singly charged ions and subexcitation electrons, $N_{ex}$ is the number of excited atoms, and $\bar{E}_i$, $\bar{E}_{ex}$ and $\bar{\varepsilon}_s$ are, respectively, the average energy to produce a singly ionized atom, average excitation energy and average kinetic energy of a subexcitation electron. Dividing both sides of Equation (2.7) by $N_i I$ we have

$$\frac{W}{I} = \frac{\bar{E}_i}{I} + \frac{\bar{E}_{ex}}{I} \frac{N_{ex}}{N_i} + \frac{\bar{\varepsilon}_s}{I} \qquad (2.8)$$

For He, Platzman (1961) gave for the various terms in (2.8) the values $1.72\dagger \simeq 1.06 + 0.85(0.40) + 0.31$. The ratio $\bar{E}_i/I$ exceeds unity due to the production of excited and multiply-charged ions. It is also worth noting that a good proportion of the total oscillator strength lies in the continuum and that an energy of $0.31I$ goes into subexcitation electron production.

For molecules, $W/I$ exceeds its value for the rare gases considerably and varies from 2.2 for $CH_4$ to $\sim 2.9$ for a number of polyatomic gases. Thus a larger portion of the particle's initial energy is dissipated in non-ionizing processes for molecules than for the rare gas atoms. It is worth noting that $W/I$ is lower for some small molecules and $CH_4$,†† while more complex asymmetrical molecules show higher values. Franck-Condon transitions to molecular potential energy curves (surfaces) with equilibrium nuclear separations markedly different from those of the ground state, multiple ionization and excitation, superexcitation that leads to molecular dissociation, and possibly larger (and widely varying) percentages of the total oscillator strength in the continuum may partly account for the variations of $W/I$ in molecules and its larger value for molecules compared to that for the rare gases.

Although the difference between $W/I$ for molecules and the rare gases can

---

† This number was obtained using $W(\text{He}) = 42.3$ eV/ip.
†† Methane behaves like the rare gases in a number of respects (see Chapter 4 on cross sections for the scattering of low-energy electrons by $CH_4$) due probably to its high symmetry.

be attributed to the factors mentioned above, which increase the *W* for molecules, the author believes that a good portion of these differences arises from a decrease in *W* for the rare gases because of increased ionization via process (2.4). Writing $(W_\beta)_{\text{molecule}}$ as

$$(W_\beta)_{\text{molecule}} = \bar{E}_i + \bar{\varepsilon}_s + \bar{E}_{ex} \frac{N_{ex}}{N_i} \tag{2.9}$$

and (for a given *I*) attributing the difference $(W_\beta)_{\text{molecule}} - (W_\beta)_{\text{rare gas}}$ to the effect of process (2.4), we obtain

$$\frac{W_{\text{rare gas}}}{W_{\text{molecule}}} = 1 - \frac{\bar{N}_i}{N_i} \tag{2.10}$$

Here $\bar{N}_i$ is the ultimate number of singly charged ions produced through process (2.4) and $N_i$ is the total number of singly charged ions produced directly or indirectly (via process (2.4)). Using $(W_\beta)_{\text{molecule}}$ as given by Equation (2.6), one obtains for the ratio $\bar{N}_i/N_i$ the values listed in Table 2.2. It is seen from this analysis that an appreciable amount of ionization in the rare gases originates via process (2.4). The values for $\bar{N}_i/N_i$ are close to the ratio $(I - E_1)/I$, where $E_1$ is the energy of the first excited state of the rare gas atom. We also note that using $\bar{E}_i + \bar{\varepsilon}_s \simeq 1.4\ I$ (Platzman, 1961), we obtain from Equations (2.6) and (2.7) the relation $\bar{E}_{ex} N_{ex}/N_i = 5.9 + 0.42\ I$.

**Table 2.2** Ratio $\bar{N}_i/N_i$ for the rare gas atoms

| Rare gas atom | $\bar{N}_i/N_i$ |
| --- | --- |
| He | 0.10[a] |
| Ne | 0.19 |
| Ar | 0.24 |
| Kr | 0.24 |
| Xe | 0.22 |

[a] Assuming $W_\beta = W_\alpha (= 46\ \text{eV/ip})$ (Bortner and Hurst, 1953, 1954).

## 2.3 *W* for Low-Energy Positive Ions

The experimental information on the total ionization produced by charged particles absorbed in gases, discussed in the previous section, has been obtained by using, to a large extent, alpha and beta particles from radioactive sources. This, naturally, limits the range of ion masses and energies. Only a small amount of information exists concerning total ionization produced in gases by slow positive ions. Phipps, Boring and Lowry (1964) studied total ionization in argon using ions from an ion accelerator with masses ranging from 1 to 40 a.m.u. (atomic mass units) and with energies in the range of 8 to 100 keV. Similar studies in nitrogen have been made by Boring, Strohl and Woods (1965). In the low-energy range of these studies the incident ion velocities are comparable to the electron orbital velocities, so that processes relatively unimportant at higher energies, such as elastic scattering and electron capture and loss by the ion, become significant compared to ionization and excitation, and contribute to the loss of energy by the primary ion. It then seems possible that $W$ might depend more strongly on the type and velocity of the incident ion than $W_\alpha$ and $W_\beta$ discussed in the previous section. The results of Boring and coworkers on argon and nitrogen are reproduced in Figure 2.2. $W$ has been plotted as a function of $v_0/v$ (where $v_0(= e^2/\hbar)$ is the electron velocity in the first Bohr orbit of the hydrogen atom) so that a comparison of the results for the various ions with equal velocity can be made. It is clear that for both argon and nitrogen, $W$ depends on the ion velocity (especially for heavy ions) and the ion mass. $W$ increases with decreasing ion velocity (except for protons) and for a given initial ion velocity, $W$ is found to increase slightly as the mass of the ion

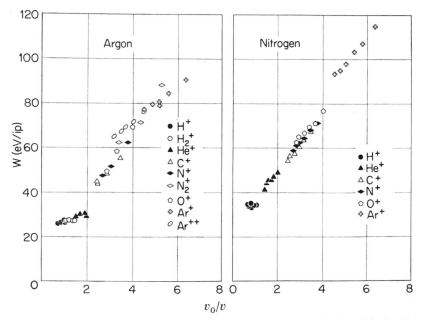

**Figure 2.2** $W$ values for different positive ions incident with varying initial energies in argon (Phipps, Boring, and Lowry 1964), and nitrogen (Boring, Strohl, and Woods, 1965). $W$ is plotted against $v_0/v$ where $v_0 = e^2/\hbar = 2 \cdot 2 \times 10^8$ cm/sec and $v$ is the incident particle velocity

increases (Figure 2.2). This latter regularity seems to hold better for nitrogen than for argon. The $W$ for argon ranges from 26.8 eV/ip for 50 keV protons to 90.9 eV/ip for 25 keV argon ions. For the lighter ions ($H^+$, $H_2^+$) $W$ appears to be approaching, at high velocities, a common value of $\sim 24.6$ eV/ip which is somewhat smaller than $W_\alpha (= 26.4$ eV/ip) for argon. For nitrogen, $W$ ranges from 34.6 eV/ip for 25 keV protons to 115.2 eV/ip for 25 keV argon ions and extrapolates to $\sim 35$ eV/ip as compared to a $W_\alpha$ of 36.4 eV/ip (see Table 2.1). Some previous values of $W$ for protons stopped in argon (Lowry and Miller, 1958; Larson, 1958) are in general agreement with the result of Boring and coworkers.

Approximate calculations by Boring, Strohl and Woods (1965) have indicated the importance of elastic losses at low ion energies. Following Lindhard and coworkers (1963), they calculated the fraction of the incident ion energy which ultimately goes into motion of the gas atoms (elastic). By subtracting the energy lost to atomic motion from the initial ion energy they obtained the energy expended by the ion to inelastic processes (mainly electronic). Combining this energy with the measured ionization they obtained $W$ values which were nearly constant for the various ions in a given gas and approximately the same as those found experimentally at high energies where the energy lost to elastic collisions is small. Thus, as a whole the effect of elastic collisions can account to a first approximation for the experimentally observed variation of $W$ with energy and type of incident ion at low energies. In a subsequent publication Boring and Woods (1968) reported $W$ values for $H^+$, $O^+$ and $Ar^+$ stopped in $CO_2$, $C_2H_4$ and $C_2H_2$. These gases exhibited similar behaviour to that of Ar and $N_2$.

## 2.4 Total Ionization in Gas Mixtures

### 2.4.1 *Normal Mixtures*

Normal mixtures will be referred to here as those for which the Jesse effect is not appreciable. The $W_{ij}$ for such a binary gas mixture usually lies between the extreme values, $W_i$ for the $i$th and $W_j$ for the $j$th component, and changes smoothly from one extreme to the other as the composition of the gas mixture is changed. It has been found for alpha particles (Bortner and Hurst, 1954; Moe, Bortner and Hurst, 1957) that $W_{ij}$ can be expressed as

$$W_{ij}^{-1} = (W_i^{-1} - W_j^{-1})Z_{ij} + W_j^{-1} \tag{2.11}$$

where

$$Z_{ij} = P_i(P_i + a_{ij}P_j)^{-1} \tag{2.12}$$

and $P_i$ and $P_j$ are the partial pressures of the components $i$ and $j$, and $a_{ij}$ is an empirical parameter which depends on the particular $ij$ pair. If we replace in (2.11) $W$ by $100/G(\text{ions})$, the appropriate expression for $G(\text{ions})$ is obtained. Replacing in expression (2.11) $a_{ij}$ by $S_j/S_i$, where $S_j$ and $S_i$ are the molecular stopping powers for the $j$th and $i$th component, respectively, the empirical expression (2.11) becomes identical to that of Huber, Baldinger and Haeberli (1949). The identification of $a_{ij}$ with $S_j/S_i$ appears to be true only for a very limited number of binary mixtures, e.g. $N_2$–$O_2$. However, expression (2.11) has been found experimentally to hold for a large number of normal mixtures, the empirical parameter $a_{ij}$ now being thought of as an 'effective stopping power ratio' (Strickler, 1961, 1963; Klots, 1966, 1967) compounded of two molecular parameters in the sense

$$a_{ij} = \frac{Z_j'}{Z_i'} \tag{2.13}$$

The quantities $Z_j'$ and $Z_i'$ are empirically determined energy partition parameters, which are generally different from the macroscopic stopping powers (for some other expressions used to represent $W_{ij}$ see Valentine and Curran, 1958).

### 2.4.2 *Dose Partition in Gas Mixtures*

The work of Strickler (1961, 1963) on the determination of the parameters $Z_i'$ has been scrutinized recently by Klots (1966, 1967). Table 2.3 lists the 'effective atomic numbers' $Z_i'$ for a number of gases. The '$f$' values given by Strickler† are related to the '$Z$' values of Klots by

$$Z_i' \propto f_i W_i \tag{2.14}$$

All '$Z$' values listed in Table 2.3 are relative to the '$Z$' value for $CH_4$ for which a value of ten has been arbitrarily assigned. The agreement between the results of Strickler and those of Klots is qualitative, and the observed differences may be due to saturation effects. It seems possible from the data listed in Table 2.3 to determine the partition of dose among the components of a gaseous (homogeneous) mixture. However, it has to be pointed out that the '$Z$' parameters hold with regard to the gross production of ionized species, and their use to calculate the yield, $G_{in}$, of an excited state, $n$, below the ionization potential for a gas component from its value in the pure form, $G_{io}$, and the '$Z$' parameters is at present a pure extrapolation.

---

† Strickler (1961, 1963) replaced $a_{ij}$ in Equation (2.12) by $W_j f_j | W_i f_i$ so that
$$W_{ij} = (W_i - W_j) Z_{ij}'' + W_j, \text{ where } Z_{ij}'' = P_i(P_i + f_{ij} P_j)^{-1} \text{ and } f_{ij} = f_j/f_i.$$
He then determined $f_{ij} = f_j/f_i = (f_j/f_k)|(f_i/f_k) = f_{kj}/f_{ki}$, where $i$, $j$ and $k$ refer to any three gases for a number of gas combinations.

**Table 2.3** 'Effective atomic numbers' for various atomic and molecular gases

| Atom or molecule | 'Effective atomic number'[a] | Reference[c] | Atom or molecule | 'Effective atomic number'[a] | Reference[c] |
|---|---|---|---|---|---|
| He | 2.24±0.15 | 1 | $C_2H_4$ | 15.45±0.15 | 2 |
| | | | | 18.2 | 3, 4 |
| Ne | 4.7±0.2 | 1 | $C_2H_6$ | 16.25±0.2 | 2 |
| | | | | 17.8 | 3, 4 |
| Ar | 6.6±0.15 | 1 | $c\text{-}C_3H_6$ | 20.5±0.3 | 2 |
| | | | | 22.4 | 3, 4 |
| Kr | 9.8±0.15 | 1 | $C_3H_6$ | 22.75±0.15 | 2 |
| Xe | 14.9 ±0.4 | 1 | $C_3H_8$ | 22.8 ±0.2 | 2 |
| | | | | 23.0 | 3, 4 |
| $H_2$ | 3.08±0.02 | 2 | $1\text{-}C_4H_8$ | 29.4 ±0.1 | 2 |
| | 3.44 | 3, 4[b] | | 32.1 | 3, 4 |
| $N_2$ | 7.02±0.09 | 2 | $2\text{-}C_4H_8$ | 31.6 | 3, 4 |
| | 7.0 | 3, 4 | | | |
| $O_2$ | 6.24±0.06 | 2 | $i\text{-}C_4H_8$ | 29.8 ±0.2 | 2 |
| | 8.00 | 3, 4 | | | |
| CO | 8.39±0.10 | 2 | $cis\text{-}C_4H_8$ | 30.1 ±0.2 | 2 |
| NO | 7.91±0.07 | 2 | $C_4H_{10}$ | 28.9 ±0.4 | 2 |
| $CO_2$ | 10.9 ±0.1 | 2 | $i\text{-}C_4H_{10}$ | 29.5 ±0.2 | 2 |
| | 11.7 | 3, 4 | | | |
| $C_2H_2$ | 14.7 ±0.3 | 2 | | | |
| | 17.3 | 3, 4 | | | |
| $CH_4$ | 10 | 2 | | | |
| | 10 | 3, 4 | | | |

[a] Relative to $CH_4$ for which a value of 10 is assigned.
[b] Values from References 3 and 4 were obtained through the transformation $Z_i' \propto f_i W_i$ (see text).
[c] References to Table:
1. C. E. Klots, *J. Chem. Phys.* **46**, 3468 (1967).
2. C. E. Klots, *J. Chem. Phys.*, **44**, 2715 (1966).
3. T. D. Strickler, *J. Phys. Chem.*, **67**, 825 (1963).
4. T. D. Strickler, Oak Ridge National Laboratory Report ORNL-3080 (1961).

### 2.4.3 *Mixtures with Pronounced 'Excited-State Effects'*

For gaseous mixtures for which electronic energy from one component A (atomic or molecular) can be transferred to another component X (atomic or molecular) with subsequent ionization, dissociation, or electronic excitation of X (below *I*), Equation (2.11) is inadequate to describe the observed total ionization yields. It may, in this case, appropriately be written as

$$W_{ij}^{-1} = [(W_i^{-1} - W_j^{-1})Z_{ij} + W_j^{-1}] + \text{'excited state effects'} \quad (2.15)$$

The main 'excited-state effect' is thought to arise from[†]

$$A^* + X \rightarrow \text{ion pairs} \tag{2.16}$$

although in the case where A is a rare gas atom, the process

$$A^* + X \rightarrow \text{neutrals} \tag{2.17}$$

may complicate the total ionization analyses because of its quenching effect on the process (2.4) mentioned earlier.

Excited-state effects have been studied for most of the rare gases—for He and Ne by Jesse and Sadauskis (1952, 1953, 1955b); for Ar by Melton, Hurst and Bortner (1954), Bortner and coworkers (1963), Hurst, Bortner and Glick (1965); for Kr by Gels (1965); and for Xe by Fellers (1966). There is also evidence for an excited-state effect in $N_2$ (Strickler, 1963; Klots, 1966). Observations of such mechanisms have been reported also for a number of molecular gases (Čermák, 1965; Olmsted, Newton and Street, 1965).

Figure 2.3 shows the change in $W_{ij}$ with contaminant gas concentration for a typical mixture: argon ($i$) and ethane ($j$). It is seen that $W_{ij}$ drops sharply from its value for argon up to a molar fraction of $\sim 0.04$ and then

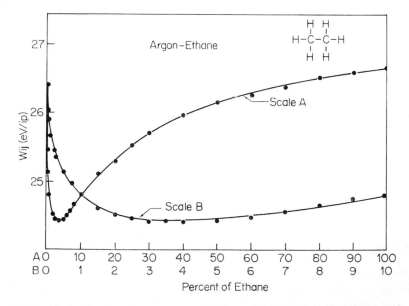

**Figure 2.3** $W_{ij}$ for argon–ethane mixtures as a function of ethane concentration (Bortner and coworkers, 1963)

[†] Ionization of X upon collision with a metastable A* has been known as the Penning effect (Chapter 3).

rises gradually to its value for ethane ($C_2H_6$). This precipitous drop in $W$ for small contaminant gas concentrations, referred to as the Jesse effect, is characteristic of a large number of contaminant gases mixed with a rare gas (see, for example, Bortner and coworkers, 1963).

A detailed study of the role of excited states in the total ionization of gaseous mixtures has been made using argon as the parent gas. The results of this study are summarized in Figure 2.4A where the minimum value of $W$, $W_{min}$, for a given Ar–X mixture is plotted as a function of the ionization potential, $I_X$, of the contaminant gas, X. The difference $W_{Ar} - W_{min}$ is greater when $I_X < 11.5$ eV, is intermediate when $11.5 \le I_X \le 14$ eV, and is insignificant when $I_X > 15.4$ eV. From Figure 2.4A alone it is suggested that two groups of long-lived electronic excited states are produced in Ar by the passage of the charged particle, one at $\sim 11.5$ eV and the other in the region between 14 to 15.4 eV. The eventual fate of these long-lived states is determined by spontaneous emission ($Ar^* \rightarrow Ar + h\nu$), collisions with ground-state argon atoms ($Ar^* + Ar \rightarrow Ar + Ar + h\nu'$), and energy transfer to $X(Ar^* + X \rightarrow Ar + X^*)$, where the excitation energy of X can be either above or below $I_X$. That excited states are produced in argon with sufficiently long lifetimes to allow collisions with ground-state argon atoms is supported by spectroscopic studies on the continuum emission from rare gases (see, for example, Tanaka, 1955; Huffman, Tanaka and Larrabee, 1962; Strickler and Arakawa, 1964) which are characteristic of Franck–Condon transitions in molecules. Also, the large rates for the production of $Ar_2^+$ (Dahler and coworkers, 1962) may be taken as further support for a long-lived $Ar^*$ involved in the production of $Ar_2^+$ (Lampe and Hess, 1964).

Energy loss spectra for 25 keV electrons through argon (Geiger, 1962) showed two peaks below the ionization potential, one at 11.7 eV and the other at 14.2 eV. No distinction between singlet and triplet states of the same total quantum number was made due to inadequate energy resolution, but the 11.7 eV loss was assigned to the resonance levels $3p^5 4s\,^3P_1$(11.62 eV) and $3p^5 4s\,^1P_1$(11.83 eV), while the loss at 14.2 eV to the levels $3p^5 5s\,^3P_1$ (14.09 eV) and $3p^5 5s\,^1P_1$(14.26 eV). Additional states at 14.9 and 15.0 eV were reported by Boersch, Geiger and Hellwig (1962), but these energies correspond to the threshold for $Ar_2^+$ (Boersch, Geiger and Hellwig, 1962; Hornbeck and Molnar, 1951; Marmet and Kerwin, 1960). The metastable states $^3P_0$ (11.72 eV) and $^3P_2$ (11.55 eV) invoked by a number of workers to explain the Jesse effect in argon are also shown in Figure 2.4A. The role of these metastables in the observed total ionization yields cannot be distinguished from that of the neighbouring resonance levels. However, since the Jesse effect is observed for $I_X > 11.7$ eV and since for the pressures used in these experiments resonance states can have long lifetimes (for optimum pressures up to $10^6$ nsec) due to the trapping of resonance radiation (Mitchell and Zemansky, 1961; Holstein, 1947, 1951), resonance states must play a substantial role in the Jesse effect for the rare gases (Hurst, Bortner and Glick, 1965). It might be mentioned that metastable states were invoked to explain the Jesse effect purely on the basis of their long lifetimes (to allow time for collisions between Ar* and X). However, since the apparent lifetime of a resonance state can be equal to or exceed that of a

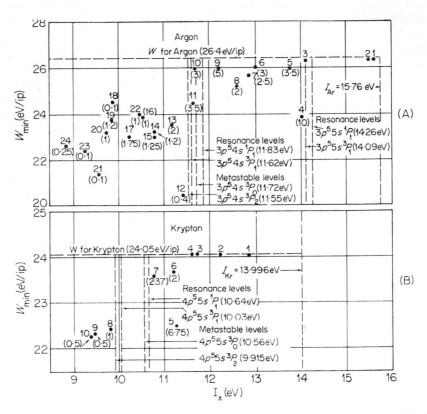

**Figure 2.4** Minimum values of $W$, $W_{min}$, plotted against the ionization potential, $I_X$, of the contaminant gas, X, for the parent gases argon (Fig. 2.4A) and krypton (Fig. 2.4B). Numbers in parentheses shown in the figure are percentages of X at which $W_{min}$ was obtained

Fig. 2.4A   $N_2$[a] (1); $H_2$[a] (2); $CO$[b] (3); $Kr$[a] (4); $CO_2$[c] (5); $CH_4$[c] (6); $N_2O$[c] (7); $H_2O$[a] (8); $O_2$[a] (9); $CCl_2F_2$[a] (10); $C_2H_6$[c] (11); $C_2H_2$[c] (12); $C_3H_8$[c] (13); i–$C_4H_{10}$[c] (14); n–$C_4H_{10}$[c] (15); $C_2H_4$[c] (16); c–$C_3H_6$[c] (17); $C_3H_6O$[b] (18); $C_3H_6$[c] (propylene) (19); l–$C_4H_8$[c] (20); $CH_3I$[a] (21); $C_2H_5OH$[a] (22); $C_6H_6$[a] (23); $C_6H_5CH_3$[a] (24)

Fig. 2.4B[d]   $N_2O$ (1); $O_2$ (2); $CCl_2F_2$ (3); $C_2H_6$ (4); $C_2H_2$ (5); $C_3H_8$ (6); i–$C_4H_{10}$ (7); $C_3H_6$ (propylene) (8); i–$C_4H_8$ (9); c–$C_4H_8$–2 (10)

[a] Melton, Hurst and Bortner (1954);   [b] Hurst, Bortner and Glick (1965);   [c] Bortner and coworkers (1963);   [d] Gels (1965).

metastable, due to the imprisonment of resonance radiation, this assumption seems less justified especially in view of the results on argon shown in Figure 2.4A. The large spreading in the points in Figure 2.4A may be the result of pronounced differences in the ionization efficiency of X.

Further support for the involvement of resonance states in the Jesse effect observed for Ar–X mixtures has been obtained recently by Klots (1968). He used a standard ionization chamber and a vacuum ultraviolet mono-chromator and measured the photoionization efficiency of a number of molecules (previously studied by Hurst, Bortner and Glick (1965) in Ar mixtures with alpha particles) at wavelengths of 1067 and 1048Å. These wavelengths correspond to the energies of the first set of resonance states in argon below which (Figure 2.4A) the Jesse effect is very pronounced. He found a good correlation between the measured photoionization efficiency and the magnitude of the observed Jesse effect. Implicit in this comparison is the assumption that the ionization efficiency of a molecule is independent of the mode of excitation.

Figure 2.4B shows similar studies for krypton binaries. Unlike argon, there is no correlation between the increase in ionization and the metastable and/or resonance states of krypton shown in the figure.

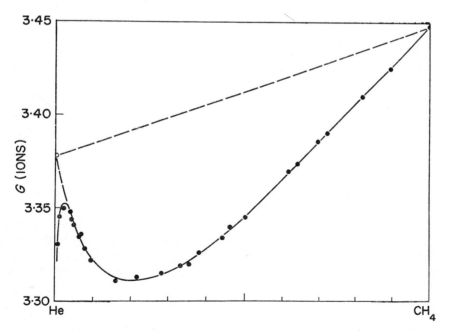

**Figure 2.5** Ion yields for He–$CH_4$ mixtures as a function of
$$P_{CH_4}/P_{CH_4} + 0.217\, P_{He} \qquad \text{(Klots, 1967)}$$

Additionally Klots (1967) discovered what seems to be a negative Jesse effect. When, for example, he added krypton to neon the ion yield increased to a high, as expected, but then it started decreasing to a minimum before it finally reached the value for pure krypton. Similar behaviour has been observed for neon–argon mixtures. Figure 2.5 shows this effect for He–$CH_4$. For this binary the effect may arise from competition between processes (2.16) and (2.17). Such a competition, obviously, cannot explain the effect for binaries consisting of two rare gases (e.g. Ne–Kr). In this case competition between (2.4) and some collision process with Kr resulting in de-excitation of Ne* without ionization may be occurring.

## 2.5 Isotope Effect in the Total Ionization of Gases

An isotope effect has been observed in the probability of molecular ionization when such an act was induced by collision with a rare gas atom (Jesse and Platzman, 1962; Jesse, 1964), by beta particles (Jesse, 1963), by alpha particles (Jesse, 1967) and by 70 eV electrons (Meyerson, Grubb and Vander Haar, 1963). A summary of these results is given in Table 2.4. In all cases investigated the ionization yield for the compound was smaller than that of its deuterated form†. Similar effects have been found in photo-ionization studies (Person, 1965a, b, 1967). The photoionization work (Weissler, 1956; Person, 1965a, b, 1967) is in accord with the data discussed in this section and will be elaborated upon in the next Chapter.

The marked difference in the isotope ratio for the hydrocarbon gases listed in Table 2.4 obtained by alpha and beta particles has been attributed (Jesse, 1967), in part, to differences in the energy spectrum of the secondary electrons in the two cases. The author believes that elastic collisions are, in part, responsible for these differences. Methane shows a small isotope effect compared to the other polyatomic gases while the ionization ratio for $D_2/H_2$ remains constant for alpha, beta and gamma excitation in contrast to the hydrocarbon gases.

The results presented in Table 2.4 and those on photoionization (Chapter 3) show an isotope effect in the total ionization cross section. They furnish evidence that ionization at least at energies within a few Rydbergs of the ionization potential $I$ occurs largely as a secondary process rather than by direct ejection of an orbital electron during the excitation act (within $\sim 10^{-16}$ sec) for a large number of molecules. In the next section we discuss this in some more detail.

---

† Recently Herce and coworkers (1968) studied the relative ion abundances in the Penning ionization of $CH_4$ and $CD_4$ on impact of $2^3S$ and $2^1S$ helium atoms. They did not observe any marked isotope effect in the relative abundancies of isotopic ions in collision with the same metastable atom.

**Table 2.4** Ionization ratios for deuterated and undeuterated gases

| Gases | Ratio for $\alpha$ particles[a] | Ratio for $\beta$ particles[b] | Ratio for $\gamma$-rays[a] | Ratio for mixtures with a rare gas[c] | Ratio for 70-eV electrons[d] |
|---|---|---|---|---|---|
| $D_2|H_2$ | 1.010 | 1.008 | 1.009 | | |
| $C_2D_2/C_2H_2$ | 1.017 | | | | |
| $CD_4/CH_4$ | 1.013 | 1.006 | | | |
| $C_2D_6/C_2H_6$ | 1.019 | 1.008 | | | |
| $C_2D_2H_2/C_2H_4$ | 1.021 | | | | |
| $C_2D_4/C_2H_4$ | 1.034 | 1.013 | | | |
| $C_2H_4$ | | | | 1.008 (in Ar) 1.010 (in Ne) | |
| n-$C_4H_{10}$ | | | | 1.007 (in Ar) 1.008 (in Ne) | |
| -1,3,5-$d_3$/Benzene unlabelled | | | | | 1.01 |
| -$d_6$/Benzene unlabelled | | | | | 1.03 |
| -$\alpha$-$d_3$/Toluene unlabelled | | | | | 1.01 |
| -ring-$d_5$/Toluene unlabelled | | | | | 1.01 |
| -$d_8$/Toluene unlabelled | | | | | 1.04 |
| -$d_{10}$/Phenanthrene unlabelled | | | | | 1.04 |

[a] Jesse (1967); $\alpha$ particles from $^{210}$Po (5.3 MeV).
[b] Jesse (1963); for the hydrocarbon gases the mean beta kinetic energy was $\sim$18 keV while for $H_2$ and $D_2$ $\sim$3 keV.
[c] Jesse and Platzman (1962). The ratio referred to here is given by $N_i + \Delta N_i^D/N_i + \Delta N_i^H$ where $N_i$ is the ionization for the pure parent gas (Ne or Ar) and $\Delta N_i^D$ and $\Delta N_i^H$ is the increase in ionization when small amounts of the deuterated and undeuterated gases are mixed with the parent gas. The ratio is taken for the plateau region.
[d] Meyerson, Grubb and Vander Haar (1963).

## 2.6 Superexcited States

As has been discussed in the first chapter, a large portion of the total oscillator strength of the valence electrons lies at energies above the lowest ionization potential $I$. A neutral atom or molecule excited by photon absorption, collision with excited atoms, or transfer of energy from a moving charged particle to an electronic state with energy $E_n > I$ is said to be in a superexcited state at $E_n$ (Platzman, 1962). For atoms, states with $E_n > I$

usually have more than one excited electron, or, alternatively, a single electron is excited with the residue in other than the lowest level of its ground-state configuration. Much is known about preionization (auto-ionization) in atoms (see, for example, Fano, 1961; Temkin, 1966; Marr, 1967 and Chapter 3) in which the process involves only a reorgani-zation of the electronic system. The probability that an atomic superexcited state will autoionize is practically unity although emission spectra from superexcited states are known. In atomic absorption spectra, configuration interaction between the discrete and continuous states causes a characteristic interference (Fano, 1961) which reduces the intensity of the underlying continuum at one side of the absorption line and enhances it on the other. The same effect is observed (Ogawa and Tanaka, 1962) with simple mole-cules in Rydberg states which converge to an ionization energy greater than $I$ (Chapter 3).

Complex molecules appear to behave differently. Platzman (1962) showed that superexcited states for molecules are distributed continuously from $I$ up to excitation energies of a few Rydbergs above $I$. Figure 2.6 shows the excitation spectrum of methane (Platzman, 1962) with inclusion of the main band of its superexcited states. Instead of the differential oscillator strength, $df/dE$, the quantity $(R/E)\,df/dE$ (where $R$ is the Rydberg energy) is plotted as a function of $E$. The mean energy of the superexcitation band shown in

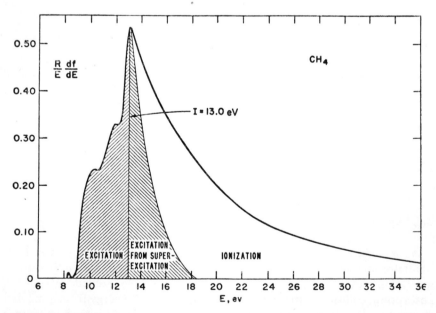

**Figure 2.6** Excitation spectrum of methane with inclusion of the main band of its superexcited states (Platzman, 1962)

the figure is approximately at 15 eV. Excitation spectra including superexcited states have been reported also for $H_2O$ and $C_2H_5OH$ (Platzman, 1967).

Molecules in a superexcited state, especially at energies within $\sim 20$ eV of $I$, have a probability or efficiency of ionization $\eta$† appreciably smaller than unity. Thus, preionization of a superexcited state competes with atomic reorganization which diverts part of the excitation energy into chemical energy or heat. We shall refer to all non-ionizing atomic reorganization processes (dissociation, predissociation, internal conversion) as 'dissociation'. The competition between indirect ionization and 'dissociation' has been shown in a number of cases.

In their original work, Jesse and Platzman (1962) assumed in discussing the data in column 5 of Table 2.4 that when energy $E(> I)$ is imparted to a molecule, two alternative pathways can be followed: (i) direct ionization with a probability $\delta$, i.e.

$$AB + energy \rightarrow AB^+ + e \qquad (2.18)$$

and (ii) superexcitation with a probability $(1 - \delta)$, i.e.

$$AB + energy \rightarrow AB^*$$

followed by

$$AB^* \nearrow^{AB^+ + e \text{ at a rate } k_1 \text{ (preionization)}}_{\searrow A + B \quad \text{ at a rate } k_2 \text{ ('dissociation')}}$$

On this simple model the total ionization efficiency $\eta$ can be written as

$$\eta = \delta + (1 - \delta) k_1 (k_1 + k_2)^{-1} \qquad (2.19)$$

If $k_1 \ll k_2$, then $\eta \simeq \delta$ and preionization and atomic reorganization would not be competitive. If, however, $\delta \ll 1$, then

$$\eta \simeq \left( \frac{k_1 + k_2}{k_1} \right)^{-1} \qquad (2.20)$$

and it measures a true competition between preionization and dissociation. Presumably, $\delta$ and $k_1$ are mass independent and the isotope dependence of the total ionization yields, discussed in Section 2.5, arises from differences in the mass dependent $k_2$. Taking $k_2 = k_2' (M)^{-1/2}$ where $M$ is an effective reduced mass for such atomic motions as to bring the molecule to a point subsequent to which preionization is not possible, expression (2.19) becomes

$$\eta = \delta + (1 - \delta)(1 + M^{-1/2} k_2'/k_1)^{-1} \qquad (2.21)$$

Although a test of (2.21) requires knowledge of $\delta$, $k_2'/k_1$ and $M$, one may

---

† In analogy to the photoionization efficiency (Chapter 3), we can define $\eta$ as the probability of AX* being ionized to the total probability of it being formed.

calculate an ionization ratio for isotopic species assuming $\delta \ll 1$ and a specific molecular splitting. Calculations of this nature have been done by Jesse and Platzman (1962) for the cases listed in column 5 of Table 2.4, assuming symmetrical splitting or dissociation of a hydrogen atom. Their results are in general agreement with the measured ionization ratios. The higher values of the ionization ratio for certain pairs in Table 2.4 may be indicative of an asymmetrical molecular fragmentation. It is finally noted that although the assumption that $\delta \ll 1$ is correct for a number of molecular systems, we now know (see Section 3.5.3) that $\eta \simeq \delta$ for others.

The isotope effects obtained by high-energy charged particles can be used to demonstrate the existence of an isotope effect in $\eta$ (see, however, Section 3.5.3 in Chapter 3). For the energies employed (see Table 2.4), most of the ionization is produced in collisions with electrons of kinetic energies $E \gg I$. Platzman (1961, 1963) calculated that in the complete absorption of 10 keV electrons in helium, 69% of the ions are produced by electrons with $E > 300$ eV and 82% with $E > 100$ eV. Thus, the observed isotope effect for energetic charged particles essentially is an isotope effect in the mean total ionization cross section, $\sigma_i(E)$, by electrons with $E \gg I$. For high electron energies $E$, $\sigma_i(E)$ is given by the Bethe asymptotic formula (Fano, 1954),

$$\sigma_i(E) = \frac{4\pi a_0^2}{E/R} M_i^2 \ln c_i E \tag{2.22}$$

where $a_0$ is the Bohr radius and $c_i$ is a collision constant, expected to be approximately constant for isotopic molecules, and[†]

$$M_i^2 = \int_I^\infty \frac{df}{dE} \frac{R}{E} \eta(E)\, dE \simeq \bar{\eta} \int_I^\infty \frac{df}{dE} \frac{R}{E}\, dE = \bar{\eta} M_I^2 \tag{2.23}$$

Thus, the isotope effect in the total ionization implies that $M_i^2$ is isotope dependent and since $M_I^2$ is mass independent, the isotope dependence of $M_i^2$ is to be found in $\bar{\eta}$.

The Bethe asymptotic formula (2.22) containing $M_i^2$ offers a way to compare the isotope effects in the total ionization of molecular gases obtained by electrons (Table 2.4) with those obtained by direct photo-absorption measurements. Writing

$$\frac{\sigma_{iD}(E)}{\sigma_{iH}(E)} = \frac{M_{iD}^2}{M_{iH}^2} \tag{2.24}$$

---

[†] Because $\bar{\eta} < 1$ for molecules, $M_i^2$ is no longer a purely optical constant. It is less than the integrated square of the dipole matrix element, $M_I^2$ (Fano, 1954). Further, $\sigma_i$ drops below that predicted by Equation (2.22) at low energies, especially near the ionization threshold.

where D refers to the deuterated and H to the undeuterated form of the species involved, one may compare the relationship between $\sigma_i$ and $M_i^2$. An effort to test (2.24) for the case of $C_6D_6$ and $C_6H_6$ has been made by Person (1965b, 1967) who used his data on the photoabsorption cross section $\sigma_p(E)$ and photoionization efficiency $\eta_p(E)$ to calculate $M_i^2$. The highest photon energy used in Person's photoabsorption experiments was 11.64 eV. Data for higher photon energies are necessary in the calculation, due to the fact that a limiting energy range weights the contribution to $M_i^2$ from the low values of $E$ at the expense of the high values. In spite of this limitation, Person's calculations gave a value of $1.04 \pm 0.02$ for the ratio $M_{iD}^2/M_{iH}^2$ (D referring to $C_6D_6$ and H to $C_6H_6$), which is to be compared with the value of $1.03 \pm 0.01$ obtained by Meyerson and coworkers (1963) for $\sigma_{iD}/\sigma_{iH}$ using 70 eV electrons.

Finally, it has to be pointed out that although the phenomenon of auto-ionization in atoms and preionization in molecules has been known for a long time and its importance to molecular states of large oscillator strength has been strongly emphasized by many workers, it is only recently that these studies have attracted intense attention (see Marr, 1967 and Chapter 3). Superexcitation is a predominant primary act in both radiation interaction and photoabsorption ($h\nu > I$). By virtue of the high energy (typically much larger than the chemical bond strengths) possessed by a molecule in a superexcited state, events subsequent to superexcitation may be quite different from those following the formation of a discrete excited state with $E_n < I$. Molecular fragmentation may lead to the production of fragment ions with large amounts of excitation energy. The formation of such excited molecular ions following superexcitation is bound to influence the sequence of ion–molecule reactions. Further, the so-called Kasha's rule (i.e. the internal (non-radiative) conversion of the excitation energy of high-lying excited states to the lowest excited state of the molecule (Chapter 3)) cannot apply to the case of superexcitation.

## 2.7 References

W. Binks, (1954). *Acta Radiol. Suppl.*, **117**, 85.
H. Boersch, J. Geiger and H. Hellwig, (1962). *Phys. Letters*, **3**, 64.
J. W. Boring, G. E. Strohl and F. R. Woods, (1965). *Phys. Rev.*, **140**, A1065.
J. W. Boring and F. R. Woods, (1968). *Rad. Res.*, **35**, 472.
T. E. Bortner and G. S. Hurst, (1953). *Phys. Rev.*, **90**, 160.
T. E. Bortner and G. S. Hurst (1954). *Phys. Rev.*, **93**, 1236.
T. E. Bortner, G. S. Hurst, M. Edmundson and J. E. Parks (1963). ORNL-3422.
V. Čermák (1965). In *Proceedings 4th International Conference on Physics of Electronic and Atomic Collisions*, Quebec, Science Bookcrafters Inc., New York, p. 75.
J. S. Dahler, J. L. Franklin, M. S. B. Munson and F. H. Field (1962). *J. Chem. Phys.*, **36**, 3332.

A. Dalgarno and G. W. Griffing (1958). *Proc. Roy. Soc.* (*London*), *A*248, 415.

G. A. Erskine (1954). *Proc. Roy. Soc.* (*London*), *A*224, 362.

U. Fano (1946). *Phys. Rev.*, **70**, 44.

U. Fano (1954). *Phys. Rev.*, **95**, 1198.

U. Fano (1961). *Phys. Rev.*, **124**, 1866.

B. R. Fellers (1966). A Study of Alpha-Particle Ionization in Xenon Mixtures, M. Sc. thesis, University of Tennessee.

J. Geiger (1962). Streuung von 25-keV Elektronen an Helium, Neon, Argon, Krypton und Xenon, doctoral dissertation, Technical University of Berlin.

G. L. Gels (1965). A Study of Alpha-Particle Ionization in Krypton Mixtures, M. Sc. thesis, University of Tennessee.

L. H. Gray (1936). *Proc. Roy. Soc.* (*London*), *A*156, 578.

L. H. Gray (1944). *Proc. Cambridge Phil. Soc.*, **40**, 72.

J. A. Herce, J. R. Penton, R. J. Cross and E. E. Muschlitz, Jr. (1968). *J. Chem. Phys.*, **49**, 958.

T. Holstein (1947). *Phys. Rev.*, **72**, 1212.

T. Holstein (1951). *Phys. Rev.*, **83**, 1159.

J. A. Hornbeck and J. P. Molnar (1951). *Phys. Rev.*, **84**, 621.

P. Huber, E. Baldinger and W. Haeberli (1949). *Helv. Phys. Acta*, **23**, Supl. III, 85.

R. E. Huffman, Y. Tanaka and J. C. Larrabee (1962). *J. Opt. Soc. Am.*, **52**, 851.

G. S. Hurst, T. E. Bortner and R. E. Glick (1965). *J. Chem. Phys.*, **42**, 713.

R. Ishiwari, S. Yamashita, K. Yuasa and K. Miyake (1956). *J. Phys. Soc. Japan*, **11**, 337.

W. P. Jesse (1961). *Phys. Rev.*, **122**, 1195.

W. P. Jesse (1963). *J. Chem. Phys.*, **38**, 2774.

W. P. Jesse (1964). *J. Chem. Phys.*, **41**, 2060.

W. P. Jesse (1967). *J. Chem. Phys.*, **46**, 4981.

W. P. Jesse, H. Forstat and J. Sadauskis (1950). *Phys. Rev.*, **77**, 782.

W. P. Jesse and R. L. Platzman (1962). *Nature*, **195**, 790.

W. P. Jesse and J. Sadauskis (1952). *Phys. Rev.*, **88**, 417.

W. P. Jesse and J. Sadauskis (1953). *Phys. Rev.*, **90**, 1120.

W. P. Jesse and J. Sadauskis (1955a). *Phys. Rev.*, **97**, 1668.

W. P. Jesse and J. Sadauskis (1955b). *Phys. Rev.*, **100**, 1755.

W. P. Jesse and J. Sadauskis (1957). *Phys. Rev.*, **107**, 766.

C. E. Klots (1966). *J. Chem. Phys.*, **44**, 2715.

C. E. Klots (1967). *J. Chem. Phys.*, **46**, 3468.

C. E. Klots (1968). *Chem. Phys. Letters*, **2**, 645.

F. W. Lampe and G. G. Hess (1964). *J. Am. Chem. Soc.*, **86**, 2952.

H. V. Larson (1958). *Phys. Rev.*, **112**, 1927.

J. Lindhard, V. Nielsen, M. Scharff and P. V. Thomson (1963). *Kgl. Danske Videnskab. Selskab. Mat. Fys. Medd.* 33, No. 10.

R. A. Lowry and G. H. Miller (1958). *Phys. Rev.*, **109**, 826.

P. Marmet and L. Kerwin (1960). *Can. J. Phys.*, **38**, 787.

G. V. Marr (1967). *Photoionization Processes in Gases*, Academic Press, New York.

C. E. Melton, G. S. Hurst and T. E. Bortner (1954). *Phys. Rev.*, **96**, 643.

S. Meyerson, H. M. Grubb and R. W. Vander Haar (1963). *J. Chem. Phys.*, **39**, 1445.

W. F. Miller (1956). *Bull. Am. Phys. Soc.*, **1**, 202.

A. C. G. Mitchell and M. W. Zemansky (1961). *Resonance Radiation and Excited Atoms*, Cambridge University Press, London.

H. J. Moe, T. E. Bortner and G. S. Hurst (1957). *J. Phys. Chem.*, **61**, 422.

M. Ogawa and Y. Tanaka (1962). *Can. J. Phys.*, **40**, 1593.

J. Olmsted, A. S. Newton and K. Street Jr. (1965). *J. Chem. Phys.*, **42**, 2321.
J. C. Person (1965a). *J. Chem. Phys.*, **43**, 2553.
J. C. Person (1965b). In *Proceedings 4th International Conference on Physics of Electronic and Atomic Collisions* Quebec, p. 419.
J. C. Person (1967). Argonne National Laboratory Report ANL-7360, p. 44.
J. A. Phipps, J. W. Boring and R. A. Lowry (1964). *Phys. Rev.*, **135**, A36.
R. L. Platzman (1960). *J. Phys. Radium*, **21**, 853.
R. L. Platzman (1961). *Int. J. Appl. Rad. and Isotopes*, **10**, 116.
R. L. Platzman (1962). *Vortex*, **23**, 372.
R. L. Platzman (1963). *J. Chem. Phys.*, **38**, 2775.
R. L. Platzman (1967). In G. Silini (Ed.), *Proc. Third Intern. Congress of Radiation Research*, North-Holland Publishing Co., Amsterdam, pp. 20–42.
T. D. Strickler (1963). *J. Phys. Chem.* **67**, 825.
T. D. Strickler (1961). ORNL-3080.
T. D. Strickler and E. T. Arakawa (1964). *J. Chem. Phys.*, **41**, 1783.
Y. Tanaka (1955). *J. Opt. Soc. Am.*, **45**, 710.
A. Temkin (1966). *Autoionization (Astrophysical, Theoretical and Laboratory Experimental Aspects)*, Mono Book Corp., Baltimore.
J. M. Valentine (1952). *Proc. Roy. Soc. (London)*, *A***211**, 74.
J. M. Valentine and S. C. Curran (1958). *Repts. Progr. in Phys.*, **21**, 1.
J. Weiss and W. Bernstein (1957). *Rad. Res.*, **6**, 603.
G. L. Weissler (1956). In S. Flügge (Ed.), *Handbuch der Physik*, Springer-Verlag, Berlin, Vol. 21, pp. 304–382.
G. N. Whyte (1963). *Rad. Res.*, **18**, 265.

# 3 Photophysical processes

## 3.1 Introduction

In this chapter we shall discuss photon–atom (molecule) and excited atom (molecule)–excited or unexcited atom (molecule) interactions. The photon energies will be assumed to be low enough (say, below 1 KeV; in most of our discussion below $\sim 100$ eV) so that the photon wavelength ($\lambda$) is much longer than the size of the atomic and molecular structures responsible for photon absorption. This ensures that the photon momentum and the attendant Compton scattering are negligible.

We shall first inquire as to how low-energy electromagnetic radiation is absorbed by the electronic system of atoms and molecules and subsequently we shall ask the question: where does the excitation energy go? That is, what are the fates of the excited electronic states produced by photon absorption? The first inquiry will lead us to a discussion of photoabsorption processes (discrete or continuum) above and below the ionization potential, and the latter to the processes of autoionization, predissociation, pre-ionization, internal conversion, radiative decay of excited electronic states and the associated emission spectra, as well as to the various energy transfer processes. At the end of the chapter a brief discussion will be made of some pertinent work on processes involving excitation other than by low-energy photons.

It is outside of our scope to present a complete coverage of the vast field of atomic and molecular photophysics. However, we shall endeavour to cover some of the recent advances, especially as they relate to radiation and life sciences. The reader is referred to a recent review article by Fano and Cooper (1968) and to the books by Marr (1967), Herzberg (1950, 1956, 1966), Jaffé and Orchin (1964), Kauzmann (1957) and Birks (1970). (See also references quoted in subsequent discussions in this chapter.)

When the photon energy is below the first ionization potential of the atom or molecule, the photoabsorption processes are not complicated by ionization products and only the direct action of the incident photon is important. The resultant spectrum, as we stated in Chapter 1, is characteristic of electronic absorption of discrete atomic and molecular structures and is referred to as the *absorption or optical spectrum*. We may note that although most of the photophysical studies on atoms and simple molecules are performed in the gaseous phase, such studies on polyatomic molecules are often performed in solution or the liquid and solid-state phases. In the former type of experiments the effect of environment on the atomic and molecular properties is negligible (especially at low pressures) compared to those which arise in the liquid and solid-state phases. There exists an extensive literature on the investigation of environmental effects on the electronic properties of atoms and molecules. Absorption studies and emission studies (both in terms of spectra and decay times) can be employed to investigate atomic and molecular interactions in high-pressure and condensed-phase experiments.

Special attention will be focused on aromatic molecules since many disciplines find interest in these systems. Aromatic molecules possess delocalized $\pi$ electron structures which are characterized by strong selective absorption in the visible and near ultraviolet region of the optical spectrum. They fluoresce, phosphoresce, and scintillate (Section 3.4). Some form basic units of structures of living matter, while others are excellent scintillating materials (Birks, 1964). Some of the higher members are carcinogenic. Knowledge of the physical properties of these systems is directly relevant to radiation and life sciences. There is still another reason why we should look into the photophysics of aromatic molecules. This stems from their relative simplicity (resulting from their $\pi$ electron structures) and also from the experimental fact that higher $\pi$ electronic states quickly (see Section 3.4) convert internally and non-radiatively to the lowest excited state of the same multiplicity. Therefore, quite generally, optical emission and energy-transfer processes proceed through the lowest excited $\pi$ electronic states (singlets or triplets) of these molecules. The importance of this experimental fact in radiation and life sciences is evident, since a great deal of information suggests that many biological and chemical reactions come about by a true excited state mechanism. The following references may be considered representative of the vast literature on the physics of these systems: Platt (1961); Kotani, Ohno and Kayama (1961); Coulson (1961); Streitwieser (1961); Roberts (1962); Jaffé and Orchin (1964); Clar (1964); Platt and coworkers (1964a, 1964b); Berlman (1965); Coulson and coworkers (1965); Clementi (1968); Birks (1970).

## 3.2 Electronic Excitation and Ionization by Photon Absorption

### 3.2.1 *Production of Excited States*

#### 3.2.1.1 Atomic photoabsorption

The energy distribution of a system of atoms and molecules in thermal equilibrium is determined by Boltzmann statistics, so that at normal temperatures practically all the atoms or molecules are in their ground (lowest) electronic or vibronic states prior to excitation. Provided the photon excitation intensity is low, the absorption is almost entirely due to atoms or molecules in their ground states.

Now, consider a parallel beam of continuous light falling onto an absorption cell containing a monatomic gas at low pressure. The transmitted intensity $I(v)$ may be represented as shown on the left-hand-side of Figure 3.1, indicating that the system possesses an absorption line at the frequency $v_0$. We may write for the transmitted intensity $I(v)$ (Lambert's law)

$$I(v) = I_0 e^{-k(v)x} \tag{3.1}$$

where $I_0$ is the intensity at $x = 0$ and $k(v)$ is the absorption coefficient expressed in $cm^{-1}$, when $x$ is expressed in cm. The variation of the absorption coefficient $k(v)$ with the frequency $v$ can be found from Equation (3.1) and the transmitted intensity as a function of $v$ (Figure 3.1). This is shown on the right-hand-side of Figure 3.1. The total width, $\Delta v$, of the line profile in Figure 3.1 at one-half of its maximum value is called the half-width of the absorption line. In the absence of other line-broadening factors (Doppler broadening, pressure broadening, etc.; see Mitchell and Zemansky (1961)), $\Delta v$ is related to the spontaneous transition probability through the uncertainty principle (Section 3.3.4). In practice, the natural line width is obscured by a number of broadening processes including the instrumental line width.

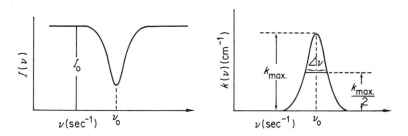

**Figure 3.1** An absorption line. Variation of the absorption coefficient with frequency in an absorption line

Electronic transitions from an initial atomic state to higher excited states yield absorption lines which form a series. The energy separation between higher states decreases and the series converges on the series limit, the ionization potential $I$. Beyond $I$ the absorption spectrum is not discrete but continuous, since the atom is ionized and the excess energy $(hv - I)$ can appear as kinetic energy of the ejected electron. Discrete transitions in the ionization continuum can and do take place (Section 3.5). A schematic energy level diagram for atomic systems is shown in Figure 3.2.

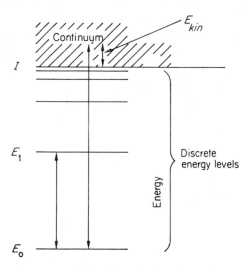

**Figure 3.2** Schematic energy level diagram
for atomic systems

For low enough pressures for which intermolecular forces can be neglected, the absorption coefficient $k(v)$ (in cm$^{-1}$) can be expressed in terms of the absorption cross section $\sigma_{ab}(v)$ (in cm$^2$), and the number density $N$ (defined here as the number of atoms $(2.69 \times 10^{19})$ per cm$^3$ at 273 °K and 760 torr), namely, $\sigma \times N = k$. If we conveniently normalize the optical path $x$ we may write for $\sigma_{ab}(v)$

$$\sigma_{ab}(v) = \frac{T}{273} \frac{760}{P} \frac{1}{x} \frac{1}{N} \ln \frac{I_0}{I} \qquad (3.2)$$

where the pressure $P$ is in torr and $\sigma_{ab}(v)$ in cm$^2$. In view of the magnitude of $\sigma_{ab}(v)$ in photoabsorption processes, it is usually given in units of megabarns (1 Mbn$\equiv 10^{-18}$ cm$^2$).

### 3.2.1.2 Molecular photoabsorption

The molecular absorption spectrum, in contrast to the atomic, is a great deal more complicated owing to the additional degrees of freedom present in molecules. A molecule can exist not only in discrete electronic energy states, but also a number of vibrational states are associated with each electronic state, and a number of rotational states are associated with each vibrational state. The molecular absorption spectra, therefore, are of three types, rotation, vibration, and electronic. The total energy, $E$, of a molecule in an energy state, aside from the translational energy, can be expressed as

$$E = E_{el} + E_{vib} + E_{rot} \tag{3.3}$$

where $E_{el}$, $E_{vib}$, and $E_{rot}$ represent the electronic, vibrational, and rotational energy, respectively.

The rotation spectrum of a molecule is associated with changes in its rotational states without simultaneous changes in the vibrational and electronic states. The separation of the rotational energy levels is of the order of a few hundredths of one eV, and this relatively small energy spacing results in absorption in the far infrared region ($\lambda > 20$ microns).

The vibration and vibration–rotation spectrum of a molecule is associated with changes which occur in the vibrational states of the molecule without simultaneous changes in the electronic state. Since changes in vibrational states are generally accompanied by changes in rotational states the vibrational spectrum has rotational fine structure. The spacing of vibrational energy levels is of the order of a few tenths of one eV and the most important vibrational spectra occur in the near and middle infrared regions ($\lambda \lesssim 30$ microns).

Electronic spectra arise from transitions between electronic states. Such transitions are accompanied with simultaneous changes in the vibrational and rotational states. The electronic spectrum of a molecular gas may assume a very complicated structure as a result of the superposition of vibrational and rotational changes on the electronic transition. Rotational structure is not resolved and vibrational structure is often barely resolved in the electronic spectra of solutions, liquids, and solids. Here the molecular electronic spectra consist of relatively broad bands which usually lie in the ultraviolet ($\lambda \lesssim 4000\text{Å}$) and visible ($4000\text{Å} \lesssim \lambda \lesssim 8000\text{Å}$) region.

The transitions just discussed are shown schematically in Figure 3.3 for a diatomic molecule. The upper state (excited) shown in Figure 3.3 is stable (attractive), i.e. it possesses a potential minimum. It could, however, have been unstable (repulsive) and the transition would have resulted in molecular dissociation. Generally, absorption of a photon with energy greater than the dissociation limit of the excited state leads to dissociation of the diatomic molecule, the excess energy appearing as kinetic and/or excitation energy of the fragments. If in a molecular system discrete energy levels overlap a continuous range of energy levels, a radiationless transition from a discrete level to the continuum (transition from a discrete state into a stable state above its dissociation limit for example) can lead to molecular dissociation. The process is called *predissociation* and, being a non-radiative intra-molecular energy transfer process, will be discussed in Section 3.4.2 in connection with other energy-transfer processes.

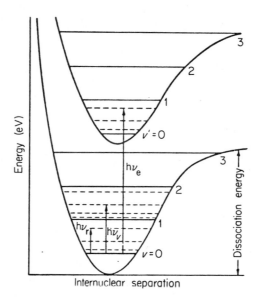

**Figure 3.3** Potential energy level diagram for a diatomic molecule. $h\nu_r$, $h\nu_v$, and $h\nu_e$ stand for pure rotational, vibrational–rotational, and electronic transitions, respectively (Marr, 1967)

In Figure 3.3 we have shown all upward transitions starting from the lowest ($v = 0$) vibrational level of the ground state. This is justified since at room temperature the largest fraction of molecules is in the lowest ($v = 0$) vibrational level. We may further note that since the ground electronic states of most molecules are singlets (electrons are paired) the excited electronic states reached by photon absorption are also singlets (to satisfy the spin selection rule ($\Delta S = 0$)). That is, photoabsorption studies yield no information on spin-forbidden transitions unless other techniques (e.g. perturbation techniques, luminescence spectroscopy, flash photolysis, and particle impacts) are employed.

When the photon energy, $hv$, exceeds the ionization potential, $I$, the photoabsorption process may lead to direct ionization or to the excitation of a superexcited state (Section 3.5). In the former case the ejected electron carries away as kinetic energy part or all of the excess energy $(hv - I)$, and in the latter case the molecule will either 'dissociate' or undergo indirect ionization. As we have mentioned in the previous chapter (Section 2.6) superexcited states of molecules constitute broad bands extending from $I$ to within a few Rydbergs of $I$. Ionization of a superexcited state, i.e., a radiationless transition from a stable molecular state to an ionic state, is called *preionization*, in direct analogy to the similar process for atoms, the process of *autoionization*. Preionization and predissociation are both non-radiative intramolecular energy transfer processes and as such they have much in common. They are mainly distinguished by the entities which result from the radiationless transition.

In closing this section we note that complex molecules are often studied in solution with transparent solvent(s). The absorption coefficient of the solution is proportional to the concentration $c$ of the compound in the solution*. When the concentration is expressed in units of moles per liter and the cell thickness $x$ in cm, Equation (3.1) is customarily written as

$$I(v) = I_0\, 10^{-\varepsilon(v)cx} \tag{3.4}$$

where $\varepsilon(v)$ is called the molar (decadic) extinction coefficient. The molar extinction coefficient (or $\log \varepsilon(v)$) is generally plotted as a function of the exciting-light wavelength (in nanometers (nm) or angstroms (Å)) or the wave number (in $cm^{-1}$).

Our discussion will be centred on electronic transitions, with occasional reference to vibrational structure. No discussion of rotational structure will be made, the omission being reasonable in our context.

### 3.2.2 *Aromatic Structures*

A great deal of our discussion in subsequent chapters will be concerned with organic molecules the structure of which possesses certain unique features. To aid our future discussions we shall now elaborate very briefly on the electronic structure of these systems.

### 3.2.2.1 $\sigma$ and $\pi$ Orbitals

Organic molecules are made up mostly of light atoms. Thus the atomic orbitals commonly involved in bonding are $s$ and $p$ orbitals. An $s$-atomic orbital is spherically symmetric, while a $p$-atomic orbital is distinctly directional. The formation, direction, and strength of a molecular bond can

---

*Care should be taken for this condition to be fulfilled experimentally (Beer's law).

be understood in terms of these types of atomic orbitals. A two-centre molecular orbital such as that involved in a diatomic molecule or that binding two atoms in a polyatomic molecule can at best have cylindrical symmetry. Such an orbital may be achieved by two $s$ orbitals (e.g. $H_2$), two $p$ orbitals acting along their axis (e.g. $Cl_2$), or an $s$ and a $p$ orbital (e.g. HCl). In all of the above cases the atomic orbitals are almost completely localized in the region between the nuclei. This type of chemical bond (covalent) is often referred to as a $\sigma$ bond and the electrons which participate are called $\sigma$ electrons.

Other orbitals have lower symmetry than that of a $\sigma$ orbital. Consider two $p$-atomic orbitals interacting with their cylindrical axes lying side-by-side. The result is a $\pi$ orbital, the bond so formed is called a $\pi$ bond, and the electrons participating are called $\pi$ electrons. Here, in contrast to the case of a $\sigma$ bond, the charge is moved away from the line joining the nuclei. Most of the known cases of $\pi$ orbital formation occur in connection with multiple-bond phenomena.

### 3.2.2.2 Hybridization

Let us now turn to the $\pi$ electronic structures of aromatic molecules. These can best be understood by considering the electronic configuration of the carbon atom, a common element of organic molecules. The ground state electronic configuration of the carbon atom ($Z = 6$) is $1s^2 2s^2 2p^2$. On the other hand, a carbon atom participating in bonding is thought of (see, for example, Coulson, 1961) as being in the configuration $1s^2 2s^1 2p^3$, i.e. one of the $2s$ electrons is considered to be excited into a $2p$ state. The $1s^2 2s^1 2p^3$ configuration has now four valence electrons, one $1s$ and three $2p$. These four atomic orbitals can be mixed or hybridized* in three alternative configurations. The resulting four hybrid orbitals allow for not only more but also stronger bonds. This makes up for the promotion energy $2s \rightarrow 2p$.

*Tetrahedral or $sp^3$ hybridization.*   Here the one $2s$ and the three $2p$ orbitals are mixed and the resultant four equivalent hybrid orbitals are directed towards the corners of a regular tetrahedron. All saturated compounds such as methane ($CH_4$) are examples of the tetrahedral hybridization, the four hybrids forming four $\sigma$ bonds.

*Trigonal or $sp^2$ hybridization.*   Here one of the three $2p$ orbitals, say $p_z$, remains unchanged and the one $2s$ and the remaining two $2p$ orbitals are

---

*The concept of hybridization or mixing of atomic orbitals involves the determination of those linear combinations of orbitals ($s$ and $p$ for the carbon atom) which might make more effective bonds than those of the individual orbitals, for a given total number of bonds. This concept helps us to understand the multivalency of the carbon atom and the structures of aromatic systems.

mixed to yield three $sp^2$ equivalent coplanar hybrid orbitals, symmetrical about their bonding axes and about the molecular plane. Their axes of symmetry are at 120 degrees from each other in the $x$–$y$ plane. The unchanged $p_z$ atomic orbital is mirror symmetric about the nodal $x$–$y$ plane. The three $sp^2$ hybrid orbitals form three equivalent $\sigma$ bonds in the plane containing the nuclei and the inner-shell electrons. The $\pi$ electronic structures of aromatic molecules are determined by $sp^2$ hybridization (see next section). Ethylene ($H_2C{=}CH_2$) is a simple example of $sp^2$ hybridization.

*Diagonal or sp hybridization.* Here two of the original $2p$ orbitals (say $p_y$ and $p_z$) remain unchanged and two equivalent hybrid $\sigma$ orbitals are produced by mixing the one $2s$ and the remaining $2p_x$ orbital. The hybrid $\sigma$ orbitals so formed are directed at 180 degrees to each other along the $x$ axis, i.e. along the line formed by the intersection of the nodal planes of the two $p$ orbitals. Triple-bonded molecules such as acetylene (H—C≡C—H) are typical examples of this type of hybridization. We note that apart from the C—H and C—C $\sigma$ bonds, the $p_y$ and $p_z$ atomic orbitals of each carbon atom are paired to produce two C—C $\pi$ bonds.

An idea as to how the different carbon orbitals overlap as a function of the internuclear distance $r_{ij}$ between the nuclei of two ($i$ and $j$) carbon atoms can be obtained from Figure 3.4 (Roberts, 1962) which is based on calculations by Mulliken (1952).

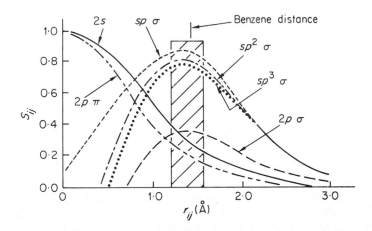

**Figure 3.4** Overlap of the different carbon-atom orbitals as a function of the internuclear distance $r_{ij}$ between the nuclei of two ($i$ and $j$) carbon atoms. $S_{ij}$ is the overlap integral defined as $\int \psi_i \psi_j \, d\tau$ where $\psi_i$ and $\psi_j$ are the wave functions for the orbitals $i$ and $j$, respectively (Roberts, 1962; based on calculations by Mulliken, 1952)

### 3.2.2.3 The $\pi$ electron structures of aromatic molecules

In aromatic structures the $\pi$ orbitals have a node (zero electron density) in the plane of the molecule while the closed-shell $\sigma$ or single-bond electrons are tightly and locally bound to the molecular skeleton. The almost complete localization of the $\sigma$ electrons and the delocalization of the $\pi$ electrons makes it possible to treat the $\pi$ electron system independently of the rest of the molecule. It is this separability of $\sigma$ and $\pi$ orbitals that makes these systems so attractive to the physicist since it introduces a relative simplicity in theoretical treatments. Let us consider a conjugated system, such as benzene. Here each of the six carbon atoms contributes one $\pi$ electron and the $\pi$ electron orbital is in contact with both of its neighbours. The six $p$-atomic orbitals (achieving maximum interaction when their nodal planes are coplanar, i.e. when the molecule is planar) interact to produce a nodal plane (that of the molecule) and to form six $\pi$-molecular orbitals which are completely delocalized. This is shown schematically in Figure 3.5

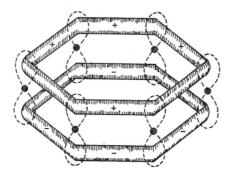

Figure 3.5 The $\pi$-molecular orbitals in benzene (Coulson, 1961)

(Coulson, 1961). The $\sigma$ hybrids interact as shown in Figure 3.6 (Coulson, 1961) forming the localized C—H and C—C $\sigma$ bonds. Similar delocalized $\pi$ electron structures characterize other aromatic and conjugated molecules. The $\pi$ electron system, possessing many-centre delocalized orbitals, is loosely bound by a potential which is defined by the shape of the planar molecular skeleton. Because of this, the $\pi$ electron states are the most closely spaced in the molecule, and transitions between $\pi$ electron states usually give absorption spectra in the visible and near ultraviolet region.

Among the theoretical approaches to determinations of the electronic structure of molecules are the valence-bond method, the method of linear combination of atomic orbitals (LCAO), and the free-electron model (due to Platt (see Platt and coworkers, 1964a, b)) for cata-condensed aromatic hydrocarbons*. Full account of all three treatments can be found in the references cited at the end of Section 3.1. A brief outline of the Platt classification of the electronic spectra of cata-condensed aromatic systems is in order since this classification had a quite general use by virtue of its simplicity.

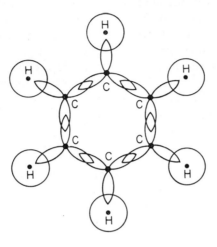

**Figure 3.6** The $\sigma$-hybrids of the carbon atoms of benzene (Coulson, 1961)

The Platt model yields less accurate but more rapid results. It assumes separability of the $\sigma$ and $\pi$ electrons and treats the $\pi$ orbitals as orbitals of free electrons travelling in a one-dimensional loop around the molecular perimeter. Approximating the molecular perimeter by a circle of circumference $l$ the allowed energy levels (corresponding to those of a plane rotator) are given by

$$E_q = \frac{q^2 h^2}{2ml^2} \tag{3.5}$$

where $q$ is called the orbital ring quantum number and describes the number of nodes of the wave function. All energy levels (except the lowest, $E_0$) are doubly degenerate since the electrons can move in either direction along the perimeter. Introduction of the periodic potential due to the atoms and the cross-links between the atoms removes this degeneracy and each state is

---

*In this group of polycyclic hydrocarbons no carbon atom belongs to more than two benzene rings, and every carbon atom is on the molecular periphery. The cata-condensed ring systems, whose general formula is $C_{4n+2}H_{2n+4}$, include some of the most important aromatics (e.g. the polyacenes (naphthalene, anthracene, etc.)), some of the chemical carcinogens, and some of the best organic scintillating compounds.

split into two components denoted by the subscripts a and b* ($q$ still determines the number of nodes of the wave function).

In the ground state of an $n$-ringed cata-condensed aromatic hydrocarbon the shell with $q = n$ is the highest filled since there are $2(2n+1)$ $\pi$ electrons all together (each carbon atom contributes one $\pi$ electron). The $q$'s of the individual electrons in the multielectron system are added algebraically to give the total ring quantum number $Q$ of the system. The nomenclature introduced by Platt as well as the configuration and states for the cata-condensed hydrocarbons are shown in Table 3.1. The letters $e, f, g, h$ describe electron shells for which $q = n-1$, $n$, $n+1$, $n+2$, respectively, and the letters A, B, C,... and K, L, M,... describe states for which $Q = 0, 1, 2, ...$ and $Q = 2n, 2n+1, 2n+2, ...,$ respectively.

The ground state of the system is the singlet $^1A$ ($f^4$ configuration ($Q = 0$) and paired electron spins). Excited states corresponding to the configuration $f^3g$ (i.e. one electron is excited from $f$ to $g$) are the lowest excited states of these systems commonly observed in their absorption spectra (see, for example, Friedel and Orchin, 1951 and Clar, 1964). There are eight excited states corresponding to the $f^3g$ configuration: two $B$ singlet states ($^1B_a$, $^1B_b$), two $B$ triplet states ($^3B_a$, $^3B_b$), two $L$ singlet states ($^1L_a$, $^1L_b$), and two $L$ triplet states ($^3L_a$, $^3L_b$). Excited states corresponding to other configurations are shown in Table 3.1†.

**Table 3.1** Configurations and states for cata-condensed hydrocarbons

| $\Delta q$ from ground state $^1A$ | Configuration | $\Delta Q$ | State |
|:---:|:---:|:---:|:---:|
| 2 | $f^3h$ | 2 | $^{1,3}C_{a,b}$ |
|   |        | $2n+2$ | $^{1,3}M_{a,b}$ |
| 2 | $e^3f^4g$ | 2 | $^{1,3}C_{a,b}$ |
|   |           | $2n$ | $^{1,3}K_{a,b}$ |
| 1 | $f^3g$ | 1 | $^{1,3}B_{a,b}$ |
|   |        | $2n+1$ | $^{1,3}L_{a,b}$ |
| 0 | $f^4$ | 0 | $^1A$ |

* The nodes in the electronic wave function occur at planes of symmetry in the molecule. When the nodal lines bisect the C—C cross-links the suffix a is used; when they pass through the C atoms the suffix b is used.
† See Jaffé and Orchin (1964) for further discussion and other types of notation.

In Figure 3.7 the polarization diagrams for the states $L_{a,b}$ and $B_{a,b}$ are shown for naphthalene. The $B_a$ and $B_b$ states correspond to a strong dipole oscillation in the molecular plane with polarization perpendicular and parallel to the long molecular axis, respectively. The states $L_a$ and $L_b$ correspond to weaker dipole oscillations; $L_a$ is polarized perpendicular and $L_b$ parallel to the long molecular axis. Since $^1A$ is singlet, transitions from this state to higher triplet states are spin-forbidden. From the intensity and polarization of the absorption bands the excited states can be classified (see Platt and coworkers, 1964a, b).

The absorption spectra of cata-condensed hydrocarbons and other aromatic and conjugated molecules contain a sequence of absorption bands in the visible and ultraviolet region. These are attributed to $\pi \rightarrow \pi^*$ transitions. Transitions involving $\sigma$ electrons (i.e. $\sigma \rightarrow \sigma^*$) lie at higher energies and

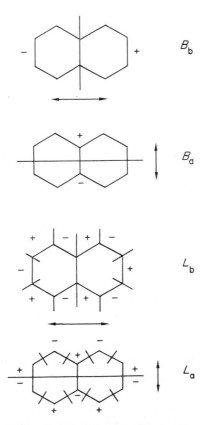

**Figure 3.7** Polarization diagrams for the $B_{a,b}$ and $L_{a,b}$ states of naphthalene

overlap high-lying $\pi \to \pi^*$ absorption bands. Generally, three (or more) $\pi$ electron absorption bands are observed in the absorption spectra of organic molecules (see Friedel and Orchin, 1951 and Clar, 1964) which correspond to transitions from the singlet ground state to singlet $\pi$ electronic states. An example is shown in Figure 3.8 for anthracene in cyclohexane; the molar extinction coefficient is plotted as a function of the wave number. We note that the absorption bands display fine (vibrational) structure. More fine structure is resolved at lower temperatures.

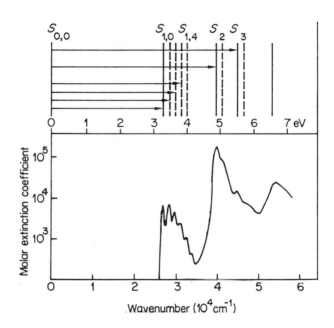

**Figure 3.8** Absorption spectrum of anthracene in cyclohexane (Birks, 1964)

A simple notation may be adopted(Birks, 1964) whereby $S_{0,0}, S_{0,1}, \ldots S_{0,n'}$ denote the ground $\pi$ electronic state in its $v = 0, 1, \ldots n'$ vibrational states, and $S_{1,0}, S_{1,1}, \ldots S_{1,n'}, S_{2,0}, S_{2,1}, \ldots S_{2,n'}, S_{n,0}, S_{n,1}, \ldots S_{n,n'}$ denote the $1^{st}$, $2^{nd}$ and $n^{th}$ excited $\pi$ singlet states in their $v = 0, 1, \ldots n'$ vibrational states, respectively. There is also a sequence of excited $\pi$ electronic triplet states which in analogy are denoted as $T_{1,0}, T_{1,1}, \ldots T_{1,n'}, T_{2,0}, T_{2,1}, \ldots T_{2,n'}, \ldots T_{n,0}, T_{n,1}, \ldots T_{n,n'}$. As a rule, for molecules which have closed electron subshells, $T_{n,n'}$ lie lower in energy than the corresponding $S_{n,n'}$. The transitions $S_{0,0} \to T_{n,n'}$ are spin-forbidden, but triplet states can be populated by other means. As is seen from Figure 3.8 at higher energies the presence of a series of $\sigma \to \sigma^*$ excited states obscures the observation of higher $\pi$ singlet states. At the upper portion of Figure 3.8 the energy level diagram which corresponds to the absorption spectrum plotted on the lower portion of the figure is shown. All transitions are shown to begin from the lowest ($v = 0$) vibrational level of the ground state ($S_{0,0}$) since at normal temperatures this is the predominantly populated level. The levels $S_{1,0}, \ldots S_{1,4}, S_{2,0}, S_{3,0}$ are also shown in Figure 3.8.

In terms of biological significance, the most important conjugated cyclic systems are the heterocyclic molecules. In these molecules another type of electronic transitions can be introduced. These are discussed in the next section.

### 3.2.2.4 $n \to \pi^*$ Transitions

We have seen in the previous section that, quite generally, bonding $\pi$ orbitals have higher energies than bonding $\sigma$ orbitals, and antibonding $\pi$ orbitals ($\pi^*$) have lower energies than antibonding $\sigma$ orbitals ($\sigma^*$), that is $\sigma \to \sigma^*$ transitions occur at higher energies (shorter wavelengths) than $\pi \to \pi^*$ transitions. In aromatic hydrocarbons all the weakly held electrons are already involved in molecular bonds. If, however, we introduce into the ring system heteroatoms such as N, O, S, etc., these atoms possess some rather weakly bound non-bonding or 'unshared-pair' electrons, often called 'lone-pair' electrons. Such electrons maintain their atomic character and are located at the heteroatom, occupying orbitals called $n$ orbitals. The non-bonding orbital ($n$ orbital) may have a higher energy than bonding $\pi$ orbitals and transitions involving $n$ and $\pi^*$ orbitals ($n \to \pi^*$) may assume a very important role in determining the principal optical and photochemical properties of the heteromolecule. Absorption bands arising from $n \to \pi^*$ transitions (i.e. transitions from non-bonding $n$ orbitals to antibonding $\pi$ orbitals ($\pi^*$)) usually lie in the near ultraviolet and visible region. Electronic transitions from non-bonding $n$ orbitals to antibonding $\sigma$ orbitals ($n \to \sigma^*$) also occur (at $\lambda \gtrsim 2000\text{Å}$).

The $n \to \pi^*$ transitions are rather weak even when they are not forbidden by symmetry. This is due to the fact that an $n$ orbital retains its atomic character and thus the overlapping of an $n$ orbital with a $\pi^*$ orbital is small. Absorption bands due to $n \to \pi^*$ transitions are characterized by certain features (which, alternatively, are used for their identification) such as low intrinsic intensity, blue shifts in solvents of high dielectric constants, disappearance of the absorption bands in acid media, and blue shifts upon introduction of conjugative electron-donating substituents (e.g. $-NH_2$, $-OH$) on the molecule (Kasha, 1961). Also, for $n \to \pi^*$ transitions there is an inherently greater spin–orbital interaction which arises from the nature of the orbitals involved. Because of this and the inherent weakness of $n \to \pi^*$ transitions, molecules with lowest singlet states corresponding to this type do not generally exhibit fluorescence emission. In rigid glass solutions

at 77 °K the total luminescence emission spectra consist of only phosphorescence (Kasha, 1961). Because of the vital role of atoms with available unshared pairs of electrons $(N, O, P, S)$ in biological reactions, $n \to \pi^*$ transitions are important in biochemical transformations. It has to be stressed, however, that when heteroatoms are inserted into a molecule, the possibility of $n \to \pi^*$ transitions is not necessarily introduced. The $n$ orbitals must retain their non-bonding atomic character and not conjugate effectively with the ring.

From the preceding discussion it is seen that molecular orbitals in organic molecules are of three main types: $\sigma$ orbitals, $\pi$ orbitals and $n$ orbitals. The lowest electronic transitions of wide occurrence are: $n \to \pi^*$, $\pi \to \pi^*$, $n \to \sigma^*$ and $\sigma \to \sigma^*$. The formaldehyde molecule ($\overset{H}{\underset{H}{>}}C{=}O$) allows the simplest basis for classification of $\sigma$, $\pi$ and $n$ orbitals. For this molecule there are three $\sigma$ bonds (C—H, C—H, C—C), one $\pi$ bond (C—O) and an $n$ orbital which is non-bonding (or at most slightly bonding) on the O atom. The $n$ orbital ($2p$ orbital on the O atom) has its axis perpendicular to the axes of the $2p$ orbitals which form the $\sigma$ and $\pi$ bonds. Figure 3.9 shows schematically the energy levels for the various types of electronic transitions in organic molecules. For some typical electronic-state sequences of heteroatomic systems see Kasha (1961).

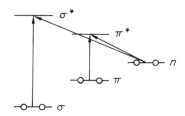

**Figure 3.9** Various types of electronic transitions in organic molecules

In concluding this section we point out again that $\sigma$ orbitals are considered, to a first approximation, to be localized on the respective bonds, but $\pi$ orbitals are delocalized over the whole conjugated system of the molecule. Hence, $\sigma \to \sigma^*$ transitions are not characteristic of the overall structure of the molecule in contrast to electronic transitions involving $\pi$ orbitals (i.e. $\pi \to \pi^*$ and $n \to \pi^*$). Also, $\sigma$ orbitals are not affected by twist of the bonds, in contrast to $\pi$ orbitals (hence $\pi \to \pi^*$ and $n \to \pi^*$ transitions) which are sensitive to changes in the geometry of the conjugated system. Divergence from planarity, for example, strongly affects the quantum yield of fluorescence of an aromatic molecule.

### 3.2.3 *Basic Relations Describing Photoabsorption Processes*

The intensity of an absorption line or band depends on the transition probability between the combined states. The transition probability is determined by a number of factors such as spin, symmetry, angular momentum and spatial overlap of the wave functions of the combined states. For an intense transition the total spin should remain unchanged. Certain symmetry rules (see Herzberg, 1966) must be fulfilled; the total angular momentum must change by certain quantized amounts (see Herzberg, 1966), and the wave functions should overlap as much as possible. Transition probabilities are available from thermodynamics, classical dispersion theory, and quantum mechanics.

#### 3.2.3.1 Einstein thermodynamic derivation of transition probability

Einstein (1917) considered the thermodynamic equilibrium between radiation and atoms and derived the fundamental equations which relate the transition probabilities for induced absorption $(A + h\nu \rightarrow A^*)$, spontaneous emission $(A^* \rightarrow A + h\nu)$, and induced (stimulated) emission $(A^* + h\nu \rightarrow A + 2h\nu)$. Consider an isotropic radiation of frequency between $\nu$, $\nu + d\nu$, and intensity $I(\nu)$ striking an assembly of atoms which are capable of being raised by absorption of this radiation from their ground state $i$ to an excited state $f$. If $B_{i \rightarrow f} I(\nu)$ is the probability per second that an atom in $i$ exposed to $I(\nu)$ will absorb a quantum of energy $h\nu$ and pass to $f$ (stimulated absorption), $A_{f \rightarrow i}$ is the probability per second that an atom in $f$ will spontaneously emit (in a random direction) a quantum of energy $h\nu$ and pass to $i$, and $B_{f \rightarrow i} I(\nu)$ is the analogous probability for induced emission, then Einstein's thermodynamic relations are:

$$\frac{A_{f \rightarrow i}}{B_{i \rightarrow f}} = \frac{2h\nu^3}{c^2} \frac{g_i}{g_f} \tag{3.6}$$

and

$$\frac{B_{f \rightarrow i}}{B_{i \rightarrow f}} = \frac{g_i}{g_f} \tag{3.7}$$

where $g_i$ and $g_f$ are the statistical weights of the ground and excited states, respectively.* For transitions between states of the same multiplicity $g_i/g_f = 1$, and from Equation (3.7) it is seen that for this case the probability

---

* Note that the coefficients $B$ have been defined in terms of the intensity of the isotropic radiation and not in terms of the radiation density; the two are related by

$$B(\text{density}) = \frac{c}{4\pi} B(\text{intensity})$$

of stimulated absorption is equal to the probability of stimulated emission. Equation (3.6) is applicable to the case where the refractive index of the surrounding medium is unity. If we allow for the refractive index of the medium containing the absorbing atoms or molecules and the consequent reduction in the velocity of light in the medium, Equation (3.6) becomes (Perrin, 1926, 1931; Lewis and Kasha, 1945)

$$\frac{A_{f \to i}}{B_{i \to f}} = \frac{2hn^3 v^3}{c^2} \frac{g_i}{g_f} \tag{3.8}$$

where $n$ is the refractive index of the medium at the frequency $v$. From Equation (3.8) the radiative or natural mean lifetime* $\tau_0$ for the excited state (i.e. the mean lifetime of the state when the only de-excitation pathway is the $f \to i$ radiative transition) is

$$\tau_0 = \frac{1}{A_{f \to i}} = \left[ \frac{2hn^3 v^3}{c^2} \frac{g_i}{g_f} B_{i \to f} \right]^{-1} \tag{3.9}$$

It is noted that the Einstein derivations apply to a two-state atomic energy-level diagram.

Now, suppose that we expose an atomic system with $N$ atoms per $cm^3$ in state $i$ to a parallel beam of light of intensity $I(v)$, and let $\delta N(v)$ of these atoms being capable of absorbing the frequency range $v$, $v + dv$ and $N'$ atoms per $cm^3$ in state $f$ of which $\delta N'(v)$ are capable of emitting this frequency range, then (Mitchell and Zemansky, 1961)

$$\int k(v) \, dv = \frac{hv_0}{4\pi} (B_{i \to f} N - B_{f \to i} N') \tag{3.10}$$

where $v_0$ is the frequency at the centre of the absorption line and the integral is taken over the absorption line. From Equations (3.6), (3.7), (3.9) and (3.10) we have (taking $n = 1$)

$$\int k(v) \, dv = \frac{\lambda_0^2}{8\pi} \frac{N}{\tau_0} \frac{g_f}{g_i} \left( 1 - \frac{g_i}{g_f} \frac{N'}{N} \right) \tag{3.11}$$

The second term in the parenthesis cannot be neglected when $N'$ is comparable with $N$. This may be the case for gases electrically excited at high current densities.† If the only agent responsible for atomic excitation is the

---

*This should be distinguished from the decay time, $\tau_M$, measured in emission studies; $\tau_0 \geqslant \tau_M$ (Section 3.4).
†If the ratio $N'/N$ exceeds unity (this can be obtained by various means such as collisions of A with other (metastable) excited species which transfer their exitation energy to A producing A*), then there will be a net emission. This possibility is used in the laser (Light Amplification by Stimulated Emission of Radiation) for the production of a high intensity beam of coherent light of very precisely defined energy.

absorption of a low intensity beam of light itself, then $N'/N$ is very small ($\lesssim 10^{-4}$) and Equation (3.11) becomes

$$\int k(v)\, dv = \frac{\lambda_0^2}{8\pi} \frac{N}{\tau_0} \frac{g_f}{g_i} \tag{3.12}$$

where $\lambda_0$ is the wavelength corresponding to $v_0$. Equation (3.12) is a fundamental one. It says that whatever physical processes are responsible for the formation of the absorption line, the integral of the absorption coefficient remains constant when $N$ is constant.

### 3.2.3.2 Classical dispersion theory

The quantity which describes the probability of a transition between two electronic states in classical dispersion theory is the oscillator strength $f$, or $f$ value for the transition. That is, classically atoms are thought of as composed of electrons oscillating harmonically at various resonant frequencies. The number of such presumed oscillators with a given frequency could be represented in terms of experimental quantities. The oscillator strengths are characteristic of the normal modes and are defined as the effective number of electrons in the atom or molecule which can participate in a (hypothetical) harmonic oscillation with the frequency of the corresponding modes. The effective number of electrons which can undergo this harmonic oscillation is not always integral. In actual fact, usually it is much less than unity. On the basis of quantum theory the $f$ value is proportional to the Einstein $A$ coefficient, or for a resonance line (i.e. a line for which the $i \rightarrow f$ transition is followed by the $f \rightarrow i$ radiative transition) it is inversely proportional to the lifetime of the resonance line.

The $f$ number and the absorption coefficient $k(v)$ are related by (Mitchell and Zemansky, 1961)

$$f_n = \frac{mc}{\pi N e^2} \int_n k(v)\, dv \tag{3.13}$$

i.e. the $f$ number of a normal mode $n$ is directly proportional to the area under the absorption line or band. In Equation (3.13) $N$ is the number of atoms per $cm^3$. If we express the frequency $v$ in wave numbers ($cm^{-1}$) indicated by $\bar{v}$ and use the molar extinction coefficient $\varepsilon(\bar{v})$, rather than $k(v)$, we obtain

$$f = 4.319 \times 10^{-9} \int \varepsilon(\bar{v})\, d\bar{v} \tag{3.14}$$

Thus, the intensity of an absorption band can be stated in terms of the $f$ value corresponding to the integrated absorption obtained graphically from a plot of $\varepsilon(\bar{v})$ against $\bar{v}(= v/c = \lambda^{-1})$.

Combining Equations (3.12) and (3.13) we have

$$\tau_0 f = \frac{mc}{8\pi^2 e^2} \lambda_0^2 \frac{g_f}{g_i} = 1.51\lambda_0^2 \frac{g_f}{g_i} \tag{3.15}$$

where the refractive index of the medium at $v_0$ was assumed equal to unity. We may note that $g_f/g_i$ is equal to one if both states are singlets and is equal to three if either of the states is triplet and the other is singlet. Equation (3.15) enables the calculation of $\tau_0$ from the $f$ value and vice versa. It is clear from Equation (3.15) that the greater the $f$ value, i.e. the stronger the transition, the shorter the natural mean lifetime or the excited state.

From the preceding discussion it is seen that either of three quantities (oscillator strength, natural lifetime, integrated absorption coefficient) can be used to measure the allowableness or the forbiddingness of an electronic transition, and therefore the intensity of a given absorption line or band.

### 3.2.3.3 Quantum mechanical transition probability

Quantum mechanically the transition probability between two states, $i$ and $f$, of wave functions $\psi_i$ and $\psi_f$, respectively, with absorption or emission of radiation of frequency $v_{i,f}$ depends on the square of the transition moment integral

$$\left| \int_{-\infty}^{+\infty} \psi_f^* Q \psi_i \, d\tau \right|^2 \tag{3.16}$$

where the integral is taken over all space for the state wave functions $\psi_f^*$ (the complex conjugate of $\psi_f$) and $\psi_i$, and $Q$ is an operator, usually the electric dipole operator $M$. For electric dipole transitions the $f,i$ matrix element of the electric-dipole operator is

$$R_{f,i} = \int_{-\infty}^{+\infty} \psi_f^* M \psi_i \, d\tau \tag{3.17}$$

where

$$|R_{f,i}|^2 = |X_{f,i}|^2 + |Y_{f,i}|^2 + |Z_{f,i}|^2$$

and

$$X_{f,i} = e \int \psi_f^* \sum_j x_j \psi_i \, d\tau \tag{3.18}$$

is the $x$ component of the transition dipole moment of the whole collection of the $j$ particles marking up the system. $R_{f,i}$ is a real physical property of the molecule and as such it must be invariant to symmetry operations. $R_{f,i}$ must be totally symmetric, i.e. it cannot change sign under a symmetry operation. If for a given pair of wave functions the integral (Equation (3.17)) is zero, the transition between the states involved is forbidden. In certain cases although the electric dipole matrix elements vanish, the transition may still take place via other moments such as the magnetic dipole or electric quadrupole. However, such transitions are weak.

The electric dipole matrix elements are related to the probability $A_{f \to i}$ for a spontaneous radiative transition from the state $f$ to the state $i$ by

$$A_{f \to i} = \frac{64\pi^4 v^3}{3hc^3} |R_{f,i}|^2 \tag{3.19}$$

where both states were assumed singlets. Similarly, the oscillator strength for a given transition can be expressed in terms of $|R_{f,i}|^2$. Knowledge of the transition moment matrix elements for a given transition allows determination of $A_{f \to i}$, oscillator strength, natural lifetime of the excited state, and natural line-width of the absorbed or emitted radiation.

### 3.2.3.4 Kuhn–Thomas sum rule

This rule states (Kuhn, 1925; Thomas, 1925)* that if $Z$ is the total number of electrons in the atom or molecule, then

$$\sum_f f_{i,f} = Z \tag{3.20}$$

The sum is taken over all final states $f$ (see further discussion in Section 3.5.2.4).

### 3.2.3.5 Differential oscillator strength

Extending the terminology in Section 3.2.3.2, we can define the differential oscillator strength as

$$\frac{\mathrm{d}f}{\mathrm{d}v} = \frac{mc}{\pi e^2} \sigma_{ab}(v) \tag{3.21}$$

Equation (3.21) has been derived from Equation (3.13) where $k$ was substituted for $\sigma_{ab}(v) N$. The Kuhn–Thomas sum rule can now be expressed as

$$\sum_n f_n + \int \frac{\mathrm{d}f}{\mathrm{d}v} \, \mathrm{d}v = Z \tag{3.22}$$

i.e. the sum of all oscillator strengths in discrete transitions plus the integral oscillator strength over the continuum equals the total number of electrons

---

*For a derivation see Kauzmann (1951), pp. 651–653. See also a general discussion of sum rules in Fano and Cooper (1968).

($Z$) of the absorbing atomic unit. For the validity of this rule and further discussion see Section 3.5.2.4.

### 3.2.4 *The Franck–Condon Principle*

In the Born–Oppenheimer approximation the ground, $\psi_i$, and excited, $\psi_f$, state wave functions each are written as a product of an electronic wave function $y$, a function of the coordinates, $x$, of the electrons and, $N$, of the nuclei, and a nuclear wave function, $\varphi$, a function only of $N$ viz.

$$\psi_i(x, N) = y_i(x, N)\varphi_i(N)$$

$$\psi_f(x, N) = y_f(x, N)\varphi_f(N) \qquad (3.23)$$

It is only within this approximation that the concept of potential energy curves and vibrational levels have a meaning.

Let us now consider expression (3.18). Since the operator is only a function of the electronic coordinates, $X_{f,i}$ can be expressed as

$$X_{f,i} = e \int y_f^* \sum_j x_j y_i \, d\tau \int \varphi_f^* \varphi_i \, d\tau' \qquad (3.24)$$

If, now, it is assumed that the first integral is but a very weak function of the nuclear coordinates $N$, then $X_{f,i}$ in the above approximation is seen to be proportional to the integral over the vibrational wave functions of the levels involved (second integral on the right-hand side of Equation (3.24)). For this integral (usually called overlap integral) to be large the nuclei in the excited and ground states must occupy the same spatial positions. Otherwise the overlap integral is small and the probability for the transition is small. Stating this differently, during an electronic transition the electron motion is so much faster compared to the nuclear motion, that the nuclei can be considered as having retained their initial positions during the electronic transition. This tendency for vertical transitions is known as the Franck–Condon principle (Franck, 1925; Condon, 1928). It is within this principle that on potential energy diagrams (see, for example, Figure 3.3) electronic transitions are represented as vertical lines. It might be noted that the wave function of the ground state is of greatest intensity (i.e. the square of the zero-point vibrational wave function is maximum) at the centre point of the two vibrational extremes. Therefore, the most probable transitions are likely to start from this centre point, i.e. at the most probable nuclear separation. Some forms of potential energy curves arising in diatomic molecules are discussed in Chapter 6 in connection with electron attachment processes.

## 3.3 Experimental Methods

Low-energy photons induce resonance transitions in atoms and molecules. Higher-energy photons may, in addition, induce superexcitation, ionization and dissociative ionization, the products of ionization and/or dissociation being left with or without excess energy. As in all collision processes, the measurement of the interaction cross section as a function of photon energy is of primary importance. Therefore, in low-energy photon–atom (molecule) collisions the location of and cross section for the various resonance states as well as the cross section for photoionization and its energy dependence should be measured. Other quantities of particular interest are the lifetimes of excited electronic states, emission spectra and quantum yields, as well as the study of intra- and intermolecular electronic energy transfer. The products of photoionization need be identified and much physics can be learned by energy and angular distribution analyses of the photoionization electrons.

The experimental techniques commonly employed for the study of photo-physical processes fall into three broad categories: (i) Techniques employed for the study of absorption and emission spectra in the visible and near ultraviolet region. (ii) Techniques employed for the study of photoabsorption processes in the vacuum ultraviolet region. These include arrangements for the measurement of total and specific ionization cross sections, identification of the ionization products, determination of the kinetic energies and angular dependences of the ejected photoelectrons, and ion-molecule reactions. (iii) Techniques employed for the measurement of the lifetimes of excited states. In this category, the flash photolysis method used for the investigation of higher triplet states of complex molecules might be included.

In each of the above techniques the experimental set-up begins with a light source and ends with a detector (and recording device). Some comments regarding the availability of these two important accessories are in order prior to elaborating briefly on the individual methods themselves.

### 3.3.1 *Light Sources and Detectors*

#### 3.3.1.1 Light sources

The study of photoabsorption processes requires intense sources of photons which have a significant portion of their radiation on the short wavelength side of the spectrum under study. Such light sources are of two kinds: those with a closely packed line spectrum and those with a continuous spectrum. In most experiments the light sources are of a steady-state nature or, as in the case of certain gas discharges, they are fired at repeated intervals. In certain studies (category (iii) above) pulsed sources are required. Quantitative detailed investigations of photoabsorption processes require light sources of reasonable constancy with time.

As a rule, line emission sources provide a greater accuracy than sources with continuous spectra. This is chiefly because the scattered background and second-order radiation can be more easily corrected for.

Line sources utilize gas discharges in which the excited and ionized (often multiply ionized) atoms give numerous strong emission lines upon de-excitation. Line spectra accompanying de-excitation of outer-shell electrons give long wavelength emission lines (the well-known 2537 and 3650 Å mercury lines, for example, are frequently used in ultraviolet and visible spectroscopy),* while the emission from multiply ionized atoms extends throughout a higher energy range. It might be noted that whenever (as in vacuum ultraviolet studies) the gas discharges operate in successive pulses, the average intensity and spectral distribution must be reproducible.

X-ray sources with spectra consisting of lines (fewer than in the case of ultraviolet line sources) superimposed on a continuum are used for photophysical studies at higher photon energies. Since in this energy region absorption measurements are made for discrete wavelengths in the middle of individual lines, the measurements cannot cover the whole spectrum continuously and do not yield an adequate analysis of fine structure of absorption spectra.

Available source continua for ultraviolet absorption studies obtained in gas discharges include those of molecular hydrogen and of the rare gases. Much work has been done to develop such continuous light sources (see, for example, Tanaka, Jursa and LeBlanc (1958); Newburgh, Heroux and Hinteregger (1962); Huffman, Tanaka and Larrabee (1962); Wilkinson and Byram (1965) and references quoted therein). In Figure 3.10 a schematic representation of the rare gas continua including part of the hydrogen continuum are shown (Tanaka, Jursa and LeBlanc, 1958). It is seen that the neon continuum is very satisfactorily covered by the helium continuum which can be used instead, since it is more intense and can be produced at a much lower pressure than that which is required with neon (Tanaka, Jursa and LeBlanc, 1958). Other discharges yielding continua to much higher energies are being developed (Garton, 1966, 1968).

Probably the most 'noble' continuous source of high-energy photons is the 'synchrotron light'. It is well known that high-energy electrons subjected to a centripetal acceleration emit electromagnetic radiation whose wavelength distribution depends on the electron energy and the orbital radius. The power radiated per unit wavelength interval into all angles for a single mono-energetic electron of energy $E$ travelling an orbit of radius $R$ is (Tomboulian

---

*For a survey of light sources used in photobiological studies, see Withrow and Withrow (1956).

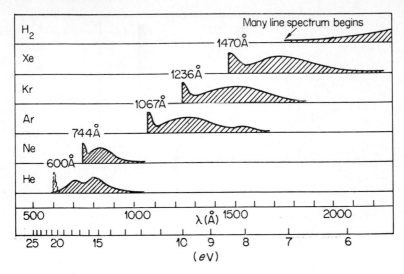

**Figure 3.10** Schematic representation of rare gas continua (Tanaka, Jursa, and LeBlanc, 1958)

and Hartman, 1956)

$$P(\lambda) = \frac{3^{5/2} e^2 c}{16\pi^2 R^3} \left(\frac{E}{mc^2}\right)^7 G\left(\frac{\lambda}{\lambda_c}\right) \tag{3.25}$$

where the so-called critical wavelength $\lambda_c$ is defined by

$$\lambda_c = \frac{4\pi R}{3} \left(\frac{mc^2}{E}\right)^3 \tag{3.26}$$

$c$ is the speed of light and $m$ the rest mass of the electron. The critical wavelength $\lambda_c$ is close to the wavelength for the maximum power. The function $G(\lambda/\lambda_c)$, the universal spontaneous power distribution for the electron synchrotron radiation, is shown in Figure 3.11 (upper portion). We note from Equations (3.25) and (3.26) that the peak of the spectral distribution moves to shorter wavelengths as $E^3$ and the power radiated at the peak increases as $E^7$. In Figure 3.11 (lower portion) the power (erg sec$^{-1}$ Å$^{-1}$) radiated into all angles per electron as a function of the wavelength for monoenergetic electrons of 120 and 180 MeV energy is shown for the National Bureau of Standards synchrotron ($R = 83.4$ cm) (Codling and Madden 1965a). For technical details on this source, see Codling and Madden (1965a) and Madden, Ederer and Codling (1967). See also a discussion in Marr (1967).

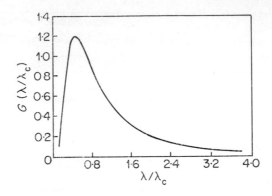

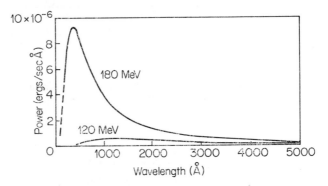

**Figure 3.11** *Upper portion:* Universal spectral distribution curve for the radiation from monoenergetic electrons (integrated over all angles) as a function of reduced wavelength $\lambda/\lambda_c$. The maximum occurs at $\lambda/\lambda_c = 0.42$ (Madden, Ederer, and Codling, 1967). *Lower portion:* Spectral distribution of the 'synchrotron light' for the NBS 180 MeV electron synchrotron. Power (ergs/sec Å) radiated per electron into all angles, as a function of wavelength, for monoenergetic electrons of 120 and 180 MeV energy (Codling and Madden, 1965a)

The use of the synchrotron light continuum, instead of the conventional line sources, in atomic photophysics enabled many new spectral features to be obtained (see Section 3.5). A number of laboratories are now equipped with such facilities and much is expected in the coming years on the photophysical processes which occur in the high vacuum ultraviolet region.

The development of intense light sources of short duration attracted considerable attention also. Much effort has been devoted recently for improving such sources (see detailed references on nanosecond pulsed discharge lamps in Birks and Munro (1967)). Such light sources are useful in fluorescence studies for the measurement of the lifetimes of excited states. They are also used for testing the time response of various photodetectors in the nanosecond region. The development of the laser light source is another outstanding advancement. The high intensity and monochromaticity of the laser beams unquestionably make them ideally suited in photophysics.

### 3.3.1.2 Detectors

Two general types of detecting devices have been in use for the measurement of light intensities: photographic plates and photomultiplier detectors (electronic detectors in general). Photographic plates are simpler devices and can be used to measure relatively low light intensities by prolonging the exposure time. They must, however, be photometrically calibrated to translate densities into intensities. Such a calibration is laborious and not usually of the desirable accuracy. For this reason photographic plates do not lend themselves well to extensive quantitative absorption measurements. For some fast photographic plates for vacuum ultraviolet work, see Madden, Ederer and Codling (1967).

Photomultipliers are inherently more accurate and flexible but require subsidiary apparatus. The output from a photomultiplier, directly proportional to the number of photons per second in the beam, can be presented on a recorder chart. Photomultipliers (or gas discharge counters such as the Geiger–Mueller counter) can be used for single photon counting. For work in the visible and near-ultraviolet region the conventional glass-windowed photomultipliers present no problems (one should, however, choose a phototube with the correct spectral response). Glass-windowed photomultipliers, however, cannot be used directly in vacuum ultraviolet studies, because the glass window absorbs, the ultraviolet radiation before it strikes the photocathode. Lithium fluoride (or calcium fluoride) windowed photomultipliers may be used down to 1100Å (quartz has a cut-off at ~1800Å). Below this wavelength windowless photomultipliers are necessary. These are connected directly to the vacuum system. Since most electronic detectors must operate either at high vacuum or at pressures different from those of the rest of the system, the need for differential pumping with windowless photomultipliers in high-vacuum ultraviolet photophysics is evident.

Some workers covered the glass window of a conventional photomultiplier with a fluorescent material which upon excitation by ultraviolet radiation emits, with a high and nearly constant (independent of the exciting wavelength) quantum efficiency, light that matches the photocathode spectral response of a glass-ended photomultiplier. One such material which has very often been used is sodium salicylate (Samson, 1964a). The overall sensitivity of such an arrangement is low, since some of the emitted photons are lost in their passage from the fluorescent material to the photocathode.

For a discussion of electric detectors see Tousey (1962) and Birks (1964). See also Weissler (1956) and Marr (1967) for other ways of monitoring light intensity.

### 3.3.2 *Techniques for the Study of Absorption and Emission Spectra in the Visible and Near-Ultraviolet Region*

Such techniques present no specific difficulties. They have been in use for a number of years. A typical single-beam arrangement for absorption studies is shown in Figure 3.12. The intensities $I_0$ and $I$ (Equations (3.1) and (3.2))

**Figure 3.12** Schematic representation of a typical single-beam arrangement for absorption studies in the visible and near ultraviolet region

are measured without and with the sample in place. Whenever a compound is studied in solution with a transparent solvent, $I_0$ is often measured with the absorption cell in place and filled with the solvent. In this case the molar extinction coefficient $\varepsilon(\bar{\nu})$ is obtained by making use of Equation (3.4). $I_0$ should be as low as possible to minimize the concentration of any photochemical products.

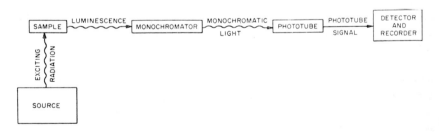

**Figure 3.13** Schematic representation of a typical single-beam arrangement for emission studies in the visible and near ultraviolet region

Rearranging Figure 3.12 as shown in Figure 3.13 studies of emission spectra can be made. The exciting radiation can be either continuous or discrete. Certain exciting lines can be isolated by the use of appropriate filters, or by the use of a second monochromator located between the source and the sample.

### 3.3.3 *Techniques for the Study of Photoabsorption Processes in the Vacuum Ultraviolet Region*

Although a number of difficulties have limited the measurement of absorption coefficients in this energy region for some time, vacuum ultraviolet atomic and molecular photophysics have come under intense study recently. Vaccum ultraviolet spectroscopy is becoming a quantitative method for the study of autoionizing, preionizing, and ionic states, as well as of the spectral distribution of optical oscillator strengths. Recent advances in light sources, detecting devices, and vacuum and optical techniques have contributed to this. The energy spread in such studies can be held considerably smaller compared with commonly used electron impact techniques. Further, photon collision processes are more restricted in their degree of freedom. There are, of course, still a number of experimental problems which severely restrict studies in certain areas (such as metallic atomic vapours, and atoms (e.g. H, N, O) normally existing in the molecular form). The use of windowless vacuum ultraviolet monochromators introduces certain difficulties, too.

Generally, in vacuum ultraviolet studies (as in all photoabsorption studies), one should keep in mind that quantitative measurements of the absorption cross section $\sigma_{ab}(v)$, require the energy width of the monochromatic exciting light to be small compared with the corresponding absorption region of the gas. Ionization and dissociation absorption continua present no problem in this connection. However, values of $\sigma_{ab}(v)$ measured at or near sharp resonance absorption lines are often subject to great uncertainties.

In photoabsorption studies the measured absorption coefficient (or $\sigma_{ab}(v)$) characterizes the attenuation of light, irrespective of whether one or more mechanisms contribute to this attenuation. In vacuum ultraviolet studies (where $hv > I$) the total absorption cross section is equal to the photoionization cross section $\sigma_i(v)$ for atomic systems for which the ionization efficiency $\eta(v)$ is unity. For more complex systems $\sigma_i(v) < \sigma_{ab}(v)$ since the absorption of the photon can lead to excitation and dissociation in addition to resulting in the production of an ion plus an electron. In this case one may define the photoionization efficiency ($< 1$) as

$$\eta(v) = \frac{\sigma_i(v)}{\sigma_{ab}(v)} = \frac{\text{number of ion pairs produced per second}}{\text{number of incident photons absorbed per second}} \qquad (3.27)$$

If $i(v)$ is the total collector current due to ions produced by absorption of monochromatic radiation of energy $hv$, the intensity of which decreased from $I_1$ (at the beginning of the absorption path) to $I_2$ (at the end of the absorption path), the right-hand side of Equation (3.27) is equal to $i(v)hv/e(I_1 - I_2)$. Here, $i(v)$ refers to the total current collected, independently of the specific kinds (or energy) of the products formed and $e$ to the electron

charge. The ionization cross section so determined is called the total
ionization cross section.

Some standard techniques employed for the measurement of the total
ionization cross section have been discussed by Weissler (1956) and more
recently by Marr (1967). In all such studies the experimental arrangement
involves a windowless vacuum ultraviolet monochromator through which
light of a particular wavelength is focused onto an exit slit behind which,
often, two parallel plate ionization chambers in tandem are located at a
known distance apart. The ionization chambers are filled with gas at
appropriate pressure. The absolute photon flux entering the chamber is also
measured at the slit. Knowing the absorption coefficient and the absolute
photon flux, one can determine the actual number of photons disappearing
along the length of one of the ionization chambers. Measurement of the ion
current (i.e. the number of ion pairs produced per sec) in connection with
the number of photons absorbed, yields the photoionization efficiency
through Equation (3.27). In such experiments the main sources of error arise
from flux, pressure, and current measurements.

Samson (1964a) constructed a double ionization chamber which has
certain advantages over other experimental arrangements of this type.

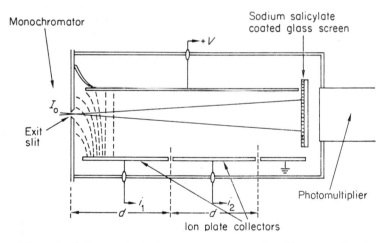

**Figure 3.14** Samson's double ionization chamber (Samson, 1964a)

Samson's double ionization chamber is shown in Figure 3.14. The two
collector plates are identical in length. The light enters at the exit slit and the
transmitted light is detected by a photomultiplier. Photons absorbed along
the length of each of the ion plate-collectors produce ions which are collected
by the respective plates giving rise to the ion currents $i_1$ and $i_2$ (proportional

to the number of photons absorbed in their respective regions). Since these currents are measured simultaneously they are independent of any fluctuations in the incident radiation intensity. In such experiments care must be taken to apply correct voltages so that only ions (and not electrons) are collected. The slits have to be held at a positive potential (which gives rise to an electric field as shown in Figure 3.14) to drive away over to the first plate ions formed in the vicinity of the slit. If all ions formed were collected by the respective plates, then using Lambert's law for the light absorbed in the light path in front of each collector plate, one obtains

$$k(v) = (\ln i_1/i_2)/d \tag{3.28}$$

where the optical path $d$ is equal to the length of one of the collector plates and is reduced to standard temperature and pressure. For the geometry shown in Figure 3.14 the photoionization efficiency $\eta(v)$ is given by (Samson, 1964a)

$$\eta(v) = \frac{(i_1)^2/e}{I_0(i_1 - i_2)} \tag{3.29}$$

Since in vacuum ultraviolet studies two or more photoionization processes are possible at a given exciting wavelength, the measured total cross section is, in general, a sum of several partial cross sections, specific to individual photoionization mechanisms. The ion(s) and electron(s) produced in such processes must be studied further separately (or in coincidence). Determination of the ionic mass identifies the ionization product(s), while measurement of the ionic charge allows knowledge of the relative abundance of singly, doubly, or multiply charged ions. The yield of specific ions as a function of the exciting wavelength need also be studied.

Determination of the state of excitation (discrete or continuum) of the residual ion is of great significance. Such information can be obtained by measuring the kinetic energy of the ejected electron which, for a fixed wavelength, is a function of the state of excitation of the residual ion. Measurements of the spectra of photoelectrons have been performed using electric or magnetic spectrometers (Frost, McDowell and Vroom, 1967a, b, c; Berkowitz, Ehrhardt and Tekaat, 1967; Turner and May, 1966; Carlson, 1967; Turner, 1968, see discussion and results in Section 3.5). Alternatively, the state of excitation of the residual ion can be determined by studying the fluorescent radiation emitted when the excited ion returns to its ground state. Such fluorescent radiation has been observed for some molecules (e.g. $N_2$ and $O_2$, Huffman, Tanaka and Larrabee (1963b,e); $O_2$, James (1956); Weissler (1962)).

Measurements of the angular distribution of the photoelectrons yield information as to the relative transition probability and phase of different degenerate channels. Such studies have been made (e.g. Berkowitz, Ehrhardt and Tekaat, 1967; see Section 3.5). Studies of ion–molecule reactions have been performed also.

The investigations just discussed require the combination of vacuum ultra-violet and mass spectroscopic techniques. The major difficulties arising in these studies stem from limited (weak) intensity currents, detection of low-energy electrons, as well as in achieving a monochromatic light beam of sufficient intensity and constancy. The cross sections for specific ions are often measured as a function of the exciting photon energy (or wavelength) in relative units and then normalized to known cross sections (see reviews by Marr (1967); Comes (1968); Tousey (1962) and Section 3.5).

### 3.3.4 *Measurement of the Lifetime of Excited Electronic States*

Knowledge of the various transition probabilities and lifetimes of excited atomic and molecular electronic states is necessary in many branches of physics, and basic for the understanding of the structure of atoms and molecules. From measurements of the transition probabilities the $f$ values and the matrix elements can be deduced, respectively, through Equations (3.15) and (3.19) in connection with Equation (3.9). Also, knowledge of the lifetime of an excited state (especially of the lowest electronic states of organic molecules) is of extreme importance in studies of molecular inter-actions, transfer or quenching of the excitation energy, and luminescence. It is of primary interest in a number of practical problems such as nuclear organic scintillation detectors.

Consider now an atomic system which at time $t = 0$ is raised by some means to an excited state $i$ from which it subsequently decays radiatively to lower energy levels $f$, and let $N_0$ be the number of atoms raised initially to this state. Due to the radiative decay to lower energy levels the number of atoms, $N(t)$, in the state $i$ at a time $t$ later will be decreased. In the absence of all external perturbations the radiative lifetime of the excited state $i$ is determined according to quantum theory by the total probability of all possible transitions to lower energy states. These transition probabilities $A_{i \to f}$ can be calculated if the wave functions of the combining levels $f$ and $i$ are known. We may write for the number of atoms $N(t)$ in the excited state $i$ after a time $t$

$$N(t) = N_0 e^{-t/\tau'_M} \qquad (3.30)$$

where

$$\tau'_M = \frac{1}{\sum\limits_f A_{i \to f}}$$

and the sum is taken over all final states $f$.

Let us now, for simplicity, assume that at time $t = 0$ the system is excited by a $\delta$ function light pulse of unit intensity. The emission (fluorescence) response function $i(t)$, proportional to the exciting light intensity, is defined as the fluorescence intensity at subsequent time $t$. Assuming a unimolecular decay, one obtains the response function which is a simple exponential of the form

$$i(t) = i_0 e^{-t/\tau_M} \tag{3.31}$$

where $i_0$ is the emission intensity immediately after the cessation of the exciting pulse at $t = 0$ and $\tau_M$ is the average decay time defined as the time required for the intensity of light emission to fall to $i_0/e$ (where $e$ is the basis of the natural logarithms). If only two states are combined and non-radiative de-excitation processes are not involved, the decay time $\tau_M$ measured in emission studies is equal to the natural lifetime $\tau_0$, determined from the integrated absorption spectrum. If the excited state $i$ can decay to states other than the initial state or if de-excitation can take place non-radiatively, the decay time $\tau_M$ measured in emission studies is smaller than the natural lifetime $\tau_0$ measured in absorption studies. The two lifetimes are related by

$$\tau_M = q_M \tau_0 \tag{3.32}$$

where

$$q_M = \frac{\text{probability for emission to the initial state}}{\text{total probability of de-excitation}} \leq 1$$

In decay-time studies, both for atomic and molecular systems, care must be taken that energy-transfer processes as well as re-absorption of emitted photons do not take place or are properly accounted for. The process of re-absorption of emitted photons (self-absorption) is extremely troublesome for resonance states (e.g. in rare gases) where the imprisonment (trapping) of resonance radiation (Holstein, 1947; Mitchell and Zemansky, 1961) results in an apparent lifetime for the resonance state often orders of magnitude longer than $\tau_0$ (depending on the pressure). To avoid these effects, measurements should be performed at low pressures and when in solution at low concentrations. In the latter case the effect of environment (solvent) on the lifetime should be investigated. The process of non-radiative energy transfer to neighbouring species (Section 3.4) and the process of collisional quenching (Section 3.4) affect (decrease) the measured lifetime $\tau_M$.

Transition probabilities between various levels of one and the same atom or molecule are of widely different magnitudes. The natural lifetime $\tau_0$ for emission of radiation is usually of the order of $10^{-8}$–$10^{-9}$ sec, but in many cases may be as long as $10^{-3}$ sec, and in exceptional cases may be even several seconds. Lifetimes of excited electronic states may be obtained from:

theory, uncertainty principle, integrated absorption, direct measurement following the decay of the excited state, and indirect determinations.

### 3.3.4.1 Theory

As has been discussed in Section 3.2.3, evaluation of the matrix elements for the transition between the states involved allows determination of the radiative transition probability between the two states, and, therefore, calculation of the natural lifetime $\tau_0$ of the excited (upper) state. The feasibility of such calculations, however, is very limited. Actually, they are possible for simple atomic systems. Some atomic transition probabilities calculated from theoretical approximations can be found in Wiese, Smith and Glennon (1966). The transition probabilities reported by these authors can be used to determine the lifetimes of the respective excited states. A comparison of experimental and theoretical lifetimes of some of the states in neutral helium has been given by Bridgett and King (1967) and has shown that the experimental and theoretical values are in good agreement (within 15%).

### 3.3.4.2 Uncertainty principle

If the total width $\Delta\bar{v}$ of the line profile at one-half of its maximum value (half-width of the absorption line; see Figure 3.1) is expressed in cm$^{-1}$ a value for the radiative lifetime of the state responsible for the absorption line can be determined through the uncertainty principle viz.

$$\Delta E\,\Delta t \geq \hbar$$

$$\Delta t = \tau' \geq \frac{\hbar}{\Delta E} = \frac{1}{2\pi c\,\Delta\bar{v}} \qquad (3.33)$$

Since the natural line width is in practice obscured (and broader) by a number of broadening processes, the lifetime $\tau'$ estimated through Equation (3.33) is always a lower limit to the true lifetime. It might be noted that although the line width is determined by the lifetimes of both states between which the transition occurs, when one of the states is the ground state its lifetime is very long and its contribution becomes negligible (Agarbiceanu and coworkers, 1963).

### 3.3.4.3 Integrated absorption

Expressions (3.12) (or (3.13) and (3.15)) may be used to evaluate $\tau_0$ and $f$ from a measurement of the integrated absorption coefficient (see details in Mitchell and Zemansky, 1961). Strictly speaking, the above-mentioned expressions are valid only for distinct narrow spectral lines, although sometimes they have been applied to molecular cases. Lifetimes and oscillator

**Table 3.2** Wavelengths, lifetimes and oscillator strengths of resonance lines of rare gases obtained by optical absorption

| Atom | $\lambda$(Å) | $\tau$(nsec) | $f$ |
|---|---|---|---|
| Xe $^1P_1^o$ | 1296[a] | 2.8 $\pm$0.20[a] | 0.270$\pm$0.02[a] |
| | — | 3.17$\pm$0.19[b] | 0.238$\pm$0.015[b] |
| Xe $^3P_1^o$ | 1470[a] | 3.74$\pm$0.30[a] | 0.260$\pm$0.02[a] |
| | — | 3.79$\pm$0.12[b] | 0.256$\pm$0.008[b] |
| Kr $^1P_1^o$ | 1164[c] | 4.55$\pm$0.2[c] | 0.135$\pm$0.01[c] |
| Kr $^3P_1^o$ | 1236[c] | 4.38$\pm$0.2[c] | 0.158$\pm$0.01[c] |

[a] P. G. Wilkinson, *J. Quant. Spectry. Rad. Transfer*, **6**, 823 (1966).
[b] D. K. Anderson, *Phys. Rev.*, **137**, A21 (1965).
[c] P. G. Wilkinson, *J. Quant. Spectry. Rad. Transfer*, **5**, 503 (1965).

strengths of resonance lines of rare gas atoms obtained from optical absorption studies are listed in Table 3.2.

Expression (3.9) has been expressed by several authors in terms of experimental parameters (Perrin, 1926; Lewis and Kasha, 1945; Mulliken, 1939; Ladenburg, 1921; Tolman, 1924). For transitions between states of the same multiplicity (e.g. singlet–singlet transitions, as in fluorescence), the relation obtained is:

$$\frac{1}{\tau_0} = \frac{8\pi \times 2303 c \bar{v}^2 n^2}{N_A} \int \varepsilon(\bar{v}) \, d\bar{v} = 2.88 \times 10^{-9} \bar{v}^2 n^2 \int \varepsilon(\bar{v}) \, d\bar{v} \quad (3.34)$$

where $c$ is the velocity of light in vacuum, $N$ is Avogadro's number, $\eta$ is the refractive index at $\bar{v}$, $\varepsilon(\bar{v})$ is the molar extinction coefficient at $\bar{v}$, and the integral is taken over the absorption band under discussion. For states of different multiplicity the right-hand side of Equation (3.34) is multiplied by $g_i/g_f$ where $g_i$ and $g_f$ are the degeneracies of the lower and upper states, respectively. Expression (3.34) is strictly applicable to sharp atomic transitions and to resonance fluorescence,* since it has been derived under the assumption that the absorption band is sharp and that both the fluorescence and absorption occur at the same energy.

Other relations between the radiative lifetime and the molecular fluorescence and absorption spectra have been derived. Förster (1951) assumed mirror symmetry between the fluorescence and absorption bands and obtained for the natural lifetime $\tau_0$, the following expression, applicable to

---

*Resonance fluorescence is referred to light emission from an excited state back to the lower state from which the absorption process took place (Section 3.4).

molecular fluorescence,

$$\frac{1}{\tau_0} = 2.88 \times 10^{-9} n^2 \int_{\bar{\nu}\,\text{min}}^{\bar{\nu}\,\text{max}} \frac{(2\bar{\nu}_0 - \bar{\nu})^3}{\bar{\nu}} \, \varepsilon(\bar{\nu}) \, d\bar{\nu} \qquad (3.35)$$

In Equation (3.35) $n$ is the mean refractive index of the medium over the fluorescence and absorption bands, $\bar{\nu}_0$ is to the first approximation the wave number at the line of symmetry between the absorption and fluorescence bands, and the integration is taken over the absorption band for the transition.

Strickler and Berg (1962) and Birks and Dyson (1963) derived a somewhat more general expression for the calculation of the natural lifetime, $\tau_0$, which does not directly involve the assumption of mirror symmetry. Birks and Dyson (1963) obtained the following expression for the value of $\tau_0$,

$$\frac{1}{\tau_0} = 2.88 \times 10^{-9} \frac{n_f^3}{n_a} \langle (\bar{\nu}_f)^{-3} \rangle_{av}^{-1} \int \frac{\varepsilon(\bar{\nu})}{\bar{\nu}} \, d\bar{\nu} \qquad (3.36)$$

where $n_f$ and $n_a$ are the mean refractive indices over the fluorescence band and over the absorption band, respectively; $\langle (\bar{\nu}_f)^{-3} \rangle_{av}^{-1}$ is the reciprocal of the mean value of $(\bar{\nu}_f)^{-3}$ in the fluorescence spectrum. The Birks and Dyson Equation (3.36) may be reduced to the Förster Equation (3.35) if a mirror symmetry between the fluorescence and absorption spectra is assumed. The expression given earlier by Strickler and Berg (1962) is identical to that of Birks and Dyson (1963) if it is assumed that the medium has no optical dispersion (i.e. $n_f = n_a$). Expressions (3.35) and (3.36) can be tested experimentally. The quantities $\tau_M$ and $q_M$ are measured for a given state and an experimental value of $\tau_0$ is evaluated through Equation (3.32). This experimental value of $\tau_0$ is then compared with the predictions of Equations (3.35) and (3.36). Comparisons of this nature have been performed by Cherkasov and coworkers (1956), Strickler and Berg (1962), Birks and Dyson (1963), Ware and Baldwin (1964), and Dawson and Windsor (1968) for a number of organic molecules. The results of these workers have been summarized by Birks and Munro (1967). A representative set of data has been collected in Table 3.3. Although the data of Cherkasov and coworkers (1956) on $q_M$ and $\tau_M$ for a number of mesoderivatives of anthracene in deaerated alcohol solution gave experimental values of $\tau_0$ which agreed more closely with the predictions of Equation (3.35) rather than with Equation (3.34), the experimental values of $\tau_0$ are found to be generally higher than the theoretical ones. This cannot be ascribed to the use of Equation (3.35) (rather than Equation (3.36)) since the absorption and fluorescence spectra of many of the compounds considered exhibited reasonable mirror symmetry. However, for a few of the compounds listed

**Table 3.3** Sample data on $\tau_M$, $q_M$, $\tau_0$(emission) and $\tau_0$(absorption) for organic molecules in solution[a,b]

| Compound | Solvent | $\tau_M$(nsec)[e] | Ref. to $\tau_M$[f] | $q_M$ | Ref. to $q_M$[f] | $\tau_0$(emission)[d] | $\tau_0$(absorption)[e] | Ref. to $\tau_0$(abs.)[f] | $\dfrac{\tau_0(\text{abs.})}{\tau_0(\text{emiss.})}$ |
|---|---|---|---|---|---|---|---|---|---|
| Benzene | Hexane | 26.0 (5.7) | 1 | 0.11 | 2 | 236 | | | |
| | | | | 0.05 | 3 | 520 | | | |
| Toluene | Hexane | 26.0 (5.8) | 1 | 0.23 | 2 | 113 | | | |
| | | | | 0.12 | 3 | 217 | | | |
| p-Xylene | Hexane | 28.0 (6.1) | 1 | 0.415 | 2 | 67.5 | | | |
| | | | | 0.20 | 3 | 140 | | | |
| Hexamethyl benzene | Hexane | 6.0 (2) | 1 | 0.037 | 2 | 162 | | | |
| Naphthalene | Hexane | 103 (8.3) | 1 | 0.38 | 2 | 271 | | | |
| Anthracene | Benzene | 4.2 | 4 | 0.24 | 2 | 17.5 | 14.3 | 6 | |
| | Benzene | 4.1 | 5 | 0.26 | 8 | 15.8 | 13.5 | 7 | |
| | Benzene | 4.0 | 6 | | | 15.3 | | | |
| | Ethanol | 5.8 | 4 | 0.27 | 9 | 16.7 | | | |
| | Ethanol (95%) | 5.2 | 6 | | | 21.5 | 17.9 | 6 | |
| | Vapour phase | 5.7 | 10 | | | 19.3 | | | |
| | Vapour phase | 4.3 | 11 | | | | | | |
| | Benzene | 4.26 | 7 | 0.27 | 12 | 15.8 | 12.9 | 12 | |
| 9,10-Diphenyl anthracene | Benzene | 8.2 | 4 | 0.86 | 13 | 9.6 | 8.5 | 6 | |
| | | | | 0.85 | 13 | | | | |
| | Benzene | 7.3 | 6 | 0.84 | 8 | 8.7 | | | |
| | Ethanol | 9.4 | 4 | | | 12.7 | | | 0.82 |

**Table 3.3** (*continued*)

| Compound | Solvent | $\tau_M$(nsec)[c] | Ref. to $\tau_M$[f] | $q_M$ | Ref. to $q_M$[f] | $\tau_0$(emission)[d] | $\tau_0$(absorption)[e] | Ref. to $\tau_0$(abs.)[f] | $\dfrac{\tau_0(\text{abs.})}{\tau_0(\text{emiss.})}$ |
|---|---|---|---|---|---|---|---|---|---|
| 9,10-Dichloro-anthracene | Benzene | 9.6 | 5 | | | 14.8 | | | |
| | Benzene | 9.98 | 7 | 0.71 | 12 | 14.1 | 10.7 | 12 | 0.76 |
| 1,2-Benzanthracene | Cyclohexane | 45.0 | 14 | 0.19 | 14 | 232 | | | |
| 5-Methyl-1,2-Benzanthracene | Cyclohexane | 43.0 | 14 | 0.20 | 14 | 215 | | | |
| 6-Methyl-1,2-Benz-anthracene | Cyclohexane | 56.5 | 14 | 0.14 | 14 | 404 | | | |
| 10-Methyl-1,2-Benz-anthracene | Cyclohexane | 54.2 | 14 | 0.25 | 14 | 217 | | | |
| Perylene | Benzene | 5.2 | 4, 5 | 0.89 | 6, 13 | 5.3 | 5.06 | 7 | |
| | Benzene | 4.9 | 6, 13 | 0.89 | | 5.5 | 5.1 | 6 | |
| | Benzene | 4.79 | 15, 16 | 0.87 | 8 | 5.4 | 4.29 | 15 | |
| | Ethanol | 6.0 | 4 | 0.99 | 8 | 6.9 | | | |
| | Benzene | 5.02 | 7 | | 12 | 5.07 | 4.76 | 12 | 0.94 |
| Pyrene | Benzene | 300 | 13 | 0.8 | 13 | 375 | | | |
| | Cyclohexane | 480 | 17 | 0.68 | 13 | 706 | | | |
| | Cyclohexane | 435 | 13 | | | | | | |
| 9-Aminoacridine | Ethanol | 15.15 | 7 | 0.81 | 12 | 18.7 | 14.6 | 4 | 0.78 |
| | | | | | | 14.0 | 15.4 | 15 | 1.10 |
| Quinine sulphate | 1N-$H_2SO_4$ | 19.4 | 7 | 0.54 | 12 | 36.0 | 17.3 | 12 | 0.48 |
| | | | | | | 37.2 | 27 | 6 | 0.73 |

*a* For more data, see Birks and Munro (1967).
*b* Measurements at room temperature.
*c* Values in parentheses are for oxygen saturated solutions; the other values are for oxygen-free solutions.
*d* $\tau_0 = \tau_M/q_M$
*e* From Equation (3.36).
*f* 1. T. V. Ivanova, P. I. Kudryashov and B. Ya. Sveshnikov, *Sov. Phys. Doklady*, **6**, 407 (1961).
   2. E. J. Bowen and A. H. Williams, *Trans. Faraday Soc.*, **35**, 44 (1939).
   3. M. D. Lumb and D. A. Weyl, *J. Mol. Spectroscopy*, **23**, 365 (1967).
   4. W. R. Ware, *J. Phys. Chem.*, **66**, 455 (1962).
   5. W. R. Ware, *J. Am. Chem. Soc.*, **83**, 4374 (1961).
   6. J. B. Birks and D. J. Dyson, *Proc. Roy. Soc.*, **A275**, 135 (1963).
   7. W. R. Ware and B. A. Baldwin, *J. Chem. Phys.*, **40**, 1703 (1964).
   8. W. H. Melhuish, *J. Phys. Chem.*, **65**, 229 (1961).
   9. C. A. Parker, *Anal. Chem.*, **34**, 502 (1962).
   10. W. R. Ware and P. T. Cunningham, *J. Chem. Phys.*, **43**, 3826 (1965).
   11. K. H. Hardtl and A. Scharmann, *Z. Naturforsch.*, **12a**, 715 (1957).
   12. W. R. Dawson and M. W. Windsor, *J. Phys. Chem.*, **72**, 3251 (1968).
   13. D. J. Dyson, quoted in Birks and Munro (1967).
   14. J. B. Birks, D. J. Dyson and T. A. King, *Proc. Roy. Soc.*, **A277**, 270 (1964).
   15. S. J. Strickler and R. A. Berg, *J. Chem. Phys.*, **37**, 814 (1962).
   16. L. Brewer, C. G. James, R. G. Brewer, F. E. Stafford, R. A. Berg, and G. M. Rosenblatt, *Rev. Sci. Instr.*, **33**, 1450 (1962).
   17. J. B. Birks, D. J. Dyson and I. H. Munro, *Proc. Roy. Soc.*, **A275**, 575 (1963).

in Table 3.3 such as perylene for which there is good mirror symmetry between their fluorescence and absorption spectra, Equation (3.36) (derived on the assumption that there are no major differences between the nuclear configurations of the excited and the ground states) is in reasonable agreement with experimental values of $\tau_0$. For the majority of the cases $(\tau_0)_{exp} > (\tau_0)_{theo}$ and the discrepancy tends to increase with departure from the mirror symmetry relation (Birks and Dyson, 1963). This may be taken to suggest a relaxation of the excited state into a more stable nuclear configuration following the Franck–Condon absorption transition. The reader is referred to Birks and Munro (1967) for a list of lifetimes and quantum efficiencies for other organic molecules in various liquid media not listed in Table 3.3.

### 3.3.4.4 Direct determination of the lifetime of excited electronic states

Two basic techniques are in use for the direct measurement of the lifetime of excited electronic states: phase and/or modulation techniques, and pulse techniques. The latter are of wider application.

A. *Phase and/or modulation techniques.* In this method, the fluorescing system is excited by a continuous light source, the intensity of which is modulated at a high frequency. The phase and/or modulation of the fluorescence emission is compared with the phase and/or modulation of the exciting light.

Consider a fluorescent system having a fluorescence response function $i(t)$ given by Equation (3.31), which is exposed to an exciting light pulse described by a function $p(t)$. The functions $i(t)$ and $p(t)$ are related to the fluorescence response function $F(t)$ by the superposition integral (Bennett, 1960; Birks, Dyson and Munro, 1963)

$$F(t) = \int_{-\infty}^{t} p(t')\, i(t-t')\, dt' \tag{3.37}$$

$$= \int_{0}^{\infty} i(t'')\, p(t-t'')\, dt'' \tag{3.38}$$

If $p(t)$ is a periodic function, it may be represented by a Fourier series

$$p(t) = c_0 \left\{ 1 + \sum_j c_j \exp\left[i(\omega_j t + \phi_j)\right] \right\} \tag{3.39}$$

where $\omega_j$, $c_j$ and $\phi_j$ are the angular frequency, relative amplitude, and phase of the $j$th component, respectively.

Using Equations (3.31) and (3.38), the fluorescence response to the exciting pulse (Equation (3.39)) is given by (Birks and Munro, 1967)

$$F(t) = \frac{c_0\, i_0}{\beta} \left\{ 1 + \sum_j \left(\frac{\beta}{\beta + i\omega_j}\right) c_j \exp\left[i(\omega_j t + \phi_j)\right] \right\} \tag{3.40}$$

where $\beta = \tau_M^{-1}$. It is seen that $F(t)$ consists of a Fourier series which is similar to $p(t)$ in which the amplitudes and phases of the various frequency components depend only on those of the same frequency components in $p(t)$ and on $i(t)$. The following equations give the phase and amplitude of the $j$th component in $F(t)$ relative to those in $p(t)$.

$$\mathbf{m}_j = m_j \exp\left(-i\theta_j\right) = \frac{\beta}{\beta + i\omega_j} \tag{3.41}$$

$$\theta_j = \tan^{-1}(\omega_j/\beta) \tag{3.42}$$

$$m_j = \frac{\beta}{(\beta^2 + \omega_j^2)^{1/2}} = \cos\theta_j \tag{3.43}$$

Equations (3.42) and (3.43) are the basic equations for the phase and modulation fluorometers, respectively. In the former case the lifetime (decay time) is determined from the phase difference $\theta_j$ between the fluorescence and the exciting light of the component of angular frequency $\omega_j$ (usually the fundamental). In the latter case the lifetime is determined from the ratio $m_j$ of the modulation amplitudes of the fluorescence and exciting light of the component of angular frequency $\omega_j$. Equations (3.42) and (3.43) apply to systems with an exponential fluorescence response function (Equation (3.31)).

Birks and Dyson (1961) described a versatile instrument in which both the phase and modulation of the fluorescence can be measured. This can be used to analyse more complex fluorescence response functions (Birks, Dyson and Munro, 1963). Modulation fluorometry has received relatively little attention. Lifetimes obtained by this method have been reported by Birks and Little (1953) and Hamilton (1957, 1961). An excellent review of the various phase and/or modulation methods has been given by Birks and Munro (1967) and the reader is referred to this work for further discussion.

B. *Pulse techniques.* These methods are of wider use than the phase and/or modulation. They often employ techniques quite familiar to the nuclear physicist. The system under study is excited by intermittent radiation pulses of short duration, and the radiation emission decay is observed directly during the interval between the excitation pulses with a high gain, fast photomultiplier. The signal is displayed and measured with the aid of an oscilloscope or other device. The width of the signal received from the photomultiplier by the oscilloscope is generally directly related to the spread in transit time of the photoelectrons collected by the photomultiplier anode. If the interdynode transit time is too long, the lifetime cannot be measured with great precision since this method requires the response function of the whole equipment to be shorter in time than the true fluorescence lifetime. Therefore, knowledge of the lifetime range serves as a guide in designing a lifetime apparatus. The nature of the apparatus limits the measurable range of the lifetime. The general features of the equipment used for the measurement of lifetimes by this technique are shown in Figure 3.15. A number of specific arrangements have been discussed by Birks and Munro (1967) (see also a review on atomic lifetimes by Stroke, 1966).

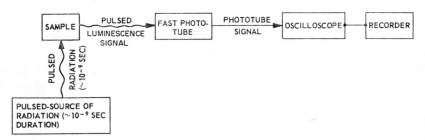

**Figure 3.15** Block diagram showing the general experimental arrangement used for the measurement of decay times

The time that elapses between excitation of an atom or molecule by a pulse of light and the subsequent emission of radiation can be measured also by delayed coincidence techniques. In this more recent method two separate detectors (rather than a single detector for direct observation of the luminescence) are used, one detector to observe the exciting source directly and the other to observe single photons from the luminescent system under study. The pulse from the first detector establishes the zero time origin and is used to gate the output from the second detector. The single photon pulses so selected are fed into a time-to-height converter and the output is displayed on a multi-channel analyser. For gas samples in the region below 1 torr (low pressures are often necessary to avoid imprisonment of radiation) the emission intensities are low and integration of light emission is necessary. The method was first proposed by Heron, McWhirter and Rhoderick (1956). Subsequently it was modified and applied to atomic lifetime studies by Bennett (1961) and Bennett, Kindlemann and Mercer (1965). Bollinger and Thomas (1961) applied the method in molecular lifetime studies. Bridgett and King (1967) used single-photon counting techniques in connection with a pulsed beam of electrons of fixed energy and sharp time cut-off to determine directly atomic-excited-state lifetimes (see also Klose, 1966, 1968.) When excited gas atoms are produced by sharp pulses of low-energy electrons, care must be taken that the excited level under study is allowed to decay freely and is not repopulated by cascading from upper (higher energy) states. Therefore, the exciting electron beam must be monoenergetic with energy as close to the excitation threshold as possible. Some recent atomic lifetime values measured by the method of delayed coincidence utilizing pulsed-electron excitation can be found in Bridgett and King (1967) and Klose (1966, 1968). It is finally noted that Imhof and Read (1969) have described a delayed coincidence method for determining lifetimes of neutral atoms and molecules in which coincidences are observed between inelastically scattered electrons which produce the excited atomic and molecular states and the subsequent decay of photons.

Recently Christophorou, Abu-Zeid and Carter (1968) have developed a pulsed-electron technique for the study of lifetimes of organic liquids under electron impact. This has been motivated by their earlier finding (Carter, Christophorou and Abu-Zeid, 1967) that the emission from organic liquids under intense electron bombardment is predominantly due to excimers (Section 3.4). The principle of the method is shown in Figure 3.16. A beam of electrons with energy $E$ passes between two parallel deflection plates $A$ and $B$, which are connected to an alternating voltage

$$V = V_0 \cos \omega t \tag{3.44}$$

A screen is placed in a plane perpendicular to the common axis of the two deflection plates. If the separation distance between the two deflection plates is $d(=0.7\text{ cm})$, the length of the deflection plates is $l(=21\text{ cm})$, and the distance from the centre of the plate to the screen is $y(=25\text{ cm})$, then the sweep velocity of the final spot of the beam on the screen is given by

$$\left(\frac{dx}{dt}\right)_R = \frac{1}{2}y\frac{l}{d}\frac{eV_0\omega}{E} \tag{3.45}$$

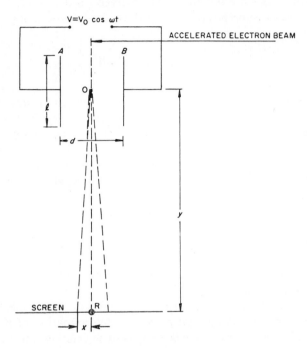

**Figure 3.16** Schematic diagram illustrating the principle of the pulsed-electron source

For a slit width of 0.05 cm centred at $R$ (see Figure 3.16) and for values of $E$, $V_0$, and $\omega$ equal to 56 keV, 1.25 kV, and 2.4 MHz, respectively, an electron pulse of $\sim 0.3$ nsec duration can be obtained as the beam sweeps across the slit.* The sample cell was located below the slit and was made of a cylindrical brass tube, 0.8 cm long and 0.6 cm in diameter, the top of which was sealed with a 0.0002 cm thick nickel foil. The cell's base was made of quartz and was located on a short quartz light guide which was sealed to the photomultiplier (Amperex type XP 1023). The electron beam,

---

*The duration of the pulse is somewhat shorter if relativistic effects are considered (Abu-Zeid, Christophorou and Carter, 1968).

deflection plates, sample cell, and phototube were placed into a vacuum chamber which was continuously evacuated by a Turbo-Molecular pump. A sampling oscilloscope was used to display the signal which was recorded on a strip-chart recorder.

Assuming that the response function of the instrument without the sample in place (Figure 3.16) is $I(t)$, and that the fluorescence response function is $i(t)$, the total response function $F(t)$ observed on the oscilloscope can be represented by the superposition integral which we now write as:

$$F(t) = \int_0^t I(t')\, i(t-t')\, dt' \qquad (3.46)$$

Equation (3.46) has been used in various ways to determine $i(t)$. We remind ourselves that the pulse method requires a pulsed source which cuts off in a shorter time than the fluorescence lifetime and a detection system which has a fast time response. The instrumental response function $I(t)$ is, in general, limited by the response of the photomultiplier rather than by the response of the light flash. When a pulsed light source is used for excitation, the function $i(t)$ can be determined easily. The pulse contour of the source is recorded first without the sample in place. This determines $I(t)$. Under the same conditions as for the determination of $I(t)$, $F(t)$ is measured. Equation (3.46) is then used to determine $i(t)$, usually with the aid of a computer. If the fluorescence response function $i(t)$ is exponential the 'method of moments' may be used to determine the lifetime from the measured functions (Brody, 1957).

When the excitation pulse is an ionizing particle, or an electron beam or x-ray beam, rather than a light pulse, the determination of $i(t)$ is more difficult. Various procedures have been worked out in these cases, especially in connection with scintillation decay times (Birks, 1964). The instrumental response function, for example, may be determined directly from first principles. This requires combining the contributions to the time dispersion from various parts of the apparatus such as the width of the exciting pulse, the spread in the transit time of the phototube, the distortion in the electrical circuit and the response of the recording system. Calculation of $I(t)$ this way is rather impractical. An alternative method may be employed, in which one observes the oscilloscope pulse caused by a material whose decay time is very much shorter than the width of the response function. Several attempts were made to use this method by obtaining materials (scintillators) with short decay times. A third method is to determine $I(t)$ using standard scintillators of known decay times.* If the response function $i(t)$ of the

---

*Smith–Saville (1968) has used Čerenkov light pulses, generated by individual ionizing particles, to determine $I(t)$ ($\sim 0.5$nsec) for a single photon counting instrument.

standard system is represented by

$$i(t) = \frac{N}{\tau_M} e^{-t/\tau_M} \tag{3.47}$$

where $N$ is a normalizing constant, through Equation (3.46) we have

$$\tau_M \frac{dF(t)}{dt} + F(t) = NI(t) \tag{3.48}$$

Equation (3.48) may be used to determine the response function $I(t)$ of the system from measurements of $F(t)$ and known $\tau_M$. A number of workers have used anthracene crystals as a standard (see Birks (1964) for problems arising when anthracene is used as a standard). Others (e.g. Christophorou, Abu-Zeid and Carter, 1968) used liquid standards.

If $I(t)$ is represented by $\exp\text{-}(t/\tau_i)^2$ where $\tau_i$ is a parameter characteristic of the width of the Gaussian function, we have for large values of $t$

$$F(t) \simeq \text{constant } e^{-t/\tau_M} \tag{3.49}$$

Equation (3.49) has been applied by a number of workers to determine $\tau_M$ from $F(t)$ directly.

### 3.3.4.5 Indirect determination of the lifetime of excited electronic states

A number of indirect determinations of the lifetime of excited electronic states have been reported based on the measurement of various quantities such as the dispersion of radiation, fluorescence depolarization, and solvent quenching. The reader is referred to Mitchell and Zemansky (1961), Stroke (1966), and Birks and Munro (1967) for references to such methods.

### 3.3.5 Flash Photolysis, Double-Photon Absorption, and Simultaneous Excitation Studies

### 3.3.5.1 Flash photolysis

As will be discussed in Section 3.4, information about the position and lifetime of the lowest triplet state for luminescent molecules can be obtained from emission studies, namely phosphorescence (Section 3.4.1). Also, perturbation techniques (Section 3.4.2) as well as electron impact (Chapter 5) and other heavy particle impact (Section 3.5) techniques provide information about the lowest triplet state of a molecule. Particle impact techniques allow location of higher triplet states as well. Another very useful technique has been shown especially suited for the observation and quantitative study of triplet states of organic molecules by allowing measurement of the triplet–

triplet absorption spectra of such molecules. This is the flash photolysis method.

In this method a strong flash of light (of several hundred joules of energy and of a few microseconds duration) produces a large enough number of molecules in the lowest triplet state (by intersystem crossing from $S_1$ (Section 3.4.2)) that the triplet–triplet absorption spectrum of the transient species produced by the high intensity photolysis flash, i.e.

$$T_1 + h\nu \rightarrow T_{n \geqslant 2} \tag{3.50}$$

can be studied. Such transitions are allowed since no change in multiplicity is involved. The transient spectrum is customarily taken by using a second flash, timed to photograph the absorption spectrum at pre-set time intervals after the photolysis flash. Once the transient spectrum is known, detailed kinetic studies can be made by using a continuous photoelectric recording of the optical density at one wavelength.

The first application of flash photolysis to the study of triplet–triplet absorption spectra in solution has been made by Porter and Windsor (1953, 1954). Since then a variety of molecules have been investigated (see, for example, Porter and Windsor (1958)). These studies allowed determination of the position of higher triplet states and extinction coefficients for triplet–triplet transitions as well as the concentration of $T_1$ as a function of time. The method is equally useful for gases, liquids, and solids (Porter, 1961). Care should be taken that the intense flash does not cause photodecomposition of the substance under study.

A recent method for measuring the quantum yield of triplet-state formation of aromatic hydrocarbons in solution has been described by Medinger and Wilkinson (1965) (see also Horrocks and Wilkinson, 1968).

### 3.3.5.2 Double-photon absorption and simultaneous excitation studies

We have seen in the previous section (3.3.5.1) the importance of pulsed light sources of high intensity in studying transient species. Other multiple excitation phenomena have recently been discovered and studied, especially in condensed media. Some biphotonic processes and transient excited species formed in condensed media or high-pressure gases will be discussed in Section 3.4. Here we wish to briefly elaborate on two distinct new effects, namely double-photon absorption by a molecule and simultaneous excitation of two molecules by a single photon.

Double-photon capture by organic and inorganic molecules has by now been well established. Double-photon absorption has been demonstrated in inorganic crystals (Kaiser and Garrett, 1961; Hopfield, Worlock and Park, 1963) and inorganic liquids and crystals (e.g. Peticolas, Goldsborough and Rieckhoff, 1963; Singh and Stoicheff, 1963; Peticolas and Rieckhoff, 1963; Hall, Jennings and McClintock, 1963; Singh and coworkers, 1965;

McMahon, Soref and Franklin, 1965; Dowley, Eisenthal and Peticolas, 1967; Fröhlich and Mahr, 1966; Weisz and coworkers, 1964; Peticolas, Norris and Rieckhoff, 1965). These studies allowed new insight into the molecular structure and have become possible by the availability of high intensity beams of coherent light (lasers) and other intense flash lamps such as xenon flash-lamp sources of suitable intensity and response (see, for example, Weisz and coworkers, 1964).

Quite generally, double-photon absorption has been observed via the prompt fluorescence decay of the excited species following excitation by a laser or other highly intense sources. A plot of the logarithm of the fluorescence intensity versus that of the exciting intensity yields a slope of two, corresponding to a square-law relationship between intensity of fluorescence and excitation, thus indicating the double-photon effect. For some theoretical considerations of two-photon radiation processes in mater, see Kleinman (1962) and Braunstein (1962). See Birks (1970) for a discussion of two-photon absorption processes in organic systems.

Simultaneous excitation in gaseous mixtures has been observed also (Kasha, 1968). The absorption spectrum of an irradiated mixture of hydrogen and carbon dioxide was reported (Kasha, 1968) to have shown lines which corresponded not only to the characteristic transitions of each molecule ($H_2$ and $CO_2$), but also to transitions corresponding to the sum of energies of transitions in both molecules. Such single-photon sharing is greater at high gas pressures. It can also be observed at low pressures as an emission process.

Double-photon and simultaneous-transition studies as well as other biphotonic and non-linear phenomena are becoming of increasing importance in studies of irradiated media, especially in the condensed phase. It might be interesting to look for the reverse process of double-photon absorption, namely, double-photon emission. Such a process is known to take place in the decay of the positronium.

### 3.3.6 *Photoelectron and Penning-Ionization-Electron Spectroscopy*

A brief discussion of these two new types of spectroscopy and some typical results are presented in Section 3.5.4.

## 3.4 The Fates of Electronic Excitation Energy

In Section 3.2 we have discussed the absorption of low-energy photons by the electronic system of atoms and molecules. The process of photo-absorption results in an increase in the total energy of the system which, being in a less stable state should return to its ground state. The time for this return varies drastically (from $\sim 10^{-12}$ sec to minutes or even for longer times).

The most obvious de-excitation (de-activation) process is that in which the excitation energy appears as radiation. Such radiative de-activation processes usually occur in times of $10^{-9}$ to $\sim 10$ sec. Thus, for the excited system to decay radiatively it has to survive other types of de-activation processes for comparable periods of time. That other, and often faster ($\sim 10^{-13}$ sec) de-excitation mechanisms occur, is clearly indicated by the fact that most molecules do not emit light (although they absorb light) and that the amount of energy re-emitted in most molecules is considerably smaller than the amount of energy absorbed. Much of our information about the radiationless processes in molecules is inferred from studies of radiative transitions which are easier to study. The various relaxation processes, that is the various mechanisms of de-excitation of excited electronic states, fall into three types as follows: (i) radiative de-excitation; (ii) internal and external non-radiative de-excitation, and (iii) transfer of excitation energy to other species. We shall discuss molecular properties characteristic of isolated or semi-isolated systems. No discussion of collective phenomena will be made. However, the environmental effects on the molecular properties under study will be briefly elaborated upon as most of the existing information on radiative and non-radiative de-activation processes is on systems which cannot be considered isolated.

### 3.4.1 *Radiative De-excitation*

Radiative de-excitation of an electronically excited system gives rise to light emission which may be referred to generally as luminescence. A more specific terminology is given later in this section (Table 3.4). The emission of radiation by an excited species is considered to be independent of the mode of excitation.* However, the intensity and spectral distribution of the emitted photons may vary with the intensity of excitation. Such changes may be introduced because the probability of formation (and quenching) of the various luminescent species possessing different spectral characteristics, depends on the intensity of the exciting radiation (see, for example, Carter, Christophorou and Abu-Zeid, 1967 and Christophorou, Abu-Zeid and Carter, 1968).

Radiative de-excitation of atoms and molecules is studied either through the emission spectra or through the measurement of the decay time of the excited state. Prior to discussing radiative transitions in molecules, a brief discussion of similar processes in monatomic molecules is indicated.

---

*Atomic and molecular excitation may be achieved by means other than photoabsorption as, for example, by ionizing radiation, slow electrons, sound and shock waves, thermally and chemically.

In atomic systems the only degrees of freedom are translational and electronic. Hence, in atoms strictly electronic transitions are the source of luminescence. Transitions between atomic electronic states give rise to characteristic series of lines in emission (as in absorption). Emission by low pressure monatomic vapours may take place by direct transition to the ground state or step wise (Pringsheim, 1949). When emission takes place from the electronic state which has been reached by absorption back to the ground state it is called resonance radiation.* Resonance spectra are observed not only from the lowest but also from higher excited atomic states.

The relative simplicity of atomic emissions at low pressure is often lost when the gas pressure is increased. Two main complexities arise at high pressures: one involves the imprisonment of resonance radiation and the other involves the interaction of excited and unexcited atoms. The former increases the apparent lifetime of the resonance line and the latter results in a continuous rather than a line emission spectrum. The continuous emission is often characteristic of Franck–Condon transitions in molecules.

The two processes mentioned above are clearly seen in the case of the rare gas atoms. Resonance emission and resonance trapping in the rare gases—indicated by the vectors to the right and to the left in the process below—

$$R^* \leftrightarrows R + h\nu_R \qquad (3.51)$$

have been the subject of many studies (see, for example, Mitchell and Zemansky, 1961; Pringsheim, 1949; Holstein, 1947). The rare gas continua have also been investigated (see, for example, Hopfield, 1930; Tanaka, 1955; Strickler and Arakawa, 1964; Huffman, Larrabee and Tanaka, 1965; Hurst, Bortner and Strickler, 1968, 1969; and Nichols and Vali, 1968). Although for exciting photon energies above the ionization potential of the rare gas atom the continuous emission may originate from excimer ions ($R_2^+*$) it may also originate from the process

$$R^* + R \rightarrow R_2^* \rightarrow R + R + h\nu_{cont.} \qquad (3.52)$$

where $R^*$ is an excited and $R$ is an unexcited rare gas atom interacting at collision. While process (3.52) is similar to the well-established mechanism of excimer formation in organic molecules in fluid and condensed media (Förster and Kasper, 1955; Birks and Christophorou, 1962a, b; 1963a, b; 1964a, b) it is different in many respects. Here the system lacks the many molecular degrees of freedom and since kinetic energy of the order of $kT$ is brought into the system by the colliding atoms, $R_2^*$ is unstable against dissociation. Hence, although photoassociation of excited and unexcited molecules in fluid media occurs at collision with a high probability (for certain

---

*This term has often been used in a wider context.

molecules $\gtrsim 0.5$ (Birks, Braga and Lumb, 1965) in the case of atomic vapours the collision complex most probably dissociates a large number of times without emission (i.e. $R^* + R \rightarrow R_2^* \rightarrow R^* + R$). There is need for a third body to stabilize $R_2^*$, although at the instant of closest approach there is always a finite probability that the system will undergo radiative decay. Whether this happens or not radiative dissociation of $R_2^*$ (process (3.52)) results in a continuous emission if the radiative transition ends on the repulsive ground state (see discussion on excimers in polyatomic molecules in Section 3.4.1.5; Figure 3.17d).

De-excitation processes in molecules differ from those in atoms. In a given range of energy the total number of degrees of freedom for molecules is very much greater, owing to the presence of rotations and vibrations. As a consequence of the large increase in the number of possible excited states the spacing of the energy levels is smaller. In addition, the processes of

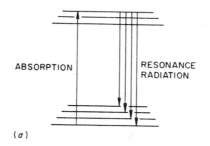

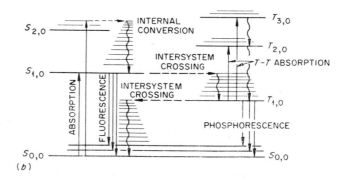

**Figure 3.17** (*a*) Resonance radiation. (*b*) Energy levels and energy transfer in a complex molecule.

molecular dissociation (direct or predissociation) may compete with radiation and other de-excitation processes. Hence, in molecules non-radiative de-excitation mechanisms (see Section 3.4.2) are very efficient, especially among the higher excited states. In spite of the many (and often fast) non-radiative de-excitation processes occurring in molecules, luminescence

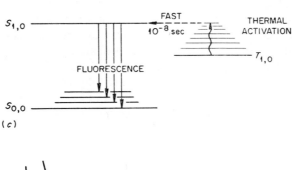

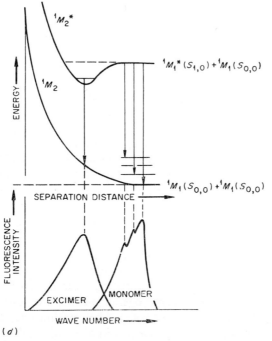

**Figure 3.17** *(continued)* (*c*) Schematic diagram showing the process of E-type delayed fluorescence. (*d*) Schematic representation of the potential energy diagrams for $^1M_3^*$ and $^1M_2$ and the $^1M_3^*$ and $^1M_3^*$ emissions

spectra (thus radiative de-activations) are a common feature of a large number of substances. It should be noted, however, that almost generally radiation is observed from the lowest excited state of either multiplicity (singlet or triplet) of a molecule. This would indicate that the non-radiative processes which compete with radiation are very much faster for the higher than the lowest excited molecular electronic states. Since the lowest excited states have smaller chance of crossing and for their energies the processes of dissociation and predissociation may not be operative, molecules for which the radiative de-excitation of the lowest excited states is fast compared to non-radiative quenching have a greater probability to emit light. As stated earlier, light emission from the lowest excited electronic states of a molecule can be exhibited under proper chemical and physical conditions in almost all forms of matter (Pringsheim, 1949). It should be noted that because the higher the initial excited state is, the greater the probability of non-radiative de-activation, luminescence due to absorption in the far ultraviolet region is highly improbable. It is not surprising, therefore, that luminescence is a rare event for molecules which absorb only in the far ultraviolet. Dissociation (direct or predissociation) following photoabsorption has been provided as an explanation for the almost general lack of luminescence in inorganic polyatomic molecules* (Pringsheim, 1949).

Radiative de-activations are environment dependent. They are seriously influenced by changes in collision rates both in terms of absolute yields and spectral distributions. Most of the work on luminescence and radiationless transitions in polyatomic molecules has been performed in the dissolved or liquid and solid states. The effects of specific environments (especially solvent effects) on the molecular properties have, in certain cases, been investigated. As a rule the efficiency of radiative de-excitation increases with restricting molecular motion. Since molecular rigidity is favourable in preventing radiationless transitions the probability of luminescence will increase with decreasing temperature.

Among the systems which have been found to luminesce efficiently in condensed media are the aromatic hydrocarbons. Owing to the presence of $\pi$ electron structures in these systems, they often have high rates for radiative de-excitation from their lowest $\pi$-singlet state to the ground state. This property explains their use as scintillating materials (Birks, 1964). The strong luminescence of certain carcinogenic members of this group such as 3,4-benzopyrene and derivatives of 1,2-benzanthracene, led to their identification in carcinogenic tars. As a rule in aromatic molecules radiative emission is observed irrespective of what $\pi$ electronic state is initially excited,

---

*The same reason may be provided for the absence of luminescence in saturated hydrocarbons and other aliphatics. However, luminescence spectra in saturated hydrocarbons have recently been reported (Hirayama and Lipsky, 1969).

and is characteristic of the lowest $\pi$ electronic state of a given multiplicity, although the radiative lifetimes of the higher $\pi$-singlet states are quite generally shorter than the lifetime of the lowest excited $\pi$-singlet state (see, however, section 3.4.1.2.). The absence of radiative decay from higher $\pi$-singlet states has been attributed to efficient, rapid ($\sim 10^{-11}$–$10^{-12}$ sec) radiationless conversion between these states, which is evident in the overlap of adjacent bands in their absorption spectra. In this group of molecules, therefore, excitation into any of the $\pi$-singlet states is rapidly brought by the non-radiative processes of internal conversion and internal degradation (see Section 3.4.2) to the lowest vibrational state of the first excited $\pi$-singlet state $S_{1,0}$, from which radiative decay may occur. Non-radiative de-excitation of $S_{1,0}$ may take place also but this process is somewhat slow in this case as a result of the larger energy gap between $S_{1,0}$ and $S_{0,0}$. A similar process occurs between the $\pi$-triplet states. A process which may be in competition with internal quenching from higher $\pi$-singlet states to $S_{1,0}$ is energy transfer to other species (see, for example, Oster and Kallmann, 1962, 1966). The important emissions which follow the radiative de-excitation of $S_{1,0}$ and $T_{1,0}$ as well as those arising from secondary interactions involving these states will now be outlined. Some references of general interest are: Pringsheim (1949); Förster (1951); Van Duuren (1963); Garlick (1949); Bowen and Wokes (1953); Parker (1968); Birks (1970). See also Feofilov (1961) for polarization of luminescence.

## 3.4.1.1 Resonance radiation

As we have mentioned earlier in this section, resonance radiation is commonly observed in atomic gases at low pressures where the time intervals between collisions are longer than the lifetime of the excited state ($\sim 10^{-8}$ sec). Such radiation is observed when the only possible radiative transition from the excited state is to the ground state, which is the case for the lowest excited level or for a higher level for which intermediate transitions are forbidden. In Figure 3.17a the resonance spectrum for a molecule is shown. It is seen that the emission is from the initial state to the various vibrational (and rotational; not shown in the Figure 3.17a) levels of the ground state. This emission is often referred to as resonance molecular fluorescence and occurs when the molecule cannot, within its radiative lifetime, dissipate vibrational energy. It is observed in low-pressure vapours.

## 3.4.1.2 Fluorescence

Molecular non-resonant emission almost always appears either from the lowest excited singlet or the lowest triplet state, the excited state losing some or all of its excess vibrational energy through collisions. The short-lived re-emission of the absorbed radiation at lower energies than the absorbed radiation is referred to as fluorescence. As a result of the fast relaxation processes of higher states down to the lowest vibrational state of the lowest electronic state of a given multiplicity a general observational rule has been formulated: the emitting level of a given multiplicity is the lowest excited level of that multiplicity. This rule would imply, then, that the character of the emission spectrum of a substance does not depend on the exciting wavelength.

Most studies on fluorescence have been made on organic molecules in the dissolved state. In these systems with rare exceptions fluorescence occurs from the lowest excited state of a given multiplicity independently of which (higher) excited state is initially excited. The only exception to this experimental generalization appears to be the azulene molecule from which fluorescence originates from the second rather than from the first excited $\pi$-singlet state.* Various suggestions have been made to explain the azulene fluorescence anomaly (Binsch and coworkers, 1967). The most generally accepted explanation is that because in azulene the $S_2 - S_1$ energy gap is large, the $S_2 \rightarrow S_0$ emission is observed due to the low internal conversion rate from $S_2$ to $S_1$.

As a consequence of the fact that the fluorescence emission in aromatic molecules takes place from the lowest vibrational level of the first excited $\pi$-singlet state and that the vibrational spacings for $S_1$ and $S_0$ are not very different, the fluorescence spectrum of aromatic molecules is usually the mirror-image of the longest wavelength absorption band. Figure 3.18 shows the spectral relationship between absorption and emission for a dilute solution of perylene in cyclohexane (Berlman, 1965). Numerous studies of this nature have been made (e.g. Birks and Dyson, 1963; Cherkasov and coworkers, 1956). Berlman (1965) has given a collection of fluorescence spectra of aromatic hydrocarbons which he has compared with the respective first absorption bands. The reader is referred to this source for further information. The vibrational structure observed in the fluorescence spectrum (Figure 3.18) is that of the ground state; the structure observed in the absorption spectrum corresponds to the vibrational levels of the excited states. Which of the vibrations is more intense in emission depends on the relative configuration of the nuclei in the upper and lower states. For most aromatic molecules the change in nuclear configurations for the two states is slight and (as in absorption) the $0 \rightarrow 0$ transition is the most intense. The following quantities characterize the fluorescence spectrum of an organic molecule in solution.

(i) The fluorescence quantum yield, $\phi_M$, defined as the number of molecules in $S_1$ which undergo radiative decay to the ground state, over the total

---

*Fluorescence from the second excited $\pi$-singlet state of pyrene and 3,4-benzopyrene in the vapour phase has recently been reported (Geldof, Rettschnick and Hoytink, 1969). A similar emission has been observed for these two molecules and 1,2-benanthracene and 1,12 benzperylene in solution (Easterly and coworkers, 1970).

number of molecules in $S_1$. This quantity is strongly dependent upon the environment and the molecular structure. For results on quantum yields see Birks and Munro (1967); Dawson and Windsor (1968) and References quoted therein.

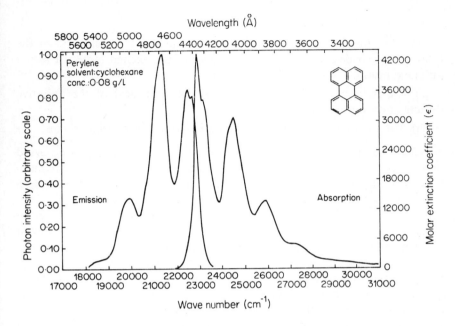

**Figure 3.18** First absorption band and fluorescence spectrum of perylene in cyclohexane (Berlman, 1965). Note the excellent mirror-image relation of absorption and emission spectra

(ii) The fluorescence quantum efficiency, $q_M$, defined as the limiting (maximum) value of $\phi_M$ at infinite dilution (Birks and Munro, 1967; Dawson and Windsor, 1968). Molecular fluorescence for most organic molecules in the dissolved state is subject to the so-called concentration quenching. The reduction in $\phi_M$ with increase in molar concentration $c$ is described by (Birks and Munro, 1967)

$$\phi_M = \frac{q_M}{1 + \dfrac{c}{c_h}} \tag{3.53}$$

where $c_h$ is the concentration at which $\phi_M = \frac{1}{2}q_M$; $1/c_h$ is the Stern–Volmer concentration quenching parameter.

(iii) The fluorescence spectrum. The relative fluorescence quantum intensity $F(\bar{v})$ at the wave number $\bar{v}$ can be normalized to absolute units by

$$q_M = \int_0^\infty F(\bar{v}) \, d\bar{v} \tag{3.54}$$

(iv) The fluorescence (decay) lifetime $\tau_M$ which is the reciprocal of the total first-order de-activation rate parameter of the excited molecule M* at infinite dilution.

(v) The radiative lifetime $\tau_0$ defined as the reciprocal of the radiative de-activation rate parameter, and

(vi) The polarization of the emission relative to the molecular axes.

When the fluorescence and absorption spectra overlap, the observed fluorescence spectrum (as well as the measured decay time and quantum yield) may differ from its molecular (true) value because of self-absorption as can be seen clearly from the spectra presented in Figure 3.19.* That portion of the fluorescence spectrum which overlaps the absorption spectrum is suppressed at high concentrations especially when the lowest absorption band has a large oscillator strength ($^1L_a$ states for aromatic molecules). The degree of overlap is determined by the relative positions of the potential energy curves (surfaces) of the ground and excited states. Numerous such examples have been reported by Christophorou (1963) (see also Birks and Christophorou, 1962a, 1963a, b, 1964b).

### 3.4.1.3 Phosphorescence

The radiative transition from the lowest triplet state of a molecule to its ground state (singlet) is known as phosphorescence (Lewis, Lipkin and Magel, 1941; Lewis and Kasha, 1944; and Kasha, 1950). This is a spin-forbidden transition and the radiative triplet lifetime is long. It may vary from $10^{-3}$ sec for molecules which contain a heavy atom, and thus promote spin–orbit coupling, to perhaps $10^2$ sec for highly symmetric systems such as benzene. The large difference in the fluorescence and phosphorescence lifetimes can help separate the fluorescence and phosphorescence emissions. The phosphorescence spectrum shows similar vibrational structure to the fluorescence spectrum, since the transition occurs to the various vibrational sublevels of the ground state. The phosphorescence spectrum is shifted, however, to lower energies (see Figure 3.17b) since, as a rule (Hund's rule),

---

*In this case the observed decrease in the emission intensity at the short-wavelength side of the spectrum with increasing concentration is due to self-absorption, but also due to the depletion of excited single molecules in $S_{1,0}$ via the process of excimer formation (see Section 3.4.1.5).

$T_{1,0}$ lies below $S_{1,0}$. Often the phosphorescence and fluorescence spectra show some degree of overlap.

In photoabsorption studies the lowest triplet state is reached by inter-system crossing usually from $S_{1,0}$. Since $S_1$ decays radiatively in most cases within $\sim 10^{-8}$ sec, intersystem crossing from $S_{1,0}$ to $T_1$ must be fast (comparable to the decay time of $S_1$) in order to compete effectively with the $S_1$ emission, and this, despite the required change in the electron spin.

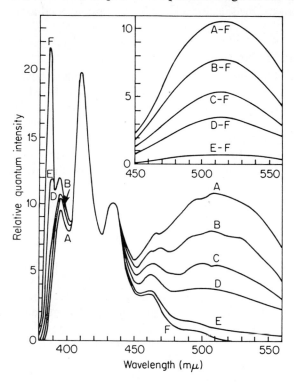

**Figure 3.19** Fluorescence spectra of 5-methyl 1,2-benzanthracene in cyclohexane at various concentrations: A, $10^{-2}$ M; B, $7.5 \times 10^{-3}$ M; C, $5 \times 10^{-3}$ M; D, $3 \times 10^{-3}$ M; E, $5 \times 10^{-4}$ M; F, $5 \times 10^{-5}$ M (M = moles per litre). The relative quantum intensities are normalized at $\lambda = 430$ mμ (Birks and Christophorou, 1962a)

Similarly, $T_1$ can be reached by intersystem crossing from higher singlet states to the triplet manifold followed by internal conversion to $T_{1,0}$. However, internal conversion from higher singlet states to the lowest singlet is a fast ($\sim 10^{-12}$ sec) process, making intersystem crossing from higher triplet states an inefficient way of populating $T_1$.

The forbiddingness of $T_{1,0} \to S_{0,n}$ radiative transition results in a long lifetime for the lowest triplet state allowing other de-activation processes to compete effectively with phosphorescence. In fluid media phosphorescence is usually completely quenched by the faster radiationless mechanisms, although phosphorescence in the gas and liquid phases has been observed in some molecules (e.g. biacetyl). Phosphorescence is a more likely process of de-activation of $T_{1,0}$ when the temperature is lowered and when the molecule is constrained, as in rigid or highly viscous media or in the adsorb state.

### 3.4.1.4 Delayed fluorescence

Two types of delayed molecular fluorescences have been distinguished: $E$-type and $P$-type. The processes leading to $E$-type delayed fluorescence are shown schematically in Figure 3.17c. The molecule in $T_{1,0}$ is lifted thermally back to $S_{1,0}$ from which it undergoes radiative decay to $S_{0,n}$. Obviously the spectrum is indistinguishable from that of normal fluorescence (Section 3.4.1.2), but it has a much longer lifetime, characteristic of $T_1$. Its intensity is naturally temperature dependent, whilst the overall process is a function of the $T_{1,0} - S_{1,0}$ energy gap.

The mechanism suggested originally by Parker and Hatchard (1962a) for the $P$-type delayed fluorescence,

$$^3M_1^* + {}^3M_1^* \to {}^1M_1^* + {}^1M_1 \to {}^1M_1 + {}^1M_1 + h\nu_{DF} \tag{3.55}$$

involves a bimolecular interaction between two molecules ($^3M_1^*$) in their lowest triplet state giving an excited molecule ($^1M_1^*$) in its lowest excited $\pi$-singlet state and another ($^1M_1$) in its ground state. The emission, $h\nu_{DF}$, is then that of $^1M_1^*$ produced through triplet–triplet interaction. The observation (Parker and Hatchard, 1962a, b) that the intensity of the delayed emission varied as the square of the intensity of the exciting light led to this suggestion. The phenomenon was identified in rigid solutions of naphthalene by Czarnecki (1961) and 3,4-benzopyrene by Muel (1962). Muel (1962) proposed that process (3.55) proceeds via an excimeric intermediate (Section 3.4.1.5) which rapidly dissociates to give $^1M_1^*$ and $^1M_1$. This suggestion was also made by Parker and Hatchard (1962a). In rigid media the interaction between the two $^3M_1^*$ molecules is non-collisional over a distance.

There exists some controversy as to the detailed mechanism of $P$-type delayed fluorescence (Muel, 1962; Parker, 1964; Naqvi, 1967; Birks, 1963, 1968; Parker and Hatchard, 1962a, b, 1963; Parker, 1963, 1966, 1967; Birks, Moore and Munro, 1966; Tanaka and coworkers, 1963; Moore and Munro, 1967; Birks, Srinivasan and McGlynn, 1968; Stevens and Ban, 1968). Birks (1968), summarizing the work on $P$-type delayed fluorescence, argued that in fluid solutions triplet–triplet interaction occurs by two parallel processes, namely through process (3.55) and through

$$^3M_1^* + {}^3M_1^* \rightarrow {}^1M_2^* \tag{3.56}$$

Mechanism (3.55) is due to electron exchange interaction and consists of a diffusion-controlled and a non-diffusion-controlled process. Mechanism (3.56) is due to excimer interaction which is diffusion-controlled. In rigid solutions the two diffusion-controlled components are inhibited. Obviously $^1M_2^*$ may dissociate to give $^1M_1^*$ and $^1M_1$ with subsequent $^1M_1^*$ emission. Otherwise, $^1M_2^*$ may directly undergo radiative dissociation to the repulsive excimeric ground state giving excimer fluorescence (Section 3.4.1.5).

Delayed fluorescence has been reported in vapours (e.g. Stevens, Hutton and Porter, 1960; Williams, 1958); fluid media (e.g. Stevens and Walker, 1963; Parker and Hatchard, 1961, 1962a,b, 1963; Birks, Moore and Munro, 1966; Tanaka and coworkers, 1963; Parker, 1963, Moore and Munro, 1967; Birks, Srinivasan and McGlynn, 1968; Stevens and Ban, 1968); rigid solutions (e.g. Azumi and McGlynn, 1963; Czarnecki, 1961; Muel, 1962), and molecular crystals (e.g. Kepler and coworkers, 1963).

### 3.4.1.5 Excimer fluorescence

Many organic molecules—as well as other systems such as rare gas atoms and other monatomic vapours (e.g. those of mercury) at high pressures (see discussion earlier in this section and Colli (1954))—are known to exhibit a second type of fluorescence in concentrated solutions or in liquid and crystal phases, known as excimer fluorescence (Förster and Kasper, 1955; Birks and Christophorou, 1962a, 1963a,b, 1964a,b). Excimer fluorescence originates from radiative dissociative de-activation of the excimer*(excited dimer)$^1M_2^*$. The excimer $^1M_2^*$ is produced by collisional interaction of excited $^1M_1^*$ and unexcited $^1M_1$ molecules (monomers).

$$^1M_1^* + {}^1M_1 \rightleftarrows {}^1M_2^* \tag{3.57}$$

where the arrow to the left indicates excimer dissociation reproducing $^1M_1^*$ and $^1M_1$. Excimer formation is typified by the behaviour of many aromatic molecules such as pyrene (Förster and Kasper, 1955; Birks and Christophorou, 1963a) and derivatives of 1,2-benzanthracene (Christophorou, 1963; Birks and Christophorou, 1962a, 1963b, 1964a,b). Figure 3.19 shows a typical example of $^1M_2^*$ emission from 5-methyl 1,2-benzanthracene (Birks and Christophorou, 1962a) in cyclohexane solutions. As the molecular concentration increases, the $^1M_2^*$ emission increases, while that of $^1M_1^*$ decreases due to self-absorption and depletion of $^1M_1^*$ via the excimer

---

*The term excimer has been introduced by Stevens (1961) to describe $^1M_2^*$ and to distinguish it from excited states of dimers which are stable in the ground state.

formation process (3.57). The ratio of the excimer to monomer quantum yields is proportional to the concentration, consistent with (3.57). The absorption spectrum is characteristic of $^1M_1$ and independent of the concentration (Christophorou, 1963) showing that the excimer has a repulsive ground state. The excimer fluorescence (inset in Figure 3.19),

$$^1M_2^* \rightarrow {}^1M_1 + {}^1M_1 + h\nu_{EF} \qquad (3.58)$$

is a structureless band shifted to longer wavelengths compared to the $^1M_1^*$ emission. For organic molecules the difference in energy between the first vibrational level in the $^1M_1^*$ emission and the maximum of the $^1M_2^*$ band is of the order of 6000 cm$^{-1} \simeq 0.75$ eV (Christophorou, 1963). The shape of the excimer band is determined principally by the degree of repulsion of the ground state $^1M_2$. Spectra, lifetimes, and energetics, as well as theoretical treatments of excimers have been reported in abundance following the generalization of the formation of excimers by Christophorou (1963) and Birks and Christophorou (1963a, b, 1964a, b). The extensive literature has been cited in Birks (1967), in which references to the reaction kinetics of excimer formation (Equation (3.57)) can be found (see also Förster (1969)).

The attractive molecular interactions involved in excimer formation are of two types: (i) charge-transfer interactions and (ii) dipole–dipole (exciton) interactions. In the case of aromatic excimers (i) is a charge resonance interaction between positive and negative molecular ions, while (ii) is an exciton interaction between transition dipoles in the two molecules. The magnitude of (i) is a function of the ionization potential as well as the electron affinity of the species involved, while that of (ii) is determined by the magnitude of the $^1L_{a,b}$ (or $^1B_{a,b}$)–$^1A$ transition dipole which ever has the largest moment among the transitions to the lowest excited states of $^1M_1$ from the singlet ground state $^1A$ (Platt, 1949). A number of theoretical treatments (Konijnenberg, 1963; Murrell and Tanaka 1964; Azumi and McGlynn, 1964, 1965; Azumi, Armstrong and McGlynn, 1964; Vala and coworkers, 1966) have shown that neither (i) or (ii) alone can adequately account for the magnitude of the aromatic excimer interaction of $\sim 0.75$ eV. Configuration interaction between the charge and exciton resonance states has been proposed as being responsible for the resultant excimer attraction. That charge transfer is important in excimer binding has been demonstrated by many experimental studies (Eisinger and coworkers, 1966; Kawaoka and Kearns, 1966; Mataga, Okada and Ezumi, 1966; Mataga, Ezumi and Okada, 1966). Recent reports on the observation of emission from $^3M_2^*$ (Christophorou, Abu-Zeid and Carter, 1968; Lim and Chakrabarti, 1967; Castro and Hochstrasser, 1966; Langelaar and coworkers, 1968) and formation of excimers via processes (see below) other than Equation (3.57) strongly indicate the importance of charge resonance. In the case of triplet state excimers $^3M_2^*$, the contribution of the exciton resonance to the excimer attraction is naturally negligible.

Process (3.57) is one of the possible mechanisms via which excimers are formed. We have seen earlier that triplet–triplet interaction (process (3.56)) also leads to excimer formation. Further, Chandross, Longworth and Visco (1965) reported that positive ($M^+$) and negative ($M^-$) ion annihilation leads to the formation of excimers, i.e.

$$M^+ + M^- \rightarrow {}^{1,3}M_2^* \tag{3.59}$$

Although originally no specific mention was made about the excited states of the excimer formed via processes (3.56) and (3.59), the excimer formed through process (3.56) can be in either a singlet ($^1M_2^*$), a triplet ($^3M_2^*$), or a quintet state ($^5M_2^*$), while that formed through process (3.59) can be in either a singlet or a triplet state.

Another process, ion recombination, occurring when ionizing radiation is used for excitation may lead to excimer formation. In the past this process has been considered as leading to singlet or triplet excited molecules,[*] i.e.

$$M^+ + e \rightarrow {}^{1,3}M^* \tag{3.60}$$

Nevertheless, ion recombination can be visualized as a three-body process or as a sequential two-body reaction process leading to excimer formation, i.e.

$$M^+ + e + M \rightarrow {}^{1,3}M_2^* \tag{3.61}$$

or

$$M + e \rightarrow M^{-*}$$

$$M^+ + M^- \rightarrow {}^{1,3}M_2^* \tag{3.62}$$

Which of the two processes ((3.61) or (3.62)) occurs will depend principally on the lifetime of $M^{-*}$ and the nature and concentration of M. Carter, Christophorou and Abu-Zeid (1967) presented experimental evidence that process (3.61) (or (3.62)) becomes important when organic liquids are excited by intense electron beams and determines through the abundant $^1M_2^*$ emission the spectral distribution of the light emission (see discussion in Section 3.4.1.6).[†]

It might be noted that process (3.56) leads to delayed excimer fluorescence. Further, the emission from $^3M_2^*$ may be termed in analogy to that of $^3M_1^*$ as excimer phosphorescence.

---

[*]Process (3.60) has been proposed in one way or another to explain the slow or delayed component in organic scintillators (Birks, 1964), but this seems not to be the case (Birks, 1964; King and Voltz, 1966).

[†]According to Badger, Brocklehurst and Russell (1967) and Badger and Brocklehurst (1968) process (3.59) (or (3.61), (3.62)) does not yield $^{1,3}M_2^*$. These workers suggested that excimers are formed via $M^+ + M \rightleftarrows M_2^+$, followed by $M_2^+ + e(M^-) \rightarrow {}^{1,3}M_2^*(+M)$.

Temporary photoassociation of two different molecules has first been demonstrated by Birks and Christophorou (1962b) for the case of two dissimilar aromatic molecules in solution. The resultant transient species has been called by Birks and Christophorou (1962b) a 'mixed excimer'. The work of Birks and Christophorou (1962b) has been confirmed and extended by a number of workers for a variety of pairs of dissimilar molecules and in different environments (Selinger, 1964; Cherkasov and Basilevskaya, 1965; Neznaiko, Obyknovennaya and Cherkasov, 1966; Kawaoka and Kearns, 1966). One of the most striking examples of 'mixed excimer' fluorescences is that exhibited by DNA and by various dinucleotides and polynucleotides in solution at low temperatures (Eisinger and coworkers, 1966). Other mixed excimers have been reported in biologically important molecules (Mataga, Okada and Ezumi, 1966; Mataga, Ezumi, and Okada, 1966; Walker, Bednar and Lumry, 1966, 1967).

Quite generally, we may write for the case of mixed excimers

$$^1M_1^* + Q \rightleftarrows (^1M_1Q)^* \tag{3.63}$$

where Q is any molecule which interacts with $^1M_1^*$ to form an excited complex $[(^1M_1Q)^*]$ which is dissociated in the ground state. Certain authors (Walker, Bednar and Lumry, 1966; Birks, 1967) proposed to call $(^1M_1Q)^*$ an exciplex ( = exci(ted com)plex).

Intramolecular excimers—excimer formation between two units of a long molecule—have also been reported (Hirayama, 1965). The importance of excimer and exciplex interactions in atomic and molecular photophysics, as well as in molecular biophysics, is self-evident. It might be noted that Christophorou and Carter (1966) presented experimental evidence that organic scintillators may be improved by the use of excimer forming solvents (see also Carter and Christophorou, 1967).

### 3.4.1.6 Dependence of the emission spectrum on the intensity of excitation

Especially in organic scintillation work the assumption is implicit that the optical emission spectrum under ultraviolet excitation is identical with the scintillation spectrum excited by ionizing radiation and independent of the ionizing radiation intensity. However, since the use of different types and intensities of excitation changes the relative abundance and quenching probabilities of the possible emitting excited species ($^1M_1^*$, $^3M_1^*$, $^1M_2^*$, $^3M_2^*$ and possibly $^5M_2^*$), the emission spectra may depend on the type and intensity of excitation. Recent studies by Carter, Christophorou and Abu-Zeid (1967) on the emission from liquid benzene and liquid benzene and naphthalene derivatives excited by electrons, x-rays, and ultraviolet light have shown that the usual emission spectrum, characteristic of the monomeric $S_{1,0} \rightarrow S_{0,n}$ radiative transitions in organic molecules under ultraviolet and weak x-ray excitation, is absent (or very weak compared to the $^1M_2^*$ emission) under intense electron bombardment. These studies have been extended to other organic liquids (Christophorou, Abu-Zeid and Carter, 1968) and the results confirmed the original findings of Carter, Christophorou and Abu-Zeid (1967). The above authors attributed the abundance of $^1M_2^*$ emission (and $^3M_2^*$ emission; see Christophorou, Abu-Zeid and Carter (1968)) to the formation of $^{1,3}M_2^*$ through process (3.61) or (3.62). It is possible, however, that the species $^{1,3}M_2^*$ are formed through dimeric positive ion–electron neutralization, i.e. $M_2^+ + e \rightarrow {}^{1,3}M_2^*$. The absence of monomer $^1M_1^*$ emission has been attributed (Carter, Christophorou and Abu-Zeid, 1967) to strong monomer quenching. This would imply a drastic difference in the ionization quenching properties of $^1M_1^*$ and $^1M_2^*$.

In Table 3.4 we summarize the various types of luminescences discussed in this section.

**Table 3.4** Types of luminescences

---

### Emissions involving single species

---

| | |
|---|---|
| Resonance fluorescence | $S_{0,0} \rightarrow S_{n,n}''$ ($n = 1$ for molecules), followed by $S_{n,n}'' \rightarrow S_{0,n}'$ |
| Fluorescence | $S_{1,0} \rightarrow S_{0,n}'$ |
| Phosphorescence | $T_{1,0} \rightarrow S_{0,n}'$ |
| $E$-type delayed fluorescence | $T_{1,0} \rightarrow S_{1,0}$, followed by $S_{1,0} \rightarrow S_{0,n}'$ |

---

### Emissions involving bimolecular interactions

---

| | |
|---|---|
| $P$-type delayed fluorescence | $^3M_1^* + {}^3M_1^* \rightarrow {}^1M_1^* + {}^1M_1 \rightarrow$ delayed monomer fluorescence $\rightarrow {}^1M_2^* \rightarrow {}^1M_1^* + {}^1M_1 \rightarrow$ delayed monomer fluorescence $\rightarrow {}^1M_1 + {}^1M_1 + h\nu_{DEF}$ (delayed eximer fluorescence) |
| Excimer fluorescence | $^1M_2^* \rightarrow {}^1M_1 + {}^1M_1 + h\nu_{EF}$ |
| Excimer phosphorescence | $^3M_2^* \rightarrow {}^1M_1 + {}^1M_1 + h\nu_{EP}$ |

---

### 3.4.2 *Internal and External Non-radiative De-excitation*

Competing with radiative de-excitation are a number of quenching processes which either proceed internally or are induced by external agents. Such processes are often faster than the process of radiative decay and are responsible for the low yields or absence of light emission.

#### 3.4.2.1 Non-radiative transitions in atoms

Owing to the lack of vibrational and rotational structure in atoms and the wide separation of atomic states, internal non-radiative de-excitation is absent in atoms when the excitation energy is below the first ionization potential $I$. It can, however, be induced externally by collision, either in the form of electronic energy transfer to other species or through the process of excimer or contact complexes discussed in Section 3.4.1. A quenching collision may also induce conversion of atomic electronic energy to translational energy.

The state of affairs is different for atomic states with energies above $I$. Atomic superexcited states overlap with the ionization continuum and may undergo autoionization. Such non-radiative processes from discrete states above $I$ into the ionization continuum are, when allowed, very fast and compete very effectively with radiative de-excitation. Descriptions of auto-ionization effects—mainly large natural breaths, line asymmetries, and peculiarities in spectral intensities—have been accumulating since the early 1930's.*

The process of autoionization is schematically illustrated in Figure 3.20 (Marr, 1967). The case of the thallium atom may help explain the transitions shown in this figure. The ground-state configuration of electrons in the outermost unfilled shell for the thallium atom is $6s^2 6p$ and the normal spectrum consists of the $s$ and $d$ series corresponding to the transitions $6s^2 6p \rightarrow 6s^2 ns$ ($n \geq 7$) and $6s^2 6p \rightarrow 6s^2 nd$ ($n \geq 6$), respectively, i.e. to excitation of valence electrons; $n$ is the principal quantum number of the excited electron. Excitation of the inner $6s$ electrons can also take place, namely $6s^2 6p \rightarrow 6s 6p np$ ($n \geq 6$). In this latter case the photon energies involved are obviously greater and the absorption lines which lie in the vacuum ultraviolet

---

*States capable of autoionization were first evidenced in x-ray absorption (Auger, 1925a, b; 1926a, b). Inner-shell atomic ionization following x-ray absorption leads to an ionized atom plus the ejected electron. Depending on the level $x( = K, L_1, L_{11}, L_{111}, \ldots$ in the usual x-ray notation) from which the electron has been ejected, the system (ionized atom plus ejected electron) is left with a corresponding positive energy $E_x$ relative to the initial state. The vacancy formed following electron ejection may be filled by an electron from a higher state of energy $E_y$, the energy $E_x - E_y$ appearing as radiation. Alternatively, following the production of an inner-shell vacancy the atom may spontaneously undergo a radiationless transition in which it reverts to a lower ionization energy $E_z(z = L_1, L_{11}, L_{111}, M_1, \ldots)$ with the concomitant ejection of an (Auger) electron having kinetic energy equal to $E_x$-$E_y$-$E_z$. Other transitions involving the ejection of less energetic Auger electrons from successive radiationless transitions to lower ionized states can occur (Section 3.5.5).

region may partly or totally overlap with the normal ionization continuum (Figure 3.20). Those states overlapping with the continuum may undergo autoionization, the probability for such transition being determined by the degree of similarity between the discrete state and the continuum state at the same energy. As a result of autoionization distinct peculiarities are observed in the ionization cross section at the positions of superexcited states (see results in Section 3.5).

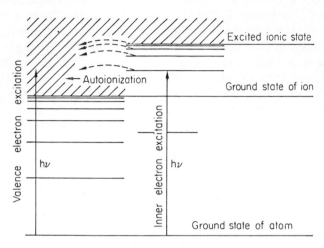

**Figure 3.20** Autoionization; inner electron transitions
(Marr, 1967)

The development of new vacuum ultraviolet techniques (Section 3.3) has allowed such optical studies to be performed. The vacuum ultraviolet work on autoionization coupled with electron impact-induced autoionization (Chapters 4 and 5) proved a very powerful tool for the understanding of highly excited atomic and molecular states. Electron impact data supplement optical data, especially on optically forbidden transitions. Optical studies, however, owing to their increased detection sensitivity enable the observation of resonances which are often too weak to detect in electron impact spectroscopy. Autoionization induced by heavy ion impacts on gases has been investigated recently also (Barker and Berry, 1966; Kessel, McCaughey and Everhart, 1966; Rudd, 1964, 1965; Rudd and Lang, 1965; Rudd, Jorgensen and Volz, 1966a, 1966b; Rudd and Smith, 1968).

In polyelectronic atoms autoionization may take place also as a result of double excitation. In this process two electrons are excited simultaneously to higher orbits than their normal one. When the excitation energy of the two electrons exceeds $I$, the atom may undergo a radiationless transition in which one of the excited electrons falls into a lower state, the de-excitation energy being used to eject the other excited electron from the atom. This process occurs very quickly following double excitation. The lifetime of the excited atom against autoionization is generally very much shorter than its radiative lifetime (compare lifetime values in Tables 3.2 and 3.6).

### 3.4.2.2 Non-radiative transitions in molecules

Here, contrary to the case of atoms, most of the non-radiative transitions investigated lie below $I$. Non-radiative transitions from a molecular superexcited state into the ionization continuum (preionization), although very important in view of the large amount of oscillator strength that goes into these states, have not been as extensively studied. For molecules the following non-radiative de-excitation processes may be distinguished: (i) preionization, (ii) predissociation, (iii) internal conversion, and (iv) externally-induced de-excitation. Processes (i) and (ii) are radiationless transitions from a discrete level to one belonging to a continuum of states. Such radiationless decompositions (as autoionization) are usually referred to as Auger processes. Process (iii) is a radiationless transition between two discrete levels.

Discrete levels overlapping with a continuum of states—as in the case of autoionization, preionization and predissociation—have a finite probability of undergoing a radiationless transition if there is a mixing of the eigen-functions of the states between which the non-radiative transition occurs. The selection rules for predissociation and preionization have been derived by Kronig (1928). Kronig's selection rules are: (i) $\Delta J = 0$; (ii) $\Delta S = 0$; (iii) $\Delta \Lambda = 0$ or $\pm 1$; (iv) $+ \leftrightarrow -$; and (v) $s \leftrightarrow a$. That is, both states must have the same total angular momentum $J$ and the same multiplicity; their $\Lambda$ values must differ by 0 or $\pm 1$, both states must be positive or negative (the $+$ and $-$ signs refer to the behaviour of the total eigenfunction with respect to inversion), and for identical nuclei the states must both have the same symmetry in nuclei ($s$ and $a$ refer to the behaviour with respect to exchange of identical nuclei). Selection rules (i), (iv), and (v) hold rigorously. Rule (iii) holds only insofar as $\Lambda$ is defined, and the spin selection rule (ii) holds approximately as it does in autoionization and radiative transitions. For further discussion on the selection rules for predissociation and preionization the reader is referred to Herzberg (1950, 1966) where ample examples pertaining to special cases are discussed in detail.

A. *Preionization.* Excitation of an atomic superexcited state followed by a radiationless transition into a state of identical energy belonging to the ionization continuum will lead to ionization. For molecules superexcitation does not necessarily lead to ionization since other processes, grouped together as 'dissociation' are possible. Predissociation, for example, due to an unstable or a stable state above its dissociation limit, may compete with indirect ionization. We note, that molecular ionization may yield an excited positive ion which subsequently may fluoresce, decaying to its own ionic ground state.

Observations of preionizing effects in molecular absorption spectra trace back to the early 1930's. The diffuse bands in $CO_2$ around 750 to 785Å, for example, were associated with preionization (Henning, 1932). Similar effects have been reported (Beutler and Jünger, 1936) for $H_2$, and recently by Codling and Madden (1965b) in $O_2$ and $N_2$. Other molecules (e.g. $N_2O$, $CS_2$) were reported to have shown preionizing effects (see Marr, 1967). We refer also to our discussion on superexcited states in Section 2.6. In Section 5 of this chapter results will be presented on the (total) photoabsorption cross section, $\sigma_{ab}(v)$, and the ionization cross section, $\sigma_i(v)$, for a variety of molecules. It will be apparent from these data that for most molecules, in contrast with atoms, $\sigma_i(v) < \sigma_{ab}(v)$ especially for photon energies close to $I$. This clearly demonstrates the competition between 'dissociation' and preionization. The difference between $\sigma_i(v)$ and $\sigma_{ab}(v)$ decreases with increasing photon energy above $I$ since superexcited states lie mostly within a Rydberg of $I$ (see Section 2.6) and for higher photon energies direct ionization dominates.

B. *Predissociation*. A diatomic molecule, given sufficient energy for dissociation, will fly apart within the time of one vibration ($\sim 10^{-14}$ to $10^{-13}$ sec) unless some other process intervenes. Polyatomic molecules do not necessarily dissociate even though they may possess sufficient energy to do so. The excitation energy in a polyatomic molecule will initially reside in accordance with the Franck–Condon principle in several degrees of freedom. If dissociation is to occur, concentration of the vibrational energy in the appropriate mode is required. This would imply that dissociation in a polyatomic molecule does not occur instantly but more leisurely following redistribution of energy.

The radiationless transition from a discrete state to a continuum of states with subsequent dissociation shortens the lifetime of the excited discrete state and thus broadens the excitation levels. Such a process naturally results in a diffuse spectrum. If the lifetime becomes as short as that of half-vibration, the energy width is of the same magnitude as the difference between two successive vibrational levels. Ordinarily a diffuse region separates the discrete and continuous parts of the spectrum. The width of the diffuse region is much smaller for diatomic as compared with polyatomic molecules. The gradual transition from discrete to continuous bands (i.e. the absence of sudden onsets for predissociation) in polyatomic molecules is a result of their many vibrational degrees of freedom.

The process of predissociation can be conveniently discussed through

Figure 3.21, in which the upper attractive state a is crossed by a repulsive (Figure 3.21a) or an attractive (Figure 3.21b) curve b. Absorption from G to a would give a discrete vibrational spectrum in accord with the Franck–Condon principle. Absorption ending at or above the crossing point c (Figure 3.21a) may result in a cross-over to state b with subsequent dissociation. The same is true for Figure 3.21b if the level reached on a has

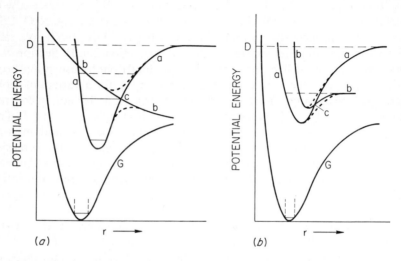

**Figure 3.21** Potential energy curves showing predissociation

at least the same energy as the asymptote of b. For such a radiationless transition to occur, the two upper curves a and b must disturb each other in the region of closest approach. The cross-over process is adiabatic and not well understood. For the transition from a to b to occur, the perturbation must be moderate. If the interaction is weak, a and b are independent. If it is strong, a and b repel each other appreciably. This is shown by the broken curves in Figure 3.21. In this case of strong interaction the spectrum for a situation as shown in Figure 3.21a is initially discrete; subsequently (in going to higher energies) it becomes diffuse and then discrete again. The first discrete portion is due to absorption in state a; the diffuse part of the spectrum is due to dissociation from curve a while the second discrete portion of the spectrum is due to absorption in curve b.

In the above discussion we explicitly used the Born–Oppenheimer approximation. The basic assumption of this approximation (nuclei move with infinitesimal velocity) is not valid for real molecules. Retaining, however, the concept of the potential energy curve as discussed above, we can think of predissociation as a result of the system undergoing adiabatic skips from one curve to another, a process probable at close approach.

Examples of predissociation are numerous in spectra of diatomic and polyatomic molecules. They have also been observed in the absorption spectra of simple aromatics. As discussed in Section 3.4.1 predissociation may be in competition with fluorescence, the absence of the latter often being indicative of the onset of predissociation.

The processes discussed so far can be summed up with the aid of Figure 3.22 where the simplified potential energy diagrams for the known states of $O_2$ and $O_2^+$ are shown (Gilmore, 1965). Absorption bands due to the process $O_2 + h\nu \rightarrow O_2^*$ can take place through the transitions $X^3\Sigma_g^- \rightarrow A^3\Sigma_u^+$ or $X^3\Sigma_g^- \rightarrow B^3\Sigma_u^-$. The former are the forbidden Herzberg transitions and the latter is the Schumann–Runge band system in the region of 2000Å. The transition to the $B^3\Sigma_u^-$ state converges to a dissociation limit at $\sim 1670$Å to the short wavelength side of which the strong dissociative continuum $O_2 + h\nu \rightarrow O(^3P) + O(^1D)$ takes place (Marr, 1967). According to Wilkinson

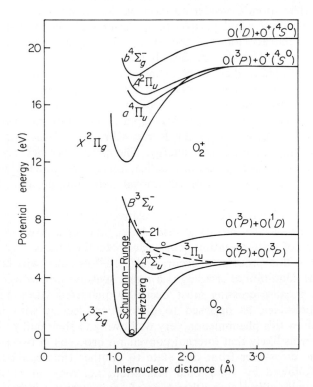

**Figure 3.22** Potential energy level diagrams for oxygen based on Gilmore (1965)

and Mulliken (1957) and Carroll (1959), the unstable state $^3\Pi_u$ (Figure 3.22) crosses the $B^3\Sigma_u^-$ state in the region of $v' = 3$ or 4. Thus, following excitation of the $B^3\Sigma_u^-$ state, the system may undergo predissociation by crossing over from the stable part of $B^3\Sigma_u^-$ to the unstable $^3\Pi_u$ from which it dissociates into $O(^3P)+O(^3P)$. Excitation to states above $I$ yield absorption bands which exhibit preionization characteristics. For example, the bands between 1026 and 860 Å are broad and Cook and Metzger (1964) suggested that they are preionized by the interaction of the $O_2^+(X^2\Pi_g)+e$ continuum. We also note that the process of direct photoionization may be represented by

$$O_2(X^3\Sigma_g^-)+hv \rightarrow O_2^+(X^2\Pi_g)+e \tag{3.64}$$

When both $O_2$ and $O_2^+$ are in their lowest states (0, 0 transition), the threshold for (3.64) corresponds to the ionization potential of $O_2$.

C. *Internal conversion.* Radiationless combination of discrete electronic states within a molecule is called internal conversion. Before elaborating on this process a word or two on what is called vibrational redistribution and vibrational relaxation. Consider an electronically and vibrationally excited molecule. The most rapid process to occur in such an excited species is probably the redistribution of the excess vibrational energy among the different vibrational modes. Although this process, as mentioned in the previous paragraph, is likely to be of major importance in dissociation and predissociation, unless it takes place in a time shorter than one collision (for a liquid this time is $\lesssim 10^{-13}$ sec), it is unlikely to be important. Vibrational relaxation is of course less rapid than vibrational redistribution and refers to the dropping to the lowest vibrational level that does not involve a change in electronic energy. The process comes about either through a collision (as in gases) or a collision-like environmental interaction (as in liquids). The processes of vibrational redistribution and vibrational relaxation are fast.

The process of internal conversion is relatively a slower ($\sim 10^{-12}$ sec) one. It relaxes the molecule into its lowest excited state of a given multiplicity. As in the case of predissociation, crossing of potential surfaces is a necessary condition for this process to occur, and selection rules will have to be fulfilled. The condition of crossing is a rather sensitive criterion for internal conversion, because crossing must occur at quantized vibrational states. Complex molecules, as opposed to diatomic molecules with their only vibration, show this phenomenon very well owing to their many vibrational modes. It seems likely that internal conversion proceeds in two steps. First an adiabatic cross-over from one state to another (this can be in either direction) followed by vibrational relaxation, i.e. removal of the excess energy through collisions or other type of environmental interaction. If a molecule were completely isolated, its energy following adiabatic cross-over would have to be lost radiatively, a very much slower process. In this case the molecule would have ample time to cross back to its original state.

The rate of internal conversion from an electronic state $S_{n,0}$ with electronic and vibrational wave functions $\psi_n$ and $\varphi_0$, respectively, to the state $S_{n-1,n'}$ with electronic and vibrational wave functions $\psi_{n-1}$ and $\varphi_{n'}$, respectively ($n'$ is a high vibrational level of $S_{n-1}$ of the same energy as $S_{n,0}$) depends on the magnitude of the perturbation matrix element $\langle \psi_n, \varphi_0 | \mathcal{H}' | \psi_{n-1}, \varphi_{n'} \rangle$; $\mathcal{H}'$ is the change in the Hamiltonian due to perturbation. The magnitude of the perturbation matrix element is proportional to the vibrational overlap integral $\langle \varphi_0 | \varphi_{n'} \rangle$ which decreases rapidly with increasing $n'$, that is with increasing energy separation of $S_n$ and $S_{n-1}$. For $n > 1$ this energy separation is small. For organic molecules, for example, for $n = 2$, $n'$ is $\sim 4$ and the rate of internal conversion is normally high ($\sim 10^{10}$–$10^{12}$ sec$^{-1}$), for $n = 1$, $n'$ is typically $\sim 16$ and the rates decrease to $\sim 10^7$ sec$^{-1}$ or less (Birks and Munro, 1967). This explains why fluorescence is observed from $S_1$ to the ground state in competition with internal conversion, but radiative transitions from $S_2$ to $S_0$ or $S_2$ to $S_1$ are not normally observed except in rare cases such as in azulene and its derivatives (Beer and Longuet-Higgins, 1955). That internal conversion from higher excited states takes place in times $\lesssim 10^{-12}$ sec can be seen from the following: From the absorption spectra of aromatic hydrocarbons one may obtain (Section 3.2.3) a radiative lifetime for the intense second absorption bands of these systems, which is of the order of $10^{-10}$–$10^{-9}$ sec. The failure to detect experimentally any emission from upper states in these systems suggests that the process of internal conversion is at least $10^2$–$10^3$ times faster than radiative emission; it thus should occur in times $\lesssim 10^{-12}$ sec. If the only process competing with radiation from higher excited states were internal conversion to the emitting state ($S_{1,0}$), the emission quantum yield would be independent of the exciting wavelength. For most organic molecules that have been studied in dilute solution this seems to be the case for excitation through the first absorption bands (Weber and Teale, 1958). However, exceptions have been reported (Ferguson, 1959; Braun, Kato and Lipsky, 1963) and have been generally attributed either to predissociation or to intersystem crossing (see later this section) to the triplet manifold. It might be noted that as a result of the high rate of internal conversion, the lowest absorption band of a given multiplicity will be better resolved vibrationally, than higher energy bands of the same multiplicity which correspond to internally converting levels. It should be noted also that the non-radiative processes discussed above are fast and not observable separately. As a result, the amount of knowledge about each process is sketchy.

In most molecules the $S_1$–$S_0$ energy separation is sufficiently large that internal conversion is inefficient in the absence of collisions or equivalent effects in fluid media. The internal conversion quenching rate of $S_1$, then, may become zero at low temperatures. For certain molecules such as anthracene and pyrene in aliphatic solvent solutions there is experimental evidence (Laposa, Lim and Kellogg, 1965; Bennett and McCartin, 1966) that internal conversion quenching rate is zero even at 300 °K.

To distinguish internal conversion between states of like multiplicity and between states of different multiplicity Kasha (1950) (see also Kasha and McGlynn, 1956) introduced the term intersystem crossing to describe the spin-dependent internal conversion, retaining the term internal conversion for radiationless transitions between states of like multiplicity (singlets or triplets). Like internal conversion, intersystem crossing is probably a two-step process. In photoabsorption studies excitation of triplet states is spin forbidden and since internal conversion to $S_{1,0}$ in the singlet manifold is fast, intersystem crossing in most cases is the radiationless transition from $S_{1,0}$ to a vibrationally excited triplet state $T_m$. If $m > 1$ then $T_m$ internally converts to $T_{1,0}$ which can undergo radiative decay to $S_0$ (phosphorescence), intersystem crossing to $S_0$, or it can lead to $E$-type delayed fluorescence or undergo triplet–triplet annihilation (see Section 3.4.1).

As all processes involving a change in total electron spin momentum, intersystem crossing would have a low probability compared with the corresponding processes in which spin momentum is conserved. The probability of a radiative spin-forbidden transition is $\sim 10^{-6}$ that of a radiative spin-allowed transition. A similar factor for the ratio of the $S$–$T$ intersystem crossing and $S$–$S$ internal conversion between states of similar energy separation has been proposed (Kasha, 1950).

Intersystem crossing becomes allowed through spin–orbit coupling, which gives the triplet level a small amount of singlet character by mixing it with one or more of the singlet levels of the molecule. It is also possible for the coupling to give the ground singlet level a small amount of triplet character, but it appears that the former mechanism is normally more important. Moreover the introduction of a high $Z$ (atomic number) atom such as iodine or bromine into the molecule will strongly increase the probability of the transition by enhancing spin–orbit coupling. The same result can be achieved by strong external electric and magnetic fields or by paramagnetic molecules (e.g. $O_2$, NO) or molecules containing heavy atoms dissolved in solution.

It is of interest to note that although the high yields of phosphorescence in some molecules require the rate constant, $k_{IC}$, for the $S_1 \rightarrow T_1$ intersystem crossing to be comparable to that of fluorescence ($\sim 10^8$ sec$^{-1}$), triplet state lifetime measurements in solution (e.g. Jackson, Livingston and Pugh, 1960; Linschitz and Pekkarinen, 1960; Porter and Windsor, 1954) show that the rate for the $T_1 \rightarrow S_0$ intersystem crossing is by a factor of $10^4$ or more slower (see an explanation in Section 3.4.2.2.D).

Kasha (1950) studied several aromatic compounds in rigid glass solution at 77 °K and observed that the total luminescence quantum yield $\phi_P + \phi_F$

was $\sim 1$; $\phi_P$, the quantum yield of phosphorescence, is defined as the ratio of the number of phosphorescence photons emitted to the number of molecules excited into $S_1$. This finding then, shows that under these conditions the radiationless transitions from $S_1$ and $T_1$ to $S_0$ are unimportant. Therefore, one can determine the rate constant $k_{IC}$ from

$$\frac{\phi_P}{\phi_F} = \frac{k_{IC}}{k_{FM}} \tag{3.65}$$

where $\phi_M$ is the quantum yield of fluorescence and $k_{FM}$ the radiative fluorescence rate constant. For a series of aromatic hydrocarbons $k_{IC}$ decreases rapidly with increase in the $S_1$–$T_1$ energy gap indicating that $k_{IC}$, like the internal conversion rate, depends on the vibrational overlap integral $\langle \varphi_0 | \varphi_{n'} \rangle$ (see further discussion in Kasha, 1960). Birks and coworkers studied the intersystem crossing rate constant $k_{IC}$ for a number of aromatic molecules and distinguished a temperature-independent component–natural intersystem crossing rate in the absence of external perturbations—and a temperature-dependent component (Birks, Braga and Lumb, 1965; Birks, Lumb and Munro, 1964; Birks and King, 1966).

In the absence of impurity or concentration quenching (see next section) $q_M$ and the fluorescence lifetime $\tau_M$ (not to be confused with the natural lifetime $\tau_0$) are given by

$$q_M = \frac{k_{FM}}{k_{FM} + k_{IM}} \tag{3.66}$$

and

$$\tau_M = \frac{1}{k_{FM} + k_{IM}} \tag{3.67}$$

where $k_{IM}$ is the rate constant for internal quenching of $S_{1,0}$.

D. *Externally induced de-excitation.* As a rule the action of the environment on the excited states of molecules is that of a passive receiver of energy. The interaction between two molecules, however, may be sufficient that the entire electronic energy of the excited molecule, $^{1,3}M_1^*$ (usually in the $S_{1,0}$ or $T_{1,0}$) may be quenched or transferred in the process.

Let us now consider the case of singlet excited molecules, i.e. $^1M_1^*$. Externally induced de-excitation may occur as a result of interactions between excited and unexcited molecules either on collision or at a distance. Those occurring in a collision and resulting in a quenching of the excitation energy are usually known as *impurity* or *concentration* quenching, when the unexcited molecule is of different Q or of the same $^1M_1$ kind, respectively. The discussion that follows is confined to the case of organic molecules, although it may be generalized in its most extent.

Let us first consider impurity quenching. Three such types can be distinguished (Birks and Munro, 1967): *static quenching* due to the formation of a metastable (e.g. charge-transfer) complex between $^1M_1$ and Q, *dynamic or collisional quenching* due to collision between $^1M_1^*$ and Q, and *energy transfer quenching* due to resonance interaction between $^1M_1^*$ and Q. The first type of quenching is thought to proceed via a metastable complex ($^1M_1Q$), while the second type (dynamic) via the transient complex ($^1M_1Q$)*. The excitation energy of ($^1M_1Q$) and that of ($^1M_1Q$)* may be dissipated by radiationless and/or radiative transitions in the complex itself or in Q*. Similarly, in the third type of quenching (energy transfer)—which is discriminated against the collisional quenching in that the energy from $^1M_1^*$ can be transferred over to Q resulting in Q* while $^1M_1^*$ and Q are at a distance—the energy transferred from $^1M_1^*$ to Q is dissipated radiatively and/or non-radiatively in Q. More on the energy-transfer quenching is in Section 3.4.3.

The collisional quenching process is obviously a short-range interaction and it is diffusion-controlled. Following Birks and Munro (1967) we modify expressions (3.66) and (3.67) as

$$q_{QM} = \frac{k_{FM}}{k_{FM} + k_{IM} + k_{QM}[Q]}$$

$$= \frac{q_M}{1 + \tau_M k_{QM}[Q]} \tag{3.68}$$

$$\tau_{QM} = \frac{1}{k_{FM} + k_{IM} + k_{QM}[Q]}$$

$$= \frac{\tau_M}{1 + \tau_M k_{QM}[Q]} \tag{3.69}$$

where [Q] is the concentration of the impurity Q, $k_{QM}$ is the impurity quenching rate parameter (proportional to [Q]), and $q_{QM}$ and $\tau_{QM}$ are the values of $q_M$ (Equation (3.66)) and $\tau_M$ (Equation (3.67)) which are now reduced due to the presence of [Q]. If $D_M$ and $D_Q$ are the diffusion coefficients and $r_M$ and $r_Q$ are the encounter radii of $^1M_1$ and Q, respectively, then we may write

$$k_{QM} = \frac{4\pi N p(D_M + D_Q)(r_M + r_Q)}{1000} \tag{3.70}$$

where $p$ is the probability that quenching occurs in a collision. Assuming that the encounter radii are equal to the respective molecular radii and using for $D$ the Stokes–Einstein expression

$$D = \frac{kT}{6\pi\eta r} \tag{3.71}$$

where $\eta$ is the solvent viscosity, Equation (3.70) becomes

$$k_{QM} = \frac{2RTp(r_M + r_Q)^2}{3000\eta r_M r_Q} \tag{3.72}$$

If it is assumed that $r_M = r_Q$, Equation (3.72) reduces to

$$k_{QM} = \frac{8RTp}{3000\eta} \tag{3.73}$$

See Stevens and Dubois (1963), Kropp and Burton (1962), and Ware(1962) on the applicability of Equation (3.73). The quantity $p$ in Equation (3.73) depends on the nature of M and Q. Birks, Dyson and Munro (1963) and Birks, Lumb and Munro (1964) have shown Equation (3.73) with $p = 1$ to be valid for encounters between excited and unexcited pyrene molecules leading to excimer formation (Equation (3.57)) in several solvents over a range of temperature and viscosity.*

Let us now return to the process of concentration quenching. As we have indicated earlier in this chapter, this is the quenching of $^1M_1^*$ by $^1M_1$. Equation (3.53) (Section 3.4.1.2) describes how, in general, the emission intensity from $^1M_1^*$ in solution is decreased at high concentrations due to this process. In many cases of organic molecules dissolved in solution the concentration quenching of the fluorescence of $^1M_1^*$ is accompanied by the structureless excimer emission at larger wavelengths produced by process (3.57). We have discussed excimer formation in Section 3.4.1.5. Here we present a somewhat detailed account of the various processes and their respective rate parameters in a solution of $^1M_1$ at a molar concentration $c$. Table 3.5 summarizes the reactions taking place in a concentrated solution

---

*In view of the large quenching effect of $O_2$ care must be taken to remove dissolved $O_2$ from exposed solutions or liquids at room temperature. Impurity quenching by $O_2$ has been and still is a major problem in photophysical studies of solutions and liquids. Various methods of removing dissolved oxygen such as bubbling of $O_2$-free-$N_2$-gas through the liquid or repeated freezing of the solution with subsequent melting while pumping (freeze-pump-thaw technique) have been used. However, Wagner, Christophorou and Carter (1969) found the above procedures to be inadequate. These authors adopted a procedure whereby high purity helium gas is passed through a porous filter into the liquid under study and the helium flow arranged such that only small-size bubbles are formed. The liquid was contained in a vacuum-tight box under $N_2$ atmosphere. Bubbling periods of up to 30 minutes were necessary to remove the dissolved oxygen for liquid benzene and naphthalene derivatives. Wagner, Christophorou and Carter (1969) reported that the lifetimes of liquid benzene, toluene and mesitylene at 25 °C are, respectively, greater than or equal to 30.3, 41.2 and 44.2 nsec.

of $^1M_1$. Considering the kinetics of the processes shown in Table 3.5 we obtain for the fluorescence quantum yield of $^1M_1^*$ (Birks, Dyson and Munro, 1963)

$$\phi_M = \frac{k_{FM}}{k_M + k_E K_e c} \qquad (3.74)$$

$$= \frac{q_M}{1 + c/c_h} \qquad (3.75)$$

where $K_e = (k_{EM})/(k_E + k_{ME})$ and

$$1/c_h = k_E/k_M K_e \qquad (3.76)$$

is the Stern–Volmer concentration quenching parameter.

**Table 3.5** Reaction scheme for excimer formation[a]

| Process | Description | Rate parameter (sec$^{-1}$) |
|---|---|---|
| $^1M_1 + h\nu \rightarrow {}^1M_1^*$ | Excitation of $^1M_1$ | — |
| $^1M_1^* \rightarrow_1 {}^1M + h\nu_M$ | Fluorescence of $^1M_1^*$ | $k_{FM}$ $\Big\}$ $k_M$ |
| $^1M_1^* \rightarrow {}^1M_1$ | Internal quenching of $^1M_1^*$ | $k_{IM}$ |
| $^1M_1^* + {}^1M_1 \rightarrow {}^1M_2^*$ | Excimer formation | $k_{EM} c$ |
| $^1M_2^* \rightarrow {}^1M_1 + {}^1M_1 + h\nu_E$ | Excimer fluorescence | $k_{FE}$ $\Big\}$ $k_E$ |
| $^1M_2^* \rightarrow {}^1M_1 + {}^1M_1$ | Internal quenching of $^1M_2^*$ | $k_{IE}$ |
| $^1M_2^* \rightarrow {}^1M_1^* + {}^1M_1$ | Dissociation of $^1M_2^*$ | $k_{ME}$ |

[a] The reaction scheme can be extended to include impurity quenching of $^1M_2^*$ and $^1M_1^*$.

The $^1M_2^*$ fluorescence quantum yield $\phi_E$ is defined as the ratio of the total number of $^1M_2^*$ fluorescence photons emitted to the total number of $^1M_1^*$ molecules excited.

Hence,

$$\phi_E = \frac{k_{FE} K_e c}{k_M + k_E K_e c} \qquad (3.77)$$

$$= \frac{q_E}{1 + c_h/c} \qquad (3.78)$$

where

$$q_E = \frac{k_{FE}}{k_E} \tag{3.79}$$

is the fluorescence quantum efficiency of $^1M_2^*$, defined as the limiting (maximum) value of $\phi_E$ at infinite concentration (Förster and Kasper, 1955). From Equations (3.74) and (3.77) we have

$$\frac{\phi_E}{\phi_M} = \frac{k_{FE}}{k_{FM}} K_e c \tag{3.80}$$

The fluorescence lifetimes $\tau_M$ of $^1M_1^*$ and $\tau_E$ of $^1M_2^*$ can be similarly defined at infinite dilution and at infinite concentration as

$$\tau_M = \frac{1}{k_M} \tag{3.81}$$

and

$$\tau_E = \frac{1}{k_E} \tag{3.82}$$

The fluorescence time characteristics of a concentrated solution depending on $c$ and all the rate constants in Table 3.5 are quite complex. Analyses of decay response functions of $^1M_1^*$ and $^1M_2^*$ following excitation by a $\delta$ function light pulse have been given by Birks, Dyson and Munro (1963).

We finally turn our attention to the concentration quenching of triplet states. Here it is essential to distinguish self-quenching of the triplet state by another triplet state (see Section 3.4.1 on triplet–triplet annihilation) and self-quenching by ground-state molecules. The latter type of quenching is often, but not always small (see, for example, Porter and Windsor, 1954; Linschitz and Sarkanen, 1958). It appears to be still not clear as to why the process $T_1 \to S_0$ is more than $10^4$ to $10^6$ times slower than the process $S_1 \to T_1$. The fact that $T_1$ (and possibly $T_2$) and $S_1$ are usually closer in energy than $T_1$ and $S_0$ are, has provided a qualitative but not necessarily satisfactory explanation. The closeness of $T_1$ (or $T_2$) and $S_1$ allows easier crossing and less energy need be changed from electronic to vibrational (see discussion on internal conversion between singlets). Other suggestions have also been made (e.g. Pariser, 1956, see also other references in Seybold and Gouterman, 1965).

Oxygen and other paramagnetic substances as well as transition metal ions are effective quenchers of $T_1$. The observed apparent viscosity dependence of the $T_1 \to S_0$ radiationless transition, contrary to the lack of such dependence for the rate $S_1 \to T_1$, seems to have been explained in terms of quenching of $T_1$ by traces of impurities such as $O_2$ (see Seybold and Gouterman, 1965), although recent evidence (Langelaar (1969)) suggests that it is mainly due to diffusion-controlled triplet concentration quenching.

Electronic energy from a molecule in the triplet state can be transferred upon collision (as in the case of a molecule in the singlet state) to another molecule. Triplet energy transfer can occur also non-collisionally due to electron-exchange interaction over a distance ($\lesssim 13\text{Å}$) in rigid solutions. The process may be represented as

$$D^*(\text{triplet}) + A(\text{singlet}) \rightarrow D(\text{singlet}) + A^*(\text{triplet}) \qquad (3.83)$$

It refers to electronic energy transfer from one triplet to form another and has received considerable attention in recent years (e.g. Bäckström and Sandros, 1958; Hammond, Turro and Leermakers, 1962; Porter and Wilkinson, 1961; Sandros and Bäckström, 1962; Terenin and Ermolaev, 1956).

### 3.4.3 Transfer of Electronic Excitation Energy

Energy exchange between atoms and molecules certainly occurs most efficiently in close collisions where electron transfer or orbital overlap takes place between degenerate states. Hence, as discussed in the previous section, in many cases of electronic energy transfer between molecules, electronic energy exchange occurs between pairs of molecules which form well-defined complexes. The transfer of excitation energy in molecular encounters is one way of energy transfer from one system to another, the rate of energy transfer being determined by material diffusion. Four other distinct mechanisms of electronic energy transfer can be conceived: (i) Radiative transfer,* (ii) Long-range non-radiative resonance transfer, (iii) Short-range electron exchange interaction, and (iv) Exciton migration.*

Mechanism (i), is an important energy transfer process in low-pressure gases and low-concentration solution. This process does not compete with electronic de-excitation but occurs after radiative de-excitation has taken place. Mechanism (ii), contrary to bimolecular processes which play little or no part, is an important mechanism in solid media. After a short discussion of radiative transfer we shall elaborate on mechanism (ii) since under certain conditions this is a very efficient process, competing with radiative de-excitation. No discussion of exciton migration (process (iv)) will be made. For a discussion of this process see Frenkel (1931), Davydov (1962), Kasha (1959), Dexter (1953), Förster (1960a, b).

Experimental studies of electronic energy transfer have been made with gases, liquids, and solutions between like and unlike molecules, the problem of intermolecular energy exchange being closely associated with that of

---

*The word transfer is used to describe exchange of electronic energy between unlike molecules and the word migration between like molecules.

intermolecular forces. The problem is of special interest in biology (e.g. photosynthesis, vision, biological polymers, macromolecules, etc.), in radiation and photochemistry, and radiation and photophysics, as well as of practical importance (e.g. scintillation detectors).

### 3.4.3.1 Radiative transfer

This is the exchange of excitation energy whereby an excited atom or molecule D* (donor) emits and another atom or molecule A (acceptor) absorbs the emitted radiation, being raised to an excited electronic state. The process depends on the overlap of the emission spectrum of D and the absorption spectrum of A. Its probability is determined by the Beer–Lambert law and depends, obviously, on the geometry of the system. In principle, the range for this kind of energy transfer is infinite since the intermediaries are real photons. It is the only process of energy transfer that occurs at intermolecular distances $\gtrsim 100$Å. It is, of course, assumed that the medium separating D and A is not capable of absorbing the emission of D*. The lifetime of the excited state of any particular atom or molecule is unaffected by this process since transfer occurs following light emission. However, the persistence of excitation in a finite system composed of many similar atoms or molecules may be increased by the process of multiple absorption and emission. Such a process we have seen to be of major significance when A and D are identical and the emitting state has a large oscillator strength (e.g. imprisonment of resonance radiation in rare gas atoms). Further, for a unitary system whose emission and absorption spectra overlap, radiative migration affects the spectral distribution of D (self absorption, Section 3.4.1; see also Figure 3.19).

Radiative transfer is probably the simplest process of energy transfer from D* to A; it is well understood and for this reason it has been referred to by some authors (Förster (1959)) as the 'trivial process'.

### 3.4.3.2 Long-range non-radiative resonance transfer of electronic excitation energy

Interatomic and intermolecular transfer of electronic excitation energy can occur non-radiatively between otherwise well-separated atoms or molecules, over distances greatly exceeding kinetic-theory collision diameters. Such a non-radiative energy-transfer process, known variously as 'inductive resonance', 'classical resonance', or 'sensitized fluorescence', is of general occurrence between atoms and molecules in the gas phase or in solution or crystalline environments. For atoms, the transfer is most probable when there is exact or nearly exact energy resonance between D* and A. For molecules, it is greatest when there is a large overlap between the emission of D* and the absorption of A and when the acceptor absorption transition is allowed.

Resonance transfer takes place within the lifetime of the excited state of D* and thus competes with radiative de-excitation. It, therefore, decreases the lifetime of D*, but it has no effect on its fluorescence spectrum $F(\bar{v})$. The process is independent of the geometry of the system, it is responsible for sensitized fluorescence of atoms and molecules, it leads to the so-called depolarization of fluorescence (i.e. the falling off of polarization from a fluorescent solution due to transfer of excitation energy between differently oriented molecules), and sometimes contributes to concentration quenching (see Section 3.4.2.2.D).

The first observation of energy transfer of this type was made in 1923 by Cario and Franck in their classical experiments of sensitized fluorescence of atomic vapours (see a number of examples in Livingston(1957)). The recent interest of the photophysicist, photochemist, and photobiologist in this mechanism of electronic energy transfer is overwhelming, and the literature contains a large body of experimental data which confirm the theoretical predictions.

For the process to occur within the short lifetimes of excited molecules ($10^{-8}$ sec—this is a long time compared to $10^{-15}$ sec of the orbital motion), there should exist some kind of resonance between D* and A. When this condition is fulfilled a comparatively weak coupling between distant atoms or molecules may be sufficient for excitation transfer. Thus for energy transfer to take place an interaction between D* and A is required. This is one between the two configurations D*A (initial) and DA* (final). Neglecting exchange interactions, the interaction energy $u$ is of the type (Förster, 1960b)

$$u = \iint \psi_{D*}(x)\,\psi_A(x')\,H(x,x')\,\psi_D(x)\,\psi_{A*}(x')\,\mathrm{d}x\,\mathrm{d}x' \qquad (3.84)$$

where $x,x'$ represent the electronic coordinates, the $\psi'$s the corresponding molecular electronic wave functions, and $H(x,x')$ the interaction Hamiltonian. The interaction energy decreases with a certain power of $R$, the intermolecular distance. It also depends on the mutual orientation of both molecules. At relatively long separations the dipole–dipole term dominates for allowed electric dipole transitions.

The detailed mechanism of transfer depends on the magnitude of $u$ as compared to certain inherent molecular energy parameters, namely (i) the Franck–Condon bandwidth $\Delta\varepsilon$, (ii) the vibrational quantum energy $\Delta E$, and (iii) the width of a single vibronic level $\Delta\varepsilon'$. Parameter (i) is the largest (typically $\sim 3000\ \mathrm{cm}^{-1}$ for aromatic molecules) and results from the difference between nuclear equilibrium configurations of the ground and excited states. Parameter (ii) is typically of the order of $1000\ \mathrm{cm}^{-1}$ while parameter (iii) is distinctly smaller, $\sim 10\ \mathrm{cm}^{-1}$ for a condensed system at room temperature. (This width corresponds to a vibrational lifetime of $\sim 10^{-12}$ sec.) Förster (1960b)* distinguished the following three different cases depending on the relative magnitudes of $|u|$, $\Delta\varepsilon$ and $\Delta\varepsilon'$: (a) $|u| \gg \Delta\varepsilon \gg \Delta\varepsilon'$ strong, (b) $\Delta\varepsilon \gg |u| \gg \Delta\varepsilon'$ medium, and (c) $\Delta\varepsilon \gg \Delta\varepsilon' \gg |u|$ weak. Condition (a) prevails in crystals and involves strong coupling. Cases (a) and (b) have been discussed among others by Simpson and Peterson (1957). The weak coupling case has been treated in detail by Förster (1946, 1948, 1959, 1960a, 1960b). For the weak coupling case the absorption and fluorescence spectra are not affected. Strong coupling would not be expected for many fluid solvents, but weak coupling could lead to energy-transfer rates in excess of those of diffusion.

The size of $\Delta\varepsilon'$ determines the efficiency of redistribution of vibrational energy. In the case of weak coupling (case (c) above), the electronic transition is slow compared with the vibrational relaxation time. Hence, once the energy has passed over to the acceptor, vibrational relaxation takes place and there is little chance of back transfer.

Let us now consider a simplified energy level diagram for a molecular case as shown in Figure 3.23. Here the superposition of electronic and vibrational levels is essential. In case (c) above relaxation within D* occurs prior to transfer, i.e. the energy available for transfer is that which would otherwise be emitted. Suppose now that the energy for one of the possible de-activating processes in D* corresponds exactly to that of absorption in A. Then, with sufficient energetic coupling between the two molecules, both processes may occur simultaneously resulting in energy transfer from D* to A. With the broad spectra of polyatomic molecules in solution there is always sufficient correspondence between the levels of D* and A, especially when the absorption of A is shifted slightly to the red compared to that of D (Figure 3.23).

Förster (1948) considering only dipole–dipole interaction gave for the rate of resonance transfer between two well-separated stationary molecules the expression

$$k_{D^* \to A} = \frac{9000 \ln 10 K^2}{128\pi^5 n^4 N_A \tau_M^D R^6} \int_0^\infty F_D(\bar{v}) \varepsilon_A(\bar{v}) \frac{d\bar{v}}{\bar{v}^4} \qquad (3.85)$$

In Equation (3.85) $\bar{v}$ is the wave number, $\varepsilon_A(\bar{v})$ the molar decadic extinction coefficient of A, $F_D(\bar{v})$ the normalized fluorescence spectrum of D (Equation 3.54), $N_A$ is Avogadro's number, $\tau_M^D$ the fluorescence lifetime of D, $n$ the refractive index of the medium, $R$ the intermolecular distance, and $K^2$ is an orientation factor ($= 2/3$) for randomly oriented molecules. Denoting by $R_0$ the distance at which transfer from D* to A and radiative decay

---

*The same classification has been used earlier by Franck and Teller (1938).

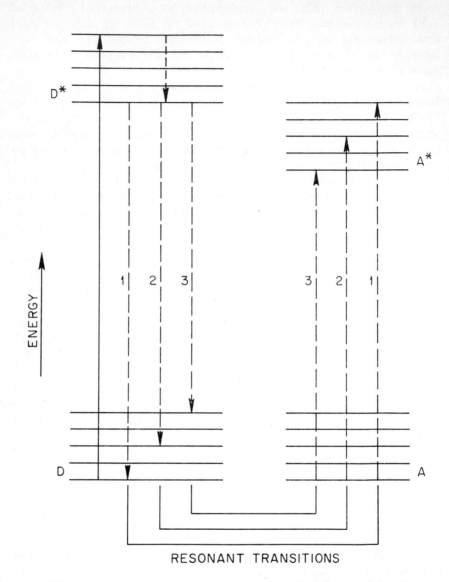

RESONANT TRANSITIONS

**Figure 3.23** Energy level diagram for a donor D and an acceptor A.
———— Radiative transitions, – – – Non-radiative transitions,
— — Resonant transitions

of $D^*$ are equally probable (i.e. $k_{D^* \to A} = 1/\tau_M^D$), we have

$$R_0^6 = \frac{9000 \ln 10 K^2}{128 \pi^5 n^4 N_A} \int_0^\infty F_D(\bar{v}) \varepsilon_A(\bar{v}) \frac{d\bar{v}}{\bar{v}^4} \qquad (3.86)$$

where $\int_0^\infty F_D(\bar{v}) \, d\bar{v} = q_M$ (Equation 3.54).

Equation (3.86) yields values of $R_0$ in the range 50–100Å for typical cases of allowed electric dipole transitions. These are generally found to be in agreement with experimental data. The values of $R_0$ increase with increasing quantum yield of $D^*$ and the overlap of the spectra determined by the integral in Equation (3.86). The formulae (3.85) and (3.86) become invalid when transfer occurs before thermal equilibrium is established. Such a situation may arise in gases at low pressures (slow thermal relaxation) or in liquids or solids when the interaction is strong and thus transfer is very fast (case $a$). It is also noted that (3.85) and (3.86) only apply to stationary molecules. If the diffusion mean free path is greater than $R_0$ energy transfer occurs more efficiently.

When in either D or A the transitions involved are not allowed, higher coulombic terms such as dipole–quadrupole must be considered. Also, when the orbitals of the donor and the acceptor overlap, exchange interaction is important. It is possible to conceive a number of electronic energy transfer processes involving singlet and triplet states of the donor and the acceptor. One should keep in mind, also, that spin-forbidden mechanisms can become important if other processes competing for the excitation energy of $D^*$ are similarly forbidden. Thus the process,

$$D^*(\text{triplet}) + A(\text{singlet}) \to D(\text{singlet}) + A^*(\text{singlet}) \qquad (3.87)$$

which is spin-forbidden by electron-exchange interaction has been reported to occur in rigid media (Ermolaev and Sveshnikova, 1963), its occurrence being a consequence of the fact that the competing process of phosphorescence in $D^*$ is also spin-forbidden.

In summary, the transfer of $\pi$-electronic singlet excitation energy from $D^*$ to A can take place by radiative transfer, non-radiative transfer due to coulombic interaction between $D^*$ and A and non-radiative transfer due to short-range electron exchange interaction. Self-absorption, non-radiative migration between donor molecules, excimer formation and dissociation, and diffusion of $D^*$ and A may influence energy transfer from $D^*$ to A. Effects of viscosity on energy transfer between aromatic molecules in fluid media have recently been discussed by Birks, Georghiou and Munro (1968).

In closing this section it is of interest to note that energy transfer within a molecule (intramolecular) where the absorbing part is different from the emitting one has been observed in a number of systems. This kind of energy transfer between different parts of a complex molecule is of special interest to the photobiologist. Stryer (1960), for example, has discussed energy transfer between aromatic amino acids in proteins.

Although our discussions in this chapter have been confined to the case where the exciting means is a low-energy photon, to their most extend are also valid for other mechanisms of primary excitation such as ionizing radiation. In the latter case, excitation occurs directly or indirectly via primary and secondary electrons. Among the many surveys and general discussions of energy-transfer processes and non-radiative transitions are those by Franck and Livingston (1949), Kasha (1950, 1960), Terenin and Ermolaev (1956), Dexter (1953), Livingston (1957), Sponer (1959), Porter and Wright (1959), Robinson (1961), Förster (1960a, 1960b), Seybold and Gouterman (1965), Wilkinson (1964) and Birks (1970).

## 3.5 Typical Experimental Results

### 3.5.1 *General Remarks on the Photoionization of Atoms and Molecules*

Photoionization of atoms and molecules is of major interest not only in the field of photophysics, but also in radiation physics, astrophysics, photochemistry, radiation chemistry and biology, as well as photobiology. In spite of its wide application and long history, it is only in recent years that the investigations in the field have become quantitative, so that cross sections for ionization and photoionization efficiencies can be reliably estimated for a large body of atoms and molecules. It is only recently, also, that a quantitative study of the products of photoionization as well as of the phenomena of autoionization and preionization has been undertaken.

In photoabsorption processes leading to ionization, as in those that do not, if the electron is ejected sufficiently quickly after photoabsorption, the nuclei would have moved insignificantly, the transition would have been vertical, and the Franck–Condon principle obeyed. The positive ion so-formed may dissociate or predissociate depending on its internal energy, or it may undergo radiative de-excitation. The sum of the kinetic energies, $E_{kin}$, of the ejected electron and the positive ion, is related to the energy $h\nu$ of the absorbed photon by

$$h\nu = I + E_{kin} \tag{3.88}$$

where $I$ is the ionization energy in eV ($I = eV_i$, where $V_i$ is the ionization potential in volts; we refer to $I$ throughout this book as the ionization potential in eV). Since the positive ion is very much heavier than the electron, nearly all of $E_{kin}$ appears as kinetic energy of the ejected electron and one neglects the kinetic energy of the positive ion. If an amount of energy, $E_{ex}$, is residing in the positive ion as excitation energy, Equation (3.88) becomes

$$hv = I + E_{kin} + E_{ex} \tag{3.89}$$

Since the allowed energy levels for the ejected photoelectron(s) are non-quantized, a continuous photoabsorption spectrum can be observed. This is clearly seen from the data on the photoionization cross section as a function of electron energy for a number of atomic systems (see Figures 3.26, 3.27 and 3.29 in Sections 3.5.2 and 3.5.3). However, quite generally, the true ionization continua are modified by the additional processes of auto-ionization and predissociation (see Sections 3.5.2 and 3.5.3).

For atoms such as those of the alkali metals, ionization can be produced by ultraviolet photons. Other metallic vapours, noble gas atoms, and molecular gases require more energetic photons in the vacuum ultraviolet or soft x-ray region. First, second and third atomic ionization potentials are given in Appendix I. Ionization potentials of molecules and free radicals and the method employed for their determination can be found in Appendix II.

### 3.5.1.1 The photoionization cross section

In Section 3.2.3 we gave expressions relating the basic quantities involved in photoabsorption processes—oscillator strength, integrated absorption coefficient, natural lifetime, and the transition moment matrix elements—in the electric dipole approximation. In the case of photoionization these relations need be extended to account for the fact that while the initial state is discrete the final state lies in the continuum. From Equation (3.21) (Section 3.2.3.5) which gives the differential oscillator strength $df/dv$, the photoabsorption cross section associated with a particular frequency $v(hv > I)$ per unit frequency interval is

$$\sigma_{ab}(v) = \frac{\pi e^2}{mc} \frac{df}{dv} \tag{3.90}$$

or in terms of the kinetic energy $\varepsilon$ of the ejected electron

$$\sigma_{ab}(\varepsilon) = \frac{2\pi^2 e^2 \hbar}{mc} \frac{df}{d\varepsilon} = C \frac{df}{d\varepsilon} \tag{3.91}$$

where $C$ has the value $8.0722 \times 10^{-19}$ cm$^2$Ry if the energy is in Rydbergs and the other constants are expressed in cgs units.

The cross section $\sigma_{ab}(v)$ is greater or equal to the photoionization cross section $\sigma_i(v)$. In the dipole approximation and under the assumption that only one electron is directly involved in the photoabsorption process,

$\sigma_i(v)$ can be written as (Bates, 1946a)

$$\sigma_i(v) = \frac{32\pi^4 m^2 e^2}{3h^3 c} \frac{1}{g_i} \sum_i \sum_f vvC_p |\psi_i^*(\sum_j \mathbf{r}_j)\psi_f(\varepsilon)\,d\tau|^2 \qquad (3.92)$$

where $g_i$ is the statistical weight of the initial bound and discrete state $i$, whose normalized wave function is $\psi_i$. $\psi_f(\varepsilon)$ is the normalized wave function of the ion plus the ejected electron whose velocity is $v$ (for various kinds of normalization of $\psi_f$ see Bates (1946b) and Marr (1967)). The position vector of the $j$th electron is represented by $\mathbf{r}_j$ and the sum $\sum_j \mathbf{r}_j$ characterizes the dipole moment. Allowance is made for the distortion of the wave function of all atomic electrons except that of the ejected one by the introduction of the factor $C_p$ which is slightly less than unity. The exact application of (3.92) may not be feasible and approximations need to be introduced.

Expression (3.92) gives $\sigma_i(v)$ in terms of the dipole length matrix element. In addition to the dipole length formula there are two other equivalent ways of defining the dipole photoionization cross section for exact wave functions for the initial and final states, namely, the dipole momentum (or velocity) formula and the dipole acceleration formula (Chandrasekhar, 1945). (See expressions for the dipole length, dipole velocity, and dipole acceleration formulae in Chapter 7.) The three formulae stress contributions from different regions of electron coordinate space. The dipole length formula emphasizes regions of large $r$, while the velocity and acceleration formulae emphasize, respectively, regions of intermediate and small $r$. For a given set of wave functions one of the three formulae is more appropriate to use. Although there exist no general rules as to which of the three formulae is more applicable, certain points need to be considered in making the decision. Theoretical predictions of $\sigma_i(v)$ in the above formulae are compared with experimental results in Sections 3.5.2 and 3.5.3. The reader is referred also to discussions by Ditchburn and Öpik (1962) and Marr (1967). In general, theoretical calculations on the photoionization of molecules and molecular ions are very much fewer and less accurate than for atoms principally because of the difficulty in obtaining accurate electronic wave functions for the former.

### 3.5.1.2 Atomic collision-induced ionization

Although most of the information on atomic energy levels has been obtained from optical studies, atomic collision techniques have extended our knowledge of atoms and molecules in recent years. The latter group of techniques is not restricted by optical selection rules.

Experimental studies of autoionizing states have been carried out using electron impact (Chapters 4 and 5) and heavy ion impact (see, for example: Berry, 1966; Kessel, McCaughey and Everhart, 1966; Rudd, 1964, 1965;

Rudd and Lang, 1965; Rudd, Jorgensen and Volz, 1966a,b; and Rudd and Smith, 1968) techniques. We note here that ionization by particle impact differs in several respects from photoionization. As will be seen in Chapter 5 in the case of ionization by particle impact the outermost electrons are preferentially ejected irrespective of the particle's energy.

Contrary to the case of photoionization where the cross section often peaks at the threshold (Section 3.5.2) in impact ionization the cross section always peaks at energies considerably above threshold (Chapter 5). Also, while in photoionization as a rule, a single electron is ejected by a given photon (see, however, Section 3.5.5) in particle impacts multiple electron ejection frequently occurs (Chapter 5). It is to be noted further that the cross sections for ionization by electron impact are greater than those by photon absorption (compare data in this and Chapter 5).

### 3.5.1.3 Absorption profile of autoionizing levels

The strength of the interaction of atomic superexcited states with the continuum of states determines the lifetime of the discrete levels. The absorption profile of resonances in atomic photoionization continua resulting from interferences of high-lying states with the continua has been formulated by Fano (1961) and Fano and Cooper (1965), who gave for the absorption cross section the expression

$$\sigma_{ab}(v) = \sigma_1 \frac{(q+\varepsilon)^2}{1+\varepsilon^2} + \sigma_2 \tag{3.93}$$

In Equation (3.93) $\sigma_1$ and $\sigma_2$ are the background cross sections associated, respectively, with the fraction of the available continua with which the discrete states interact and with the fraction of the continua which does not enter into the interaction; $\varepsilon$ is an energy parameter which measures the displacement from some energy within the resonance (in units of half-width, $\Gamma/2$, of the discrete state) and $q$, referred to as the profile index, is determined by the autoionization and transition matrix elements and it can be assumed constant over a given resonance. Far from the resonance $\sigma_{ab} = \sigma_1 + \sigma_2$. The quantities $q$ and $\rho^2 \left( = \frac{\sigma_1}{\sigma_1 + \sigma_2} \right)$ determine the resonance profile type as is shown in Figure 3.24 (Codling, Madden and Ederer, 1967). A change in the sign of $q$ reverses the asymmetry of the resonance.

### 3.5.1.4 Cross sections for photoionization, ionization by electron impact and radiative capture

Photoionization cross sections $\sigma_i(v)$ can be used to estimate cross sections for ionization by electron impact $\sigma_i(\varepsilon)$ and also cross sections for radiative

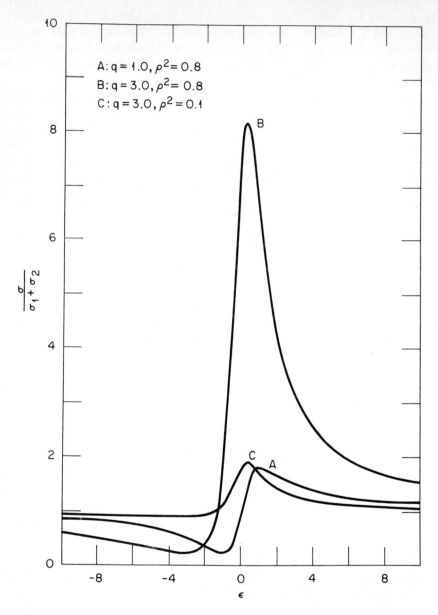

**Figure 3.24** Theoretical resonance profiles, showing the effect on the line shape of: (i) increasing profile index ($q$) and (ii) reducing the autocorrelation coefficient ($\rho^2$). The cross section is normalized to unity in the wings and plotted as a function of a reduced energy variable $\varepsilon$ (Codling, Madden and Ederer, 1967)

capture. In the former case if $\sigma_i(v)$ is known for two systems at the threshold and $\sigma_i(\varepsilon)$ is also known for one of the systems at the threshold to higher energies, $\sigma_i(v)$ for the other from the threshold to higher energies can be obtained approximately by a comparison procedure (Seaton, 1959; McDowell and Peach, 1961). In the latter case the principle of detailed balancing is applied to relate the photoionization cross section to the cross section, $\sigma_c(v)$, of radiative capture of electrons of velocity $v$ by positive ions (the inverse process of photoionization), namely (McDaniel, 1964)

$$\frac{\sigma_c(v)}{\sigma_i(v)} = \frac{e}{2mc^2} \frac{g_i}{g_f} \frac{V_p^2}{300V_e} \tag{3.94}$$

where $g_i$ and $g_f$ are, respectively, the statistical weights of the initial state and the residual system in the final state. $V_p$ and $V_e$ are the potentials (in volts) corresponding to the energies of the incident photon and the ejected electron. Expression (3.94) applies only if the same atomic state is involved in the forward and the reverse processes. However, radiative recombination generally occurs to highly excited states while the photoionization cross sections are usually for the ground state. Approximate values of $\sigma_i(v)$ and $\sigma_c(v)$ may be obtained if either is known.

### 3.5.2 *Atomic Photoionization and Autoionization*

Atomic photoabsorption cross sections are essentially equal to the photo-ionization cross sections. Naturally, for molecules this is not so due to the multiplicity of the available absorption mechanisms.

In measurements of $\sigma_i(v)$ for atomic vapours, care must be taken that the system is not present in the molecular form or any such content is properly accounted for. Even a small concentration of molecules in the atomic gas under study can cause serious errors owing to the differences between molecular and atomic photoabsorption cross sections.

Atomic photoionization cross sections peak at photon energies corresponding to the various photoionization thresholds (absorption edges), rising abruptly as each of the edges is approached from below (low-energy side). The photoionization cross section decreases gradually as the photon energy increases above the corresponding threshold (see, for example, $\sigma_i(v)$ for $H$ in Figure 3.25, Na and Rb in Figure 3.26 and He in Figure 3.27). This behaviour, however, cannot be generalized. For some atoms (e.g. Ne, Figure 3.27) $\sigma_i(v)$ continues to rise and then falls off with increasing energy above threshold.

As we have stated in the previous section, most of the experimental values of $\sigma_i(v)$ refer to photoionization of the atomic system from its ground state. These are the ones we shall be presenting. Few observations and predictions

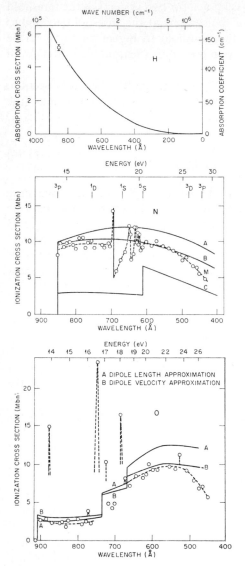

**Figure 3.25** Photoionization cross sections for atomic hydrogen, nitrogen and oxygen. H: Solid curve is computed values; Φ is the measurement of Beynon and Cairns (1965) (Marr, 1967). N: Solid lines represent calculations by Dalgarno and Parkinson (1960) in the dipole velocity approximation (Curve C) and calculations by Henry (1968) in the dipole length (Curve A) and dipole velocity (Curve B) approximation. M is the experimental curve of Comes and Elzer (1967, 1968) (after Comes (1968)). O: Solid lines represent calculations by Dalgarno, Henry, and Stewart (1964): Dipole length (Curve A) and dipole velocity (Curve B) approximation. Experimental results are those of Comes, Speier, and Elzer (1968) (Comes, 1968) (see text for discussion of structure)

on photoionization from excited states have been made (Marr, 1967). In this case the atomic system is raised to a long-lived state (metastable) and the photoionization cross section for the absorption of radiation from this excited state is measured.

### 3.5.2.1 Atomic hydrogen, nitrogen and oxygen

In Figure 3.25 the photoionization cross sections for atomic hydrogen, nitrogen, and oxygen are presented. The figure allows a direct comparison between the theoretical and experimental results. A discussion of the photoionization of atomic nitrogen and oxygen has been given by Comes (1968) in the light of recent experimental and theoretical results. References to earlier work can be found in Marr (1967).

The experimental results of Comes and Elzer (1967, 1968) for atomic nitrogen, obtained by a crossed-beam technique, from the ionization limit down to 432Å are compared in Figure 3.25 with the recent calculations of Henry (1968) and an earlier calculation by Dalgarno and Parkinson (1960). Apart from the experimentally detected autoionizing Rydberg series the calculation by Henry (1968), in the dipole velocity approximation, is in very good agreement with experiment. It should be noted that Henry's calculation did not consider ionization due to discrete absorption. Below 500Å the experimental values decrease more rapidly than the theoretical calculations predict. The calculated cross sections by Dalgarno and Parkinson (1960) are seen to be far below the experimental values. At wavelengths below 700Å a Rydberg series of levels belonging to a discrete spectrum has been observed (Carroll and coworkers, 1966; Comes and Elzer, 1967, 1968). Comes (1968) noted that the scattering of the experimental values of the cross section near the ionization threshold may be indicative of some structure in this wavelength region.

The recent experimental cross sections of Comes, Speier and Elzer (1968) (see also Comes, 1968) for the photoionization of atomic oxygen (also obtained by a crossed-beam technique) are compared in Figure 3.25 with the theoretical calculations of Dalgarno, Henry and Stewart (1964). The experimental results show the participation of autoionization processes in ionization and except for wavelengths where autoionization occurs the experimental and theoretical cross section values seem to be in fair agreement up to $\sim 500$Å. Below this wavelength value the theoretical results are higher. The dipole length approximation result (curve A) fits the experimental data better in the long wavelength region whereas for the short wavelength region the dipole velocity approximation (curve B) seems to give better results. Comes (1968) noted that all but two peaks in the ionization cross section correspond to known absorption peaks and can be attributed to auto-ionization.

Information on autoionization processes for atomic oxygen has also been provided by the recent absorption measurements of Huffman, Larrabee and Tanaka (1967) and by the particle-impact work of Rudd and Smith (1968). The latter workers bombarded oxygen gas with 100 keV $H^+$ and $He^+$ ions

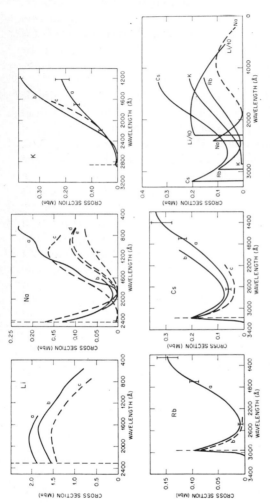

**Figure 3.26** Atomic photoionization cross sections for alkali metal vapours (Marr and Creek, 1968). Li: (*a*) Marr (1963), amended to bring into line with Honig (1962); (*b*) Hudson and Carter (1965a, 1967); (*c*) Burgess and Seaton (1960), Peach (1967) (quantum-defect-method (q.d.m.) calculations). Na: (*a*) Hudson (1964), Hudson and Carter (1967), experimental; (*b*) Ditchburn, Jutsum, and Marr (1953), experimental; (*c*) Cooper (1962), unrelaxed core calculation; (*d*) Boyd (1964), Hartree–Fock calculations, dipole length; (*e*) Boyd (1964), Hartree–Fock calculations, dipole velocity; (*f*) Peach (1967), q.d.m. K: (*a*) Marr and Creek (1968); (*b*) Hudson and Carter (1965b); (*c*) Ditchburn, Tunstead and Yates (1943). Rb: (*a*) Marr and Creek (1968); (*b*) Mohler and Boeckner (1929) amended to bring into line with Honig (1962) vapour pressure data. Cs: (*a*) Marr and Creek (1968); (*b*) Braddick and Ditchburn (1934) amended to bring into line with Honig (1962) vapour pressure data; (*c*) Mohler and Boeckner (1929) amended to bring into line with Honig (1962) vapour pressure data. Last figure shows the best values of the photoionization cross sections for the alkali metals. The cross section data for lithium have been reduced by a factor of ten for convenience of display. The dashed curve represents the Hartree–Fock calculations of Boyd (1964)

and presented experimental and theoretical data showing the energy spectra of autoionizing electrons from atomic oxygen. Finally, a recent calculation by Henry (1967) on the photoionization cross section for atomic oxygen for wavelengths down to 25Å yielded lower values than the experimental data of Cairns and Samson (1965).

### 3.5.2.2 Alkali metals

The onset of ionization for these systems takes place in the near ultraviolet region and owing to the tight binding of the electrons in the outermost closed shell, autoionizing transitions lie well beyond the first ionization potential. With the exception of some very recent work (e.g. Berkowitz and Lifshitz, 1968) the data on photoionization and autoionization of the alkali metals have been discussed by Marr (1967) and by Marr and Creek (1968). Figure 3.26 shows some of the recent data on the photoionization of the alkali metals as they have been summarized by Marr and Creek (1968). It is seen that for all the alkali-metal-vapours but lithium the photoionization cross section shows a minimum to the short wavelength side of the ionization threshold which is predicted from theoretical calculations. Although theoretically this is a zero minimum since it depends on the change of sign in the matrix element, the experimental data on Rb and Cs exclude the possibility of such a zero minimum: those on Na are not inconsistent with it (Figure 3.26) (see discussion in Marr and Creek, 1968).

A comparison between theoretical and experimental results can be made easily on the basis of Figure 3.26. The last set of curves in Figure 3.26 represents the most probable photoionization values for all the alkali-metal-vapours as reported by Marr and Creek (1968). The combined data for each element have been obtained by these authors through smoothed weighted mean values of the experimental curves for all of the vapours except sodium.

### 3.5.2.3 Alkaline earth metals and other elements

The photoionization and autoionization data on these systems have been summarized and discussed by Marr (1967).

### 3.5.2.4 Rare gases

A. *Photoionization cross sections and autoionizing levels.* Figure 3.27 shows the photoabsorption cross sections for the rare gases. The sources of origin for the plotted cross sections are given in the figure caption. It is seen that the rare gas absorption continua extend into the soft x-ray region and close the energy gap between the vacuum ultraviolet and the soft x-ray energy regions. The rare gases, especially He and Ne, have been the subject

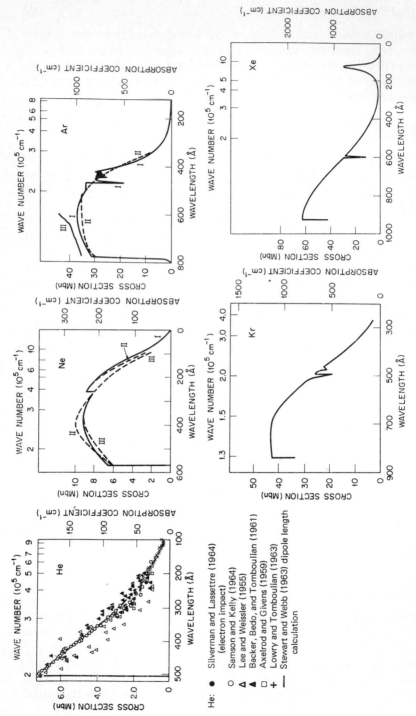

**Figure 3.27** Photoionization cross sections for the rare gases (after Marr, 1967). Ne: (I) Composite experimental curve from Samson and Kelly (1964) and Ederer and Tomboulian (1963). (II) Dipole length and (III) dipole velocity calculations by Sewell (1965). (See other references in Samson and Kelly (1964)). Ar: (I) Samson (1964b). (II) Rustgi (1964). (III) Huffman, Tanaka and Larrabee (1963c). Kr and Xe: Samson and Kelly (1964)

He:
● Silverman and Lassettre (1964) (electron impact)
○ Samson and Kelly (1964)
△ Lee and Weissler (1955)
▲ Backer, Bedo, and Tomboulian (1961)
□ Axelrod and Givens (1959)
+ Lowry and Tomboulian (1963)
— Stewart and Webb (1963) dipole length calculation

of many experimental and theoretical investigations and offer a comparison of both.

Recent photoabsorption studies, especially those employing synchrotron light as an excitation source, revealed new autoionizing states due to single and double electron excitation. (See a review by Madden and Codling (1966) and references cited therein.) The observed resonances, superimposed on the absorption continua, tie up well with electron impact-induced single and double excitation (Chapters 4 and 5). Recent studies on the photoelectron spectra of rare gases (Section 3.5.4) provided additional information on these systems.

From the photoionization cross sections shown in Figure 3.27 for He it is seen that there is an excellent agreement between the best experimental curve of Samson and Kelly (1964) and the calculations of Stewart and Webb (1963). In agreement with the conclusions of Stewart and Webb (1963), the dipole length formula is the best fit at the spectral head, while the dipole velocity formula gives better results at shorter wavelengths. Bell and Kingston (1967) have attempted to make a more complete allowance for the effect of polarization by using the method of polarized orbitals to determine continuum wave functions. Madden and Codling (1963, 1965), using the continuous radiation from the NBS 180 MeV synchrotron, discovered a number of absorption lines around 200 Å. They attributed the observed series to a double electron excitation process of the type $1s^2(^1S_0) \rightarrow 2s\ 2p(^1P_1)$ (and $1s^2 \rightarrow 2pns$ and $1s^2 \rightarrow 2pnd$).

The latest experimental data for Ne, shown in Figure 3.27, substantially agree with the calculations of Sewell (1965). Resonances in the photoionization continuum have also been reported (see, for example, Codling, Madden and Ederer (1967)). Discrete structure has been observed in three distinct energy ranges: between the $2p^5\ ^2P_{3/2,\,1/2}$ limits near 22 eV, between 44 and 60 eV, and near 80 eV. The structure near 22 cV involves resonances due to the transitions $2s^2 2p^6 \rightarrow 2s^2 2p^5\ ns$ and $nd$ analogous to those observed by Beutler (1935) for Ar, Kr and Xe (Figure 3.28). The structure between 44 and 60 eV has been classified (Codling, Madden and Ederer, 1967) as being due to two types of processes: (i) excitation of a subshell $2s$ electron, and (ii) simultaneous excitation of two outer $2p$ electrons; that near 80 eV has been attributed to simultaneous excitation of a subshell $2s$ and a $2p$ electron (see detailed discussion in Codling, Madden and Ederer, 1967).

For Ar, the theoretical treatment of Cooper (1962) does not agree in detail with the experimental data of Samson (1964b) and Samson and Kelly (1964), which agree well with those of Rustgi (1964), apart from the autoionizing transitions. The autoionizing levels in the region between the $^2P_{3/2,\,1/2}$ ionization thresholds are not shown in Figure 3.27. These are shown in Figure 3.28 (Huffman, Tanaka and Larrabee, 1963c) in connection with those for Kr and Xe. Although for Ar no line was resolved under the resolution of Huffman, Tanaka and Larrabee (as it has been in the case of Kr and Xe; see Figure 3.28) the autoionized levels shown should be members of the two series $3s^2 3p^6(^1S_0) \rightarrow 3s^2 3p^5(^2P_{1/2})nd$ and $ns$. The ionization continuum underlying these autoionizing levels and extending to higher energies is, of course, due to the direct ionization process $3s^2 3p^6 \rightarrow 3s^2 3p^5 + e$. The structure seen in Figure 3.27 for Ar around 500–400Å is due to inner-shell absorption, namely

$$3s^2 3p^6(^1S_0) \rightarrow 3s 3p^6 np(^1P_1) \ n \geq 4$$

(Madden and Codling, 1963, 1966). Additional resonances due to double-electron excitation (e.g. two $3p$ electrons) have been also observed (Madden, Ederer and Codling, 1969).

For krypton, the only theoretical treatment, that of Cooper (1962), does not agree well with the latest experimental data of Samson and Kelly (1964). The experimental cross sections of Samson and Kelly (1964) for both krypton and xenon evidence the distinct features of various autoionizing

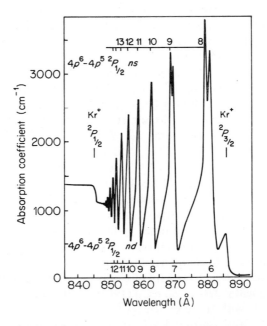

**Figure 3.28** Absorption coefficients of Kr autoionized lines in the region 840–886 Å (Huffman, Tanaka and Larrabee, 1963d)

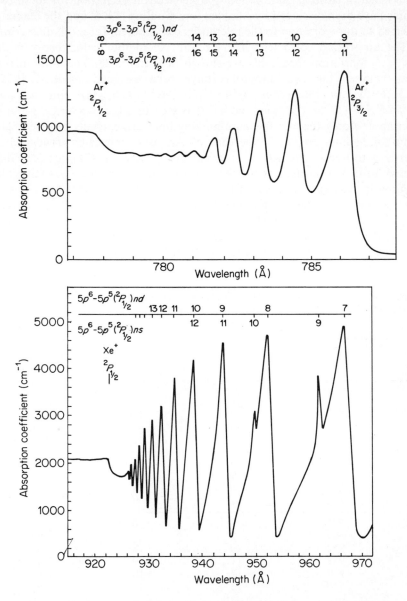

**Figure 3.28** *(continued)* Absorption coefficients of Ar autoionized lines in the 776–788 Å region (Huffman, Tanaka and Larrabee, 1963c), and of Xe autoionized lines in the region 923–972 Å (Huffman, Tanaka and Larrabee, 1963c)

transitions (see references to other experimental work in Marr, 1967). Inner-shell electron excitation and double-electron excitation for these two atoms have been discussed by Madden and Codling (1966) and the reader is referred to this work and to the references cited therein for a full discussion.

The absorption coefficients for Kr and Xe in the region between the $^2P_{3/2, 1/2}$ ionization thresholds are shown in Figure 3.28. The autoionizing lines between the two thresholds have been assigned (Beutler, 1935), respectively, to the series $4s^2 4p^6(^1S_0) \rightarrow 4s^2 4p^5(^2P_{1/2})ns$ and $nd$ and $5s^2 5p^6(^1S_0) \rightarrow 5s^2 5p^5(^2P_{1/2})ns$ and $nd$. It is seen that the two series, $s$ and $d$, overlap and only the first few autoionizing lines are partially resolved. The autoionized lines are, as in the case of Ar, asymmetrically broadened in accord with Fano's prediction. It is seen from the data of Figure 3.28 that the width of the resonances for the autoionizing series decreases (thus the lifetime increases) with higher series members. Table 3.6 lists estimates of the lifetimes of some of the autoionizing levels in the region between $^2P_{3/2, 1/2}$ for Ar and Xe (Huffman, Tanaka and Larrabee, 1963c) and Kr (Huffman, Tanaka and Larrabee, 1963d). These values are seen to be very much shorter (about a factor of $10^5$) than those for electronic states

**Table 3.6** Lifetimes for autoionized lines occurring between the $^2P_{3/2}$ and $^2P_{1/2}$ thresholds for Ar and Xe (Huffman, Tanaka and Larrabee, 1963c) and Kr (Huffman, Tanaka and Larrabee, 1963d)[a]

|            | $\lambda(\text{Å})$ | $ns$ | $nd$ | $\tau(10^{-14}\text{sec})$ |
|------------|---------------------|------|------|-----------------------------|
| Ar         | 786.6               | 11   | 9    | 5.6                         |
|            | 784.9               | 12   | 10   | 6.8                         |
|            | 783.6               | 13   | 11   | 7.6                         |
|            | 782.7               | 14   | 12   | 9.1                         |
| Kr[b]      | 880                 | 8    | 6    | 1.5                         |
|            | 869                 | 9    | 7    | 2.0                         |
|            | 863                 | 10   | 8    | 3.0                         |
|            | 858                 | 11   | 9    | 3.9                         |
| Xe[c]      | 995                 | 8    | 6    | 0.52                        |
|            | 966                 | 9    | 7    | 0.62                        |
|            | 952                 | 10   | 8    | 1.3                         |
|            | 943                 | 11   | 9    | 2.4                         |
|            | 938                 | 12   | 10   | 2.9                         |
|            | 935                 | 13   | 11   | 4.6                         |

[a] The authors quote an accuracy of ~25%.
[b] $\tau$'s more appropriate for d-series; $\lambda = \lambda_{av}$.
[c] Wavelengths indicate approximate values of d lines.

below *I* for the same systems (see Table 3.2). This clearly demonstrates the fastness of the autoionization process as compared to the process of radiative de-excitation.

B. *Photoionization efficiencies.* Matsunaga, Jackson and Watanabe (1965) measured the absolute photoionization yield of Xe in the region 860–1022Å and found that 85 percent of the measured yields fell within the range $100 \pm 5$ percent. Thus, within experimental error the Rydberg lines are completely autoionized, although it must be noticed that in Figure 3.28 lines due to higher members of the Rydberg series are much sharper than the broad lines at longer wavelengths which may indicate that for these lines the ionization efficiency may actually be $< 100\%$. Samson (1964a) too, has reported that the rare gases have the same relative photoionization yield at many wavelengths and concluded that the most probable value for the ionization yield of the rare gases is unity even in the autoionizing regions. Based on Samson's work and their own, Matsunaga, Jackson and Watanabe (1965) concluded that the photoionization yield for Ne, Ar, Kr and Xe is unity at least down to 450Å. In view of this constancy of the photoionization yields of the rare gases, they provide a convenient method for absolute intensity measurements (Samson, 1964a; Matsunaga, Jackson and Watanabe, 1965).

C. *Sum rules.* The extensive data on the photoionization cross sections for some of the rare gases over a wide range of energies allows computation of the various moments of the oscillator strength for these systems. From the expression for the oscillator strength given in Section 3.2.3 we can write (Lowry, Tomboulian and Ederer, 1965).

$$f_{gn} = \tfrac{1}{3}(E_g - E_n) |\langle g | \sum_i^z \mathbf{r}_i | n \rangle|^2 \tag{3.95}$$

where $E_g$ and $E_n$ are the binding energies (in Rydbergs) of the atom in the ground, $g$, and excited, $n$, states. Similarly, we may write for the differential oscillator strength

$$df/d\varepsilon = 0.1238 \times 10^{19} \sigma_{ab}(\varepsilon) \tag{3.96}$$

where $\varepsilon$ refers to the kinetic energy (in Rydbergs) of the ejected electron and $\sigma_{ab}(\varepsilon)$ is in $cm^2$. The non-relativistic formulations for the oscillator strength sum rules may be formally expressed as (see discussion and derivation in Dalgarno and Lynn (1957) and Piech and Levinger (1964); also Fano and Cooper (1968) and references cited therein)

$$\mu_q = \sum_n f_{gn}(E_g - E_n)^q \tag{3.97}$$

where $q = \pm 2, \pm 1$, or 0. In Equation (3.97) the summation involves an integration of Equation (3.96) from $\varepsilon = 0$ to $\varepsilon = \infty$ and a summation over all discrete excited states. Observable physical quantities such as the polarizability, diamagnetic susceptibility, refractive index, and total binding energy of the atom are related to the various oscillator-strength-sum-rules.

The most important and general of the sum rules is the well-known Kuhn–Thomas sum rule (see also Section 3.2.3) obtained from Equation (3.97) for $q = 0$ which then yields $\mu_0 = Z$; $Z$ is the total number of atomic electrons. This rule has been tested experimentally for He and Ne. Thus, Samson and Kelly (1964) using their data on the absorption coefficients for He obtained through Equation (3.13) a value for $f$ equal to 1.54 by evaluating the integral in Equation (3.13) from the ionization threshold down to 0.01 Å. Since the contribution between 4 and 0.01 Å amounted to an $f$ value of $\sim 0.0001$, they assumed the contribution to the total oscillator strength below 0.01 Å to be negligible. A theoretical value of 0.45 was obtained by both Dalgarno and Stewart (1960) and by Salpeter and Zaidi (1962) for the oscillator strength for the discrete transitions in He. Thus the total oscillator strength for helium is found equal to 1.99 which is in excellent agreement with the Kuhn–Thomas sum rule. In agreement with Samson and Kelly (1964), Lowry, Tomboulian and Ederer (1965) reported a value of $2.05 \pm 0.15$ for the total oscillator strength for He. They obtained this value from their ionization data in connection with theoretical estimates of the discrete oscillator strength.

For Ne, Samson and Kelly (1964) estimated a value of 10.03 for the oscillator strength due to the ionization continuum. Ederer and Tomboulian (1964) reported a value of $10.2 \pm 0.4$. Adding to the above values a contribution of 0.4 due to discrete transitions (coming from theoretical calculations; see Ederer and Tomboulian (1964)), the total oscillator strength for Ne, from the two experimental sources is found equal to 10.43 and $10.6 \pm 0.4$, respectively, which is in excellent agreement with the value of 10 predicted by the Kuhn–Thomas sum rule.

For Ar, Samson and Kelly (1964) obtained a contribution of the ionization continuum to the $f$ values for the $K$, $L$ and $M$ shells equal to 1.77, 7.51 and 5.69, respectively. Theoretically, the $f$ values (discrete plus continuum) for the $K$, $L$ and $M$ shells is 2, 8 and 8, respectively. Samson and Kelly, considering the contribution to the $f$ value from discrete transitions for the $K$ and $L$ shells obtained total $f$ values close to two and eight, but no such estimation could be made for the $M$ shell. The finding that the $K$ and $L$ shells contribute, respectively, two and eight may be suggestive that the Kuhn–Thomas sum rule holds for each shell separately. For Kr and Xe, no reliable estimates of the total oscillator strength have been made.

## 3.5.2.5 Other atomic systems

A discussion of experimental and theoretical photoionization cross sections of other atomic systems, such as alkaline earth metals, zinc, cadmium and mercury can be found in Marr (1967).

### 3.5.3 *Molecular Photoabsorption, Photoionization and Preionization*

Most of the recent work on molecular photoionization and preionization has been summarized and discussed by Marr (1967). Much of the available data relevant to photoionization effects in large molecules have been reviewed by Vilesov (1964), while experimental data on the absorption, photoionization and fluorescence of gases which are of importance in upper atmosphere studies have been gathered together by Cook and Ching (1965). Typical photoabsorption and photoionization cross sections are shown in Figures 3.29 and 3.30. It is clearly seen from these data that a significant fraction of the absorbed photons does not result in ionization even at energies considerably in excess of the molecular ionization potential.

Preionization effects are also clearly evident, especially for $H_2$ in the region within 1 to 1.5 eV of the ionization threshold. The ionization efficiency for $H_2$ approaches unity at $\sim 700$Å. Similar features are apparent for $D_2$ (Cook and Metzger, 1964b) with the exception of isotopic wavelength shifts. A mass spectroscopic study of the photoionization of $H_2$, HD and $D_2$ in the energy range 15.2 to 17.7 eV has been made by Dibeler, Reese and Krauss (1965). The distinct structure in the energy range just above threshold for $H_2$ is quite generally agreed to be due to the formation of preionizing vibrationally excited Rydberg states. It has been discussed by a number of workers within the description given by Fano (1961).

Considerable theoretical interest has been shown recently (e.g. Berry, 1966; Bardsley, 1967 and Hernandez, 1968) for molecular photoionization near threshold. In particular, Berry (1966) and Bardsley (1967) have discussed molecular ionization near the ionization threshold with specific reference to the work of Dibeler, Reese and Krauss (1965) on $H_2$, HD and $D_2$. Two of the major results reached by Berry are: (i) preionization occurs most readily when transitions with $\Delta v = 1$ are energetically allowed, and the rate decreases with increasing values of $\Delta v$ ($\Delta v \equiv v_i - v_f$), and (ii) the preionization rate increases rapidly with increasing principal quantum number $n$. Because of the second result, it was concluded by Berry that Rydberg states of $H_2$ with $n > 7$ have very short lifetimes such that they cannot contribute structure to the cross section. This could explain the absence of structure in $H_2$ for energies greater than $\sim 1$ eV of the threshold. Bardsley (1967) argued that although Berry's first conclusion is correct the second is false, and that the absence of structure in the experimental cross sections for energies corresponding to states with principal quantum numbers greater

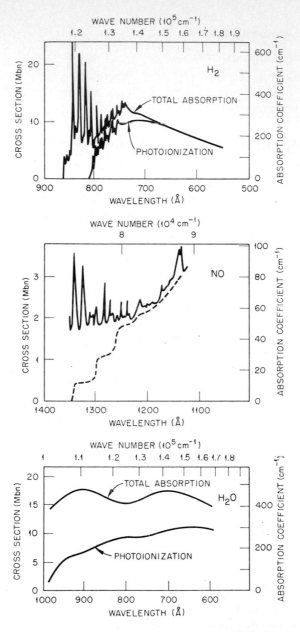

**Figure 3.29** Total absorption and photoionization cross sections (Marr, 1967). $H_2$: Data of Cook and Metzger (1964b); see Cook and Metzger (1964a) for data on $O_2$ and $N_2$ in the region between 600-1000Å. NO: Data of Watanabe (1958); the photoionization cross section is shown by the broken line and the steps in the ionization continuum correspond to vibrational levels of NO in the ground state. $H_2O$: Data of Metzger and Cook (1964)

than seven or eight is not due to large preionization rates, but due to poor energy resolution. Owing to the small spacing of higher Rydberg levels, a better energy resolution is required. It has to be borne in mind, however, that high Rydberg states may, in addition to preionization, decay by the competing process of predissociation which competes effectively with preionization, especially for high vibrational levels.

Relative photoionization and absorption cross sections for $H_2$ (ordinary and para), HD, and $D_2$ from 745 to 810Å at 300 and 78 °K have been measured recently by Chupka and Berkowitz (1968). They found that for para-

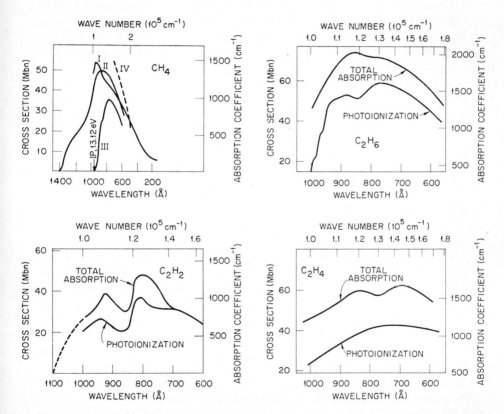

**Figure 3.30** Total absorption and photoionization cross sections for hydrocarbons (Marr, 1967). $CH_4$: (I) Data from Metzger and Cook (1964)—total absorption cross section. (II) Data from Ditchburn (1955) and Rustgi (1964)—total absorption cross section. (III) Photoionization cross-sectional data from Metzger and Cook (1964). (IV) Theoretical calculation by Dalgarno (1952). It is noted that the peak of the total absorption cross section does not coincide with the maximum of the ionization cross section. The total ionization cross section rises slowly at threshold and irregularly to a maximum at $\sim 800$ Å. $C_2H_6$, $C_2H_2$ and $C_2H_4$: Data of Metzger and Cook (1964)

hydrogen at 78 °K (where nearly all of the molecules are in their ground rotational state) most of the lines in the photoionization curve could be members of Rydberg series converging to vibrationally excited states of the $H_2^+$ ion. The preionization efficiencies of some states were found to vary greatly; they can be very small and be affected by Berry's rule (i). The finding, however, that the preionization lifetimes for Rydberg states with principal quantum numbers, $n$, up to eight are between $\sim 0.3$ to $1 \times 10^{-12}$ sec and greater than $1 \times 10^{-12}$ sec for $n > 9$ is in support of Bardsley's correction of Berry's calculation (conclusion (ii)).

Some other recent work on molecular photoionization includes that of Villarejo, Stockbauer and Ingrham (1968) on $CO_2$, Chupka and Lifshitz (1968) on $CH_3$, the mass spectroscopic study of the photoionization of $H_2S$ and $SO_2$ by Dibeler and Liston (1968), and the work of Cook, Metzger and Ogawa (1968) on $N_2O$. A theoretical discussion of low-energy photoionization of large molecules was given by Johnson and Rice (1968).

We finally note the work of Person (1965) on isotope effects in the photoionization efficiency of complex molecules. Person reported for the ratio, $\eta_D/\eta_H$, of the photoionization efficiency $\eta_D$ of $C_6D_6$ and $\eta_H$ of $C_6H_6$ a value equal to 1.07 to 1.08 for photon energies between 9.21 to 11.64 eV. This result compares well with values of similar ratios for a number of pairs of deuterated and undeuterated compounds given in Chapter 2 (Table 2.4). Although these findings have been advanced as evidence for the existence of true competition between preionization and 'dissociation', i.e. as being due to a 'competitive isotope effect', recent work by Person and Nicole (1967, 1968) on ethylene and $n$-butane suggests that the observed isotope effects in photoionization studies is not simply the competitive isotope effect. The finding by Person and Nicole (1968) that for $C_2H_4$ and $C_2D_4$, $\sigma_{iH} \simeq \sigma_{iD}$ over an energy range of $\sim 1$ eV above threshold is very suggestive that although a competitive isotope effect could be present, the predominant photoionization mechanism in this case is either direct ionization or preionization of states in which the ionization processes are very much faster than those of atomic rearrangement. More experimental work is needed to elucidate these points and their inference to the earlier data on the isotope effect in ionization efficiencies discussed in Chapter 2.

Ionization and superexcitation of simple hydrocarbons have been discussed also by Ehrhardt, Linder and Tekaat (1968).

### 3.5.4 Photoelectron- and Penning-Ionization-Electron Spectra

#### 3.5.4.1 Photoelectron spectroscopy

Photoelectron spectroscopy (Vilesov, Kurbatov and Terenin, 1961; Turner, 1962; Al-Joboury and Turner, 1963) is a relatively new method in which the kinetic energies and angular distributions of photoelectrons, ejected from gaseous atoms and molecules by vacuum ultraviolet radiation of a given wavelength, are studied. Photoelectron spectroscopy provides a simple means for determining inner-ionization potentials of atoms and molecules, location of atomic and molecular ionic energy levels, specific photoionization cross sections and Franck–Condon factors as well as relative electronic transition probabilities (see, for example, Turner (1968); Al-Joboury and Turner (1963, 1964, 1967); Al-Joboury, May and Turner (1965a, b); Frost, McDowell and Vroom (1967a, b, c); Schoen (1964); Clark and Frost (1967); Blake and Carver (1967); Berkowitz, Ehrhardt and Tekaat (1967); and Samson and Cairns (1968); Baker and Turner (1968)).

The principle of the method is as follows: A photon of energy $hv$ is absorbed and an electron of energy $E_j$ is ejected in the photoionization process. Due to the large ratio of the ion to electron mass, the energy carried away by the positive ion can be neglected and thus the electron energy is given by

$$E_j = hv - I_j \tag{3.98}$$

where $I_j$ is the $j$th ionization potential of the atom or molecule. Knowing $hv$, a measurement of $E_j$ allows evaluation of $I_j$ through Equation (3.98). If the number of electrons $n_j$ of energy $E_j$ and the total number of electrons $n_0 = \sum n_j$ are measured the specific photoionization cross section $\sigma_j$ for the production of an ion in the $j$th state can be obtained from

$$\sigma_j = (n_j/n_0)\sigma_t \tag{3.99}$$

where $\sigma_t$ is the total photoionization cross section.

Several types of electron energy analysers using retarding potential, magnetic deflection, and electrostatic deflection can be employed for the measurement of the photoelectron energy spectra. In most of these energy analysers the total number of the ejected photoelectrons cannot be measured, but instead a fraction of the photoelectrons is sampled. In this case care must be taken that the analyser does not discriminate against electrons of different energy or any such discrimination is known. This is essential for an accurate determination of the ratio $n_j/n_0$. A description of spherical retarding potential analysers can be found in Samson and Cairns (1968) and Frost, McDowell and Vroom (1967a). In most experiments of this kind the resolution is such that photoelectron spectra show effects due to vibrational excitation of the positive ion.

3.5.4.2 Photoelectron energy spectra; partial photoionization cross sections

In Figure 3.31 the photoelectron stopping curves as obtained by Frost, McDowell and Vroom (1967a) using a spherical grid analyser and 584Å radiation are shown for the case of Ar. The vertical arrows indicate the spectroscopic values for the ground state doublet separation. Their

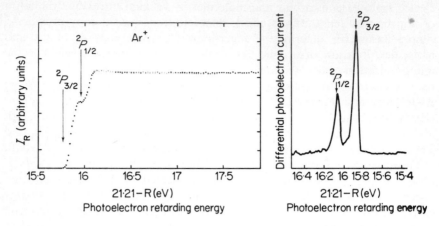

**Figure 3.31** Left: Photoelectron retarding curve for Ar. Right: A tracing of the first differential retarding curve for Ar (Frost, McDowell and Vroom, 1967a)

instrument's resolution can be seen from the differential photoelectron current curve shown on the right portion of the figure. Similar studies have been performed by other workers. Thus, Samson and Cairns (1968) used a spherical retarding potential analyser and measured the kinetic energy of the ejected photoelectrons as a function of exciting wavelength from the $^2P_{1/2}$ ionization threshold down to $\sim 400\text{Å}$ for Ne, Ar, Kr and Xe. They found that the ratio of the number of ions produced in the ground $^2P_{3/2}$ state to the number produced in the excited $^2P_{1/2}$ state is constant with exciting wavelength, and using Equation (3.99) they obtained specific ionization cross sections for these atoms. Table 3.7 gives the ratio $\sigma_{3/2}/\sigma_{1/2}$ of the specific photoionization cross sections for producing ions in their $^2P_{3/2}$ and $^2P_{1/2}$ states, as obtained by various workers.

**Table 3.7** The ratio $\sigma_{3/2}/\sigma_{1/2}$ of the specific photoionization cross sections for producing ions in their $^2P_{3/2}$ and $^2P_{1/2}$ states

| Atom | Samson and Cairns (1968) | Turner and May (1966) | Frost, McDowell and Vroom (1967a) | Comes and Sälzer (1964) |
|------|--------------------------|------------------------|-----------------------------------|--------------------------|
| Ne | 2.18 | — | — | — |
| Ar | 1.98 | — | $1.96 \pm 0.02$ | 2.14 |
| Kr | 1.79 | 1.79 | $1.73 \pm 0.02$ | 1.69 |
| Xe | 1.60 | 1.69 | $1.68 \pm 0.02$ | 1.66 |

Partial photoionization cross sections for $O_2$, $N_2$ and $H_2O$ have been obtained by Blake and Carver (1967) using a photoelectron energy spectrometer and incident monochromatic radiation in the 580–900Å region. The photoelectron peaks have been identified with particular transitions on the basis of the photoelectron energies. An example is shown in Figure 3.32 for the photoelectron spectra from $O_2$ for 584Å exciting radiation. Such measurements illustrate the competition between the various modes of decay which are energetically possible for incident radiation of a given wavelength.

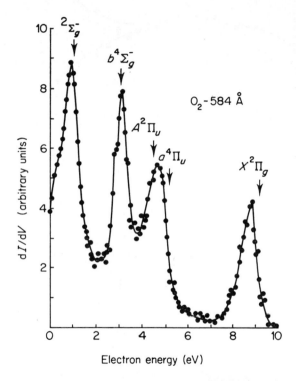

**Figure 3.32** Photoelectron-energy spectra from molecular oxygen for monochromatic radiation of wavelength 584 Å. The arrows indicate the energies corresponding to the thresholds of the various electronic states of the molecular ion (Blake and Carver, 1967)

Once the branching ratios for transitions to particular ionic states are obtained, they can be used in connection with the total photoionization cross section to deduce partial ionization cross sections. The total cross sections and partial cross sections obtained by Blake and Carver (1967) for the

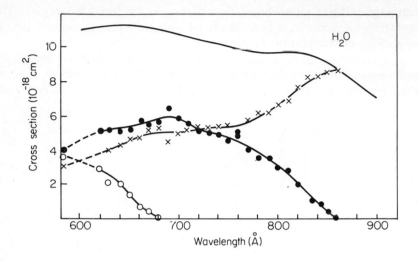

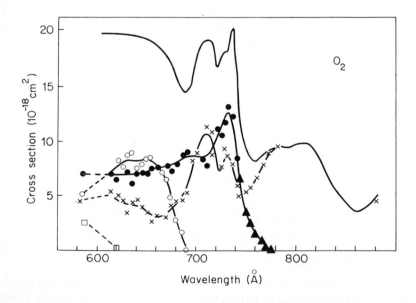

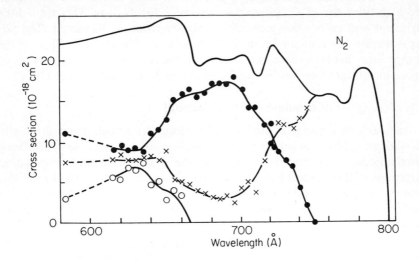

$N_2$ : Production of $N_2^+$ in the $X^2\Sigma_g^+(\times)$, $A^2\Pi_u(\bullet)$, and $B^2\Sigma_u^+(\circ)$ states.

$O_2$ : Production of $O_2^+$ in the $X^2\Pi_g(\times)$, $a^4\Pi_u(\blacktriangle)$, $a^4\Pi_u + A^2\Pi_u(\bullet)$, $b^4\Sigma_g^-(\circ)$, and $^2\Sigma_g^-(\square)$ states

$H_2O$: Production of $H_2O^+$ in the $^2B_1(\times)$ and $^2A_1(\bullet)$ states, and for dissociative ionization processes $(\circ)$.

**Figure 3.33** Partial photoionization cross sections for $N_2$, $O_2$ and $H_2O$ (Blake and Carver, 1967). The upper curve is the total ionization cross section of $N_2$ obtained by averaging the results of Cook and Metzger (1964a) over 10-Å intervals. The upper curve is the total photoionization cross section of $O_2$ obtained by averaging the results of Cook and Metzger (1964a) over 10-Å intervals. The upper curve is the total photoionization coss section of $H_2O$ obtained by averaging the results of Metzger and Cook (1964) over 10-Å intervals

formation of the ground state and various excited states of $N_2^+$, $O_2^+$ and $H_2O^+$ are shown in Figure 3.33. It is very interesting to note that the partial cross sections for $H_2O$, in contrast to those for $N_2$ and $O_2$, show little structure and each cross section rises slowly from its threshold. The $^2B_1$ cross section has a minimum at the threshold for the $^2A_1$ state showing the competition between these two modes of decay. Photoelectron spectra of linear unsaturated molecules can be found in Baker and Turner (1968).

### 3.5.4.3 Angular distribution of photoelectrons

Angular distributions of atomic and molecular photoelectrons are useful in obtaining information about the initial states of such systems. Such distributions, in general, correlate with the polarization and propagation directions of the incident light. In the early years of quantum mechanics this subject received considerable experimental and theoretical attention. The early experimental studies were limited, however, to photoelectrons ejected by x-rays.

Outside the x-ray range, experimental measurements of distributions of photoelectrons are very recent (Berkowitz and Ehrhardt, 1966; Berkowitz, Ehrhardt and Tekaat, 1967; see also similar measurements on photo-detachment from negative ions by Hall and Siegel, 1968). These measurements and recent theoretical considerations (Cooper and Zare ,1968; Tully, Berry and Dalton, 1968) have established that atomic and molecular photo-ionization (both from neutrals and negative ions) in the dipole approximation yields an angular distribution of the form $\tilde{\alpha} + \tilde{\beta} \cos^2 \Theta$, where $\Theta$ measures the angle between the direction of the ejected electron and the polarization of the incident light (see also Bethe (1933)). Tully, Berry and Dalton (1968) derived expressions for the non-relativistic differential cross sections for photoionization of randomly oriented diatomic molecules giving explicit equations for the coefficients $\tilde{\alpha}$ and $\tilde{\beta}$ in terms of the transition matrix elements. Theoretically, the ejection of an $s$ electron should have an angular distribution which is proportional to $\cos^2 \Theta$.

The angular distribution of photoelectrons from Ar and Xe obtained by Berkowitz and Ehrhardt (1966) using the 584Å resonance line of helium is shown in Figure 3.34. It is seen that the angular distribution of $p$ electrons ejected from Xe is more rapid than for Ar for which the angular distribution is more isotropic.

Angular distribution of photoelectrons corresponding to the simultaneous formation of specific electronic states of the ion have been measured by Berkowitz, Ehrhardt and Tekaat (1967) in the range 30–130° for $H_2$, $N_2$, $O_2$, CO, NO and several alkanes. Figure 3.35 shows some of their data for $N_2^+$ ($A^2\Pi_u$ and $X^2\Sigma_g^+$ states), $O_2^+(X^2\Pi_g$ state) and $NO^+(X^1\Sigma^+$ state). A preference for electron ejection in the direction of light propagation is seen in the formation of the electronic ground states of $NO^+$ and $O_2^+$. In both cases an electron from a $\pi_g$ orbital is ejected.

The angular dependence of electrons detached by a laser beam from $C^-$, $O^-$ and $H^-$ has been measured by Hall and Siegel (1968) and is discussed in Chapter 7. It should be noted that careful consideration should be given

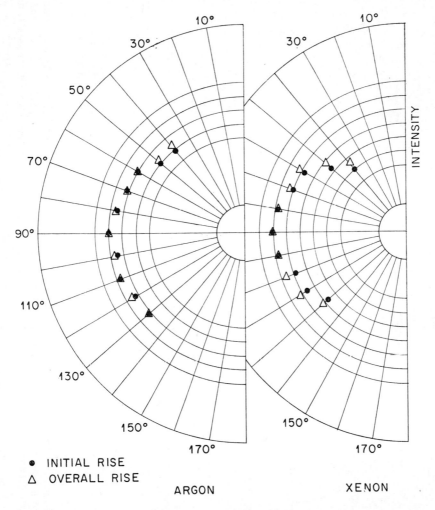

● INITIAL RISE
△ OVERALL RISE

ARGON                    XENON

**Figure 3.34** Angular distribution of photoelectrons from Ar and Xe (with respect to the direction of light propagation). The incident light was the 584 Å helium resonance line. The points (●) are based on the initial rising portion of the $^2P_{3/2}$ state and the points (△) are based on the overall rise of the retarding potential curve due to contributions of $^2P_{3/2}$ and $^2P_{1/2}$ (after Berkowitz and Ehrhardt, 1966)

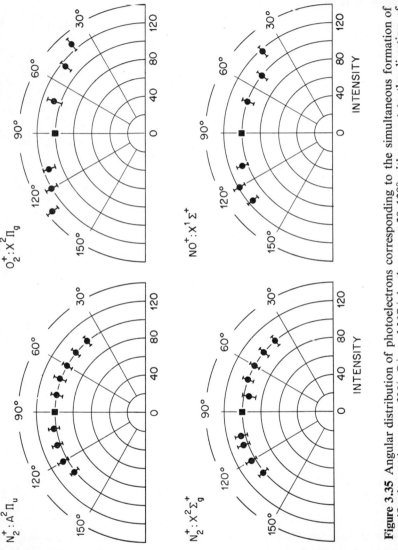

**Figure 3.35** Angular distribution of photoelectrons corresponding to the simultaneous formation of specific electronic states of $N_2^+$, $O_2^+$, and $NO^+$ in the range 30–130° with respect to the direction of light propagation (after Berkowitz, Ehrhardt, and Tekaat, 1967). See this reference for similar data on other species

to measurements which are sensitive to the form of the photoelectron angular distribution.

### 3.5.4.4 Penning-ionization electron spectra

Information about the energy of ionic states can be obtained by measuring the kinetic energy of electrons ejected in the Penning ionization process, i.e. in the ionization of an atom (B) or a molecule (XY) by collision with neutral particles in electronically excited metastable states $A^*$.[†] The process may be represented as[‡]

$$A^* + B \rightarrow A + B^+ + e \qquad (3.100)$$

and

$$A^* + XY \rightarrow XY^+ + A + e \qquad (3.101)$$

The first experimental investigations of what is now known as the 'Penning-ionization electron spectroscopy' have been made by Čermák (1966, 1968). Penning-ionization electron spectroscopy, like photoelectron spectroscopy, is a relatively new method; both are equivalent as far as measuring techniques are concerned, but there exist important differences between them. As in the case of photoionization, the energy relation for process (3.101) is

$$\varepsilon_e = E(A^*) - [I(XY) + E_{v,r}(XY^+) + E_{kin}(XY^+ + A)] \qquad (3.102)$$

where $E(A^*)$ is the excitation energy of the metastable $A^*$, $I(XY)$ is the ionization potential of the molecule XY, $E_{v,r}(XY^+)$ is the vibrational and rotational energy of the molecular ion $XY^+$ and $E_{kin}(XY^+ + A)$ is the translational energy of the products $XY^+ + A$. If $E_{kin}(XY^+ + A) = 0$ and $E_{v,r}(XY^+) = 0$, then measurements of the photoelectron energy $\varepsilon_e$ provide the first or higher ionization potentials of the molecule XY. Similarly, the energy relation for (3.100) is

$$\varepsilon_e = E(A^*) - I(B) \qquad (3.103)$$

Actually, Equation (3.103) can be written as

$$\varepsilon_e = E(A^*) - I(B) + \Delta E \qquad (3.104)$$

where $\Delta E$ is either positive or negative. If $\Delta E$ is positive and greater than the relative kinetic energy of the neutral system at infinity, then associative ionization viz.

$$A^* + B \rightarrow AB^+ + e \qquad (3.105)$$

---

†Such species can be produced by electron impact or in gaseous discharges.
‡When a metastable species $A^*$ interacts with a target atom or molecule, several types of reactions may ensue. Consider a metastable atom $A^*$ colliding with a molecule $X_2$. In addition to (3.101) other reactions can occur such as Penning ionization accompanied by molecular dissociation $(A^* + X_2 \rightarrow A + X^+ + X + e)$ and chemi-ionization or Hornbeck-Molnar reactions $(A^* + X_2 \rightarrow AX_2^+ + e$ or $AX^+ + X + e)$ (J. Hornbeck and J. P. Molnar, *Phys. Rev.*, **84**, 621 (1951)).

must have taken place. If $\Delta E$ is negative, processes (3.100) and (3.105) are possible and measurement of the energy distribution of the electrons can yield information about processes (3.100) and (3.105) themselves.

In the Penning-ionization electron spectroscopy, as opposed to photo-electron spectroscopy, spin-forbidden transitions (i.e. $A^* + B \rightarrow B^{+*} + A$ and $A^* + XY \rightarrow XY^{+*} + A$) can occur, but the former method considerably lacks the energy resolution achieved in the latter. Further work is necessary to improve the method, and Penning-ionization electron spectroscopy, as well as photoelectron spectroscopy, coupled with optical and electron impact studies will allow a better understanding of atomic and molecular processes.

Results on Penning ionization of a number of molecules can be found in Čermák (1968). Fuchs and Niehaus (1968) reported Penning-ionization electron energy distributions for collision pairs such as $A^*(=\text{He, Ne, Ar, Kr})$ and $B(=\text{Ar, Kr, Xe, Hg, C}_2\text{H}_2)$. Also Penton and Muschlitz (1968) reported that the cross section for the production of $H_2^+$, $HD_2^+$ and $D_2^+$ on impact with metastable helium atoms increased in this order. They have explained this isotope effect as being a result of competition between preionization and dissociation.

### 3.5.5 Double Electron Ejection in the Photoabsorption Process

In Section 3.5.2 we discussed double electron excitation of outer electrons in the rare gases produced by photoabsorption. Similar transitions induced by electron impact are discussed in Chapter 4. Double electron ejection, produced when two electrons share the energy of the absorbed photon, has been observed also and has been systematically studied recently (see detailed references in Krause, Carlson and Dismukes (1968)). If $\varepsilon_1$ and $\varepsilon_2$ are the kinetic energies of the ejected electrons 1 and 2, and $I_1$ and $I_2$ are the respective ionization potentials, we may write for this double-photoelectric effect process

$$\varepsilon_1 + \varepsilon_2 = hv - I_1 - I_2 \qquad (3.106)$$

Hence, when orbital electrons are ejected by monochromatic x-rays, the energy spectrum of the ejected electrons will exhibit: (i) one discrete line which is characteristic of single electron ejection ($\varepsilon_2$ and $I_2 = 0$ in Equation (3.106)), (ii) discrete lines due to excitation of a second electron ($\varepsilon_2 = 0$ and $I_2 \rightarrow E_{\text{excitation}}$ in Equation (3.106)), and (iii) continuous distributions $f(\varepsilon_1)$ and $f(\varepsilon_2)$ due to the emission of two electrons. Krause, Carlson and Dismukes (1968) performed an energy analysis of the electrons emitted in the photoabsorption process and determined directly the probability of double ionization, the probability of simultaneously ejecting one electron and promoting another to an excited discrete level, and the energy spectra of the continuum electrons. They have bombarded Ne with $\text{MgK}_\alpha$ and $\text{AlK}_\alpha$

x-rays and Ar with $TiK_\alpha$ x-rays and studied those electrons that originated from $K$, $KL$ and $KM$ shells. The energy spectrum of electrons ejected from neon atoms by 1.25 keV $MgK_\alpha$ x-rays obtained by these workers is shown in Figure 3.36. The line at 387 eV corresponds to the emission of a single photoelectron from the neon $K$ shell. The small peak on the high-energy side of the photoline is due to $MgK_\alpha$ satellite lines. The rise near channel 72 has been attributed by these workers to $NeK(MgK_\alpha)$ photoelectrons which have lost energy in collisions with neutral Ne atoms. The structure below channel 70 has been attributed to a process in which a $2p$ electron and a $2s$ or several $L$ electrons are excited or ionized in addition to a $K$ electron.

To understand the details of Figure 3.36 let us consider the energy balance for two-electron transitions:

$$\text{for excitation: } E_1 = h\nu - E_K - E_{2p(K)*} \tag{3.107}$$

$$\text{for ionization: } E_1 + E_2 = h\nu - E_K - E_{2p(K)} = \text{const} \tag{3.108}$$

In the above relations $E_{2p(K)*}$ and $E_{2p(K)}$ are the excitation and ionization energies of a $2p$ electron in a Ne atom lacking one $K$ electron and $E_K$ is the binding energy of the Ne $K$ electron. The energy distribution $f(\varepsilon_1)$ of the so-called complementary shakeoff* electrons is shown in the inset of Figure 3.36. This spectrum resulted from subtraction of the average background and inelastic loss spectrum from the recorded distribution. Krause, Carlson and Dismukes did not observe the energy distribution $f(\varepsilon_2)$ of the slower electron i.e. of the shakeoff electron. According to Equation (3.108), one distribution is the mirror image of the other. In Figure 3.36 the line appearing at $\Delta E = -28(\pm 2)$ eV (measured from the photoline; inset of Figure 3.36) indicates the $2p \to 3p$ excitation of a second electron. At $\Delta E = -42 \pm 3$ eV a continuous spectrum begins which indicates the simultaneous emission of an $L$ and a $K$ electron.

From spectra of the kind shown in Figure 3.36, Carlson and coworkers have derived for various systems $h\nu + A(A = He, Ne \text{ and } Ar)$ probability values for removing two electrons from their orbitals. These are summarized in Table 3.8. The results collected in Table 3.8 clearly show that the observed intensities of double electron ejection are well accounted for by the theory of electron shakeoff (Carlson and Krause, 1965; Krause, Carlson and Dismukes, 1968) if the electrons come from two different shells. The same

---

*Carlson and coworkers employed the sudden approximation or shakeoff theory (Levinger (1953); see other references in Carlson and Krause (1965) to interpret their results. This theory attributes the emission of a second electron to the sudden perturbation of the atomic potential at the time of the departure of the first. They referred to electrons emitted in single electron transitions as 'photoelectrons', and they used the terms 'shakeoff electron' and 'complementary shakeoff electron' to describe, respectively, the slower and faster electron of a simultaneously excited pair in two-electron transitions.

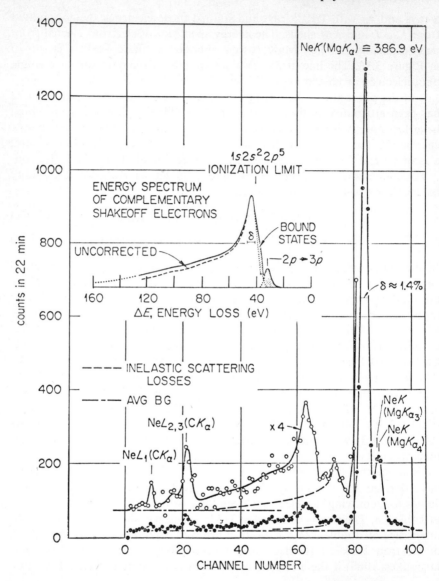

**Figure 3.36** Energy spectrum of electrons ejected from Ne by Mg K$_\alpha$ x-rays. Photoline NeK (MgK$_\alpha$) indicates single-electron emission; continuous distribution indicates emission of $K$ and $L$ electrons. Inset figure (dashed line) is the recorded spectrum after subtraction of inelastic scattering losses and background. Solid contour represents fully corrected spectrum. Channel numbers are substituted in insets by energy scale whose origin is at location of photopeak (after Krause, Carlson and Dismukes, 1968)

theory gives also a fair representation of the energy distribution of the continuum electrons (Krause, 1967). The shakeoff theory, however, underestimates the probability of two-electron interactions when the electrons originate both from the outer shell (see Table 3.8). In this case, electron correlation (interaction between the two electrons in question) is important and need be considered in calculating the matrix elements for the transitions. Inclusion of correlation effects gave good agreement between theory and experiment for the photoexcitation and photoionization of He, as can be seen from the data shown in Table 3.8.

**Table 3.8** Probabilities of double electron interaction of photons (probabilities per photon interaction (in percent))

| Shells | Excitation | | Ionization | | Total | |
|---|---|---|---|---|---|---|
| | Exp. | Theor. | Exp. | Theor. | Exp. | Theor. |
| Ne$KL^a$ | $2\pm1^b$ | $5^c$ | $16.5\pm1.5^b$ | — | $18.5\pm1.0^b$ | $18.1^c$ |
| Ar$KL^a$ | — | — | — | — | $2.5\pm0.8^b$ | $2.03^c$ |
| Ar$KM^d$ | $3^b$ | — | $18.2^b$ | — | $20.7\pm1.4^b$ | $20.53^c$ |
| He$KK^e$ | 6 | $1.8^c/6.3^f$ | 4 | $< 1^c/3.7^g$ | 13 | — |
| Ne$LL^e$ | $\leq 10$ | — | 14 | — | $\leq 24$ | $4.5^c$ |
| Ar$MM^e$ | $\sim 10$ | — | 16 | — | $\sim 26$ | $3.8^c$ |

[a] Simultaneous excitation and ionization of one or more electrons from $L$ shell of Ne (Ar) accompanying $K$ photoionization.
[b] From Krause, Carlson and Dismukes (1968).
[c] Shakeoff theory (see Carlson and Krause (1965) and Krause, Carlson, and Dismukes (1968)).
[d] Simultaneous excitation and ionization of one or more electrons from $M$ shell of Ar accompanying K photoionization.
[e] From Krause (1967).
[f] Electron correlation considered (Salpeter and Zaidi, 1962).
[g] Electron correlation considered (Byron and Joachain, 1966, 1967).

In concluding this section the importance of such multiple phenomena in studies of atomic structure should be stressed. Such direct and specific information is also important for refinement of existing theories.

### 3.6 References

M-E. M. Abu-Zeid, L. G. Christophorou and J. G. Carter (1968). Oak Ridge National Laboratory Report ORNL- TM-2219.
I. Agarbiceanu, I. Kukurezianu, I. Popesku and V. Vasiliu (1963). *Opt. Spectr.*, **14**, 8.

M. I. Al-Joboury and D. W. Turner (1964). *J. Chem. Soc.*, 4434.

M. I. Al-Joboury, D. P. May and D. W. Turner (1965a). *J. Chem. Soc.*, 616.

M. I. Al-Joboury, D. P. May and D. W. Turner (1965b). *J. Chem. Soc.*, 6350.

M. I. Al-Joboury and D. W. Turner (1963). *J. Chem. Soc.*, 5141.

M. I. Al-Joboury and D. W. Turner (1967). *J. Chem. Soc.*, 373.

P. Auger (1926a). *Ann. Phys.* (*Paris*) **6**, 183.

P. Auger (1926b). *C. R. Acad. Sci.* (*Paris*), **182**, 215, 773.

P. Auger (1925a). *J. Phys. Radium*, **6**, 205.

P. Auger (1925b). *C. R. Acad. Sci.* (*Paris*), **180**, 65.

N. Axelrod and M. P. Givens (1959). *Phys. Rev.*, **115**, 97.

T. Azumi, A. T. Armstrong and S. P. McGlynn (1964). *J. Chem. Phys.*, **41**, 3839.

T. Azumi and S. P. McGlynn (1963). *J. Chem. Phys.*, **38**, 2773.

T. Azumi and S. P. McGlynn (1964). *J. Chem. Phys.*, **41**, 3131.

T. Azumi and S. P. McGlynn (1965). *J. Chem. Phys.*, **42**, 1965.

D. J. Backer, D. E. Bedo and D. H. Tomboulian (1961). *Phys. Rev.*, **124**, 1471.

H. L. J. Bäckström and K. Sandros (1958). *Acta Chem. Scand.*, **12**, 823.

B. Badger and B. Brocklehurst (1968). *Nature*, **219**, 263.

B. Badger, B. Brocklehurst and R. D. Russell (1967). *Chem. Phys. Letters*, **1**, 122.

C. Baker and D. W. Turner (1968). *Proc. Roy. Soc.* A**308**, 19.

J. N. Bardsley (1967). *Chem. Phys. Letters*, **1**, 229.

R. B. Barker and H. W. Berry (1966). *Phys. Rev.*, **151**, 14.

D. R. Bates (1946a). *Mon. Not. Roy. Astron. Soc.*, **106**, 423.

D. R. Bates (1946b). *Mon. Not. Roy. Astron. Soc.*, **106**, 432.

M. Beer and H. C. Longuet-Higgins (1955). *J. Chem. Phys.*, **23**, 1390.

K. L. Bell and A. E. Kingston (1967). *Proc. Phys. Soc.*, **90**, 31.

R. G. Bennett (1960). *Rev. Sci. Instr.*, **31**, 1275.

R. G. Bennett and P. J. McCartin (1966). *J. Chem. Phys.*, **44**, 1969.

W. R. Bennett, Jr. (1961). In J. R. Singer (Ed.), *Advances in Quantum Electronics*, Columbia University Press, New York, pp. 28–43.

W. R. Bennett, Jr., P. J. Kindlemann and G. N. Mercer (1965). *Appl. Opt. Suppl.* **2**, *Chemical Lasers*, 34.

J. Berkowitz and H. Ehrhardt (1966). *Phys. Letters*, **21**, 531.

J. Berkowitz, H. Ehrhardt and T. Tekaat (1967). *Z. Physik*, **200**, 69.

J. Berkowitz and C. Lifshitz (1968). *J. Phys. B* (*Proc. Phys. Soc.*), **1**, (series 2) 438.

I. B. Berlman (1965). *Handbook of Fluorescence Spectra of Aromatic Molecules*, Academic Press, New York.

R. S. Berry (1966). *J. Chem. Phys.*, **45**, 1228.

H. Bethe (1933). Quantemechanik der Ein- und Zwelelektronprobleme. In *Handbuch der Physik*, Bd. 24/1, Springer, Berlin.

H. Beutler (1935). *Z. Physik.*, **93**, 177.

H. Beutler and H. O. Jünger (1936). *Z. Physik*, **98**, 181.

J. D. E. Beynon and R. B. Cairns (1965). *Proc. Phys. Soc.* (*London*), **86**, 1343.

G. Binsch, E. Heilbronner, R. Jankow and D. Schmidt (1967). *Chem. Phys. Letters*, **1**, 135.

J. B. Birks (1968). *Chem. Phys. Letters*, **2**, 417.

J. B. Birks (1963). *J. Phys. Chem.*, **67**, 2199.

J. B. Birks (1967). *Nature*, **214**, 1187.

J. B. Birks (1964). *Theory and Practice of Scintillation Counting*, MacMillan Co., New York.

J. B. Birks (1970). *Photophysics of Aromatic Molecules*, Wiley-Interscience, London.

J. B. Birks, C. L. Braga and M. D. Lumb (1965). *Proc. Roy. Soc.* (*London*), A**283**, 83.

J. B. Birks and L. G. Christophorou (1962a). *Nature*, **194**, 442.
J. B. Birks and L. G. Christophorou, (1962b). *Nature*, **196**, 33.
J. B. Birks and L. G. Christophorou (1964a). *Nature*, **197**, 1064.
J. B. Birks and L. G. Christophorou (1963b). *Proc. Roy. Soc.*, **274**, A552.
J. B. Birks and L. G. Christophorou (1964b). *Proc. Roy. Soc.*, **277**, A571.
J. B. Birks and L. G. Christophorou (1963a). *Spectrochim. Acta*, **19**, 401.
J. B. Birks and D. J. Dyson (1961). *J. Sci. Instr.*, **38**, 282.
J. B. Birks and D. J. Dyson (1963). *Proc. Roy. Soc.*, **A275**, 135.
J. B. Birks, D. J. Dyson and I. H. Munro (1963). *Proc. Roy. Soc.*, **A275**, 575.
J. B. Birks, S. Georghiou and I. H. Munro (1968). *J. Phys. B* (*Proc. Phys. Soc.*), **1** (series 2) 266.
J. B. Birks and T. A. King (1966). *Proc. Roy. Soc.*, **A291**, 244.
J. B. Birks and W. A. Little (1953). *Proc. Phys. Soc.*, **A66**, 921.
J. B. Birks, M. D. Lumb and I. H. Munro (1964). *Proc. Roy. Soc.*, **A280**, 289.
J. B. Birks, G. F. Moore and I. H. Munro (1966). *Spectrochim. Acta*, **22**, 323.
J. B. Birks and I. H. Munro (1967). In G. Porter (Ed.), *Progress in Reaction Kinetics*, vol. 4, Pergamon Press, Oxford, pp. 239–303.
J. B. Birks, B. N. Srinivasan and S. P. McGlynn (1968). *J. Mol. Spectr.*, **27**, 266.
A. J. Blake and J. H. Carver (1967). *J. Chem. Phys.*, **47**, 1038.
L. M. Bollinger and G. E. Thomas (1961). *Rev. Sci. Instr.*, **32**, 1044.
E. J. Bowen and F. Wokes (1953). *Fluorescence of Solutions*, Longmans Green, London.
A. H. Boyd, (1964). *Planet. Sci.*, **12**, 729.
H. J. J. Braddick and R. W. Ditchburn (1934). *Proc. Roy. Soc.*, **143A**, 472.
K. A. Bridgett and T. A. King (1967). *Proc. Phys. Soc.*, **92**, 75.
S. S. Brody (1957). *Rev. Sci. Instr.*, **28**, 1021.
R. Braunstein (1962). *Phys. Rev.*, **125**, 475.
C. L. Braun, S. Kato and S. Lipsky (1963). *J. Chem. Phys.*, **39**, 1645.
A. Burgess and M. J. Seaton (1960). *Mon. Not. Roy. Astr. Soc.*, **120**, 121.
F. W. Byron and C. J. Joachain (1966). *Phys. Rev. Letters*, **16**, 1139.
F. W. Byron and C. J. Joachain (1967). *Phys. Letters*, **24A**, 616.
R. B. Cairns and J. A. R. Samson (1965). *Phys. Rev.*, **139**, A1403.
G. Cario and J. Franck (1923). *Z. Physik*, **17**, 202.
T. A. Carlson (1967). *Phys. Rev.*, **156**, 142.
T. A. Carlson and M. O. Krause (1965). *Phys. Rev.*, **140**, A1057.
P. K. Carroll (1959). *Astrophys. J.*, **129**, 794.
P. K. Carroll, R. E. Huffman, J. C. Larrabee and Y. Tanaka (1966). *Astrophys. J.*, **146**, 553.
J. G. Carter and L. G. Christophorou (1967). *J. Chem. Phys.*, **46**, 1883.
J. G. Carter, L. G. Christophorou and M-E. M. Abu-Zeid (1967). *J. Chem. Phys.*, **47**, 3879.
G. Castro and R. M. Hochstrasser (1966). *J. Chem. Phys.*, **45**, 4352.
V. Čermák (1968). Penning Ionization Electron Spectroscopy. In E. Kendrick (Ed.), *Advances in Mass Spectometry*, Vol. 4, Institute of Petroleum, London, p. 697.
V. Čermák (1966). *J. Chem. Phys.* **44**,, 3774.
S. Chandrasekhar (1945). *Astrophys. J.*, **102**, 223.
E. A. Chandross, J. W. Longworth and R. E. Visco (1965). *J. Am. Chem. Soc.*, **87**, 3259.
A. S. Cherkasov and N. S. Basilevskaya (1965). *Izv. Akad. Nauk SSSR, Ser. Fiz.*, **29**, 1284.
A. S. Cherkasov, V. A. Molchanov, T. M. Vember and K. G. Voldaikina (1956). *Soviet Phys. Doklady*, **1**, 427.

L. G. Christophorou (1963). Ph. D. Thesis, Manchester.

L. G. Christophorou and J. G. Carter (1966). *Nature*, **209**, 678.

L. G. Christophorou, M-E. M. Abu-Zeid and J. G. Carter (1968). *J. Chem. Phys.*, **49**, 3775.

W. A. Chupka and J. Berkowitz (1968). *J. Chem. Phys.*, **48**, 5726.

W. A. Chupka and C. Lifshitz (1968). *J. Chem. Phys.*, **48**, 1109.

E. Clar (1964). *Polycyclic Hydrocarbons*, Vols. I and II, Academic Press, Inc., London.

I. D. Clark and D. C. Frost (1967). *J. Am. Chem. Soc.*, **89**, 244.

E. Clementi (1968). *Chem. Revs.*, **68**, 341.

K. Codling and R. P. Madden (1965a). *J. Appl. Phys.*, **36**, 380.

K. Codling and R. P. Madden (1965b). *J. Chem. Phys.*, **42**, 3935.

K. Codling, R. P. Madden and D. L. Ederer (1967). *Phys. Rev.*, **155**, 26.

L. Colli (1954). *Phys. Rev.*, **95**, 892.

F. J. Comes (1968). In E. Kendrick, *Advances in Mass Spectrometry*, Vol. 4, Institute of Petroleum, London, pp. 737–753.

F. J. Comes and A. Elzer (1967). *Phys. Letters*, **25A**, 334.

F. J. Comes and A. Elzer (1968). *Z. Naturforsch.*, **23a**, 133.

F. J. Comes and H. G. Sälzer (1964). *Z. Naturforsch.*, **19a**, 1230.

F. J. Comes, F. Speier and A. Elzer (1968). *Z. Naturforsch.*, **23a**, 125.

E. U. Condon (1928). *Phys. Rev.*, **32**, 858.

G. R. Cook and B. K. Ching (1965). *Aerospace Rept.* TDR-469 (92660-01)-4.

G. R. Cook and P. H. Metzger (1964a). *J. Chem. Phys.* **41**, 321.

G. R. Cook and P. H. Metzger (1964b). *J. Opt. Soc. Am.*, **54**, 968.

G. R. Cook, P. H. Metzger and M. Ogawa (1968). *J. Opt. Soc. Am.*, **58**, 129.

J. W. Cooper (1962). *Phys. Rev.*, **128**, 681.

J. Cooper and R. N. Zare (1968). *J. Chem. Phys.*, **48**, 942.

C. A. Coulson (1961). *Valence*, 2nd ed. Oxford University Press, Oxford.

C. A. Coulson, A. Streitwieser, Jr., M. D. Poole and J. I. Brauman (1965). *Dictionary of π-electron Calculations*, W. H. Freeman and Co., San Francisco.

S. Czarnecki, quoted in Birks (1968).

A. Dalgarno (1952). *Proc. Phys. Soc. (London)*, **A65**, 663.

A. Dalgarno, R. J. W. Henry and A. L. Stewart (1964). *Planet Space Sci.*, **12**, 235.

A. Dalgarno and R. Lynn (1957). *Proc. Phys. Soc.*, **A70**, 802.

A. Dalgarno and D. Parkinson (1960). *J. Atmospheric Terrest. Phys.*, **18**, 335.

A. Dalgarno and A. L. Stewart (1960). *Proc. Phys. Soc.*, **76**, 49.

A. S. Davydov, (1962). *Theory of Molecular Excitons*, English translation from the first Russian edition by M. Kasha and M. Oppenheimer, Jr., McGraw-Hill Book Co., New York.

W. R. Dawson and M. Windsor (1968). *J. Phys. Chem.*, **72**, 3251.

D. L. Dexter, (1953). *J. Chem. Phys.*, **21**, 836.

V. H. Dibeler and S. K. Liston (1968). *J. Chem. Phys.*, **49**, 482.

V. H. Dibeler, R. M. Reese and M. Krauss (1965). *J. Chem. Phys.*, **42**, 2045.

R. W. Dichburn (1955). *Proc. Roy. Soc. (London)*, **A229**, 44.

R. W. Ditchburn, P. J. Jutsum and G. V. Marr (1953). *Proc. Roy. Soc.*, **219A**, 89

R. W. Ditchburn and U. Öpik (1962). Photoionization Processes. In D. R. Bates (Ed.), *Atomic and Molecular Processes*, Academic Press, New York, p. 79.

R. W. Ditchburn, J. Tunstead and J. G. Yates (1943). *Proc. Roy. Soc.*, **181A**, 386.

M. D. Dowley, K. B. Eisenthal and W. L. Peticolas (1967). *J. Chem. Phys.*, **47**, 1609.

C. E. Easterly, L. G. Christophorou, R. P. Blaunstein and J. Carter (1970). *Chem. Phys. Letters*, **6**, 579.

D. L. Ederer and D. H. Tomboulian (1964). *Phys. Rev.*, **133**, A1525.

D. L. Ederer and D. H. Tomboulian (1963). Tech. Rept. 10, Contract No. Nonr-401(37).

H. Ehrhardt, F. Linder and T. Tekaat (1968). In E. Kendrick (Ed.), *Advances in Mass Spectometry*, Institute of Petroleum, London, 705.

A. Einstein, (1917). *Z. Physik*, **18**, 121.

J. Eisinger, M. Guéron, R. G. Schulman and T. Yamane (1966). *Proc. US Nat. Acad. Sci.* **55**, 1015.

V. L. Ermolaev and E. B. Sveshnikova (1963). *Soviet Phys. Doklady*, **8** (No. 4) 373.

U. Fano, (1961). *Phys. Rev.*, **124**, 1866.

U. Fano and J. W. Cooper (1965). *Phys. Rev.*, **137**, A1364.

U. Fano and J. W. Cooper (1968). *Rev. Mod. Phys.*, **40**, 441.

P. P. Feofilov (1961). *The Physical Basis of Polarized Emission*, English Translation, Consultants Bureau, New York.

J. Ferguson (1959). *J. Mol. Spectr.*, **3**, 177.

T. Förster (1948). *Ann. Physik*, **2**, 55.

T. Förster (1960b). In M. Burton, J. S. Kirby-Smith and J. L. Magee (Eds.), *Comparative Effects of Radiation*, John Wiley and Sons, New York. p. 300.

T. Förster (1959). *Discussions Faraday Soc.*, **27**, 7.

T. Förster (1951). *Fluoreszenz Organischer Verbindungen*, Vandenhoek and Ruprecht, Göttingen.

T. Förster (1946). *Naturwissenschaften*, **33**, 166.

T. Förster (1960a). *Radiation Res. Suppl.*, **2**, 326.

T. Förster (1969). *Angen. Chem. (Intern. Edition)*, **8**, 333.

T. Förster and K. Kasper (1955). *Z. Elektrochem.*, **59**, 976.

J. Franck (1925). *Trans. Faraday Soc.*, **21**, 536.

J. Franck and R. Livingston (1949). *Rev. Mod. Phys.*, **21**, 505.

J. Franck and E. Teller (1938). *J. Chem. Phys.*, **6**, 861.

J. Frenkel (1931). *Phys. Rev.*, **37**, 17, 1276.

R. A. Friedel and M. Orchin (1951). *Ultraviolet Spectra of Aromatic Compounds*, John Wiley and Sons, New York.

D. Fröhlich and H. Mahr (1966). *Phys. Rev. Letters*, **16**, 895.

D. C. Frost, C. A. McDowell and D. A. Vroom (1967a). *Proc. Roy. Soc. (London)*, **296A**, 566.

D. C. Frost, C. A. McDowell and D. A. Vroom (1967b). *Can. J. Chem.*, **45**, 1343.

D. C. Frost, C. A. McDowell and D. A. Vroom (1967c). *J. Chem. Phys.*, **46**, 4255.

V. Fuchs and A. Niehaus (1968). *Phys. Rev. Letters*, **21**, 1136.

G. F. J. Garlick (1949). *Luminescent Materials*, Oxford Clarendon Press, London, pp. 201–237.

W. R. S. Garton, (1968). quoted in Fano and Cooper.

W. R. S. Garton (1966). *Advan. Atomic Mol. Phys.*, **2**, 93.

P. A. Geldof, R. P. H. Rettschnick and G. J. Hoytink (1969). *Chem. Phys. Letters*, **4**, 59.

F. R. Gilmore, (1965). *J. Quant. Spectry. Rad. Trans.*, **5**, 369.

J. L. Hall and M. W. Siegel (1968). *J. Chem. Phys.*, **48**, 943.

J. L. Hall, D. A. Jennings and R. M. McClintock (1963). *Phys. Rev. Letters*, **11**, 364.

D. S. Hamilton (1957). *Proc. Phys. Soc.*, *B***70**, 144.

D. S. Hamilton (1961). *Proc. Phys. Soc.*, *B***78**, 743.

G. S. Hammond, N. J. Turro and P. A. Leermakers (1962). *J. Phys. Chem.*, **66**, 1144.

H. J. Henning (1932). *Ann. Phys.*, **13**, 599.

R. J. W. Henry (1967). *Planet Space Sci.*, **15**, 1747.

R. J. W. Henry (1968). *J. Chem. Phys.*, **48**, 3635.

J. P. Hernandez (1968). *Phys. Rev.*, **167**, 108.

S. Heron, R. W. P. McWhirter and E. H. Rhoderick (1956). *Proc. Roy. Soc.*, **234A**, 565.

G. Herzberg (1950). *Molecular Spectra and Molecular Structure. I. Spectra of Diatomic Molecules*, 2nd ed., D. Van Nostrand Co., New York.

G. Herzberg (1956). *Molecular Spectra and Molecular Structure. II. Infrared and Raman Spectra of Polyatomic Molecules*, D. Van Nostrand Co., New York.

G. Herzberg (1966). *Molecular Spectra and Molecular Structure, III. Electronic Spectra and Electronic Structure of Polyatomic Molecules*, D. Van Nostrand Co., New York.

F. Hirayama (1965). *J. Chem. Phys.*, **42**, 3163.

F. Hirayama and S. Lipsky (1969). *J. Chem. Phys.*, **51**, 3616.

T. Holstein (1947). *Phys. Rev.*, **72**, 1212.

R. Honig (1962). *R. C. A. Rev.*, **23**, 567.

J. J. Hopfield, J. M. Worlock, K. Park (1963). *Phys. Rev. Letters*, **11**, 414.

J. J. Hopfield (1930). *Astrophys. J.*, **72**, 133.

A. R. Horrocks and F. Wilkinson (1968). *Proc. Roy. Soc.*, **306A**, 257.

R. D. Hudson (1964). *Phys. Rev.*, **135A**, 1212.

R. D. Hudson and V. L. Carter (1967). *J. Opt. Soc. Am.*, **57**, 651.

R. D. Hudson and V. L. Carter (1965a). *Phys. Rev.*, **137A**, 1648.

R. D. Hudson and V. L. Carter (1965b). *Phys. Rev.*, **139A**, 1426.

R. E. Huffman, Y. Tanaka and J. C. Larrabee (1963a). *Appl. Opt.*, **2**, 617.

R. E. Huffman, Y. Tanaka and J. C. Larrabee, (1963d). *Appl. Opt.*, **2**, 947.

R. E. Huffman, Y. Tanaka and J. C. Larrabee (1963c). *J. Chem. Phys.*, **39**, 902.

R. E. Huffman, Y. Tanaka and J. C. Larrabee (1963b). *J. Chem. Phys.*, **39**, 910.

R. E. Huffman, Y. Tanaka and J. C. Larrabee (1963e). *J. Chem. Phys.*, **38**, 1420.

R. E. Huffman, Y. Tanaka and J. C. Larrabee (1962). *J. Opt. Soc. Am.*, **52**, 851.

R. E. Huffman, J. C. Larrabee and Y. Tanaka (1965). *Appl. Opt.*, **4**, 1581.

R. E. Huffman, J. C. Larrabee and Y. Tanaka (1967). *J. Chem. Phys.*, **46**, 2213.

G. S. Hurst, T. E. Bortner and T. D. Strickler (1968). *J. Chem. Phys.*, **49**, 2460.

G. S. Hurst, T. E. Bortner and T. D. Strickler (1969). *Phys. Rev.*, **178**, 4.

R. E. Imhof and F. H. Read (1969). *Chem. Phys. Letters*, **3**, 652.

G. Jackson, R. Livingston and A. C. Pugh (1960). *Trans. Faraday Soc.*, **56**, 1635.

H. H. Jaffé and M. Orchin (1964). *Theory and Applications of Ultraviolet Spectroscopy*, John Wiley and Sons, New York.

J. F. James (1956). In E. B. Armstrong and A. Dalgarno (Eds.), *The Airglow and the Aurorae*, Pergamon Press, London, p. 273.

P. M. Johnson and S. A. Rice (1968). *Chem. Phys.*, **49**, 2734.

W. Kaiser and C. G. B. Garrett (1961). *Phys. Rev. Letters*, **7**, 229.

M. Kasha (1950). *Discussions Faraday Soc.*, **9**, 14.

M. Kasha (1961). In W. D. McElroy and B. Glass (Eds.), *Light and Life*, The Johns Hopkins Press, Baltimore, pp. 31–68.

M. Kasha (1960). *Radiation Res. Suppl.*, **2**, 243.

M. Kasha (1968). Quoted by J. E. Turner and R. W. Wood, *Phys. Today*, **21**, No. 11, 107.

M. Kasha (1959). *Rev. Mod. Phys.*, **31**, 162.

M. Kasha and S. P. McGlynn (1956). *Ann. Rev. Phys. Chem.*, **7**, 403.

W. Kauzmann (1957). *Quantum Chemistry*, Academic Press, New York.

K. Kawaoka and D. R. Kearns (1966). *J. Chem. Phys.*, **45**, 147.

R. G. Kepler, J. C. Caris, P. Avakian and E. Abramson (1963). *Phys. Rev. Letters*, **10**, 400.

Q. C. Kessel, M. P. McCaughey and E. Everhart (1966). *Phys. Rev. Letters*, **16**, 1189.

T. A. King and R. Voltz (1966). *Proc. Roy. Soc.* (*London*), **289A**, 424.

D. A. Kleinman (1962). *Phys. Rev.*, **125**, 87.

J. Z. Klose (1966). *Phys. Rev.*, **141**, 181.

J. Z. Klose (1968). *J. Opt. Soc. Am.*, **58**, 1509.

E. Konijnenberg (1963). Thesis, University of Amsterdam.

M. Kotani, K. Ohno and K. Kayama (1961). In S. Flügge (Ed.), *Encyclopedia of Physics*, Vol. 37 (part 2), Springer-Verlag, Berlin, pp. 1–172.

M. O. Krause (1967). 15th Annual Conference on Mass Spectometry and Allied Topics, Denver, Colorado, 1967. See also Oak Ridge National Laboratory Report ORNL-P-3204.

M. O. Krause, T. A. Carlson and R. D. Dismukes (1968). *Phys. Rev.*, **170**, 37.

R. de L. Kronig (1928). *Z. Physik*, **50**, 347.

J. L. Kropp and M. Burton (1962). *J. Chem. Phys.*, **37**, 1752.

W. Kuhn (1925). *Z. Physik*, **33**, 408.

R. Ladenburg (1921). *Z. Physik*, **4**, 451.

J. Langelaar (1969). Ph. D. thesis, Amsterdam.

J. Langelaar, R. P. H. Rettschnick, A. M. F. Lambooy and G. J. Hoytink (1968). *Chem. Phys. Letters*, **1**, 609.

J. D. Laposa, E. C. Lim and R. E. Kellogg (1965). *J. Chem. Phys.*, **42**, 3025.

P. Lee and G. L. Weissler (1955). *Phys. Rev.*, **99**, 540.

J. S. Levinger (1953). *Phys. Rev.*, **90**, 11.

G. N. Lewis and M. Kasha (1944). *J. Am. Chem. Soc.*, **66**, 2100.

G. N. Lewis and M. Kasha (1945). *J. Am. Chem. Soc.*, **67**, 694.

G. N. Lewis, D. Lipkin and T. T. Magel (1941). *J. Am. Chem. Soc.*, **63**, 3005.

E. C. Lim and S. K. Chakrabarti (1967). *Mol. Phys.*, **13**, 293.

H. Linschitz and L. Pekkarinen (1960). *J. Am. Chem. Soc.*, **82**, 2411.

A. Linschitz and R. Sarkanen (1958). *J. Am. Chem. Soc.*, **80**, 4826.

R. Livingston (1957). *J. Phys. Chem.*, **61**, 860.

J. F. Lowry and D. H. Tomboulian (1963). Tech. Rept. 3 ARO(D) Project No. 2180p.

J. F. Lowry, D. H. Tomboulian and D. L. Ederer (1965). *Phys. Rev.*, **137A**, 1054.

R. P. Madden and K. Codling (1965). *Astrophys. J.*, **141**, 364.

R. P. Madden and K. Codling (1966). In A. Temkin (Ed.), *Autoionization* (*Astrophysical, Theoretical, and Laboratory Experimental Aspects*), MonoBook Corp., Baltimore, pp. 129–151.

R. P. Madden and K. Codling (1963). *Phys. Rev. Letters*, **10**, 516.

R. P. Madden, D. L. Ederer and K. Codling (1967). *Appl. Opt.*, **6**, 31.

R. P. Madden, D. L. Ederer and K. Codling (1969). *Phys. Rev.*, **177**, 136.

G. V. Marr (1967). *Photoionization Processes in Gases*, Academic Press, New York.

G. V. Marr (1963). *Proc. Phys. Soc.*, **81**, 9.

G. V. Marr and D. M. Creek (1968). *Proc. Roy. Soc.*, **A304**, 233.

N. Mataga, K. Ezumi and T. Okada (1966). *Mol. Phys.*, **10**, 201.

N. Mataga, T. Okada and K. Ezumi (1966). *Mol. Phys.*, **10**, 203.

F. M. Matsunaga, R. S. Jackson and K. Watanabe (1965). *J. Quant. Spectr. Radiative Transfer*, **5**, 329.

E. W. McDaniel (1964). *Collision Phenomena in Ionized Gases*, John Wiley and Sons, New York.

M. R. C. McDowell and G. Peach (1961). *Phys. Rev.*, **121**, 1383.

D. H. McMahon, R. A. Soref and A. R. Franklin (1965). *Phys. Rev. Letters*, **14**, 1060.

T. Medinger and F. Wilkinson (1965). *Trans. Faraday Soc.*, **61**, 620.

P. H. Metzger and G. R. Cook (1964). *J. Chem. Phys.*, **41**, 642.

A. C. G. Mitchell and M. W. Zemansky (1961). *Resonance Radiation and Excited Atoms*, Cambridge University Press, London.

F. L. Mohler and C. J. Boeckner (1929). *J. Res. Nat. Bur. Stand.*, **3**, 303.

G. F. Moore and I. H. Munro (1967). *Spectrochimica Acta*, **23A**, 1291.

B. Muel (1962). *C. R. Acad. Sci. Paris*, **255**, 3149.

R. S. Mulliken (1939). *J. Chem. Phys.*, **7**, 14.

R. S. Mulliken (1952). Record of Chemical Progress, Summer 1952, p. 67.

J. N. Murrell and J. Tanaka (1964). *Mol. Phys.*, **7**, 363.

K. R. Naqvi (1967). *Chem. Phys. Letters*, **1**, 497.

R. G. Newburgh, L. Heroux and H. E. Hinteregger (1962). *Appl. Opt.* **1**, 733.

N. F. Neznaiko, I. E. Obyknovennaya and A. S. Cherkasov (1966). *Opt. Spectr.*, **21**, 23; **21**, 285.

L. L. Nichols and W. Vali (1968). *J. Chem. Phys.*, **49**, 814.

G. K. Oster and H. Kallmann (1966). In N. Riehl and H. Kallmann (Eds.), International Symposium on Luminescence; the Physics and Chemistry of Scintillators, 1965, V. K. Thiemig, Munich, p. 31.

G. K. Oster and H. Kallmann (1962). *Nature*, **194**, 1033.

R. Pariser (1956). *J. Chem. Phys.*, **24**, 250.

C. A. Parker (1963). *Nature*, **200**, 331.

C. A. Parker (1966). *Spectrochim. Acta*, **22**, 1677.

C. A. Parker (1964). *Trans. Faraday Soc.*, **60**, 1998.

C. A. Parker (1967). *The Triplet State*, Cambridge University Press, London, p. 353. (Symposium on triplet state, Amer. Univ. of Beirut, Lebanon, 1967).

C. A. Parker (1968). *Photoluminescence of Solutions*, Elsevier Publishing Co., London.

C. A. Parker and C. G. Hatchard (1962a). *Proc. Roy. Soc. (London)*, **A269**, 574.

C. A. Parker and C. G. Hatchard (1962b). *Proc. Chem. Soc.*, 147.

C. A. Parker and C. G. Hatchard (1961). *Trans. Faraday Soc.*, **57**, 1894.

C. A. Parker and C. G. Hatchard (1963). *Trans. Faraday Soc.*, **59**, 284.

G. Peach, (1967). *Mem. Roy. Astron. Soc.*, **71** (Part I), 13.

J. R. Penton and E. E. Muschlitz, Jr. (1968). *J. Chem. Phys.*, **49**, 5083.

M. F. Perrin (1931). *Fluorescence*, Hermann, Paris.

M. F. Perrin (1926). *J. Phys. Radium*, **7**, 390.

J. C. Person (1965). *J. Chem. Phys.*, **43**, 2553.

J. C. Person and P. P. Nicole (1967). Argonne National Laboratory Report, ANL-7360, p. 44.

J. C. Person and P. P. Nicole (1968). *J. Chem. Phys.*, **49**, 5421.

W. L. Peticolas, J. P. Goldsborough and K. E. Rieckhoff (1963). *Phys. Rev. Letters*, **10**, 43.

W. L. Peticolas, R. Norris and K. E. Rieckhoff (1965). *J. Chem. Phys.*, **42**, 4164.

W. L. Peticolas and K. E. Rieckhoff (1963). *J. Chem. Phys.*, **39**, 1347.

K. R. Piech and J. S. Levinger (1964). *Phys. Rev.*, **135**, A332.

J. R. Platt (1961). In S. Flügge (Ed.), Vol. 37 (part 2), Springer-Verlag, Berlin, pp. 173–281.

J. R. Platt (1949). *J. Chem. Phys.*, **17**, 484.

J. R. Platt and coworkers (1964a). *Systematics of the Electronic Spectra of Conjugated Molecules: a source book*, John Wiley and Sons, New York.

J. R. Platt, K. Ruedenberg, C. W. Scherr, N. S. Ham, H. Labhart and W. Lichten (1964b). *Free-Electron Theory of Conjugated Molecules: a source book*, John Wiley and Sons, New York.

G. Porter (1961). In W. D. McElroy and B. Glass (Eds), *Light and Life*, The Johns Hopkins Press, Baltimore, pp. 69–77.

G. Porter and F. Wilkinson (1961). *Proc. Roy. Soc. (London)*, **A264**, 1.
G. Porter and M. Windsor (1954). *Discussions Faraday Soc.*, **17**, 178.
G. Porter and M. Windsor (1953). *J. Chem. Phys.*, **21**, 2088.
G. Porter and M. Windsor (1958). *Proc. Roy. Soc.*, **A245**, 238.
G. Porter and M. R. Wright (1959). *Discussions Faraday Soc.*, **27**, 18.
P. Pringsheim (1949). *Fluorescence and Phosphorescence*, Interscience Publishers, New York.
J. D. Roberts (1962). *Notes on Molecular Orbital Calculations*, W. A. Benjamin, New York.
G. W. Robinson (1961). In W. D. McElroy and B. Glass (Eds.), *Light and Life*, Johns Hopkins Press, Baltimore, pp. 11–30.
M. E. Rudd (1964). *Phys. Rev. Letters*, **13**, 503.
M. E. Rudd (1965). *Phys. Rev. Letters*, **15**, 580.
M. E. Rudd, T. Jorgensen, Jr. and D. J. Volz (1966a). *Phys. Rev. Letters*, **16**, 929.
M. E. Rudd, T. Jorgensen, Jr. and D. J. Volz (1966b). *Phys. Rev.*, **151**, 28.
M. E. Rudd and D. V. Lang (1965). In *Proceedings of the IVth International Conference on the Physics of Electronic and Atomic Collisions*, Hastings-on-Hudson, Science Bookcrafters, New York, p. 153.
M. E. Rudd and K. Smith (1968). *Phys. Rev.*, **169**, 79.
O. P. Rustgi (1964). *J. Opt. Soc. Am.*, **54**, 464.
E. E. Salpeter and M. H. Zaidi (1962). *Phys. Rev.*, **125**, 248.
J. A. R. Samson (1964a). *J. Opt. Soc. Am.*, **54**, 6.
J. A. R. Samson (1964b). *J. Opt. Soc. Am.*, **54**, 420.
J. A. R. Samson (1965). *J. Opt. Soc. Am.*, **55**, 935.
J. A. R. Samson (1967). *Appl. Opt.*, **6**, 403.
J. A. R. Samson and R. B. Cairns (1965). *J. Opt. Soc. Am.*, **55**, 1035.
J. A. R. Samson and R. B. Cairns (1968). *Phys. Rev.*, **173**, 80.
J. A. R. Samson and F. L. Kelly (1964). G. C. A. Tech. Rept. No. 64-3-N-(1964)
K. Sandros and H. L. J. Bäckström (1962). *Acta Chem. Scand.*, **16**, 958.
R. Schoen (1964). *J. Chem. Phys.*, **40**, 1830.
M. J. Seaton (1959). *Phys. Rev.*, **113**, 814.
B. K. Selinger (1964). *Nature*, **203**, 1062.
K. G. Sewell (1965). *Phys. Rev.*, **138**, A418.
P. Seybold and M. Gouterman (1965). *Chem. Revs.*, **65**, 413.
S. M. Silverman and E. N. Lassettre (1964). *J. Chem. Phys.*, **40**, 1265.
W. T. Simpson and D. L. Peterson (1957). *J. Chem. Phys.*, **26**, 588.
S. Singh, W. J. Jones, W. Siebrand, B. P. Stoicheff and W. G. Schneider (1965). *J. Chem. Phys.*, **42**, 330.
S. Singh and B. P. Stoicheff (1963). *J. Chem. Phys.*, **38**, 2032.
Smith-Saville (1968). Ph.D. thesis, University of Manchester.
H. Sponer (1959). *Radiation Res. Suppl.*, **1**, 558.
B. Stevens (1961). *Nature*, **192**, 725.
B. Stevens and M. I. Ban (1968). *Mol. Cryst.*, **4**, 173.
B. Stevens and J. T. Dubois (1963). *Trans. Faraday Soc.*, **59**, 2813.
B. Stevens, E. Hutton and G. Porter (1960). *Nature*, **185**, 917.
B. Stevens and M. S. Walker (1963). *Proc. Chem. Soc. (London)*, 181.
A. L. Stewart and T. G. Webb (1963). *Proc. Phys., Soc.*, **82**, 532.
A. Streitwieser, Jr. (1961). *Molecular Orbital Theory for Organic Chemists*, John Wiley and Sons, New York.
T. D. Strickler and E. T. Arakawa (1964). *J. Chem. Phys.*, **41**, 1783.
S. J. Strickler and R. A. Berg (1962). *J. Chem. Phys.*, **37**, 814.
H. H. Stroke (1966). *Phys. Today*, **19** (No. 10), 55.

L. Stryer (1960). *Radiation Res. Suppl.*, **2**, 432.
C. Tanaka, J. Tanaka, E. Hutton and B. Stevens (1963). *Nature*, **198**, 1192.
Y. Tanaka (1955). *J. Opt. Soc. Am.*, **45**, 710.
Y. Tanaka, A. S. Jursa and F. J. LeBlanc (1958). *J. Opt. Soc. Am.*, **48**, 304.
A. Temkin (Ed.) (1966). *Autoionization (Astrophysical, Theoretical and Laboratory Experimental Aspects)*, Mono Book Corp., Baltimore.
A. Terenin and V. L. Ermolaev (1956). *Trans. Faraday Soc.*, **52**, 1042.
W. Thomas (1925). *Naturwissenschaften*, **13**, 627.
R. C. Tolman (1924). *Phys. Rev.*, **23**, 693.
D. H. Tomboulian and P. L. Hartman (1956). *Phys. Rev.*, **102**, 1423.
R. Tousey (1962). *Appl. Opt.*, **1**, 679.
J. C. Tully, R. S. Berry and B. J. Dalton (1968). *Phys. Rev.*, **176**, 95.
D. W. Turner (1962). *J. Chem. Phys.*, **37**, 3007.
D. W. Turner (1968). *Chem. Brit.*, **4**, 435.
D. W. Turner and D. P. May (1966). *J. Chem. Phys.*, **45**, 471.
M. T. Vala Jr., I. H. Hillier, S. A. Rice and J. Jortner (1966). *J. Chem. Phys.*, **44**, 23.
B. L. Van Duuren (1963). *Chem. Rev.*, **63**, 325.
F. I. Vilesov (1963). *Usp. Fiz. Nauk.*, **81**, 669. (*Soviet Phys. Usp.*, **6**, 888 (1964)).
F. I. Vilesov, B. L. Kurbatov and A. N. Terenin (1961). *Soviet Phys. Doklady*, **6**, 490; **6**, 883.
D. Villarejo, R. Stockbauer and M. G. Inghram (1968). *J. Chem. Phys.*, **48**, 3342.
O. E. Wagner, L. G. Christophorou and J. G. Carter (1969). *Chem. Phys. Letters*, **4**, 224.
M. S. Walker, T. W. Bednar and R. Lumry (1966). *J. Chem. Phys.*, **45**, 3455.
M. S. Walker, T. W. Bednar and R. Lumry (1967). *J. Chem. Phys.*, **47**, 1020.
W. R. Ware (1962). *J. Phys. Chem.*, **66**, 455.
W. R. Ware and B. A. Baldwin (1964). *J. Chem. Phys.*, **40**, 1703.
K. Watanabe (1958). *Advan. Geophys.*, **5**, 157.
G. Weber and F. W. J. Teale (1958). *Trans. Faraday Soc.*, **54**, 640.
G. L. Weissler (1956). In S. Flügge (Ed.), *Handbuch der Physik*, Vol. XXI, Springer-Verlag, Berlin, pp. 304–382.
G. L. Weissler (1962). *J. Quant. Spectry. Radiative Trans.*, **2**, 383.
S. Z. Weisz, A. B. Zahlan, J. Gilreath, R. C. Jarnagin and M. Silvers (1964). *J. Chem. Phys.*, **41**, 3491.
W. L. Wiese, M. W. Smith and B. M. Glennon (1966). *Atomic Transition Probabilities*, Vol. 1, Nat. Bur. Std. NSRDS-NBS4 (Washington: U. S. Govt. Print. Off.).
F. Wilkinson (1964). In W. A. Noyes, Jr., G. Hammond and J. N. Pitts, Jr. (Eds), *Advances in Photochemistry*, Vol. 3, Interscience Publishers, New York, pp. 241–268.
P. G. Wilkinson and E. T. Byram (1965). *Appl. Opt.*, **4**, 581.
P. G. Wilkinson and R. S. Mulliken (1957). *Astrophys. J.*, **125**, 594.
R. Williams (1958). *J. Chem. Phys.*, **28**, 577.
R. B. Withrow and A. P. Withrow (1956). In A. Hollaender (Ed.), *Radiation Biology*, Vol. III, McGraw-Hill Book Co., New York, pp. 125–258.

# 4 Elastic scattering of low-energy electrons

## 4.1 Basic Background

Electron transport in gaseous media is of great theoretical and practical interest. It provides basic information on the scattering potentials and the electronic structures of the atoms and molecules which determine them. Radiation dosimetry, stopping power calculations, plasma and upper atmosphere physics, and the wide field of gaseous electronics, find unique interest in electron transport studies, especially in the low-energy region involving subionization and subexcitation electrons. Subexcitation electrons have kinetic energies below the first electronic excitation potential of the medium through which they travel. Such electrons lose energy through elastic collisions, and through inelastic collisions leading to vibrational and rotational excitation. Subexcitation electrons may lose energy also through the various types of electron attachment processes (Chapter 6), and through the excitation of compound negative ion states. The latter are unstable with respect to autoionization and can decay purely elastically, give vibrationally excited molecules or lead to dissociative attachment, or all three processes may be in competition viz.

$$e + AX \overset{\sigma_0}{\to} AX^{-*} \underset{p_{da}}{\overset{p_e}{\underset{p_{in}}{\to}}} \begin{array}{l} AX + e \text{ elastic (a)} \\ AX^* + e' \text{ inelastic (b)} \\ A(A^*) + X^- \text{ dissociative attachment (c)} \end{array} \qquad (4.1)$$

where the asterisk indicates excitation energy and the prime notation $(e')$ indicates that the scattered electron does not have the same energy as the incident one; $p_e$, $p_{in}$ and $p_{da}$ are, respectively, the probability for elastic scattering, inelastic scattering and dissociative attachment. If the auto-ionization lifetime of $AX^{-*}$ is long compared to the time between collisions, a parent negative ion $AX^-$ may be formed when the electron affinity of $AX$ is positive. Christophorou and Stockdale (1968) discussed dissociative electron attachment to polyatomic molecules (Chapter 6) and pointed out that the probability for channel (b) increases greatly when the compound negative ion state $AX^{-*}$ lies at or above the lowest excited electronic state of the neutral molecule yielding upon decay electronically excited $AX^*$ molecules. This, as will be discussed in Chapter 6 results in increased auto-ionization and thus in a decrease in the cross section for channel (c) (process 4.1). It will be shown later that the decay of the transient negative ion to vibrationally excited $AX^*$ molecules provides a very efficient mechanism for the otherwise unlikely process of *direct* vibrational excitation of molecules by slow electrons. Thus in many cases a subexcitation electron does not have as hard a time as has been implied by some authors in losing its energy and attaining thermal equilibrium with the gaseous medium.

Knowledge of the cross sections for the above processes (expression 4.1) for various molecules as a function of electron energy is of primary interest in radiation physics. Processes (b) and (c) will be discussed in subsequent chapters. In this chapter we shall focus attention on the elastic electron scattering process which proceeds either directly or via an intermediate $AX^{-*}$ (channel (a)) and which is especially important at subexcitation energies.

### 4.1.1  *Types of Collisions*

The collisions of electrons with atoms and molecules can be divided into three classes: elastic, inelastic and superelastic.

### 4.1.1.1  Elastic collisions

These are collisions in which the electron loses only that amount of energy which is necessary for the conservation of momentum in the collision process. In such a collision, therefore, only kinetic energy between the impinging electron and the target is exchanged, the internal state of the target remaining unchanged.

Let us now make the simple assumption that both the electron and the molecule can be represented by smooth elastic spheres. Although the picture of hard smooth spheres of constant cross sectional area is inappropriate—more so in electron–molecule than in molecule–molecule collisions since in the former case the interaction is a strong function of the electron energy; the same molecule is seen differently by a slow and a fast electron — still it might be instructive to make this assumption and calculate classically the average energy loss in such a collision. Conservation of energy and momentum yields for the fraction of energy lost $\Delta(\Theta)\left(\equiv \dfrac{\Delta E}{E}\right)$

$$\Delta(\Theta) = \frac{2mM}{(m+M)^2}\,(1 - \cos\Theta) \tag{4.2}$$

where $\Theta$ is the common scattering angle in the centre of mass (CM) system. For electron–atom and electron–molecule collisions $m/M \ll 1$ and expression (4.2) can be approximated by

$$\Delta(\Theta) \simeq \frac{2m}{M} (1 - \cos \Theta) \tag{4.3}$$

Further, for $m \ll M$ the angle $\Theta$ in the CM system is about equal to the angle $\Theta$ in the laboratory (LAB) system and the CM and LAB systems are almost identical. Even for the lightest atoms $M \gg m$ and the average fractional energy loss is very small. For example, $(\Delta(\Theta))_{max}$ is $\sim 5.5 \times 10^{-4}$ and $\sim 5.5 \times 10^{-5}$ for the helium and the argon atoms, respectively. Thus, the average energy transferred to the target molecule in an elastic electron–molecule collision is small. However, the electron may transfer considerable momentum in making large angle scattering. The transfer of momentum, then, is the focal point when considering elastic scattering of electrons.

The simple expressions (4.2) and (4.3) are very helpful when performing electron swarm experiments. They provide an estimate of the relaxation time, i.e. the time it takes for a swarm of electrons to reach a steady-state condition in a gas under a uniform electric field, in the absence of inelastic collisions (see discussion later in this chapter).

### 4.1.1.2 Inelastic collisions

In an inelastic collision the target (atom or molecule) gains internal energy at the expense of the kinetic energy of the impinging electron. The impact results in excitation or ionization of the target, and the amount of energy lost by the electron is determined by the quantum states of the target. These collisions will be discussed in Chapter 5. If in an inelastic collision the whole or part of the energy lost by the electron is emitted as electromagnetic radiation, the collision is called *radiative*. Such collisions, involving emission of radiation generally have small cross sections, but they may play an important role in certain phenomena such as radiative electron attachment (Chapter 6).

### 4.1.1.3 Superelastic collisions

These are collisions in which internal energy of the target atom or molecule is imparted to the electron resulting in an increase of its kinetic energy. Collisions of this type are important at low-electron energies (close to thermal) and in collisions of electrons with metastable atoms and molecules. They require the target atom or molecule to possess energy in excess of that of its lowest energy configuration.

### 4.1.2 *Collision Cross Sections*

A collision is said to have taken place between two particles when any physical change can be detected after the distance between the two particles has first been decreased and then increased. The probability that a given type of collision (reaction) will occur under given conditions is usually expressed in terms of the collision cross section. A brief review of various collision cross sections will now be made.

#### 4.1.2.1 Microscopic elastic scattering cross section

Let $N_e$ be the number of electrons per cm$^2$ per sec in a parallel mono-energetic beam directed along the $z$ axis (Figure 4.1) and $N_t$ be the number of target atoms at rest around the origin $O$. If the number $N_t$ of the target particles is sufficiently small, none is shielded by the others and no electron is scattered more than once. This is a condition characteristic of typical single-collision beam experiments. The number $N_s(\theta, \phi)$ of elastically

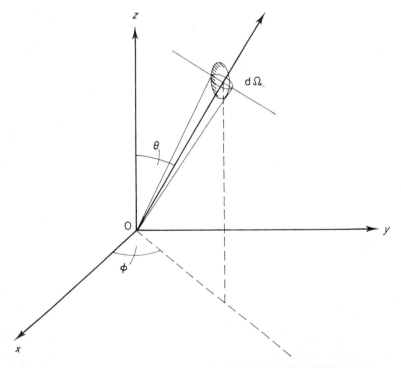

**Figure 4.1** Laboratory spherical polar and Cartesian coordinate system. The projectile approaches the target (at the origin) along the negative $z$ axis

scattered electrons per sec into the solid angle $d\Omega$ (Figure 4.1) is

$$N_s(\theta, \phi) \, d\Omega = I_s(\theta, \phi) N_e N_t \, d\Omega \qquad (4.4)$$

or

$$N_s(\theta, \phi) \, d\Omega = N_e N_t \, dI_s(\theta, \phi)$$

where $I_s(\theta, \phi) \, d\Omega \equiv dI_s(\theta, \phi)$ is the *differential microscopic elastic scattering cross section* in the LAB system and may be considered as the area presented by each target particle for scattering of the electrons into the element of solid angle $d\Omega_{LAB}$. The *total microscopic elastic scattering cross section* $\sigma_e$ can now be defined as

$$\sigma_e = \int dI_s(\theta, \phi) = \int_0^{\pi} \int_0^{2\pi} I_s(\theta, \phi) \sin\theta \, d\theta \, d\phi \qquad (4.5)$$

$\sigma_e$ represents the area presented by each of the target particles for scattering into the solid angle $4\pi$-steradians and is expressed in $cm^2$ or in

$$\pi a_0^2 = 0.88 \times 10^{-16} \, cm^2 \quad (a_0 = \text{Bohr radius}).$$

### 4.1.2.2 Momentum transfer cross section

The cross section, $\sigma_m$, for transfer of momentum to a target by an electron scattered at an angle $\theta$ from its original direction (Figure 4.1) is

$$\sigma_m = \int_0^{\pi} \int_0^{2\pi} I_s(\theta, \phi)(1 - \cos\theta) \sin\theta \, d\theta \, d\phi \qquad (4.6)$$

Equation (4.6) is the same as Equation (4.5) but with a weighting factor $(1 - \cos\theta)$. $\sigma_m$ is called the *momentum transfer or diffusion cross section*. In view of Equation (4.6), we may write for the mean fractional energy lost by the electron in an elastic collision

$$\langle \Delta(\theta) \rangle \simeq \frac{2m}{M} \int_0^{\pi} \int_0^{2\pi} (1 - \cos\theta) \sin\theta \, p(\theta) \, d\theta \, d\phi = \frac{2m}{M} \frac{\sigma_m}{\sigma_e} \qquad (4.7)$$

where $p(\theta)$ is the probability of scattering into the solid angle $d\Omega (= \sin\theta \, d\theta \, d\phi)$ and is equal to

$$\frac{I_s(\theta)}{\sigma_e}$$

The momentum transfer cross section is useful in discussions of diffusion of neutral and charged particles in gases, and the mobility of gaseous ions. As will be discussed later in this chapter, when a large number of electrons is swarming about in a gas and both the electrons and the gas atoms (or molecules) have sufficiently low energies so that internal excitation of the target gas is not pronounced, one is dealing with mostly elastic collisions and use is made of Equation (4.6) for the cross section. Small angle scattering produces only a small contribution to $\sigma_m$.

The momentum transfer cross section $\sigma_m$ differs appreciably from the total microscopic elastic scattering cross section $\sigma_e$ for distinctly anisotropic scattering. If $I_s(\theta, \phi)$ is independent of $\theta$, $\sigma_m = \sigma_e$. If a forward scattering dominates $\sigma_m < \sigma_e$; when a backward scattering predominates $\sigma_m > \sigma_e$. Both experiment and calculations show that electrons are not scattered isotropically and that large angle scattering of slow electrons is important.

### 4.1.2.3 Macroscopic total scattering cross section

Consider $N$ stationary target particles per $cm^3$ and an incident mono-energetic beam of $I_0$ particles per $cm^2$ per sec. Let $I$ be the intensity of the unscattered component at depth $x$ and let $N$ be small enough that $I \simeq I_0$. Then

$$I = I_0 e^{-N\sigma_T x} = I_0 e^{-Q_T x} \tag{4.8}$$

where $Q_T = N\sigma_T$ is the *total macroscopic or bulk scattering cross section* expressed in $cm^{-1}$. Note that $\sigma_T$ is expressed in $cm^2$. The probability that a given electron will survive scattering until it penetrates the medium at least a depth $x$ is $e^{-Q_T x}$ while $Q_T\,dx = N\sigma_T\,dx$ is the probability that it will be scattered in travelling the distance from $x$ to $x+dx$. The average distance, $\bar{x}$, between the scattering sites of a given projectile is $1/Q_T$, and the *mean free path, L*, for (elastic) scattering is

$$L = \bar{x} = \frac{1}{Q_T} = \frac{1}{N\sigma_T} \tag{4.9}$$

### 4.1.2.4 Probability of collision, collision frequency, and collision mean free time.

The probability of collision $p$ is defined here as the average number of collisions (suffered by the electron) per unit path (cm), per unit pressure (torr) at $273°K$. Hence,

$$p = \frac{1}{L} = \sigma_T N = Q_T \tag{4.10}$$

where all quantities are for $273°K$ and 1 torr. At temperatures different from $T = 273°K$ use is made of the reduced pressure $P_0$ which, under ideal gas-law conditions, is

$$P_0 = P\frac{273}{T} \tag{4.11}$$

$P$ is the actual pressure at a temperature $T$. The collision frequency $v_c$ can be defined as

$$v_c = \frac{v}{L} = vQ_T \tag{4.12}$$

and the collision mean free time, $\tau$, i.e. the average time between successive collisions experienced by a single electron as $(v_c)^{-1}$.

### 4.1.2.5 Cross sections for reactions other than elastic scattering

Let $\sigma_T$ and $Q_T$ be, respectively, the microscopic and macroscopic cross sections for an electron to undergo some reaction regardless of type in traversing a gaseous medium. Then the gross (total )microscopic cross section is written as

$$\sigma_T = \sigma_e + \sigma_{ex} + \sigma_{ion} + \sigma_a + \sigma_{oth} \tag{4.13}$$

and the gross (total) macroscopic cross section as

$$Q_T = Q_e + Q_{ex} + Q_{ion} + Q_a + Q_{oth} \tag{4.14}$$

The subscripts e, ex, ion, a and oth characterize the particular types of collisions, namely, elastic, excitation, ionization, attachment, and other processes, respectively. For subexcitation electrons, $\sigma_{ion}$ is zero and $\sigma_{ex}$ refers only to vibrational and rotational excitation. Equation (4.14) may also be written as

$$Q_T = (p_e + p_{ex} + p_{ion} + p_a + p_{oth})Q_T \tag{4.15}$$

where the various $p$'s refer to the probabilities that the collisions will result in the particular processes denoted by the subscripts.

### 4.1.2.6 Collision (reaction) rates

Consider an assembly of particles, X, at a density $N_X$. The rate, $R_X$, at which the particle density is altered by collision (reaction) is given by

$$R_X = \frac{\partial N_X}{\partial t} = -v_X N_X \tag{4.16}$$

where $v_X$ is the collision (reaction) frequency. In case of an assembly composed of two types, X, Y, of particles, the rate $R_{X,Y}$ of X, Y collisions is

$$R_{X,Y} = \frac{\partial N_X}{\partial t} = -C_{X,Y} N_X N_Y \tag{4.17}$$

where $N_Y$ is the density of component Y and $C_{X,Y}$ is the two-body rate coefficient. The coefficient $C_{X,Y}$ is related to the collision (reaction) cross section $\sigma_{X,Y}$ for the process by

$$C_{X,Y} = \langle v_{X,Y}\sigma_{X,Y}\rangle \tag{4.18}$$

where $v_{X,Y}$ is the relative velocity of X and Y and the averaging is carried over the distribution in velocity of all particles X and Y. Thus, the reaction frequency, $v_{X,Y}$, for two-body reaction processes is

$$v_{X,Y} = C_{X,Y}N_Y = N_Y\langle v_{X,Y}\sigma_{X,Y}\rangle \tag{4.19}$$

For a three-component $(X, Y, Z)$ system, the three-body reaction rate is

$$R_{X,Y,Z} = \frac{\partial N_X}{\partial t} = -C_{X,Y,Z}N_X N_Y N_Z \tag{4.20}$$

where $C_{X,Y,Z}$ is called the three-body rate coefficient and is related to the reaction frequency of component X by

$$v_{X,Y,Z} = C_{X,Y,Z}N_Y N_Z \tag{4.21}$$

When the species X,Y,Z are involved in a collision, we may visualize the X,Y,Z collision as occurring in two steps, namely, a collision of X and Y to form a complex XY, which subsequently collides with Z. The rate $R_{XY,Z}$ at which XY collides with Z can be written as

$$R_{XY,Z} = \langle v_{X,Y}\sigma_{X,Y}\rangle N_X N_Y \frac{\langle v_{XY,Z}\sigma_{XY,Z}\rangle N_Z}{\langle v_{XY,Z}\sigma_{XY,Z}\rangle N_Z + \beta} \tag{4.22}$$

where $v_{X,Y}$ is the relative velocity of X and Y, $\sigma_{X,Y}$ is the cross section for the formation of XY (the averaging is taken over the distribution in velocity of X and Y), $v_{XY,Z}$ is the relative velocity of XY and Z and $\sigma_{XY,Z}$ is the corresponding cross section (the averaging is taken over all velocities of XY and Z), and $\beta$ is the rate of decomposition of XY: all in c.g.s. units. We can distinguish two extreme cases:

(i) $\beta \gg \langle v_{XY,Z}\sigma_{XY,Z}\rangle N_Z$. In this case the rate of decomposition of XY is much faster than the rate of XY,Z collisions and Equation (4.22) reduces to

$$R_{XY,Z} = \langle v_{X,Y}\sigma_{X,Y}\rangle \langle v_{XY,Z}\sigma_{XY,Z}\rangle \frac{N_X N_Y N_Z}{\beta} \tag{4.23}$$

(ii) $\beta \ll \langle v_{XY,Z}\sigma_{XY,Z}\rangle N_Z$. In this case Equation (4.22) reduces to the two-body reaction rate

$$R_{XY,Z} = \langle v_{X,Y}\sigma_{X,Y}\rangle N_X N_Y \tag{4.24}$$

Application of (i) or (ii) will depend principally on $\beta$ and $N_Z$. For uncharged particles, case (i) generally holds, but for electron–molecule interactions (e.g. for electron attachment processes where XY may be a temporary negative ion $(Y^{-*})$ with a long lifetime (small $\beta$)) special attention should be given to distinguishing between (i) and (ii) (Chapter 6).

### 4.1.3 *Scattering Potentials*

The magnitude of the cross section for scattering of a neutral or charged particle from an atom or a molecule is a measure of the interaction potential which, in turn, is determined by the details of the atomic and molecular structure. If the structure of the scattering target could be determined accurately by appropriate wave functions, solution of the Schrödinger equation would provide reliable values for the various cross sections. Thus, in the case of low-energy electrons $\sigma_m$ could be obtained and calculation of the electron transport parameters such as the diffusion coefficient $D$ and the electron drift velocity $w$ could be made. However, the scattering of slow electrons even from the simplest atoms is difficult to treat by solution of the Schrödinger equation. Even in the calculation of the cross section for elastic scattering of electrons from the hydrogen atom the Born approximation is often used, which leads to serious errors at low energies (see a discussion in Moiseiwitsch (1962)). In the case of low-energy electron scattering from more complex atoms (and more so for molecules) all problems encountered for the hydrogen atom are magnified by further approximations to the inter-action potential of the electron in the field of the atom or the molecule. In low-energy scattering problems the wave equation for an assumed potential function is solved usually by the method of partial waves (Section 4.1.4).

Many different potential functions are used to describe the interaction between various types of particles. Usually the theory of elastic scatering assumes central force fields, i.e. angle independent functions depending only on the distance between the centres of the interacting particles. Table 4.1 lists some of the most common forms of spherically symmetric potential functions. Other interaction potentials depend on the relative angular orientation of the particles as well as the separation distance. Potentials involving an explicit dependence on the relative particle velocity are occasionally required to describe certain types of collision processes. These are more difficult to handle mathematically. For a discussion of non-central force fields see Hirschfelder, Curtiss and Bird (1954).

An important expression representing the typical potential function of a diatomic molecule was proposed by Morse (1929):

$$V(r - r_e) = D_e[1 - e^{-\beta(r - r_e)}]^2 \tag{4.25}$$

where $r_e$ is the nuclear separation at which the potential minimum occurs, $D_e$ is the dissociation energy (including the zero-point energy), and $\beta$ is a constant expressed as (Herzberg, 1950)

$$\beta = \left(\frac{2\pi^2 c M_r}{D_e h}\right)^{1/2} \omega_e = 1.2177 \times 10^7 \omega_e \left(\frac{M_r}{D_e}\right)^{1/2} \tag{4.26}$$

**Table 4.1**  Some commonly used spherically symmetric potential functions[a]

| Name | Potential function | Explanation of symbols | Comments | Graphical representation |
|---|---|---|---|---|
| Smooth elastic spheres | $V(r) = \begin{cases} \infty & (r < D) \\ 0 & (r > D) \end{cases}$ | $r$ = separation distance <br> $D = r_1 + r_2$ | Model consists of two rigid spheres of radii $r_1$ and $r_2$ | |
| Point centres of attraction or repulsion | $V(r) = \begin{cases} -\dfrac{a}{r^n} & \text{(attraction)} \\[2mm] \dfrac{a}{r^n} & \text{(repulsion)} \end{cases}$ | $a$ = positive constant <br> $n$ = index of attraction or repulsion | When $n = 1$ this is the Coulomb potential. Potential functions of this kind, based on macroscopic parameters such as the static polarizability, $\alpha$, or the permanent electric dipole moment, $\mu$, are often used to describe the scattering of electrons or ions from neutral molecules | |
| Square well | $V(r) = \begin{cases} \infty & (r < D) \\ -V_0 & (D < r < D_0) \\ 0 & (r > D_0) \end{cases}$ | $D$, $D_0$, and $V_0$ (see figure in column 5) | This is a relatively easy to use potential. Here we have a short-range repulsion and a long-range attraction | |

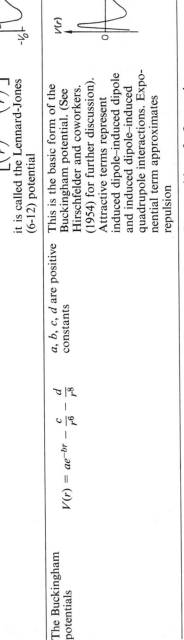

| | | |
|---|---|---|
| potential | $-\dfrac{a}{r^{n}}\ (r > D)$ | another according to an inverse power law. This potential gives for $n = 6$ the van der Waals interaction |
| The Lennard-Jones potential | $V(r) = \dfrac{a}{r^{m}} - \dfrac{b}{r^{n}}$<br><br>$a$ and $b$ are positive constants | When this potential takes the form $$V(r) = 4V_0\left[\left(\dfrac{D}{r}\right)^{12} - \left(\dfrac{D}{r}\right)^{6}\right]$$ it is called the Lennard-Jones (6-12) potential |
| The Buckingham potentials | $V(r) = ae^{-br} - \dfrac{c}{r^{6}} - \dfrac{d}{r^{8}}$<br><br>$a, b, c, d$ are positive constants | This is the basic form of the Buckingham potential. (See Hirschfelder and coworkers. (1954) for further discussion). Attractive terms represent induced dipole–induced dipole and induced dipole–induced quadrupole interactions. Exponential term approximates repulsion |
| The 12-6-4 potential | $V(r) = \dfrac{a}{r^{12}} - \dfrac{b}{r^{6}} - \dfrac{c}{r^{4}}$<br><br>$a, b, c$ are positive constants | Superposition of a Lennard-Jones (6-12) interaction and the interaction of a point charge with a polarizable molecule. Reasonable potential for interaction of an ion with a molecule |

[a] Based on a discursion given by McDaniel (1964)

The reduced mass $M_r$ in Equation (4.26) is expressed in atomic units and $D_e$ in $cm^{-1}$. The vibrational constant $\omega_e$ is related to the vibrational energy levels very approximately by (Herzberg, 1950)

$$E_v = hc\omega_e(v+\tfrac{1}{2}) - hc\omega_e\chi_e(v+\tfrac{1}{2})^2 + \text{smaller terms}$$

for vibrational quantum number $v$. Although the Morse function is of no direct interest in electron scattering, it is very important in aiding the understanding of certain processes which accompany low-energy electron–molecule collisions, electron attachment for example. The Morse function is consistent, but not uniquely so in that when substituted into the wave function it will yield the correct vibrational levels which are governed by the two constants $\omega_e$ and $\omega_e\chi_e$. Klein (1932) and Rydberg (1932, 1933) and more recently Schweinler (1967) gave a method for constructing the potential energy curve of a diatomic molecule, point by point, from the observed vibrational and rotational levels without assuming an analytical expression for it. The results are surprisingly close to the Morse function (e.g. for $H^{81}Br$ (Schweinler, 1967)).

The potential function which describes the interaction between an ion of charge $e$ and an atom of polarizability $\alpha$ is of the inverse fourth power form $(-\alpha e^2/2r^4)$; that between an ion and a molecule contains terms which depend on the relative orientation of the molecule. A discussion of the role of macroscopic potentials (based on the static polarizability, $\alpha$, and the permanent electric dipole moment, $\mu$) in the scattering of slow electrons from polar and non-polar but polarizable molecules will appear in Section 4.4. As will be seen, macroscopic potentials based on $\mu$ describe reasonably well the scattering of slow electrons from highly polar molecules. However, phenomena of interference, short-range exchange and overlap, and in general the details of the molecular structure need be considered for low $\mu(\to 0)$. That is, the target ceases to be structureless and rigid except when long-range forces dominate the scattering.

### 4.1.4 *Theory of Elastic Collisions*

In classical approaches to the scattering problem, one assumes a certain potential and determines the trajectories of each particle in terms of its initial velocity and the impact parameter. However, classical theory fails, in general, to give an accurate description of atomic collision processes, the complete understanding of which necessitates quantum calculations. On the other hand, the exact solution of the wave equation for more than two bodies is very difficult and for this reason a variety of approximate techniques are in use. It is essential to clearly understand the regions of validity of each of these approximations, most of which apply only to simple systems such as H and He. Indeed, much effort has been devoted both theoretically and experimentally to test the validity of the various quantum-mechanical approximations for the simplest of the atoms. For more complex atoms and for molecules, these techniques are difficult to apply. One, then, often has to seek guidance from more general approximations and semi-empirical rules. It is far from our purpose to discuss here the general theory of atomic collisions. The reader is referred to the following books and review articles for complete coverage of the various methods and approximations used in atomic collisions: Massey (1956); Schiff (1955); Landau and Lifschitz (1965); Bates (1962); McDaniel (1964); Hasted (1964); and Mott and Massey (1965).

A brief discussion of the method of partial waves as applied to atomic collisions will now be presented.

### 4.1.4.1 *The method of partial waves*

Let us assume a spherically symmetric force field and a monoenergetic, homogeneous and infinitely wide beam of structureless particles approaching along the $-z$ axis a structureless scattering centre fixed at the origin of the coordinates. The beam intensity is assumed to be uniform across any plane normal to the $z$ axis and constant with time. Let each incoming particle have a reduced mass $M_r = mM/(m+M)$; $m$ and $M$ are the masses of the projectile and target (assumed dissimilar), respectively. The incident beam of particles, travelling along the $z$ axis with a velocity $v_0$, is taken to be a plane de Broglie wave of wavelength $\lambda$ described by the time-independent wave function

$$\psi_{\text{inc}} = C e^{ikz}$$

where $k = 2\pi/\lambda = M_r v_0/\hbar$ and $C$ is the amplitude of the wave. The wave function for the scattered wave is presumed to have the asymptotic form,

$$\psi_{\text{sc}} \simeq \frac{C}{r} f(\Theta) e^{ikr}$$

where $f(\Theta)$ is usually called the scattered amplitude and is related to the differential elastic scattering cross section per unit solid angle, $I_s(\Theta)$, by

$$I_s(\Theta) = |f(\Theta)|^2$$

The symmetry around the $z$ axis has removed the dependence of $I_s$ on the azimuthal angle. $\Theta$ is the scattering angle in the centre of mass system.

Let us now solve the wave equation for the scattered amplitude by the method of partial waves. The time-independent wave equation is

$$\nabla^2 \psi + \frac{2M_r}{\hbar^2} [E - V(r)] \psi = 0$$

or

$$\nabla^2 \psi + [k^2 - U(r)]\, \psi = 0 \tag{4.27}$$

where $k^2 = \dfrac{2M_r E}{\hbar^2}$ and $U(r) = \dfrac{2M_r}{\hbar^2} V(r)$. We seek solution to Equation (4.27) which is everywhere bounded, continuous, single-valued, and has the asymptotic form

$$\psi(r, \Theta) = \psi_{\text{inc}} + \psi_{\text{sc}} \simeq e^{ikz} + \frac{e^{ikr}}{r} f(\Theta) \tag{4.28}$$

where the amplitude of $\psi_{\text{inc}}$ was set equal to unity for convenience. Since by assumption $V$ is only a function of $r$, the wave equation may be separated into two equations, one of which contains only the variable $r$ and the other only the variable $\Theta$. The scattering problem can be solved by expanding the wave function in a series of Legendre polynomials. We then write for $\psi(r, \Theta)$

$$\psi(r, \Theta) = \frac{1}{r} \sum_{l=0}^{\infty} A_l P_l(\cos \Theta) \phi_l(r) \tag{4.29}$$

where $l$ is a measure of the angular momentum of the projectile about the fixed scattering centre. For this reason $l$ is called *the angular momentum quantum number* and one speaks of $s$, $p$, $d$, etc., waves corresponding to $l = 0, 1, 2$, etc. In certain cases it is sufficient to consider only the $l = 0$ wave; higher $l$ values are necessary for greater accuracy.

The solution of Equation (4.27) (McDaniel, 1964) leads to the following expression for elastic scattering:

$$I_s(\Theta) = \left| \frac{1}{2ik} \sum_{l=0}^{\infty} (2l+1)(e^{2i\eta_l} - 1) P_l(\cos \Theta) \right|^2 \tag{4.30}$$

Integration of $I_s(\Theta)$ over the complete solid angle gives for the elastic scattering cross section for the $l$th partial wave

$$(\sigma_e)_l = \frac{4\pi}{k^2} (2l+1) \sin^2 \eta_l \tag{4.31}$$

where $\eta_l$ is the phase shift experienced in the scattering process by the partial wave of $l$. The total elastic scattering cross section $\sigma_e$ is represented as the sum of the partial wave cross sections $(\sigma_e)_l$, viz.

$$\sigma_e = \sum_{l=0}^{\infty} (\sigma_e)_l \tag{4.32}$$

It is seen that the solutions of scattering problems are largely determined by calculating the phase shifts. When these are small, the cross section will on the whole be small. The maximum cross sections arise when the phase shift is an odd integral multiple of $\pi/2$. From Equation (4.31), $[(\sigma_e)_l]_{max}$ is $(4\pi/k^2)(2l+1)$. This is twice the maximum possible classical cross section $(2\pi/k^2)(2l+1)$. The difference is ascribed to the diffraction of the de Broglie waves representing the impinging beam.

### 4.1.4.2 The Born approximation for elastic scattering

One of the most useful methods employed for the calculation of atomic collision cross sections is that due to Born (1926). This method permits the cross sections to be calculated with much less labour than is required by the method of partial waves. The basic assumption made in this approximation is that the effect of the scattering potential is small, so that the interaction between the particles may be treated as a perturbation. This approximation, then, holds for sufficiently fast particles (for electrons $e^2/\hbar v \ll 1$), although in some cases it can account for the general features of the cross section at lower energies.

To derive the Born expression for the scattered amplitude, we have to solve Equation (4.27) seeking a solution that has the asymptotic form given by Equation (4.28). If $K$ is the magnitude of the momentum transferred during the collision divided by $\hbar$, for weak scattering (fast particles), we have*

$$f(\Theta) = -\frac{2M_r}{\hbar^2} \int_0^\infty \frac{\sin Kr}{Kr} V(r) r^2 \, dr \tag{4.33}$$

Equation (4.33) is the Born approximation for the scattered amplitude $f(\Theta)$ and is often known as the first Born approximation. Higher approximations can be made by an iterative procedure. A large number of calculations concerning elastic and inelastic electron scattering have been carried out with the aid of the Born approximation. In Chapter 5 we shall discuss the approximation as it applies to inelastic collisions.

## 4.2 Experimental methods

Elastic scattering and more generally electron-atom (molecule) interactions are studied by two types of experimental techniques: the electron swarm and the electron beam. The former includes the drift velocity, the diffusion and the microwave methods. The latter includes, single-beam methods yielding directly the total scattering cross section ($\sigma_e$, $\sigma_T$, or $Q_T$), single-beam methods for direct determination of the differential elastic

---

*See Mott and Massey (1965) for derivation.

scattering cross section $(I_s(\theta))$, and crossed-beam methods for the study of unstable species. As will be seen in subsequent discussions there exist basic differences between the electron swarm and the electron beam methods, and some unique advantages in their combination. Electron beam experiments are more direct, they deal with nearly monoenergetic electrons and are, quite generally, performed under single-collision conditions. They,however, are difficult to perform at low energies ($\lesssim 1$ eV), due principally to contact potentials, stray fields, and mutual electrostatic repulsion, and yield in most cases relative rather than absolute values of the measured quantities. In beam experiments the gas is, as a rule, studied under dynamic and field free conditions.

On the other hand, in swarm experiments the distribution of energies is wider, and certain desirable quantities (see later this section) are not obtained directly but have to be deduced from the measured parameters through sometimes elaborate and not straightforward procedures. Knowledge of the distribution of energies in an electron swarm is a major problem and indeed a very important one. In swarm experiments, the gas is studied under static conditions and under the influence of a uniform electric field. However, in swarm experiments, unlike in beam experiments, one can work with very low energies without having necessarily low fields. This is a great advantage of the swarm method which makes swarm experiments uniquely suited at energies below $\sim 1$ eV. They are performed at high pressures and thus easily allow studies of the pressure dependence of collision processes. Swarm experiments also yield certain parameters absolutely which can be used to normalize relative beam data. The discussions that follow will enable full evaluation of the potentialities of each method and the degree, as well as the limit, of their application.

### 4.2.1 Electron Swarm Methods

The foundation of the electron swarm method traces back to the pioneering work of Townsend at the beginning of the century (Townsend, 1900, 1901, 1908a,b, 1928; Townsend and Tizard, 1912; Townsend and Bailey, 1921). This early work has been followed in the 1930's by that of Cravath (1929), Loeb (1935), Bradbury and Nielsen (1936), and Nielsen (1936). The early work has been summarized by Healey and Reed (1941), Massey and Burhop (1952) and Loeb (1955). Recent investigations will be discussed in the sections that follow.

#### 4.2.1.1 Electron swarms

In electron swarm experiments electrons undergo many collisions and diffuse in the gaseous medium through which they drift under the influence of an externally applied electric field. An equilibrium energy distribution is attained where the gain from the field is balanced by the numerous but small fractional energy losses due to the dominant elastic collisions. Large angle scattering as well as low-energy inelastic scattering has an important effect on the distance travelled by an electron in a given swarm. If $d$ is the drift distance and $x'$ is the actual distance travelled by an electron in the swarm, usually $x' \gg d$. Further, if $u$ is the random (agitation) velocity and $w$ is the average velocity in the field direction called the *electron swarm drift velocity* in the field direction, then usually $u \gg w$. Thus, if $t(=d/w)$ is the time required for the swarm to drift the distance $d$ (Figure 4.2), the actual distance travelled $x'$ is equal to $ut = (u/w)d$, where the ratio $u/w$ is typically $\gtrsim 10^2$.

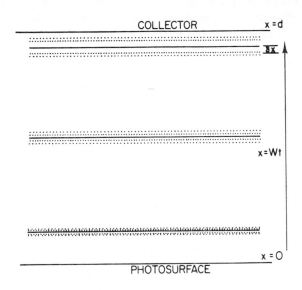

**Figure 4.2** Schematic of electron swarm at constant $E/P$

When conducting electron swarm experiments, the electron swarm or pulse should reach a steady-state condition and attain an equilibrium energy distribution quickly, that is from the beginning of the drift space. This introduces some problems as to the backscattering of electrons and the effects of boundaries in solving the Boltzmann transport equation. The relaxation time, i.e. the time it takes for an electron swarm to reach a steady-state condition in a gas must be much shorter than the drift times involved. In the absence of inelastic collisions the relaxation time is roughly equal to (McDaniel, 1964) $M/2m\bar{v}_c$ where $\bar{v}_c$ is the average electron collision frequency. Approximate values at $E/P = 0.1$ volts/cm torr are: $5/P$, $100/P$ and $500/P \,\mu$sec for He, Ne and Ar, respectively ($P$ is expressed in torr).

Estimates of the relaxation time for molecular gases must include the effect of vibrational and rotational relaxation. Average values for $H_2$ and $N_2$ are $\sim 0.1$ that of He (McDaniel, 1964, p. 557).

In Figure 4.2, we show schematically the drift and diffusion of an electron swarm which is produced by an ultraviolet pulsed source at the plane $x = 0$ (at $t = 0$) and drifts along the $x$ direction under the influence of a uniform external electric field. As the electrons drift toward the collector located at $x = d$, the swarm remains well-defined in space about its center of mass (dense line; Figure 4.2) but it spreads out due to diffusion. The velocity of the centre of mass in the field direction has been referred to as the drift velocity $w$. From the spreading, $\delta x$, of the electron swarm the longitudinal diffusion coefficient $D_l$ can be determined (Section 4.2.1.4).

The energy $\varepsilon$ of the electrons in the swarm has a considerable spreading which is characterized by a function $f(\varepsilon, E/P)$, defined by $f(\varepsilon, E/P)\,d\varepsilon =$ fraction of electrons in an energy range $d\varepsilon$ about $\varepsilon$; $E/P$ is the 'pressure reduced electric field' commonly expressed in volts/cm torr at a specified temperature $(T)$ usually indicated as a subscript to the pressure $P$. In swarm experiments, the measured quantities are averaged over $f(\varepsilon, E/P)$ and are recorded as functions of $E/P$. The parameter $E/P$ was first introduced by J. S. Townsend who pointed out that the energy distribution of electrons and ions in a gas is, in most cases, predominantly determined by the ratio of electric field strength, $E$, to the gas number density $N$. It is thus necessary to specify the temperature of the experiment and often the true value of $E/P$ is converted to an equivalent value at some standard or reference temperature by making use of Equation (4.11).

Huxley, Crompton and Elford (1966) proposed to call the unit $E/N$ the 'Townsend' (in honour of J. S. Townsend) and to make its magnitude such that

$$1 \text{ Townsend} = 10^{-17} \text{ volts cm}^2$$

The relation between $E/N$, in Townsend, and $E/P$, in volts/cm torr is

$$(E/N)_{\text{Townsend}} = [(1.0354 \times 10^{-2} T) E/P] \text{ volts/cm torr}$$

In the present discussions, we shall use either $E/N$ or $E/P$ (at specified $T$).

Two basic parameters are measured in swarm experiments: (i) the drift velocity $w$ (cm/sec) (or the electron mobility $\mu$ which is related to $w$ by $w = \mu E$) of electrons in the direction of a uniform external electric field $E$, and (ii) the diffusion coefficient $D(\text{cm}^2/\text{sec})$ lateral, $D_L$, or longitudinal $D_l$. The lateral diffusion coefficient (i.e. diffusion at right angles to the direction of the drift) cannot be measured alone, but as the ratio $w/D_L$. An independent measurement of $w$ at the same $E/P$ yields $D_L$. The longitudinal diffusion coefficient (i.e. diffusion in the direction of the drift) can be measured directly by the time-of-flight method (Section 4.2.1.4). Further, as will be discussed in Chapter 6 swarm experiments provide an accurate determination of the absolute rate of electron attachment.

From the measured basic quantities ($w$(or $\mu$); $w/D_L$(or $\mu/D_L$ or $D_l$)) as a function of $E/P$, the following physical parameters can be derived through computational techniques: (i) mean free path of the electrons $L$ at unit pressure; (ii) mean electron energy of the swarm $\langle \varepsilon \rangle$ as a function of $E/P$; and (iii) mean energy loss per collision $n$ as a function of $E/P$ or other convenient parameter such as the mean electron energy.

Also the following cross sections can be obtained as functions of the mean energy: (i) momentum transfer cross section $\sigma_m$; (ii) inelastic scattering cross sections $\sigma_{in}$ (see later this chapter and Chapter 5); and (iii) electron attachment cross section $\sigma_a$ (Chapter 6).

The parameters measured by the swarm method are averaged over $f(\varepsilon, E/P)$ and the deduced cross sections may differ from those obtained by monoenergetic electron beams. However, at low energies ($\lesssim 1$ eV) the swarm determined cross sections are close to their true values and shapes (see Section 4.4 and Chapter 6).

### 4.2.1.2 Relations between the basic swarm parameters

The quantity $eD/\mu$ has the dimensions of energy and is often called the characteristic energy of the electrons in the swarm. It is related to $kT$, where $k$ is Boltzmann's constant by

$$\frac{D}{\mu} = k_1 \frac{kT}{e}$$ (4.34)

or

$$\frac{w}{D} = \frac{N_A e}{RT} \frac{E}{k_1} = \frac{e}{kT} \frac{E}{k_1}$$ (4.35)

where $N_A$ is Avogadro's number, $e$ is the electronic charge, $R$ is the gas constant, $E$ is the electric field strength in volts/cm, and $T$ is the absolute temperature in °K. The right-hand side of Equation (4.35) is equal to $40.3E/k_1$ for $T = 288°$K. The parameter $k_1$ is related to the Townsend energy factor $k_T$, defined as the ratio of the mean energy of agitation of electrons to the mean energy of the gas molecules, by

$$k_1 = Ak_T$$ (4.36)

where $A$ is a constant that depends on the form of the energy distribution function. Both $D/\mu$ and $k_1$ are functions of $E/P$ and $T$. Thus for a specified $E/P$ and $T$ and a known (or assumed) $f(\varepsilon, E/P)$, measurements of $D/\mu$ yield the mean energy of the electron swarm $\langle \varepsilon \rangle$. For a Maxwell distribution function $A = 1$ and $k_1 = k_T$; for a Druyvesteyn form of the distribution function $k_1 = 0.875 k_T$ (Table 4.2). At low $E/P$ $(\to 0)$ $k_T \to 1$ and Equation (4.34) becomes

$$\frac{D}{\mu} = \frac{kT}{e} \tag{4.37}$$

Equation (4.37) can be used to test experimental determinations of $D/\mu$ (Section 4.3) which should converge to this value (determined by $T$ alone) as $E/P \to 0$.

In Table 4.2 a number of simple formulae which relate various swarm parameters are listed (Cochran and Forester, 1962). The numerical factors were computed for $w$ in cm/sec, $E/P$ in volts/cm torr, and $T = 298°K$ and for two forms of the energy distribution function: Maxwell and Druyvesteyn (see next section). For other relations see Huxley and Zaazou (1949) and Huxley and Crompton (1962).

More generally, the measured electron swarm drift velocity, $w$, and the diffusion coefficient, $D$, are related to $f(\varepsilon, E/P)$ and the microscopic cross section for momentum transfer, $\sigma_m(v)$, by (Allis, 1956)

$$w = -\frac{4\pi}{3} \frac{e}{mN} \frac{E}{P} \int_0^\infty \frac{v^2}{\sigma_m(v)} \frac{df_0}{dv} dv \tag{4.38}$$

and

$$D = \frac{4\pi}{3} \frac{1}{N} \int_0^\infty f_0 \frac{v^3}{\sigma_m(v)} dv \tag{4.39}$$

where $f_0$ is the spherically symmetric term in the expansion of the electron velocity distribution function, $N$ is the number of gas molecules per $cm^3$ at 1 torr at a specified $T$, and $P$ is the gas pressure in torr at the specified $T$.*

The momentum transfer cross section $\sigma_m$ is related to the momentum transfer frequency $v_m$ by

$$v_m = N\sigma_m v \tag{4.40}$$

where $v$ is the relative velocity between the electron and gas atoms (or molecules), or because $m \ll M$, $v$ is the electron velocity.

Swarm experiments which otherwise are ideally suited for the extreme low-energy region present serious difficulties in the deduction of cross sections for elastic and inelastic processes from the measured swarm parameters. A detailed analysis of swarm data requires solution of the appropriate Boltzmann transport equation for time independent electric fields and some knowledge of the energy dependence of the cross sections

---

*See an extension of the momentum transfer cross section to inelastic electron scattering by Crawford, Dalgarno and Hays (1967).

Table 4.2 Some relations between swarm parameters

| | Velocity distribution | |
|---|---|---|
| | Maxwell | Druyvesteyn |
| Townsend energy factor, $k_T$ | $k_T = k_1$ | $k_T = 0.875 k_1$ |
| Root-mean-square velocity, $\langle u^2 \rangle^{\frac{1}{2}}$ | $\langle u^2 \rangle^{\frac{1}{2}} = 1.16 \times 10^7 k_1^{\frac{1}{2}}$ | $\langle u^2 \rangle^{\frac{1}{2}} = 1.09 \times 10^7 k_1^{\frac{1}{2}}$ |
| Mean electron velocity, $\langle u \rangle$ | $\langle u \rangle = 1.07 \times 10^7 k_1^{\frac{1}{2}}$ | $\langle u \rangle = 1.04 \times 10^7 k_1^{\frac{1}{2}}$ |
| Mean free path at unit pressure, $L$ | $L = 7.20 \times 10^{-9} w k_1^{\frac{1}{2}}/(E/P)$ | $L = 7.47 \times 10^{-9} w k_1^{\frac{1}{2}}/(E/P)$ |
| Mean energy loss per collision, $n$ | $n = 1.74 \times 10^{-14} w^{\frac{1}{2}}/k_1$ | $n = 2.14 \times 10^{-14} w^2/k_1$ |
| Gas kinetic cross section, $\sigma$ | $\sigma = 4.26 \times 10^{-9}(E/P)/w k_1^{\frac{1}{2}}$ | $\sigma = 4.14 \times 10^{-9}(E/P)/w k_1^{\frac{1}{2}}$ |

for the elastic and inelastic processes. Frost and Phelps (1962) and Engelhardt and Phelps (1963) have analysed $w/D_L$ data through equation (4.41) where the power input to the electron swarm by the field, $ewE$, is balanced by that dissipated by the electrons in elastic and inelastic energy loss processes:

$$eEw = \left(\frac{2}{m}\right)^{1/2}\frac{2m}{M}\underbrace{\int_0^\infty \varepsilon^2\, N\sigma_m(\varepsilon)\left[f(\varepsilon)+kT\frac{\mathrm{d}f(\varepsilon)}{\mathrm{d}\varepsilon}\right]\mathrm{d}\varepsilon}_{\text{Elastic}} \quad +$$

$$+\left(\frac{2}{m}\right)^{1/2}\sum_j \varepsilon_j\underbrace{\int_0^\infty \varepsilon f(\varepsilon)\left[N\sigma_{+j}(\varepsilon)-N\sigma_{-j}(\varepsilon)\right]\mathrm{d}\varepsilon}_{\text{Inelastic}} \qquad (4.41)$$

In Equation (4.41), $f(\varepsilon)$ is the electron energy distribution function normalized by $\int_0^\infty \varepsilon^{1/2} f(\varepsilon)\,\mathrm{d}\varepsilon \equiv 1$, $\sigma_m(\varepsilon)$ is the cross section for momentum transfer as a function of electron energy $\varepsilon$, and $\sigma_{\pm j}(\varepsilon)$ are the cross sections for an electron to take up $(+j)$ or lose $(-j)$ energy $\varepsilon_j$. From Equation (4.41) one can obtain (Frost and Phelps, 1962; Engelhardt and Phelps, 1963) a distribution function $f(\varepsilon)$ and a set of energy-dependent momentum transfer and inelastic cross section curves which are consistent with the measured swarm parameters (see following section).

### 4.2.1.3 Electron swarm energy distribution functions

The energy distribution in an electron swarm is of fundamental importance in determining the macroscopic parameters measured in swarm experiments, such as the diffusion coefficient and drift velocity, and in deriving from these quantities microscopic cross sections for elastic and inelastic scattering, and electron attachment. Measurements of the above quantities by the swarm method represent average values and have to be used in conjunction with statistical theory to deduce atomic and molecular cross sections for the various types of collision processes.

The distribution function $f$ is defined by the Boltzmann transport equation,

$$\frac{\partial f}{\partial t} = \frac{e\mathbf{E}}{m}\cdot\nabla_v f - \mathbf{v}\cdot\nabla_r f + \left(\frac{\partial f}{\partial t}\right)_{\text{coll}} \qquad (4.42)$$

where $\nabla_v$ and $\nabla_r$ are velocity- and space-gradient operators, respectively, and $e\mathbf{E}/m$ is the acceleration due to the applied electric field $\mathbf{E}$. To obtain $f$, the cross sections for the various elastic and inelastic collision processes are required. The primary source of such information comes from measurements of the electron mobility $\mu$ and the ratio $D/\mu$. Healy and Reed (1941), Massey and Burhop (1952), Loeb (1955, 1956), and Craggs and Massey (1959) have summarized the efforts which have been made to analyse swarm experiments in terms of cross sections for elastic and inelastic electron scattering. Accurate solutions to the Boltzmann equation are difficult due mainly to a lack of knowledge as to the energy loss processes.

In most electron swarm analyses a Maxwell or a Druyvesteyn distribution form of $f$ has been assumed. In a Maxwell distribution function,

$$f(v)\,dv = N_t \left(\frac{2}{\pi}\right)^{1/2} \left(\frac{m}{kT}\right)^{3/2} v^2 e^{-\frac{mv^2}{2kT}}\,dv \qquad (4.43)$$

gives the number of electrons with velocities between $v$ and $v+dv$. $N_t$ is the total number of electrons in the swarm, $m$ is the electron mass, $T$ is the absolute temperature, and $k$ is the Boltzmann constant. The mean velocity $\langle v \rangle$ and the root-mean-square velocity $v_R$ are given by

$$\langle v \rangle = \frac{1}{N_t}\int_0^\infty v f(v)\,dv = \left(\frac{8kT}{\pi m}\right)^{1/2} \qquad (4.44)$$

and

$$v_R = \left[\frac{1}{N_t}\int_0^\infty v^2 f(v)\,dv\right]^{1/2} = \left(\frac{3kT}{m}\right)^{1/2} \qquad (4.45)$$

while the most probable velocity of the distribution $v_P$ is obtained by setting $df(v)/dv$ equal to zero and solving for $v$; this gives

$$v_P = \left(\frac{2kT}{m}\right)^{1/2} \qquad (4.46)$$

It then follows that

$$v_P/\langle v \rangle/v_R = 1/1.1284/1.2248$$

If we know $f(v)$, we may easily calculate the distribution for energies, $f(\varepsilon)$, the relation between the two distributions being

$$N_t f(v) mv\,dv = N_t f(\varepsilon)\,d\varepsilon \qquad (4.47)$$

Thus the number of electrons with energies in the range between $\varepsilon$ and $\varepsilon + d\varepsilon$ is

$$f(\varepsilon)\,d\varepsilon = N_t \frac{2}{\pi^{1/2}(kT)^{3/2}} e^{-\frac{\varepsilon}{kT}} \varepsilon^{1/2}\,d\varepsilon \qquad (4.48)$$

Druyvesteyn (1930) (see also Druyvesteyn and Penning (1940)) considered only elastic energy losses, assumed that the electron mean free path is independent of energy, and gave for the energy distribution $f(\varepsilon)$ the expression

$$f(\varepsilon) = C\varepsilon^{1/2} e^{-\frac{3\delta\varepsilon^2}{2L^2 e^2 E^2}} \qquad (4.49)$$

where $C$ is a constant, $\delta = 2m/M$, $m$ is the electron mass, $M$ is the molecular mass, $L$ is the electron mean free path, $e$ is the electron charge, and $E$ is the electric field intensity. The Druyvesteyn distribution differs from the Maxwell distribution in that it depends on the negative exponential fourth power of the electron velocity and not on the negative exponential second power of the velocity as the Maxwell distribution. Thus the Druyvesteyn distribution function has a sharper high-energy cutoff, i.e. the number of fast electrons is much larger in the Maxwell distribution. Since the Maxwell distribution is asymptotically less sharp, its high-energy tail would allow for ionization processes which are not actually observed in experiments. A comparison of the Druyvesteyn and Maxwell distribution functions for the same mean electron energy is given in Figure 4.3.

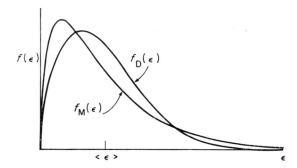

**Figure 4.3** Maxwell, $f_M(\varepsilon)$, and Druyvesteyn, $f_D(\varepsilon)$, energy distribution functions for the same mean electron energy $\langle\varepsilon\rangle$

No attempt will be made to discuss the many other distribution functions which have been derived in order to describe the energy of slow electrons in gases. An extensive coverage of the work prior to 1955 can be found in Loeb (1955). General theoretical discussions have been given by Chapman and Cowling (1952) and Allis (1956). A discussion of Margenau's (1946) calculation can be found in McDaniel (1964) also. A brief reference will be made now to some pertinent recent work which allowed great improvements in the analyses of electron swarm data, especially on electron attachment.

In the case of a rare gas when the electron energy is below the first excitation potential of the atom (quite high for the rare gases) neglect of inelastic collisions can be made. Ritchie and Whitesides (1961) have calculated equilibrium energy distributions for electrons drifting under the influence of a uniform electric field in argon. Let $f'(\theta,\ E/P,\ \varepsilon)$ be the solution of the space-and-time independent Boltzmann equation for the equilibrium energy distribution of electrons undergoing only elastic collisions with a gas of infinite extent in a uniform electric field $E$, and let its integral over all directions of motion of the electrons be

$$f(E/P,\varepsilon) = \int_0^\pi f'(\theta, E/P, \varepsilon) 2\pi \sin \theta \, d\theta \qquad (4.50)$$

where $\theta$ is the angle between the field direction and the velocity vector of an electron in the swarm. Ritchie and Whitesides (1961) (see also Chapman and Cowling, 1952) gave for $f(E/P,\ \varepsilon)$ the expression

$$f(E/P,\varepsilon) = A\varepsilon^{1/2} \exp\left\{ -\frac{6m}{M} \int_0^\varepsilon \frac{\sigma_m^2(\varepsilon)\varepsilon \, d\varepsilon}{[(eE/P)^2 + 6m/M\varepsilon\sigma_m^2(\varepsilon)kT]}\right\} \qquad (4.51)$$

where $f(E/P,\ \varepsilon)\, d\varepsilon$ is the fraction of electrons with energies in the interval between $\varepsilon$ and $\varepsilon + d\varepsilon$ for a given $E/P$, and $A$ is a normalization constant such that

$$\int_0^\infty f(\varepsilon, E/P) \, d\varepsilon \equiv 1 \qquad (4.52)$$

Originally Ritchie and Whitesides (1961) used the data of Barbiere (1951) on $\sigma_m(\varepsilon)$ for argon and evaluated numerically $f(\varepsilon, E/P)$ for values of $E/P$ low enough that electronic excitation does not occur. These distributions have been revised recently using more recent electron transport data (Nelson and Whitesides, 1968; Christophorou, Chaney and Christodoulides, 1969). Bowe (1960) used a form for the distribution function essentially the same as that of (4.51) and has shown that good agreement may be obtained between measured and calculated values of the electron swarm drift velocity in the rare gases. Normalized distribution functions for argon are plotted in Figure 4.4 for four different values of $E/P$. It is seen that by changing $E/P$ from 0.1 to 2 volts/cm torr, $f(\varepsilon, E/P)$ can be set to peak at any energy from about 1 to 6 eV. Such distribution functions are very important in experiments where the molecular gas under investigation is mixed in small proportions with another gas whose role is to determine $f(\varepsilon, E/P)$. In electron attachment studies, for example, a molecule is often mixed with a non-electron attaching gas, such as argon, in small proportions so that $f(\varepsilon, E/P)$ is characteristic of the non-electron attaching gas itself. This will allow accurate analyses of swarm data (Chapter 6) especially when the functional

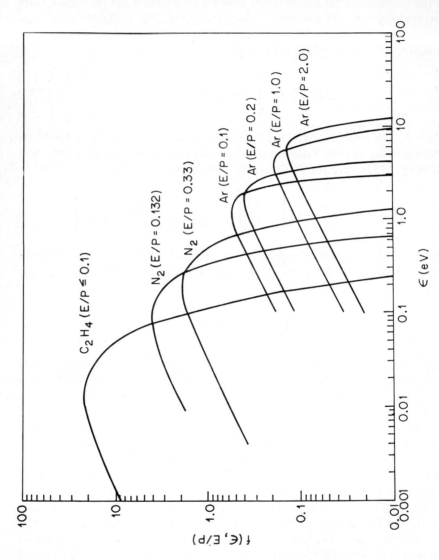

**Figure 4.4** Energy distribution in electron swarms for different gases and $E/P_{300}$ (see text)

form of the attachment cross section is known. In this respect it would be desirable to know $f(\varepsilon, E/P)$ for a molecular non-electron attaching gas for which $f(\varepsilon, E/P)$ peaks at energies $< 2$ eV, since for a large number of poly-atomic molecules electron attachment resonances peak in this energy region. Carleton and Megill (1962) performed such calculations. They obtained numerically computed solutions to the Boltzmann equation for $N_2$ considering electron energy losses by elastic scattering and by rotational, vibrational and electronic excitation and using experimental cross sections for these processes. The distribution functions calculated by Carleton and Megill (1962) for $N_2$ as well as those of Ritchie and Whitesides (1961) for argon have been used in analyses of swarm data on electron attachment to molecules (Chapter 6).

The Boltzmann transport equation has been solved numerically for the distribution $f$ of electron energies taking into account both elastic and inelastic collisions for $H_2$ and $N_2$ (Frost and Phelps, 1962), $H_2$ and $D_2$ (Engelhardt and Phelps, 1963) and $N_2$ (Engelhardt, Phelps and Risk, 1964). These workers considered a swarm of electrons drifting through a parent neutral gas at a temperature $T$ and under the influence of a uniform dc electric field $E$. Extending the work of Holstein (1946) and Margenau (1946) to include collisions of the second kind, they computed the steady-state distribution function $f(\varepsilon)$ by solution of the Boltzmann equation expressed (Engelhardt and Phelps, 1963) as

$$\frac{\mathrm{d}}{\mathrm{d}\varepsilon}\left(\frac{e^2 E^2 \varepsilon}{3N\sigma_m}\frac{\mathrm{d}f}{\mathrm{d}\varepsilon}\right) + \frac{2m}{M}\frac{\mathrm{d}}{\mathrm{d}\varepsilon}\left[\varepsilon^2 N\sigma_m\left(f + kT\frac{\mathrm{d}f}{\mathrm{d}\varepsilon}\right)\right]$$

$$+ \sum_j \left[(\varepsilon + \varepsilon_j)f(\varepsilon + \varepsilon_j)N\sigma_j(\varepsilon + \varepsilon_j) - \varepsilon f(\varepsilon)N\sigma_j(\varepsilon)\right]$$

$$+ \sum_j \left[(\varepsilon - \varepsilon_j)f(\varepsilon - \varepsilon_j)N\sigma_{-j}(\varepsilon - \varepsilon_j) - \varepsilon f(\varepsilon)N\sigma_{-j}(\varepsilon)\right] = 0 \qquad (4.53)$$

In Equation (4.53) $N$ is the neutral molecule density, $\sigma_m(\varepsilon)$ is the momentum transfer cross section, $\sigma_j(\varepsilon)$ is the rotational, vibrational, or electronic excitation cross section with an energy loss $\varepsilon_j$, and $\sigma_{-j}(\varepsilon)$ is defined as the cross section for a collision in which the electron gains energy $\varepsilon_j$; both $\sigma_j(\varepsilon)$ and $\sigma_{-j}(\varepsilon)$ include the fractional population and statistical weight factors of the initial molecular state. The distribution function $f$ is normalized by

$$\int\limits_0^\infty \varepsilon^{1/2} f(\varepsilon)\, d\varepsilon \equiv 1$$

If Equation (4.53) is multiplied by $(2/m)^{\frac{1}{2}}\varepsilon\, d\varepsilon$ and integrated over all energies, the energy balance equation (4.41) is obtained (Engelhardt and Phelps, 1963). As stated earlier, in expression (4.41), the power input to the electron swarm from the field, $eEw$, is balanced by the power dissipated by the electrons in elastic (first term on the right-hand side of (4.41) and inelastic (second term on the right-hand side of (4.41)) collisions. The energy distribution function $f(\varepsilon)$ and a self-consistent set of cross sections for the various elastic and inelastic processes can be obtained through Equations (4.53) and (4.41) as follows: Based on any available information, a set of cross sections for the elastic and inelastic processes is initially assumed. These are substituted in Equation (4.53) which is solved for $f(\varepsilon)$ with the aid of a computer. The distribution function so obtained is then used with the momentum transfer cross section to calculate a set of transport coefficients which are compared with their experimentally determined values. On the basis of the observed discrepancies, the initial estimates of the cross sections are adjusted and the electron energy distribution function and the transport coefficients are recalculated and compared with their experimental values. This procedure is repeated until the calculated transport coefficients best agree with their experimental values. A set of elastic and inelastic cross section functions and $f(\varepsilon)$ is ultimately found for which the two sides of Equation (4.41) are equal (say, within one part per $10^4$). This represents a set of elastic and inelastic collision cross sections and $f(\varepsilon)$, which is not unique but is self-consistent. In their treatment, Frost and Phelps (1962) and Engelhardt and Phelps (1963) solved Equation (4.53) for three energy regions: (i) The thermal region where the characteristic energy $\varepsilon_k$ ($\equiv De/\mu$) ranges from its thermal value to an energy where vibrational excitation begins to be important. At sufficiently low $E/P$ the electron energies approach a thermal distribution $(D/\mu \rightarrow kT/e)$, and it is necessary to consider the energy gained by the electrons in collisions with gas molecules. The thresholds and shapes of the assumed cross sections for rotational excitation considered were based on the theories of Gerjuoy and Stein (1955). (ii) The region where $\varepsilon_k$ varies from the energy where vibrational excitation is significant to that where dissociation becomes important. Neglect was made in this region of superelastic collisions. (iii) The final region where elastic scattering, vibrational and electronic excitation, and ionization were considered.

In Figure 4.4 we have plotted the normalized distribution functions calculated for $N_2$ by Phelps and coworkers (1968) for two values of $E/P$.

For $E/P = 0.132$ the distribution function is seen to peak at $\sim 0.10$ eV, while for an $E/P = 0.33$ the distribution is shifted to higher energies peaking at $\sim 0.2$ eV. In Figure 4.4 a Maxwell distribution function ($T = 298°$K) has been plotted also. This is characteristic of electron energy distribution functions in polyatomic gases at low $E/P$. Thus, such gases (e.g. $C_2H_4$, $CO_2$) can be used in swarm experiments to provide a swarm of electrons with a Maxwell energy distribution function. The $E/P$ value below which the energy distribution remains thermal in a gas depends on the molecule itself and varies quite significantly (Section 4.3) from molecule to molecule. For $C_2H_4$ electrons remain in thermal equilibrium with the gas for $E/P \lesssim 0.1$ volts/cm torr (Stockdale and Hurst, 1964; Christophorou and coworkers, 1965), but for polar molecules this $E/P$ region is often much higher. For water, for example, electrons were found to remain in thermal equilibrium with the gaseous medium up to an $E/P \simeq 6$ volts/cm torr (Pack, Voshall and Phelps, 1962; Lowke and Rees, 1963b; Christophorou and Christodoulides, 1969).

Knowledge of the energy distribution functions discussed in this section aided considerably the analysis of electron transport data, especially those on electron attachment (Chapter 6). Although the energy distribution functions for Ar as calculated by various workers are in good agreement, those for $N_2$ are not. The calculations of Carleton and Megill (1962) and those of Phelps and coworkers are compared in Figure 4.5 for two values of $E/P$. It is seen that the two sets of data differ both in over-all shape and peak energy. Christophorou, Chaney and Christodoulides (1969) tested experimentally the two sets of functions and concluded that Phelps's distributions for $N_2$ are consistent with those for Ar. For further discussion see Chapter 6 (Section 6.3).

### 4.2.1.4 Measurement of $w$, $w/D_L$ and $D_l$

A. *Measurement of $w$.*   Three experimental methods are credited for the largest body of information on the drift velocity of electrons through gases: the Townsend method, the electrical shutter method, and the pulse method(s). These will now be discussed briefly. For a more complete discussion see Loeb (1955). The drift velocities measured by the use of the methods mentioned above are generally in satisfactory agreement, the greatest discrepancies occurring in short drift chambers and small values of $E/P$. It might be pointed out that errors in pressure measurement and gas purity account for a substantial portion of the existing inaccuracies, For early data on $w$, the reader is referred to Healey and Reed (1941), Townsend (1947) and Loeb (1955). Recent data are presented in Section 4.3.

(i) The Townsend method.   A modified version of the original Townsend

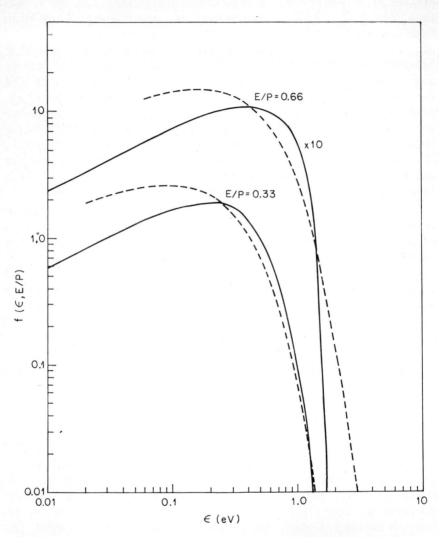

**Figure 4.5** Comparison of the calculated electron energy distributions for $N_2$ for two values of $E/P$. Solid lines are the results of Phelps and coworkers and broken lines those of Carleton and Megill

apparatus used by Huxley and his colleagues is shown in Figure 4.6. Electrons from the filament $F$ pass through the small hole $S$ and drift in the $z$ direction under a uniform electric field $E$ (maintained by guard rings) and are collected at the anode, which is split into two semicircular segments.

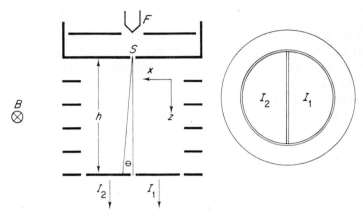

**Figure 4.6** A modified version of the original Townsend apparatus for the measurement of electron drift velocities

A magnetic field can be applied at right angles to the drift direction. The currents $I_1$ and $I_2$ (Figure 4.6) received by the two segments of the anode are equal with the magnetic field off, but become unequal with the magnetic field applied. The imbalance in the two currents is associated with the lateral deflection of the electron stream through an angle $\theta$ given by

$$\tan \theta = \frac{w_x}{w_z} \qquad (4.54)$$

where $w_x$ is the component of $w$ in the direction normal to both $E$ and $B$ ($B$ = flux density of the magnetic field) and $w_z$ is the component of $w$ along $E$ (drift direction). The quantity $\tan \theta$ can be obtained from a measurement of $I_1/I_2$ with the magnetic field applied and $B$. The drift velocity $w$ ($=w_z$) may be obtained from the equation

$$w = C\frac{E}{B}\tan \theta \qquad (4.55)$$

where $C$ is a dimensionless factor which depends on the form of the energy distribution function. For a Maxwell distribution of energies the constant $C$ is equal to 0.85 and for a Druyvesteyn equal to 0.943 (Huxley and Crompton, 1962; Loeb, 1955). Results obtained by this method will be presented in Section 4.3.

(ii) Electrical shutter method. In this method, introduced by Tyndall and his colleagues in the 1920's, electrical shutters are used to time the transit of electrons (or other charge carriers) through the gas. Figure 4.7 shows the electrical shutter arrangement of Pack, Voshall and Phelps (1962) which is based on the apparatus employed by Bradbury and Nielsen (1936). A steady-state light source or a pulsed light source is used to liberate phofto electrons from the cathode. These drift to the collector under a uniorm-

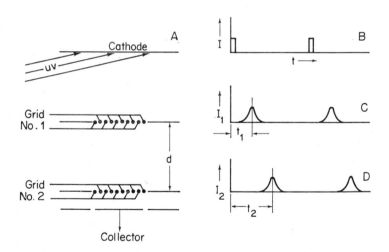

**Figure 4.7** The electrical shutter arrangement of Pack, Voshall and Phelps (1962) (shown in the conventional pulsed mode of operation) for the measurement of electron swarm drift velocities. The pulses shown are schematic illustrations and their widths are not to scale

electric field. When a steady-state light source is used, the first grid provides a time-varying electron current (i.e. it is used to inject a pulse of electrons into the drift space) and the second grid is used to measure the transit time of the electrons from the first grid. The drift velocity is determined from this time measurement and the drift distance $d$. The variation of the drift distance (distance between the two grids) may be used to correct for end effects. When a pulsed light source is used, either of the two grids can be employed for the measurement of the transit time of the electrons from the cathode. The time required for the electron pulse to travel between the two grids is equal to the difference in transit times from the cathode to the respective grids and should be independent of end effects at the cathode and effects occurring equally at both grids. In one mode of operation, 'the conventional' (Pack and Phelps, 1961; Pack, Voshall and Phelps, 1962), a potential is applied between the two halves of grid No. 1 (Figure 4.7) such that the transmitted electron current is reduced (usually by $\sim 5\%$ of its value with zero bias), while grid No. 2 is passive. The transmitted electron current can be maximized with the aid of a rectangular voltage pulse applied to each half of the grid at a certain time, $t_1$, after the light pulse. By varying the time delay between the voltage pulse applied to grid No. 1 and the light pulse, the variation with time of the collector current resembles Figure 4.7C. Likewise, making grid No. 1 passive and grid No. 2 active the variation with time of the collector current is similar to Figure 4.7D. The drift velocity is then determined from

$$ w = \frac{d}{t_2 - t_1} $$

In an alternative mode of operation, the 'zero-bias', due to Pack and Phelps (1961), a voltage pulse is used to collect some of the electrons near the grid. When the gate is closed (reducing transmission) simultaneously with the arrival of the electrons at the grid, the collector current goes through a minimum. From this, the transit time is measured and hence $w$. In this mode of operation the perturbing effects of voltages applied to the grids are believed to be reduced (Pack and Phelps, 1961).

(iii) The pulse method(s). There have been several variations of the pulse method which may be classified according to the mode of production of the electron pulses (x-rays (Herreng, 1942a,b; 1943); $\alpha$ particles (Klema and Allen, 1950; Bortner, Hurst and Stone, 1957); fission fragments (Bell, 1951); electrons (Errett, 1951); pulsed ultraviolet light (Hurst and coworkers, 1963; Christophorou, Hurst and Hadjiantoniou, 1966; Christophorou, Hurst and Hendrick, 1966; Christophorou and Christodoulides, 1969)). In all cases electrons are created suddenly and the drift velocity is determined from a measurement of the time it takes for the free electrons to travel a known distance to the collector (or detector). An amplifier and an oscilloscope are used to observe the collector current as a function of time.

Notable among the pulse methods is the single-electron detection time-of-flight method. This method has been extensively used at Oak Ridge National Laboratory and is credited for a large body of information on $w$ (Sections 4.3 and 4.4). Figure 4.8 shows the arrangement employed by Christophorou, Hurst and Hadjiantoniou (1966). The main part of the apparatus is a gold-plated, cylindrical, parallel-plate ionization chamber. The two parallel

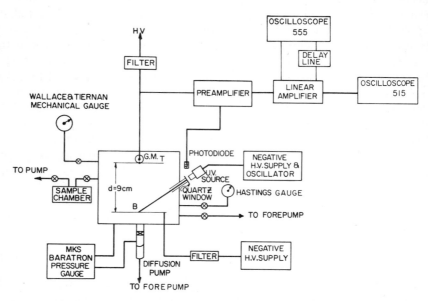

**Figure 4.8** Single electron detection time-of-flight method for the measurement of electron swarm drift velocities (Christophorou, Hurst and Hadjiantoniou, 1966)

plates, $B$ and $T$, are 9 cm apart. The ultraviolet light emitted by a mercury pulsed flash tube (Type FX–6A, Edgerton, Germeshausen and Grier, Inc.) which is operated at a rate of 160 flashes per second and has a time width of 0.5 $\mu$sec, strikes the centre of the lower plate $B$ (held at a negative potential) and releases photoelectrons. These electrons quickly come into equilibrium with the molecules of the gas and, through the action of the applied electric field, travel the drift space $BT$ and ultimately trigger the gas-discharged tube (Geiger–Müller) mounted behind the upper plate $T$, (held at zero potential) which constitutes the upper end of the drift space. The lower plate $B$ forms the lower end of the drift space $BT$.

The ultraviolet pulse which generates the electron swarm also triggers a photodiode (Type SD–100, Edgerton, Germeshausen and Grier, Inc.), and the output pulse is fed through the preamplifier and the linear amplifier to an oscilloscope. At some time $\Delta t$ after the display of the light pulse, the pulse from the gas-discharged tube is also seen on the oscilloscope. The time lapse $\Delta t$ between the two pulses is the electron swarm drift time. The number of ultraviolet flashes and the detector aperture are such that the number of pulses counted per second is small compared to the number of ultraviolet flashes. Thus the probability that more than one electron from a given swarm will enter the detector is small. This condition must be fulfilled for the counting to obey Poisson statistics, necessary to ensure random and not preferential detection of electrons from the swarm. The dimensions indicated in Figure 4.8 were found to be adequate for accurate measurement of the drift times and were chosen such that the same arrangement could be used for electron attachment studies. The equipment could be heated up to $\sim 200\,°C$. The large body of information obtained by the use of this method on polar and polarizable molecules in the pure form and in mixtures with ethylene will be presented and discussed in Sections 4.3 and 4.4. If in addition to an oscilloscope a time-of-flight multichannel analyser is used, the distribution in the time-of-arrival of the electron swarms at the detector can be obtained with a great precision. From the maximum of such distribution the drift time and thus the drift velocity is determined while from the width of the distribution the longitudinal diffusion coefficient can be calculated (see next section).

B. *Measurement of $w/D_L$ and $D_l$.* There are basically two methods for the measurement of electron diffusion coefficients in gases: (i) the Townsend method which allows determination of $w/D_L$ and (ii) the time-of-flight method which yields $D_l$. The principle of the two methods is shown in Figure 4.9. The Townsend method uses a steady-state point source and a plane detector. In contrast, the time-of-flight method utilizes a pulsed plane source and a point detector. Further, in the Townsend method one measures $w/D_L$ while in the time-of-flight method $D_l$. We shall see later in this chapter that $D_L$ and $D_l$ as measured by the two methods are not compatible for a number of gases. It might be pointed out that the time-of-flight method offers some advantages over the Townsend method in that it requires solution of the Boltzmann transport equation in one dimension rather than in three dimensions as in the Townsend method, and in that it measures the distribution in time-of-arrival of the electron swarm at a point—this time distribution can be measured with an arbitrarily large number of points with modern multichannel analysers—in contrast to the Townsend method where only two currents are measured. Both methods require extreme care in a number of problems which have been fully discussed in the references cited in the discussion below:

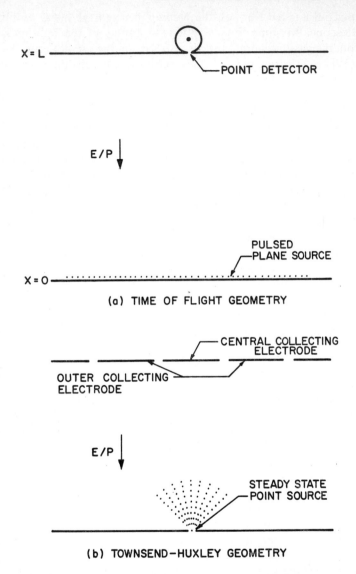

(a) TIME OF FLIGHT GEOMETRY

(b) TOWNSEND−HUXLEY GEOMETRY

**Figure 4.9** The principle of the Townsend method for the measurement of $w/D_L$ and the time-of-flight method for the measurement
of $D_l$ (see text)

(i) The Townsend lateral diffusion method.   Figure 4.10 shows a modern version of the Townsend apparatus and is due to Huxley, Crompton and their colleagues. Electrons are emitted continuously from the filament, they

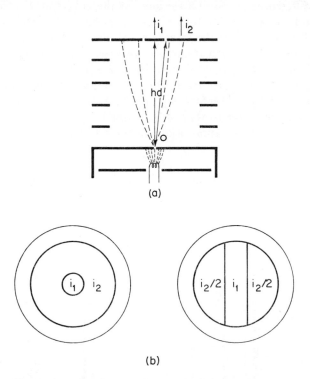

(a)

(b)

**Figure 4.10** Modern version of the Townsend apparatus for the measurement of $w/D_L$ due to Huxley, Crompton and their colleagues

pass through the hole O ($\sim 1$ mm in diameter) and drift to the anode diffusing in their passage through the gas chamber. The electron currents are kept low ($\lesssim 10^{-11}$ amps) to avoid space charge effects. The anode is segmented as shown on the right-hand-side of the lower portion of Figure 4.10 and the divergence of the electron stream in crossing the chamber is determined by the ratio $R = i_1/(i_1 + i_2)$ where $i_1$ and $i_2$ are the currents collected by the two interior segments of the anode (see Figure 4.10). Solution of the appropriate three dimensional time-independent diffusion problem yields (Huxley and Crompton, 1962) for the ratio $R$

$$R = 1 - \frac{h}{d} e^{-\lambda(d-h)} \tag{4.56}$$

where $h$ and $d$ are distances as shown in Figure 4.10 and $\lambda = w/2D_{\mathrm{L}}$. Then, $w/D_{\mathrm{L}}$ can be determined from a measurement of the ratio $R$. Use of Equation (4.35) yields $k_1$. A number of problems arising from boundaries, potential differences across the surfaces of the collecting electrodes, space charge effects in general, uniformity of the electric field, and a number of other factors influencing the accuracy of the measurements have been discussed in detail by the Australian workers (see, for example, Huxley,1940; Huxley and Crompton, 1955; Crompton and Sutton, 1952; Crompton and Jory, 1962; Huxley and Crompton, 1962; Crompton, Elford and Gascoigne, 1965; Hurst and Liley, 1965; Liley, 1967). Warren and Parker (1962a,b) used a similar apparatus to that of Townsend–Huxley and also considered the effects of boundaries appropriate to their apparatus.

(ii) The time-of-flight method.   Figure 4.11 shows the arrangement of the time-of-flight (TOF) method, as used by Wagner, Davis and Hurst (1967).

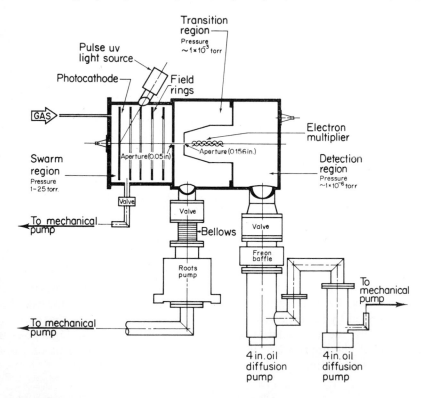

**Figure 4.11** The Wagner, Davis and Hurst (1967) arrangement of the time-of-flight method for the measurement of $D_l$

This arrangement is an improvement over that employed earlier by Hurst and coworkers (1963) in that differential pumping is used and that an electron multiplier is used as a single-electron detector rather than a Geiger–Müller tube. With the arrangement of Wagner, Davis and Hurst (1967), there is basically no limitation as to the kind of the gas to be studied. This, obviously, is not the case with a Geiger–Müller detector.

Let us now return to the upper portion of Figure 4.9 which shows the principle of the TOF method. Electrons are produced uniformly over a plane at $x = 0$ (and $t = 0$) and drift to the detector under a uniform electric field. Assuming no electron capture and ion-pair production, the time-dependent unidirectional differential equation applicable to the geometry of Figure 4.9 is

$$\frac{\partial n(x,t)}{\partial t} = D_l \frac{\partial^2 n(x,t)}{\partial x^2} - w \frac{\partial n(x,t)}{\partial x} \tag{4.57}$$

where $n(x,t)$ is the volume density of electrons as a function of time at any point $0 \leqslant x \leqslant L$. Neglecting the plane boundaries at $x = 0$ and $x = L$, a solution of Equation (4.57) is (Hurst and coworkers, 1963)

$$n(x,t) = N(4\pi D_l t)^{-1/2} \exp\left[-\frac{(x-wt)^2}{4D_l t}\right] \tag{4.58}$$

where $N$ is the area density of electrons when $x = 0$ and $t = 0$. If the area of the detector aperture is $a$ and the detector is located at a distance $L$ from the plane source, the number of electrons entering the detector between $t$ and $t + \Delta t$ is $E(t)\Delta t$, where $E(t)$ is given by

$$E(t) = Naw(4\pi D_l t)^{-\frac{1}{2}} \exp\left[-(L-wt)^2/4D_l t\right] \tag{4.59}$$

Measurement of $E(t)$ allows determination of both $w$ and $D_l$ from the quantities $t_m$ and $\delta t$ defined by

$$[\partial E(t)/\partial t]_{t=t_m} = 0$$

and

$$\delta t \equiv |t_1 - t_m|$$

where $t_1$ is determined from $E(t_1) = (1/e)E(t_m)$; $e$ is the basis of natural logarithms. Using the approximation that $D_l/Lw \ll 1$ the drift velocity is given by (Hurst and coworkers, 1963)

$$w \simeq \frac{L}{t_m} \tag{4.60}$$

Also, under the assumption that $2(D_l/Lw)^{\frac{1}{2}} \ll 1$, the diffusion coefficient $D_l$ is given by

$$D_l \simeq \frac{L^2(\delta t)^2}{4t_m^3} \tag{4.61}$$

Thus, both $w$ and $D_l$, to a first approximation, can be determined independently by fitting $E(t)$. Figure 4.12 shows schematically the function $E(t)$ and the quantities $t_m$ and $\delta t$.

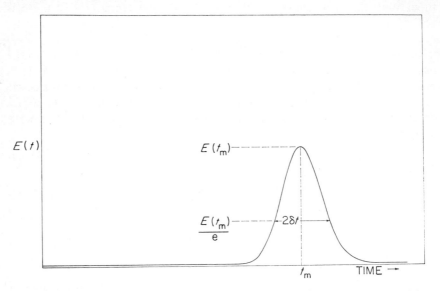

**Figure 4.12** Schematic representation of the function $E(t)$ and the quantities $t_m$ and $\delta t$

In determining $w$, and especially $D_l$, with the time-of-flight method attention should be given to properly accounting for the broadening of $E(t)$ (resulting in larger values of $D_l$) due, primarily, to the finite width of the light pulse, inadequate time-resolution, and instrumental fluctuations. Hurst and Parks (1966) and Wagner, Davis and Hurst (1967) satisfactorily accounted for these problems. They avoided any Poisson distortion by ensuring single-electron counting and dealt with all other fluctuations analytically. They measured the error distortion function and the time-of-flight distribution, and obtained the true time-of-flight distribution by a deconvolution analysis (see original references for details). A more complete account of the Poisson distortion due to the counter or analyser dead times was given by Wagner, Davis and Hurst (1967) who gave the following more precise equations for $w$ and $D_l$.

$$w = \frac{L}{t_m}\,(1+\beta^2 - \ldots)^{-1} \tag{4.62}$$

$$D_l = \frac{L^2}{t_m}\,\beta^2(1+2\beta^2 + \ldots)^{-1} \tag{4.63}$$

where
$$\beta^2 = \left(\frac{\delta t}{2t_m}\right)^2 \simeq \frac{D_l}{wL}$$

The results of these workers on $w$ agree well with those of other investigators. However, their results on $D_l$ differ in certain cases appreciably from those obtained by the Townsend lateral diffusion method (Section 4.4) and this discrepancy seems to be inherent in the two methods.

Measurements of the diffusion coefficient at very low $E/P\,(\rightarrow 0)$ are difficult to make with great accuracy due mainly to difficulties in removing the effects of stray fields, space charges and boundaries. A direct determination of the thermal value of the diffusion coefficient requires special attention in ensuring a truly thermal electron energy distribution. Such direct measurements have been made (e.g. Lloyd, 1960; Cavalleri, Gatti and Principi, 1964; Cavalleri, Gatti and Interlenghi, 1965; Nelson, 1968).

### 4.2.1.5 Microwave methods

A number of microwave methods have been developed for the study of electron collision processes in the energy range extending from thermal to several electron volts. In these methods it is essential to ensure equilibrium conditions. The often large disagreement among microwave methods themselves and also with the results obtained by other methods is disturbing. No discussion will be made here of the various microwave methods which have been summarized by Brown (1959), and have been briefly discussed by Biondi (1963), McDaniel (1964) and Hasted (1964). A number of microwave studies of electron–atom and electron–molecule scattering processes have been made (e.g. Phelps, Fundingsland and Brown, 1951; Chen and Raether, 1962). Momentum transfer cross sections, electron attachment cross sections, and rates for negative-ion molecule reactions obtained by microwave techniques will be discussed in the appropriate chapters.

### 4.2.2 *Beam Methods for the Measurement of the Elastic Electron Scattering Cross Section*

These are single-collision scattering experiments utilizing monoenergetic electron beams and fall into three categories: those which yield the total elastic scattering cross section, those which allow determination of the differential elastic scattering cross section, and those employed for the study of unstable species. A typical example from each category will now be described. For other types of experimental arrangements see Massey and Burhop (1952) and McDaniel (1964).

### 4.2.2.1 Beam methods for the measurement of the total elastic scattering cross section

The prototype of these methods is the arrangement of Ramsauer (1921a, b), the principle of which is illustrated in Figure 4.13. Electrons are produced photoelectrically at the plate P and are accelerated to the desired energy by a potential difference between P and the first slit $S_1$. A magnetic field perpendicular to the plane of the paper allows electrons which do not suffer collision to describe a circular path through the slits $S_1$–$S_7$ and enter the Faraday cage F. Those electrons which undergo elastic scattering fail to pass through the slits. Also electrons which suffer inelastic collisions (even without deflection) fail to pass through the succeeding slits and thus are not collected in F, since they move in new circular paths of smaller radii.

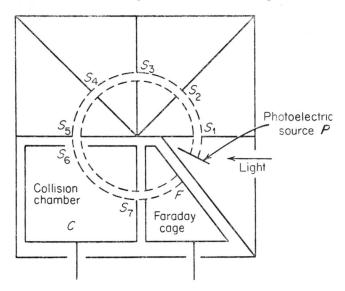

**Figure 4.13** Ramsauer's arrangement for the measurement
of total electron scattering cross sections

Let us now consider those electrons (monoenergetic) which do not suffer scattering prior to entering the collision chamber C. These can either experience collision in travelling the path $S_6$–$S_7$ and be collected in C, or reach $S_7$ and be collected in F. Thus if the arc length $S_6$–$S_7$ and the gas pressure are known, the total collision cross sections $\sigma_T$ and $Q_T$ can be determined through Equation (4.8) from measurements of the currents in C and F. If the product of the target gas pressure and the path $S_6$–$S_7$ is small enough that the probability of even a single collision is small, the probability of an electron undergoing multiple collisions in travelling the distance $S_6$–$S_7$ is negligible. Then, provided that the slits are narrow so that even small angle scattering is detected, the method gives the true total collision cross section. If the electron energy is lower than the energy for excitation of any of the levels of the target gas, the measured cross section is equal to the total elastic scattering cross section. Otherwise inelastic collisions contribute to the cross section. If the scattered electrons are energy analysed by retarding potential techniques or deflection in an electric or magnetic field, the elastically scattered electrons can be separated and collected, and both $\sigma_e$ and $\sigma_{inel}$ can be determined.

Various versions of the Ramsauer method have been used during the last 40 years. Among the most recent experiments using the Ramsauer technique to measure electron scattering cross sections from atoms and simple molecules is that of Golden and Bandel (1965a; 1966). (See further discussion in Section 4.5 and in Chapter 5 where recent electron beam techniques, such as double electrostatic analysers, are discussed.)

### 4.2.2.2 Beam methods for the measurement of the angular distribution of elastically scattered electrons

Figure 4.14 shows the 'zone apparatus' of Ramsauer and Kollath (1932) who were among the first to study the angular dependence of the elastic scattering of low-energy electrons for several simple molecules (see results

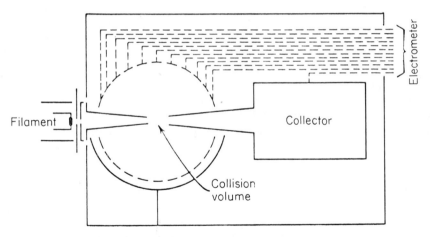

**Figure 4.14** Zone apparatus for the measurement of the angular distribution of scattered electrons (Ramsauer and Kollath, 1932)

in Section 4.5). Electrons from the filament enter the collision volume—kept at a low pressure ($\lesssim 10^{-3}$ torr)—at the centre of the sphere defined by the curved metal plate. There is cylindrical symmetry about the axis of the electron beam. The unscattered beam is collected at the collector, while the angular scattering distribution is determined by measuring the electron currents to each of the 11 plates. In the case of inelastic scattering, the elastic contribution to the total scattering may be separated out by energy selection techniques (Massey and Burhop, 1952). In the arrangement of Figure 4.14 measurements to very large angles are prevented due to the presence of the electron gun. This, however, can be overcome (see, for example, Gagge (1933)). Investigations of the angular distribution of elastically scattered electrons from unstable targets (H atoms, for example) can be achieved by the use of crossed-beam techniques (see, for example, Gilbody, Stebbings and Fite (1961)). Angular distribution data obtained during the 1930's appear in Massey and Burhop (1952) (see also Section 4.5).

### 4.2.2.3 Crossed-beam methods for the study of collisions between electrons and chemically unstable species

Such techniques have been in use since the 1930's (see Massey and Burhop (1952)). A schematic diagram of the crossed-beam apparatus of Neynaber and coworkers (1961a, b) employed for the measurement of the total scattering cross section of electrons from H and O, and with a slightly modified version, from N (Neynaber and coworkers, 1963) is shown in Figure 4.15. A neutral molecular beam emerges from the first chamber with the radio-frequency (RF) discharge source off, while a partially dissociated beam (typically $\sim 30\%$ dissociated) emerges with the RF source on. The molecular or mixed beam through a slit enters an intermediate chamber which provides vacuum isolation from the source. In this chamber any charged particles present in the beam are removed by means of a pair of deflection plates. Subsequently the beam is chopped and collimated, and it emerges into a scattering chamber where it intersects the electron beam, at right angles, scattering electrons at the chopping frequency. Due to this scattering the electron beam suffers an attenuation which is measured as an ac signal. The neutral beam ultimately enters a mass spectrometer where it is partially ionized to determine the degree of dissociation. From the scattering data for the molecular and partially dissociated beam and the dissociation measurements, the ratio of the atomic to molecular scattering cross sections are obtained as a function of the electron beam energy. To obtain absolute atomic cross sections knowledge of the molecular cross sections, obtained independently, is necessary.

### 4.3 Experimental Data on Electron Swarm Drift Velocities and Electron Diffusion

As stated in Section 4.2 electron swarm experiments provide two basic transport coefficients: the electron swarm drift velocity $w$ and the electron

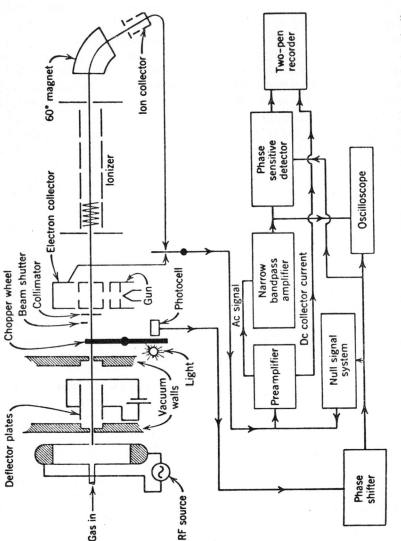

**Figure 4.15** Crossed-beam apparatus for the study of collisions between electrons and chemically unstable species (Neynaber and coworkers, 1961a)

diffusion coefficient $D$. These two coefficients characterize the motion of electrons drifting through a gas under a uniform electric field. For a specified $E/P$ (or $E/N$) their magnitude is determined by the cross sections for elastic and inelastic electron scattering. Thus, experimental values of $w(E/P)$ and $D(E/P)$ can be used to obtain information on the elastic and inelastic processes which occur in the passage of a swarm of electrons through a gaseous medium. Although elastic scattering predominates at low energies, inelastic processes leading to excitation of rotational, vibrational, and compound-negative-ion states is also important in describing the motion of slow electrons through molecular gases. The importance of rotational and vibrational excitation in swarm experiments has long been recognized (e.g. Massey and Burhop, 1952; Craggs and Massey, 1959). The importance of inelastic processes due to excitation of compound negative ion states, has recently been recognized also (e.g. Cottrell and Walker, 1965; 1966; 1967; Christophorou and Compton, 1967; Phelps, 1968; Christophorou and Carter, 1968). In spite of the fact that experimental studies of rotational and vibrational excitation are difficult (owing to the small amount of energy lost by the electrons, and to the relatively poor (fractional) energy resolution; also due to the absence (or low-intensity) of radiation from excited states), considerable knowledge has been accumulated during the recent years. In this section we will discuss the information on elastic electron scattering as obtained from electron swarm experiments and in the next section that obtained from electron beam experiments. Inelastic electron scattering will be the topic of Chapter 5.

### 4.3.1 *Electron Swarm Drift Velocities for Slow Electrons Travelling Through Gases Under the Influence of a Uniform Electric Field*

The drift velocity of electrons travelling through gases is measured as a function of $E/P$ (or $E/N$).* According to well-known theories of electron transport through gases (Loeb, 1955; Allis, 1956) no gas-density dependence of the drift velocity is expected. Indeed, numerous investigators reported no such dependence. However, recently very accurate measurements have shown that $w$ does vary with neutral gas density. Lowke (1963) found that for nitrogen at $77.6°K$ and an $E/P = 0.005$, $w$ decreased by $\sim 3\%$ with increasing pressure from 500 to 2000 torr. Similarly, Grünberg (1967) found $w$ in nitrogen and hydrogen to decrease with increasing gas density and Lehning (1968) and Huber (1968) found a strong dependence of $w$ on pressure for $CO_2$ and $C_2H_6$, respectively. Further, Crompton, Elford and

---

*When reporting $w$ as a function of $E/P$, it is important to quote the temperature of the experiment since the parameter of interest is the ratio of the electric field strength $E$ to the number density of the gas $N$, $E/N$.

McIntosh (1968) reported a density dependence for the drift velocity of electrons in deuterium (at 77°K) for which they could not, as Lowke (1963), account for by diffusion. The reciprocal of these density-dependent drift velocities varied linearly with the gas density (Frommhold, 1968). The reason for the observed density dependence of $w$ is not clear, although Frommhold (1968) suggested that electron trapping by some low-energy resonance states (e.g. rotational resonances in electron–molecule scattering at or near thermal energies) may be responsible for this delay in the electron motion at high gas densities.

The effect of temperature on the drift velocity has also been studied by a number of workers. The drift velocity is sensitive to gas temperature, $T$, and the variation of $w$ with $T$ is useful for obtaining information on the velocity dependence of the cross section for momentum transfer.

In Figures 4.16 to 4.27 representative sets of data on $w$ (and, when available, at different temperatures) for a number of gases are plotted. These data can easily be divided into three groups: those on (i) monatomic gases (Figures 4.16 through 4.19), (ii) diatomic gases (Figures 4.20 through 4.23), and (iii) polyatomic gases (Figures 4.24 through 4.27).

Although recent work appears to be more accurate than the early work of the 1920's and 1930's, a number of early measurements have been included in the graphs. In certain cases the early measurements were the only available information. Included in the graphs are also certain data the accuracy of which is believed to be inferior. This has been done since, often, these data have been used, one way or another, in various calculations, and their inclusion in the graphs offers a direct comparison with more accurate measurements. It might be noted that the accuracy of the drift velocity measurements appears to be independent of the principle of the method employed and that some of the most accurate measurements of $w$ have been conducted by Phelps and coworkers and Crompton and coworkers. The reader is directed to the references quoted in each figure caption for experimental details and for a detailed evaluation of the individual data.

For polar polyatomic gases the drift velocity, especially in the low-energy region, is a strong function of the magnitude of the molecular electric dipole moment. For non-polar molecules the drift velocity depends on the magnitude of the static polarizability as well as on the magnitude of higher molecular moments. Thus, much information can be obtained on molecular structure from drift velocity data.

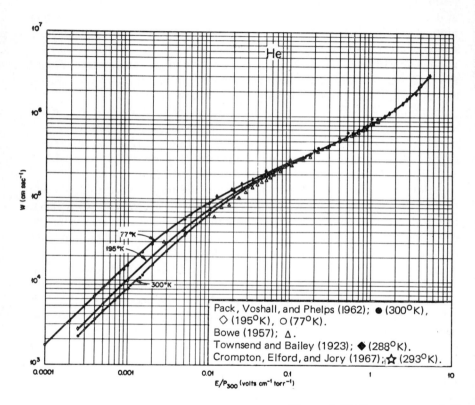

**Figure 4.16** Electron swarm drift velo

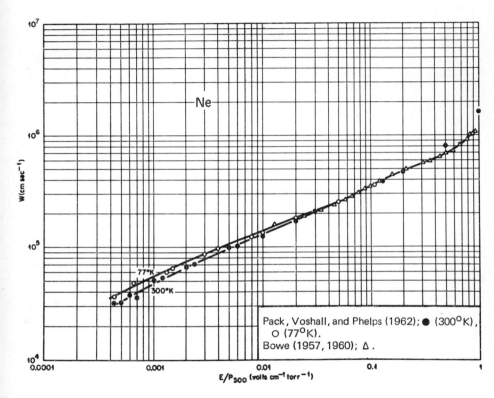

1 Ne of $E/P_{300}$ for He as a function

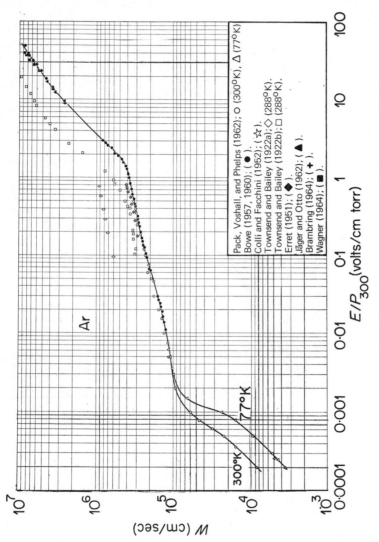

**Figure 4.17** Electron swarm drift velocities as a function of $E/P_{300}$ for Ar

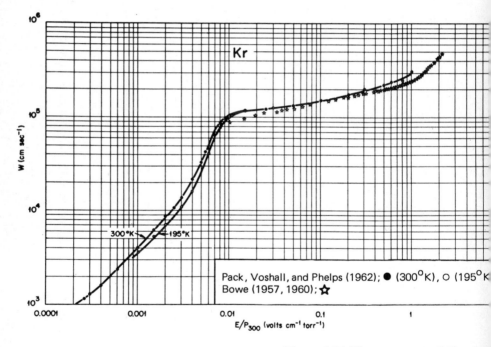

**Figure 4.18** Electron swarm drift veloc

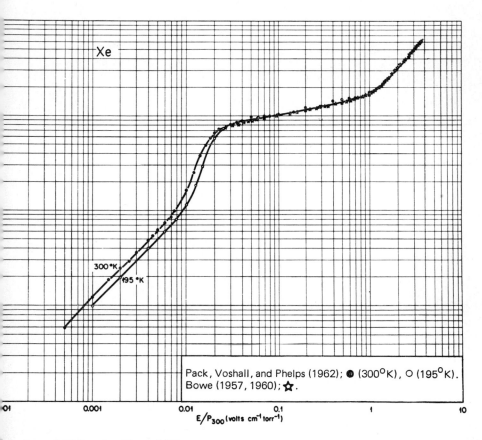

Xe

300 °K

195 °K

Pack, Voshall, and Phelps (1962); ● (300°K), ○ (195°K).
Bowe (1957, 1960); ☆.

E/P₃₀₀ (volts cm⁻¹ torr⁻¹)

unction of $E/P_{300}$ for Kr and Xe

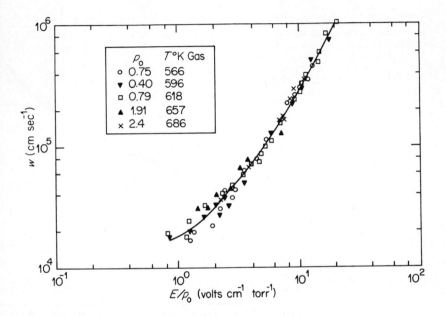

**Figure 4.19** Electron swarm drift velocities as a function of $E/P_0$ ($P_0 = 273 \, P/T$; $566 \leqslant T \leqslant 725°K$) for Cs (Chanin and Steen, 1964)

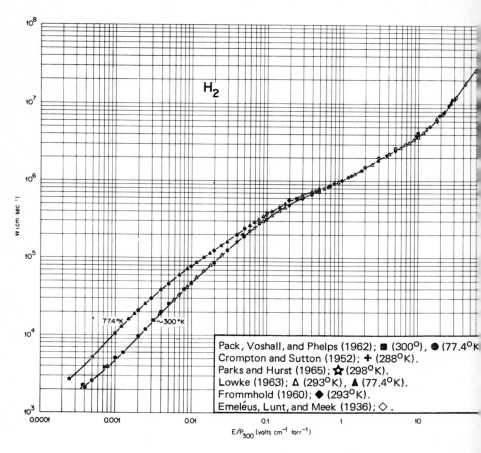

**Figure 4.20** Electron swarm drift v
For electron drift and diffusion
and

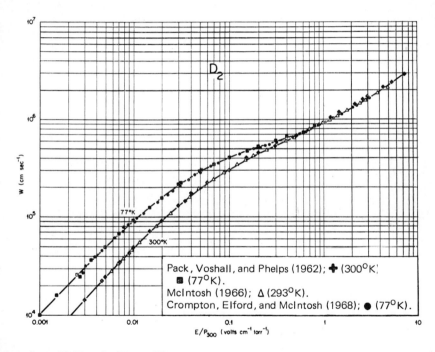

Pack, Voshall, and Phelps (1962); ✚ (300°K);
▧ (77°K).
McIntosh (1966); Δ (293°K).
Crompton, Elford, and McIntosh (1968); ● (77°K).

s a function of $E/P_{300}$ for $H_2$ and $D_2$.
rahydrogen at 77°K see Crompton
(1968)

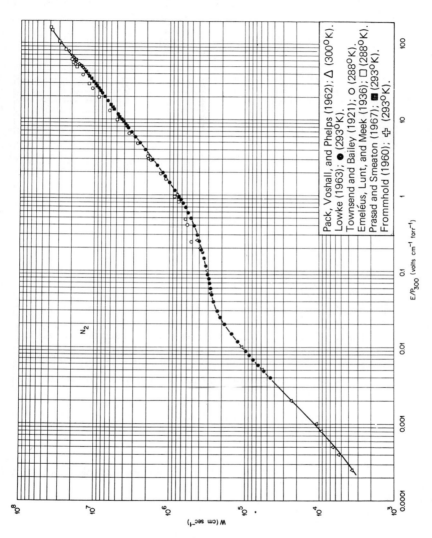

Pack, Voshall, and Phelps (1962): Δ (300°K).
Lowke (1963); ● (293°K).
Townsend and Bailey (1921); O (288°K).
Emeléus, Lunt, and Meek (1936); □ (288°K).
Prasad and Smeaton (1967); ■ (293°K).
Frommhold (1960); ✛ (293°K).

$E/P_{300}$ (volts cm⁻¹ torr⁻¹)

**Figure 4.21** Electron swarm drift velocities as a function of $E/P_{300}$ for $N_2$

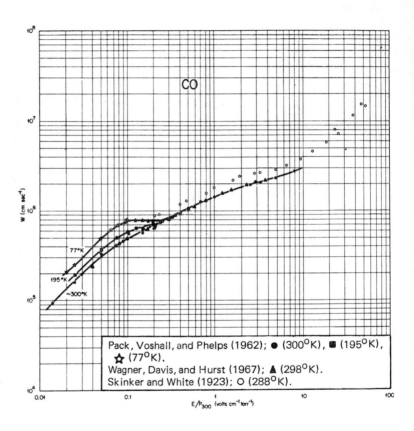

Figure 4.22 Electron swarm drift

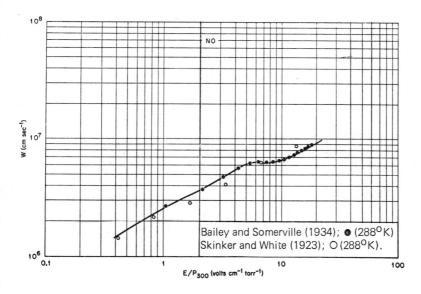

as a function of $E/P_{300}$ for CO

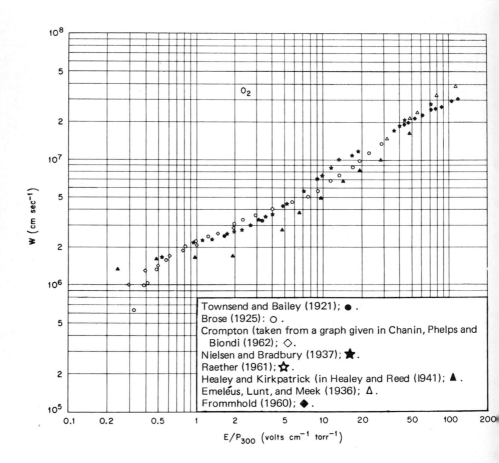

Figure 4.23 Electron swarm drift
Br

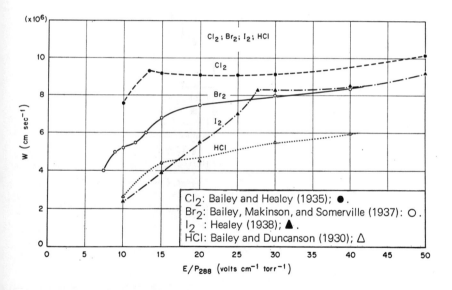

s as a function of $E/P_{300}$ for $O_2$, $Cl_2$,
HCl

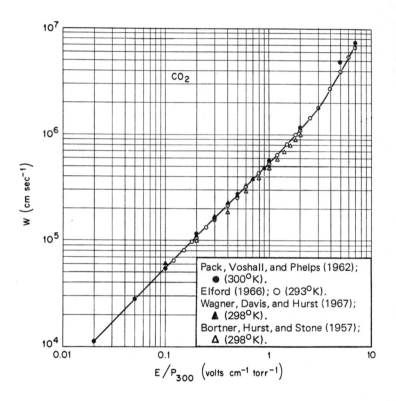

**Figure 4.24** Electron swarm
for

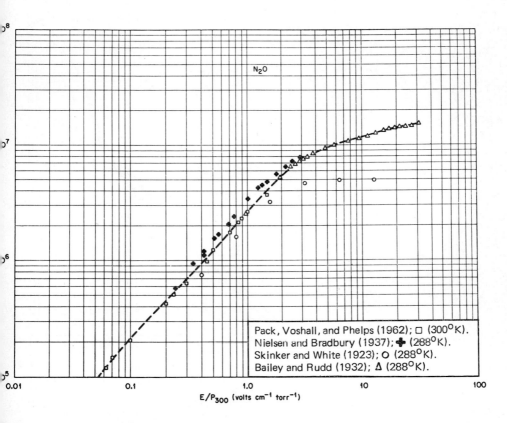

ocities as a function of $E/P_{300}$
d $N_2O$

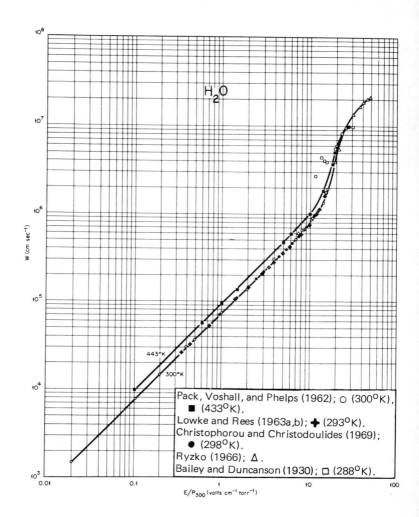

Figure 4.25 Electron swarm dr
N

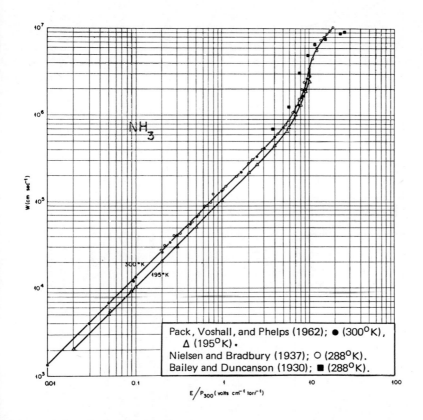

velocities as a function of $E/P_{300}$ for
nd $H_2O$

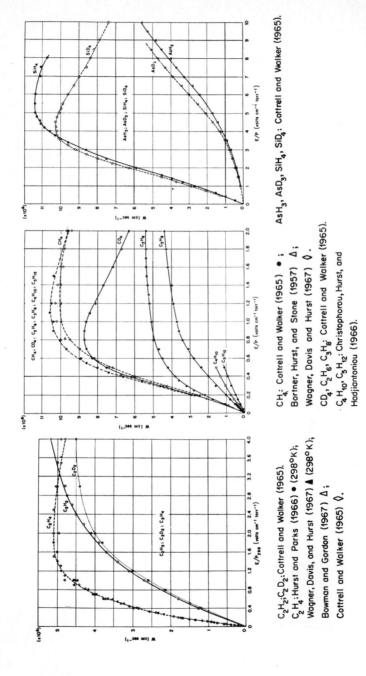

**Figure 4.26** Electron swarm drift velocities as a function of $E/P$ for hydrocarbons and other polyatomic molecules. Data quoted as Cottrell and Walker (1965) were communicated to the author in a tabular form by Dr. I.C. Walker

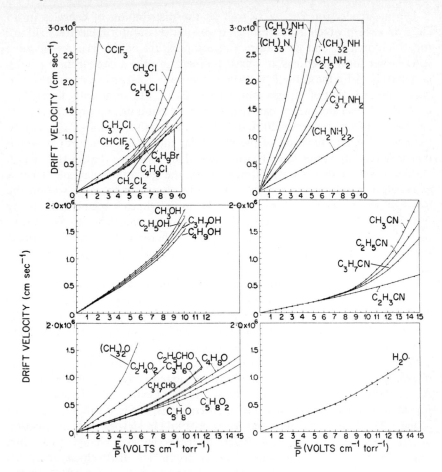

**Figure 4.27** Electron swarm drift velocities as a function of $E/P$ for polar molecules (Christophorou and Christodoulides, 1969). The temperature was $298°K$ and the identification of the compounds is as follows: Chlorotrifluoromethane ($CClF_3$), chloromethane ($CH_3Cl$), chloroethane ($C_2H_5Cl$), 1-chloropropane ($C_3H_7Cl$), chlorodifluoromethane ($CHClF_2$), dichloromethane ($CH_2Cl_2$), 1-chlorobutane ($C_4H_9Cl$), and 1-bromobutane ($C_4H_9Br$). Trimethylamine ($(CH_3)_3N$), diethylamine ($(C_2H_5)_2NH$), dimethylamine ($(CH_3)_2NH$), Ethylamine ($C_2H_5NH_2$), n-propylamine ($C_3H_7NH_2$), and ethylenediamine ($(CH_2NH_2)_2$). Methanol ($CH_3OH$), ethanol ($C_2H_5OH$), 1-propanol ($C_3H_7OH$), and 1-butanol ($C_4H_9OH$). Acetonitrile ($CH_3CN$), propionitrile ($C_2H_5CN$), butyronitrile ($C_3H_7CN$), and acrylonitrile ($C_2H_3CN$). Methyl ether ($(CH_3)_2O$), methyl formate ($C_2H_4O_2$), butyraldehyde ($C_3H_7CHO$), propionaldehyde ($C_2H_5CHO$), acetone ($C_3H_6O$), 2-butanone ($C_4H_8O$), cyclopentanone ($C_5H_8O$), and acetylacetone ($C_5H_8O_2$). Water ($H_2O$): Christophorou and Christodoulides (1969) (△); Pack, Voshall, and Phelps (1962) (○); Lowke and Rees (1963b)□

The mean electron energy as well as the distribution of energies in an electron swarm which drifts through a gas under a uniform electric field are functions of $E/P$, and the nature and temperature of the gas. As we have seen earlier for a given gas and fixed temperature the mean electron energy is determined by the ratio $E/P$. When this ratio is large the mean electron energy exceeds that of the gas molecules $(3/2kT)$. For sufficiently low values of $E/P$ the elect on swarm can reach thermal equilibrium with the gas. It can be seen from Figures 4.20 to 4.27 that there exists an initial $E/P$ region (which is a strong function of the particular molecular structure) in which the drift velocity varies linearly with $E/P$. This indicates that in this region electrons are in thermal equilibrium with the gas molecules and have a Maxwell distribution of energies. As will be shown in the next section the slope of the $w$ versus $E/P$ curves in the initial (linear) region is a strong function of the molecular electric dipole moment, and thus can be used to provide information about the scattering of thermal electrons from polar molecules. With increasing $E/P$ inelastic collisions play an increasingly important part.

The drift velocity of slow electrons through a gas has been found to depend critically on the presence of small amounts of impurities or additives. The effect is a strong function of the nature of the additive as well as the nature of the contaminated gas. In the special case of rare gases the effect of molecular additives is very pronounced since it strongly moderates the electron energy and affects the electron energy distribution. Argon-molecule mixtures have been investigated quite extensively (e.g. Klema and Allen, 1950; Nagy, Nagy and Dési, 1960; Levine and Uman, 1964; Kirshner and Toffolo, 1952; Colli and Leonardis, 1953; Bortner, Hurst and Stone, 1957; Hurst, O'Kelley and Bortner, 1961) in view of their application to radiation detectors. Figure 4.28 demonstrates the large changes in $w$ when minor amounts of water are added in pure argon. The drift velocity increases as a result of a decrease of the mean agitation energy. The effect on the electron energy distribution and average electron energy in argon resulting from the addition of small amounts of hydrogen or carbon dioxide is clearly shown in Figure 4.29 (calculations by Uman (1964)). One part of $H_2$ or $CO_2$ in $10^4$ parts of argon is sufficient to alter appreciably the electron energy distribution and the electron average energy from the values they would have in pure argon. The effect is more pronounced for $CO_2$ since a greater fraction of an electron's energy is lost in a collision with $CO_2$ than $H_2$. This effect should be considered carefully in electron attachment studies involving mixtures of molecules with rare gases. It might be interesting to note also that the effect of trace impurities on the drift velocity of electrons in pure argon has been exploited for the detection of trace concentrations of permanent gases in gas chromatography (Smith and Fidiam, 1964).

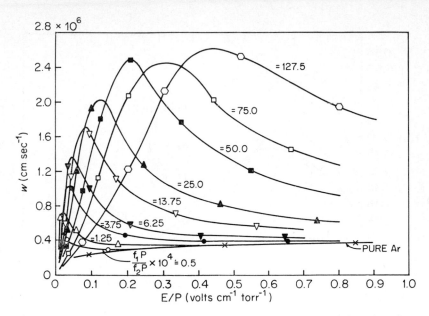

**Figure 4.28** Electron swarm drift velocities in $H_2O$–Ar mixtures (400 torr) (Hurst, O'Kelly and Bortner, 1961). $f_1P/f_2P$ is the ratio of water to argon pressure

The change in the drift velocity when a polar molecule is added in small proportions to a non-polar gas has been exploited as a means of deriving information on the scattering of thermal and near-thermal energy electrons by polar molecules. In this respect, a number of polar molecules have been mixed with ethylene (a non-polar gas) and the drift velocity for the mixtures has been measured as a function of the partial pressures (Hurst, Stockdale O'Kelly, 1963; Christophorou, Hurst and Hadjiantoniou, 1966; Hamilton and Stockdale, 1966; Christophorou and Christodoulides, 1969). A typical set of data is shown in Figure 4.30 which for the gas proportions used can be described by the relationship

$$w_M^{-1}\left(\frac{E}{P}\right) = w_E^{-1}\left(\frac{E}{P}\right) + S\left(\frac{E}{P}\right)\frac{P_I}{P_E} \tag{4.64}$$

where $w_M$ is the drift velocity for the mixture, $w_E$ that for ethylene, $P_I/P_E$ is the ratio of the partial pressures of the impurity to that of ethylene and $S(E/P)$ is the slope of the curves which is a function of $E/P$. Similar effects have been observed when a non-polar molecule of higher static polarizability has been added into another of lower polarizability (Christophorou, Hurst and Hadjiantoniou, 1966).

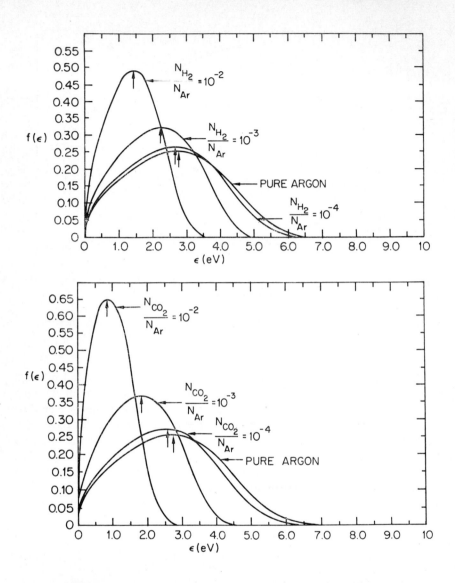

**Figure 4.29** Upper curves: Electron energy distribution function versus electron energy for several values of $H_2$ density to argon density ratio with $E/P$ = 0.5 volts/cm torr at $0\,^\circ$C. Vertical arrows mark maxima of curves. Lower curves: Electron energy distribution function versus electron energy for several values of $CO_2$ density to argon density ratio with $E/P$ = 0.5 volts/cm torr at $0\,^\circ$C. Vertical arrows mark maxima of curves (Uman, 1964)

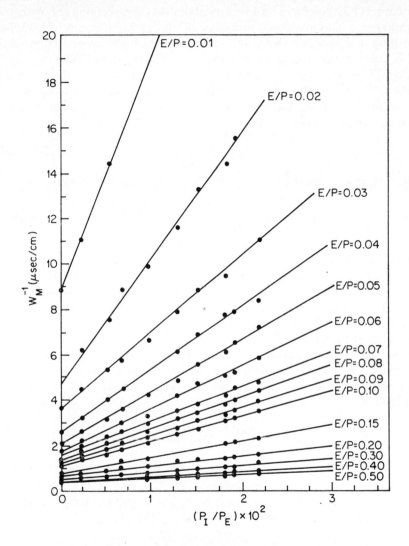

**Figure 4.30** $w_M^{-1}$ versus $P_I/P_E$ for chlorobenzene in ethylene (Christophorou, Hurst and Hadjiantoniou, 1966). $P_I$ and $P_E$ are, respectively, the partial pressures of chlorobenzene and ethylene

The experimental determination of $w(E/P)$ can be used in connection with Equation (4.38) to obtain information about the cross section for momentum transfer $\sigma_m$. This requires knowledge of the electron energy distribution function and also knowledge of the way in which the cross section varies with electron velocity. Both of these requirements are difficult to satisfy. However, in the low $E/P$ region where the drift velocity varies linearly with $E/P$ the distribution function is Maxwellian and for a known (or assumed) form of $\sigma_m$ Equation (4.38) can yield $\sigma_m(v)$. Let us then assume that $\sigma_m$ can be expressed in powers of electron velocity, i.e.

$$\sigma_m(v, \alpha) = A_\alpha/v^\alpha \tag{4.65}$$

where $A_\alpha$ is a constant depending only on $\alpha$. We may show that

$$w = \frac{4}{3(2\pi)^{1/2}} \frac{e}{mN} \frac{E}{P} \frac{2^{\alpha/2}}{A_\alpha} \Gamma(\tfrac{1}{2}\alpha + 2) \left(\frac{m}{kT}\right)^{\frac{1}{2}(1-\alpha)} \tag{4.66}$$

where $\Gamma(x)$ is the gamma function of $x$, and $\alpha$ can be any quantity greater than $-4$. The temperature-dependent $w(T)$ follows immediately from Equation (4.66). For ethylene, from studies of the temperature dependence of $w$ it has been found that $\alpha = 0$ for thermal and near thermal electron energies ($\lesssim 0.07$ eV) (Christophorou, Hurst and Hendrick (1966)). Similar studies have shown that for helium too $\alpha = 0$ (Pack and Phelps, 1961). For a polar molecule, however, $\alpha$ is of the order of two, in agreement with the Born approximation prediction (Section 4.4). Other forms of $\sigma_m$ have been assumed. For example, Christophorou and Pittman (1969) used the expression

$$\sigma_m(v) = \frac{A}{v} + \frac{B}{v^2} \tag{4.67}$$

where $A$ and $B$ are positive constants, while Pack and Phelps (1961) have assumed a power series representation of $N/v_m$ ($v_m$ is the frequency of momentum transfer collisions and $N$ is the gas density) in terms of electron velocity (Phelps, Fundingsland and Brown, 1951), i.e.

$$\frac{N}{v_m} = (\sigma_m v)^{-1} = \sum_j b_j v^j \tag{4.68}$$

Expression (4.68) has been used by a number of workers (e.g. Bowman and Gordon, 1967) to derive information on $\sigma_m$ from measurement of $w(T)$.

The finding by Christophorou, Hurst and Hendrick (1966) that the cross section for momentum transfer in ethylene, $\sigma_E$, is constant over the range of thermal and near thermal energies ($\lesssim 0.07$ eV) and equal to

$(4.37 \pm 0.26) \times 10^{-16}$ cm$^2$ aided the analysis of the drift velocity data for mixtures of polar molecules with ethylene.* Christophorou, Hurst and Hendrick (1966) and Christophorou and Christodoulides (1969) assumed that the cross section for momentum transfer, $\sigma_M(v)$ for a mixture of a polar molecule I, with ethylene E, is given by

$$\sigma_M(v) = \frac{P_I}{P_T} \frac{A_I}{v^2} + \frac{P_E}{P_T} \sigma_E$$

where $P_I$, $P_E$ and $P_T$ are the partial pressures of I and E and the total pressure, respectively, and found that the ratio of the drift velocity in pure ethylene, $w_E$, to the drift velocity in the mixture, $w_M$, is

$$\frac{w_E}{w_M} = \frac{P_E}{P_T}(z^2 e^z \varepsilon_1(z) - z + 1)^{-1} = \frac{P_E}{P_T} f(z) \qquad (4.69)$$

In Equation (4.69), $z = \frac{m}{2kT} \frac{P_I A_I}{P_E \sigma_E}$ and $\varepsilon_1(z)$ is the exponential integral

$$\varepsilon_1(z) = \int_z^\infty \frac{e^{-u} du}{u}$$

The constant $A_I$ is given by

$$A_I = \frac{1}{\beta} \frac{2kT}{m} \sigma_E \bar{S}$$

where $\beta$ is the slope of the $f(z)$ versus $z$ function and $\bar{S} = (w_E w_M^{-1} - 1)/(P_I/P_E)$. The slope $\beta$ is $\leqslant 1$; it changes from 1 as $z \to 0$ to 0.5 as $z \to \infty$. Christophorou and Christodoulides (1969) found that for the few molecules they investigated in the pure form and in mixtures with ethylene the $A_I$ values obtained from the mixtures compared very well with the corresponding ones obtained from the pure gas.

Knowledge of $A_I$ allows determination of an average cross section for momentum transfer over a known energy distribution function. If we define

---

*The investigation of certain molecules in mixtures with another gas is often necessary when due to experimental difficulties (e.g. low vapour pressure, high electron attachment cross sections) the molecule under consideration cannot be studied in the pure form.

an average of the quantity $\sigma_m(v)$ over the normalized Maxwell distribution function $4\pi f_0(v)v^2$ as*

$$\langle\sigma_m(v)\rangle \equiv 4\pi \int_0^\infty \sigma_m(v) f_0(v) v^2 \, dv \tag{4.70}$$

we have $\left(\text{for } \sigma_m(v) = \dfrac{A_1}{v^2}\right)$,

$$\langle\sigma_m(v)\rangle = \frac{m}{kT} A_1$$

### 4.3.2 *Lateral and Longitudinal Diffusion of Slow Electrons Drifting Through Gases Under the Influence of a Uniform Electric Field*

In this section, we present the available data on $D/\mu$ (ratio of the electron diffusion coefficient to electron mobility) in a graphical form (Figures 4.31–4.36), which provides in itself a comparison between the work of the various investigators. When available, data at various temperatures were included. At low $E/P$, $D/\mu$ converges to its thermal value $kT/e$.

The experimental data on electron diffusion, in certain cases, appear to be less consistent than the drift velocity data. Although the measurement of the drift velocity is independent of the principle of the method employed, the measurement of $D/\mu$ by lateral techniques (Townsend method) and by the time-of-flight method differ, often, appreciably. Electron diffusion coefficients as measured by the time-of-flight method are generally smaller in magnitude than those measured with the Townsend method. This difference is found to depend on $E/P$ and also to depend strongly on the

---

*Caution must be taken to define the kind of averaging when evaluating an average of a quantity over a function. Large errors can be introduced by the use of improper averages. For example, for a Maxwell function

$$\frac{1}{\langle v\rangle}\bigg/\left\langle\frac{1}{v}\right\rangle = \frac{\pi}{4}$$

and an error of about 20% is made if one uses the average speed to obtain the cross section. One also finds that

$$\frac{1}{\langle v\rangle^2}\bigg/\left\langle\frac{1}{v^2}\right\rangle = \frac{\pi}{8}$$

and that

$$\left\langle\frac{1}{v^2}\right\rangle\bigg/\frac{1}{\langle v^2\rangle} = 3$$

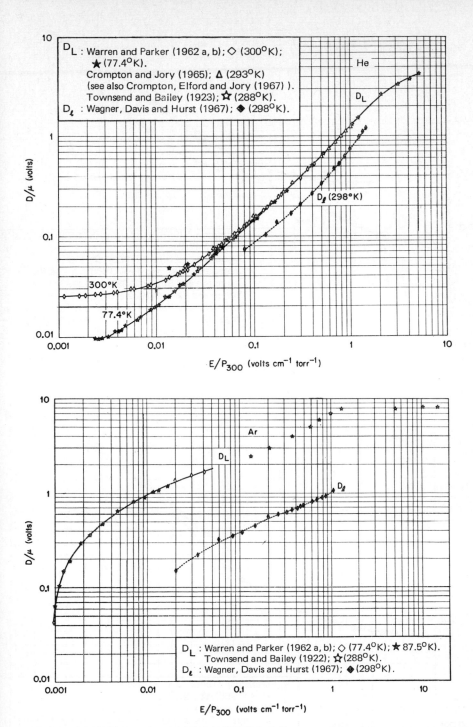

**Figure 4.31** Diffusion coefficient to mobility ratio as a function of $E/P_{300}$ for He and Ar

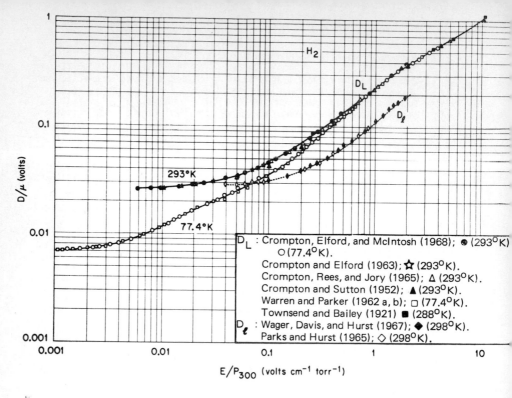

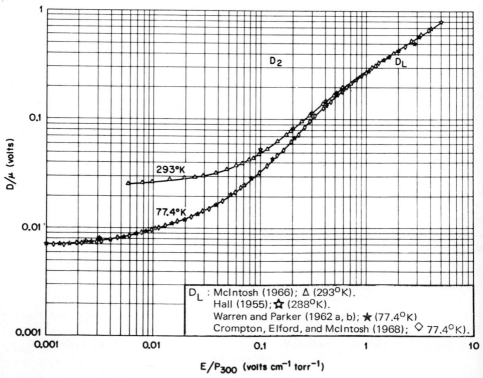

**Figure 4.32** Diffusion coefficient to mobility ratio as a function of $E/P_{300}$ for $H_2$ and $D_2$

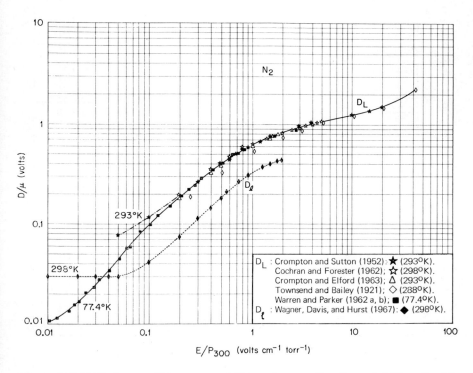

**Figure 4.33** Diffusion coefficient to mobility ratio as a function of $E/P_{300}$ for $N_2$

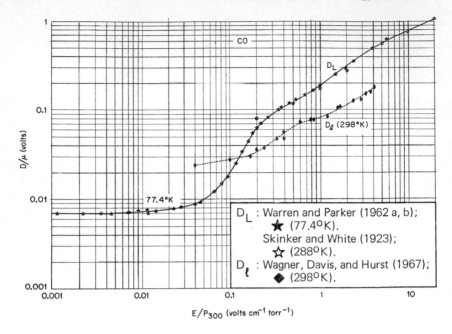

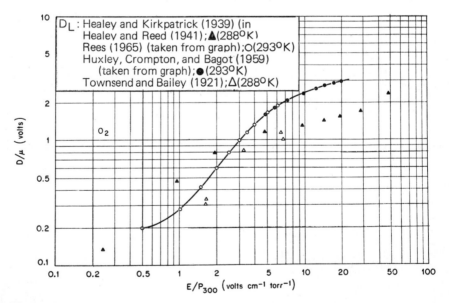

**Figure 4.34** Diffusion coefficient to mobility ratio as a function of $E/P_{300}$ for CO and $O_2$

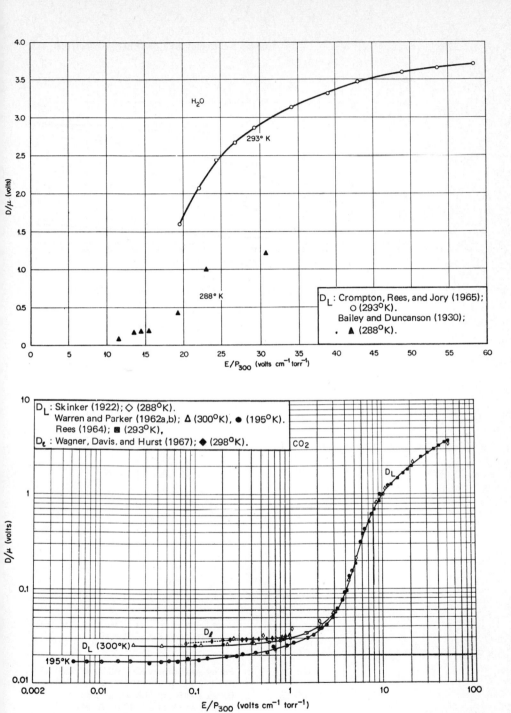

**Figure 4.35** Diffusion coefficient to mobility ratio as a function of $E/P_{300}$ for $H_2O$ and $CO_2$

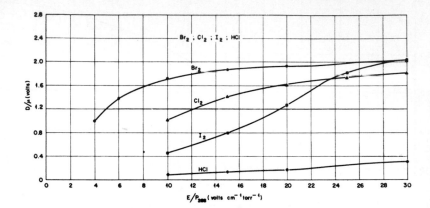

$Br_2:D_L$ : Bailey, Makinson, and Somerville (1937); ● (288°K).
$Cl_2:D_L$ : Bailey and Healey (1935); ▲ (288°K).
$I_2:D_L$ : Healey (1938); ■ (288°K).
$HCl:D_L$ : Bailey and Duncanson (1930); ✦ (288°K).

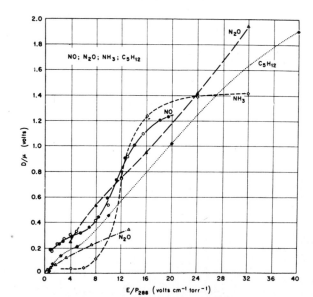

$NO: D_L$ : Bailey and Somerville (1934); ● (288°K).
              Skinker and White (1923); ○ (288°K).
$N_2O:D_L$ : Bailey and Rudd (1932); ▲ (288°K).
              Skinker and White (1923); Δ (288°K).
$NH_3:D_L$ : Bailey and Duncanson (1930); ◇ (288°K).
$C_5H_{12}:D_L$ : McGee and Jaeger (1928); ◆ (288°K).

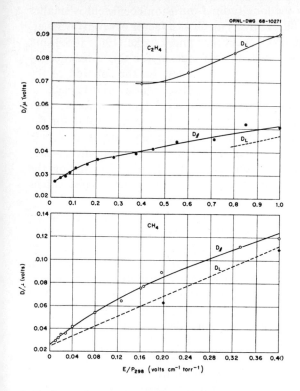

$C_2H_4:D_L$ :  Cochran and Forester (1962); ◇
　　　　　　Bannon and Brose (1928); - - - - - - -.
　　　$D_\ell$ :  Wagner, Davis, and Hurst (1967); ● (298°K).
$CH_4:D_L$ :  Cochran and Forester (1962); ◆
　　　　　　Cottrell and Walker (1965); - - - - - - -.
　　　$D_\ell$ :  Wagner, Davis and Hurst (1967); O (298°K).

re 4.36 Diffusion coefficient to mobility ratio as
a function of $E/P$ for various molecules

nature of the gas; for argon the longitudinal diffusion coefficient is smaller than the lateral diffusion coefficient by as much as a factor of seven (Wagner, Davis and Hurst, 1967; Figure 4.31). Parker and Lowke (1968, 1969) and Lowke and Parker (1969) provided an explanation for this difference by considering the effect of electron density gradients on the solution of the Boltzmann equation representing an electron pulse under a uniform electron field. Their treatment gives an energy distribution which is a function of the position within the electron pulse. This, results in a new value for the diffusion coefficient as compared to the unperturbed case (transverse). Expressing the ratio $D_l/D_L$ as integrals involving the momentum transfer cross section and the unperturbed energy distribution and including the effect of inelastic collisions in their analysis, they have determined values of $D_l$ in reasonable agreement with experiment.

The quantity $De/\mu$ has been referred to by Frost and Phelps (1962) as 'the characteristic energy', $\varepsilon_k$. This quantity is measurable and is directly related to the mean electron energy. The latter quantity can be determined as a function of $E/P$ when the electron energy distribution function is known. Assuming a Druyvesteyn or a Maxwell energy distribution function and making use of the data on $D/\mu$ summarized in this section, the mean electron energy can be determined as a function of $E/P$ by using the simple expressions listed in Table 4.2. Once the mean electron energy is obtained, mean values of a number of physical parameters describing the motion of the electrons through a gaseous medium can be calculated using simple relations such as those listed in Table 4.2. Since the quantity $D/\mu$ is a measure of the average electron energy which, in turn, is sensitive to the different cross sections, the $D/\mu$ versus $E/P$ curves reflect the gross features of the cross sections involved. Inversely, much information on the elastic and inelastic scattering cross sections can be obtained from the data on $w$ and $D/\mu$.

### 4.3.3 Energy Lost by Slow Electrons in Collisions with Atoms and Molecules

The simplest quantity which measures the energy lost in a collision of a low-energy electron with an atom or a molecule is the mean fractional energy loss per collision, $n$ (Table 4.2). Making use of the expression for $n$ given in Table 4.2, Christophorou and Carter (1968) determined $n_D$ as a function of the mean electron energy, assuming a Druyvesteyn energy distribution function. Typical data for monatomic, diatomic, and polyatomic gases are shown in Figure 4.37. The values of $n_D$ are much higher than can be accounted for by considering elastic energy losses and thus it is apparent that in a small proportion of a large number of encounters an electron loses a large proportion of its energy. The majority of encounters, however, are elastic (see Figure 4.38 and compare $v_u/N$ and $v_m/N$ at low energies). We

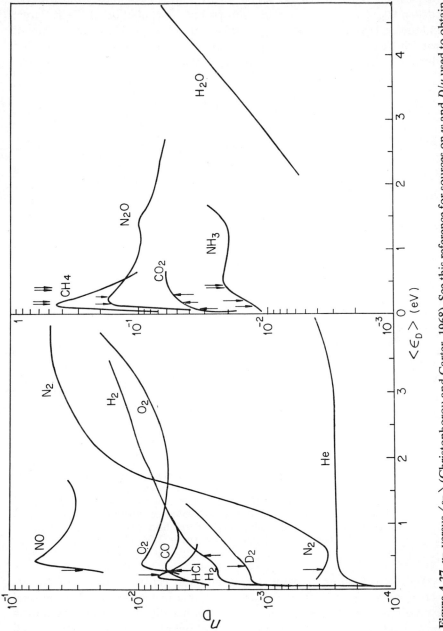

**Figure 4.37** $n_D$ versus $\langle\varepsilon_D\rangle$ (Christophorou and Carter, 1968). See this reference for sources on $w$ and $D/\mu$ used to obtain $n_D$ and $\langle\varepsilon_D\rangle$

note from the results presented in Figure 4.37 that $n_D$ for He in the energy range from ~0.3 to ~4 eV, is ~$2.9 \times 10^{-4}$ which is almost exactly equal to $2m/M_{He}$ ($\simeq 2.7 \times 10^{-4}$). This shows that $\theta$ (Equation (4.7)) is large, indicating large-angle scattering in this energy region. This is in agreement with single collision experiments using monoenergetic electrons (Massey and Burhop, 1952). Similarly, for argon $n_D$ is very close to $2m/M_{Ar}$ at ~4 eV.

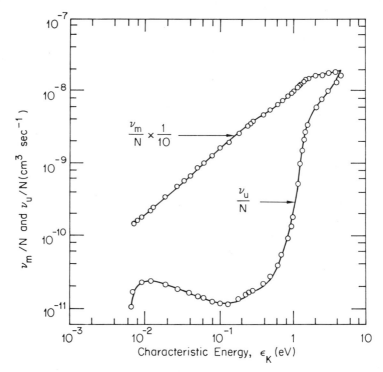

**Figure 4.38** Effective momentum transfer $\nu_m$ and energy exchange $\nu_u$ collision frequencies in $N_2$ at 77°K normalized to the neutral particle density $N$. The smooth curves represent an average of the best available experimental data and the points represent theoretical calculations by Engelhardt, Phelps, and Risk (1964) (see this reference as to the cross sections used) using no polarization correction and a positive quadrupole moment = $1.04\ ea_0^2$

The large fractional energy losses and the general features of the $n_D$ versus $\langle \varepsilon_D \rangle$ curves—note the distinct maxima in the region of vibrational excitation—can be explained in terms of either direct molecular excitation or indirect via the decay of a low-energy resonant state (Christophorou and Carter, 1968). In the latter case the location of the resonance rather than the excitation thresholds determine the position of the maximum. Energies of particular vibrational modes are indicated in Figure 4.37 by the arrows.

The quantity $n$ is composed of an elastic and an inelastic part. A convenient way to separate the effects of elastic and inelastic collisions has been suggested by Frost and Phelps (1962). They defined two different collision frequencies, which describe separately the effects of momentum transfer (elastic) and inelastic collisions. The first they called the effective frequency for momentum transfer or elastic collisions, $\nu_m$, and is defined by

$$\frac{\nu_m}{N} = \frac{e}{m\mu N} = \frac{e}{m} \frac{1}{w} \frac{E}{N} \tag{4.71}$$

The second, the energy exchange collision frequency $\nu_u$, is defined by

$$\frac{\nu_u}{N} = \frac{wE/N}{D/\mu - kT/e} = \frac{weE/N}{\varepsilon_k - kT} \tag{4.72}$$

From Equations (4.71) and (4.40) we see that $\mu N$ is inversely proportional to a function of $\sigma_m$. Thus, we expect $\nu_m/N$ to be proportional to the magnitude of $\sigma_m$. When $\nu_m/N$ is plotted as a function of the characteristic energy it is found to be very nearly independent of the details of the electron energy distribution function for gases in which $\sigma_m$ and the inelastic collision cross sections do not vary too rapidly with electron energy (Frost and Phelps, 1962). On the other hand, the quantity $\nu_u$ (Equation 4.72) is the ratio of the power input per electron to the excess energy of the electron over its thermal value, and since the power input is equal to the rate of energy loss, $\nu_u$ is a measure of the frequency of collisions in which the electrons lose their excess energy. For gases in which the inelastic cross sections do not vary too rapidly with electron energy, $\nu_u/N$ was found (e.g. Frost and Phelps, 1962; Engelhardt and Phelps, 1963; Engelhardt, Phelps and Risk, 1964; Hake and Phelps, 1967) to be a good measure for the cross sections for inelastic collisions. It is thus convenient to plot the energy exchange collision frequency as a function of the characteristic energy. An example of the variation of $\nu_u/N$ and $\nu_m/N$ with the characteristic energy $\varepsilon_k$ is shown in Figure 4.38 (Engelhardt, Phelps and Risk, 1964) for $N_2$. Similar plots have been reported for $H_2$ (Frost and Phelps, 1962), $H_2$ and $D_2$ (Engelhardt and Phelps, 1963), $N_2$ (Frost and Phelps, 1962; Engelhardt, Phelps and Risk, 1964) and $O_2$, CO and $CO_2$ (Hake and Phelps, 1967); also on rare gases by Phelps and coworkers and on some non-polar polyatomic molecules by Pollock (1968).

## 4.4 Scattering Cross Sections Derived from Swarm Experiments

### 4.4.1 *Momentum Transfer Cross Sections*

Expressions (4.65), (4.67) and (4.68) have been assumed by a number of investigators to enable deduction of $\sigma_m$ from electron swarm transport data. Using expression (4.65) for $\sigma_m$, Christophorou and Pittman (1970) obtained the power $\alpha$ of $v$ from measurements of the drift velocity as a function of gas temperature $T$ at thermal and epithermal energies. Christophorou and Pittman (1970) restricted themselves to values of $E/P$ low enough that $w$ varied linearly with $E/P$ and from their $w(T)$ measurements they determined the slope $S(\equiv w/E/P)$ as a function of $T$. From the $S(T)$ data they obtained the power $\alpha$ of $v$ from Equation (4.65) using Equation (4.38), by assuming a Maxwell distribution function for the electron energies, and applying a least-squares analysis whereby an optimum value of $\alpha$ (and $A_\alpha$) was found for which the differences between the calculated and the experimental $S$ as a function of $T$ were minimized. They then evaluated $\langle\sigma_m(v)\rangle_{v^\alpha}$ using Equations (4.65) and (4.70), the determined optimum values of $\alpha$ and $A_\alpha$, and a Maxwell distribution function for the electron energies. Their results on $A_\alpha$, $\alpha$, and $\langle\sigma_m(v)\rangle_{v^\alpha}$ are listed in Table 4.3. for the fourteen molecules studied. In Table 4.3 $\langle\sigma_m(v)\rangle_{v^2}$ for $\alpha = 2$, and $\langle\sigma_m(v)\rangle_A$ as determined from the Born approximation calculation of Altshuler (1957) (see Section 4.4.3.2) are given for comparison.

It is seen from the data listed in Table 4.3 that the values of $\alpha$ are around its Born approximation value of 2, showing that most of the scattering can be accounted for by considering the direct effect of the magnitude of the electric dipole moment. However, the magnitude of $\langle\sigma_m(v)\rangle_{v^\alpha}$ and its variation with $\mu$ (see Table 4.3 and Figure 4.45 later in this chapter) show the inadequacy of the Born approximation for pure dipole scattering to explain in detail the experimental results (see further discussion of this topic in Section 4.4.3.2). $P$ calculations of $\sigma_m$ have also been made using expression (4.67) (Christophorou and Pittman, 1969) and (4.68) (Pack, Voshall and Phelps, 1962). An elaborate analysis of electron transport data by Engelhardt and Phelps (1963) and Frost and Phelps (1964) yielded accurate momentum transfer cross sections as a function of electron energy for a number of simple molecules. The results of these workers on $H_2$ and $N_2$ and $O_2$, CO and $CO_2$ are compared with other swarm and beam data in Figures 4.39 and 4.40, respectively. The large cross sections for $CO_2$ may be due to its sizeable quadrupole moment. For this molecule $\sigma_m$ varies as $\varepsilon^{-\frac{1}{2}}$ at very low energies (Hake and Phelps, 1967). Some direct evidence for the deep minimum in $\sigma_m$ at $\sim 1$ eV can be found in the scattering data of Ramsauer and Kollath (1930, 1932). The peak at $\sim 4$ eV coincides with a maximum observed by Boness and Schulz (1968) in an energy loss experiment, and may be associated with a compound negative ion resonance

**Table 4.3** Values of the electric dipole moment $\mu$, the constants $A_\alpha$ and $\alpha$ for $\sigma_m (\equiv A_\alpha / v^\alpha)$, and the cross sections $\langle m\sigma(v)\rangle_{v\alpha'}$, $\langle\sigma_m(v)\rangle_{v^2}$ and $\langle\sigma_m(v)\rangle_A$ (see text), for polar molecules (Christophorou and Pittman, 1970)

| Molecule | $\mu^a$ (debye) | $A_\alpha$ (cm²(cm/sec)$^{(\alpha)}$) | $\alpha$ | $\langle\sigma_m(v)\rangle_{v^\alpha} \times 10^{14}$ b (cm²) | $\langle\sigma_m(v)\rangle_{v^2} \times 10^{14}$ b (cm²) | $\langle\sigma_m(v)\rangle_A \times 10^{14c}$ (cm²) |
|---|---|---|---|---|---|---|
| Dimethylamine | 1.0 | 13.79 | 1.79 | 3.89 | 5.16 | 3.808 |
| Chlorobenzene | 1.61 | 3944 | 2.375 | 34.2 | 18 | 9.871 |
| Butyl alcohol | 1.63 | 3.07 | 1.955 | 13.42 | 14.45 | 10.12 |
| n-Propyl alcohol | 1.66 | 0.966 | 1.887 | 11.82 | 13.85 | 10.49 |
| Methyl alcohol | 1.7 | 37.98 | 2.113 | 15.26 | 13.00 | 11.006 |
| Ethyl alcohol | 1.7 | 8.923 | 2.025 | 13.62 | 13.43 | 11.006 |
| Water | 1.85 | 175.6 | 2.17 | 30.44 | 23.67 | 13.034 |
| Chloromethane | 1.88 | 1545 | 2.315 | 31.67 | 18.39 | 13.459 |
| 1-Chlorobutane | 2.08 | 419 | 2.232 | 28.96 | 20.22 | 16.476 |
| Propionaldehyde | 2.74 | 985 | 2.27 | 39.5 | 26.49 | 28.589 |
| Acetone | 2.87 | 1466 | 2.285 | 46.83 | 29.5 | 31.369 |
| Cyclopentanone | 3.3 | 96.1 | 2.115 | 37.67 | 31.00 | 41.47 |
| Acrylonitrile | 3.89 | 7.71 | 1.946 | 38.59 | 42.71 | 57.627 |
| 1-Butyronitrile | 4.06 | 417 | 2.184 | 59.1 | 44.2 | 62.774 |

[a] See detailed values in Christophorou and Christodoulides (1969).

[b] For $T = 298°$K.

[c] $\langle\sigma_m(V)\rangle_A = 3.8083 \times 10^{-14} \mu^2 (cm^2 (\mu$ in debye)).

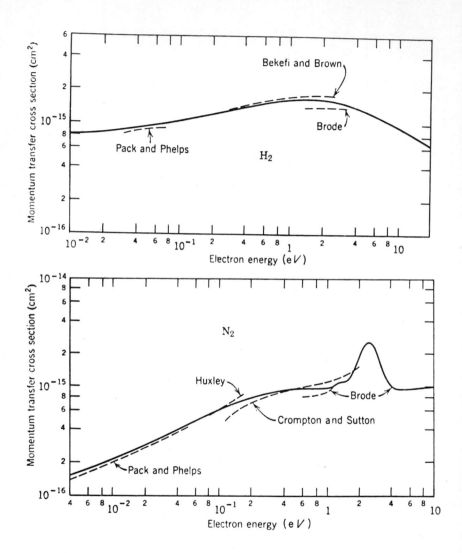

**Figure 4.39** Momentum transfer and total scattering cross sections for $H_2$ and $N_2$. $H_2$: Frost and Phelps (1962): solid line; $\sigma_m$. Also shown are the data of Pack and Phelps (1961) and Bekefi and Brown (1958) on $\sigma_m$ and data on $\sigma_T$ (Brode, 1933). $N_2$: Frost and Phelps (1962): solid line; $\sigma_m$. Also shown are the data of Pack and Phelps (1961), Huxley (1956, 1959) and Crompton and Sutton (1952) on $\sigma_m$, and data on $\sigma_T$ (Brode, 1933)

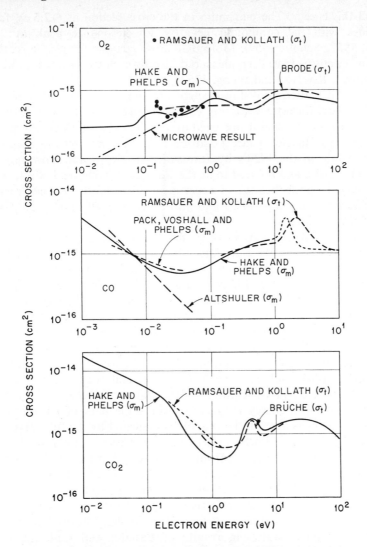

**Figure 4.40** Momentum transfer and total scattering cross sections for $O_2$, CO and $CO_2$ (Hake and Phelps, 1967). $O_2$: Hake and Phelps (1967): $\sigma_m$. Ramsauer and Kollath (1930): $\sigma_T$. Brode (1933): $\sigma_T$. Also shown is $\sigma_m$ derived from microwave experiments (see discussion in Hake and Phelps (1967)). CO: Hake and Phelps (1967): $\sigma_m$. Pack, Voshall, and Phelps (1962): $\sigma_m$. Ramsauer and Kollath (1930): $\sigma_T$. Altshuler (1957) (theory): $\sigma_m$. $CO_2$: Hake and Phelps (1967): $\sigma_m$. Ramsauer and Kollath (1930): $\sigma_T$. Brüche (1927a): $\sigma_T$

around this energy. The maximum in the cross section at $\sim 2.5$ eV for $N_2$ coincides with the compound negative ion resonance peak at 2.3 eV (Haas, 1957; Schulz, 1959). Molecular cross sections for elastic scattering for slow electrons have been measured also by cyclotron resonance absorption (Bayes, Kivelson and Wong, 1962).

The elastic scattering of slow electrons by the rare gas atoms is of special interest. Here, inelastic scattering can be neglected up to several electron volts owing to the high-lying excitation potentials of these atoms. The transparency of the heavy rare gas atoms to low-energy electrons, discovered independently in the 1920's by Ramsauer (1921a,b) and Townsend and Bailey (1922a,b), and referred to as the Ramsauer–Townsend effect, offered a challenge to both theory and experiment. This effect has been reviewed by Massey and Burhop (1952) and will be elaborated upon briefly at the end of this section.

A number of determinations of $\sigma_m$ for He, Ar, Kr and Xe from analyses of transport coefficients and data obtained from microwave and total scattering beam* experiments have been surveyed by Frost and Phelps (1964). More recently Crompton, Elford and Jory (1967) reinvestigated the elastic scattering of slow electrons by the helium atom using the swarm method. Similar studies have been made by Golden (1966b), and Golden and Bandel (1965a, 1966) for helium and argon using the Ramsauer technique. In Figure 4.41 the data on the elastic scattering of slow electrons from helium are summarized (Crompton, Elford and Jory, 1967) (see also Frost and Phelps, 1964; Golden, 1966b). Curves A and B are, respectively, the swarm-deduced $\sigma_m$ of Crompton, Elford and Jory (1967) and Frost and Phelps (1964). The energy range over which the data of Crompton and coworkers are considered to be reliable is indicated by the full curve. Also, as stated by Frost and Phelps (1964), their analysis is not expected to yield satisfactory data in the range from 5 to 10 eV, because at these energies inelastic collisions cannot be neglected. Curve C in Figure 4.41 is the momentum transfer cross section obtained by Golden (1966b) using O'Malley's (1963) modified effective range theory, while curve D was obtained by Crompton and coworkers from Golden's beam data using the theoretical angular scattering results of LaBahn and Callaway (1966). Curve E was obtained by Crompton, Elford and Jory (1967) using the total scattering cross section of Golden and Bandel (1965a) and the differential scattering cross sections of Ramsauer and Kollath (1932). The latter data have been used by Barbiere (1951) also, who calculated $\sigma_m$ from the total scattering cross section data of Ramsauer and Kollath (1929) (curve F,

---

*The reader is reminded that to compare the total scattering cross section $\sigma_T$ obtained from single-collision beam experiments with the momentum transfer cross section $\sigma_m$ derived from swarm data, the differential scattering cross section $I_s(\theta)$ must be known.

Figure 4.41). The agreement between these curves is reasonable, and the differences can be due to any of the measured quantities ($\sigma_T$, $I_s(\theta)$, swarm transport data) or the theory. It might be noted that the early beam cross section data of Ramsauer and Kollath (1929) and Normand (1930) for helium show some fine structure in the energy range from 0.5 to 2 eV. A resonance in helium at 0.45 eV has also been reported by Schulz (1965). However, neither the recent beam work of Golden and Bandel (1965a) and Bullis and coworkers (1967), nor the swarm data of Crompton, Elford and Jory (1967) showed evidence for such structure. It is generally agreed that the structure in the helium cross section function at low energies is not real.

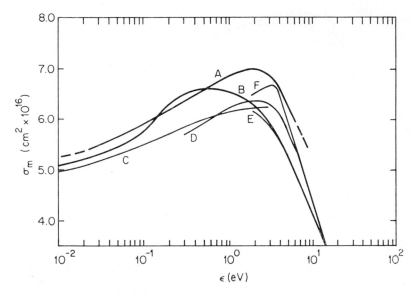

**Figure 4.41** He: A comparison of the momentum transfer cross section found from swarm experiments with the cross section deduced from single-scattering experiments (Crompton, Elford, and Jory, 1967). Curve A, Crompton, Elford, and Jory (1967); Curve B, Frost and Phelps (1964); Curve C, Golden (1966a) – O'Malley (1963); Curve D, Golden and Bandel (1965a) – LaBahn and Callaway (1966); Curve E, Golden and Bandel (1965a) – Ramsauer and Kollath (1932), Crompton and coworkers (1967); Curve F, Ramsauer and Kollath (1929, 1932) – Barbiere (1951). (See text)

Theoretical calculations on electron collision cross sections in helium have been discussed, among others, by Moiseiwitsch (1962). A recent comparison between experimental and theoretical momentum transfer cross sections is shown in Figure 4.42 (Crompton, Elford and Jory, 1967). The four curves shown for LaBahn and Callaway (1966) are each based on a particular set of assumptions concerning the interaction forces between the scattered electron and the helium atom. The curve marked adiabatic-exchange-d was obtained by LaBahn and Callaway (1964) using the adiabatic approximation, including electron exchange and considering only the dipole term of the polarization interaction. The curve marked adiabatic-exchange-T was obtained as the earlier one but with inclusion of higher order terms of the

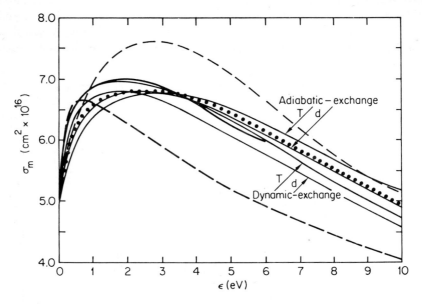

**Figure 4.42** He: A comparison between theoretical and experimental momentum transfer cross sections (Crompton, Elford, and Jory, 1967). —— Crompton, Elford, and Jory (1967); – – Frost and Phelps (1964); – · – Bauer and Browne (1964); —— LaBahn and Callaway (1966), ••• Williamson and McDowell (1965). (See text)

polarization interaction. The curves marked dynamic-exchange-d and dynamic-exchange-T were obtained by taking into account velocity-dependent interaction terms; in the former case only the dipole term of the polarization interaction was included but higher terms of the polarization interaction were included in the latter case (LaBahn and Callaway, 1966). Williamson and McDowell (1965) and Bauer and Browne (1964) used the adiabatic approximation (Figure 4.42). The data of Crompton, Elford and Jory (1967) agree very well with the curve marked dynamic-exchange-T.

The variation with energy of $\sigma_m$ (and $\sigma_T$) for neon does not differ substantially from that for helium (Ramsauer and Kollath, 1929). However, the situation is distinctly different for the heavier of the rare gases. In Figure 4.43 $\sigma_m$ (and $\sigma_T$) for Ar, Kr and Xe are plotted, and they clearly

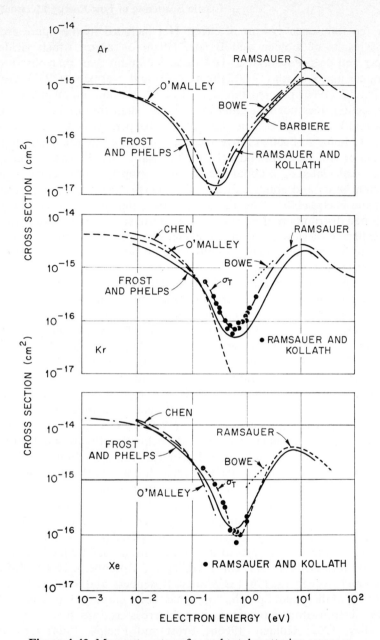

**Figure 4.43** Momentum transfer and total scattering cross sections for Ar, Kr, and Xe (Frost and Phelps, 1964). O'Malley (1963): $\sigma_m$. Frost and Phelps (1964): $\sigma_m$. Ramsauer and Kollath (1932): $\sigma_T$. Barbiere (1951): $\sigma_m$. Bowe (1960): $\sigma_m$. Ramsauer (1921): $\sigma_T$. Chen (1963): $\sigma_m$

show the Ramsauer–Townsend effect. Not included in the figure are the recent results of Golden and Bandel (1966) on argon which yielded a sharper and deeper Ramsauer–Townsend minimum than that observed by Ramsauer and Kollath (1929) (Figure 4.43) and Normand (1930). Golden and Bandel (1966) pointed out that although the minimum in their total cross section does not lie significantly deeper than the minimum in $\sigma_m$ of Frost and Phelps (1964), it is significantly sharper. Also, Golden (1966b), using O'Malley's effective range theory, deduced $\sigma_m$ at low energies from the data of Golden and Bandel (1966). The Ramsauer–Townsend minimum of the so calculated $\sigma_m$ is about a factor of 3.7 deeper than the corresponding minimum in $\sigma_T$ and about a factor of 4.4 deeper than that of $\sigma_m$ given by Frost and Phelps (1964). The difference in the latter case has been attributed to a broad distribution of energies in the swarm experiments. Cross sections for scattering of slow electrons by argon obtained using microwave techniques by Daiber and Waldron (1966) are in fair agreement with those of drift measurements.

For krypton we note (Figure 4.43) that there is a considerable discrepancy at energies below the Ramsauer–Townsend minimum. At energies $\gtrsim 2$ eV the accuracy of $\sigma_m$ derived from drift velocity data, as in the case of argon, is relatively low because of the rapid variation of $\sigma_m$ with energy. Here it is also seen that $\sigma_m$ lies well below $\sigma_T$* (Figure 4.43). For xenon the differences between the $\sigma_m$ of Frost and Phelps (1964) and the $\sigma_T$ of Ramsauer and Kollath (1932) are considerably smaller than in the case of krypton. Note, however, the differences (a factor of about two) between the data of Frost and Phelps (1964) and Bowe (1960). Recent determinations of $\sigma_m$ for Kr and Xe by Hoffmann and Skarsgard (1969) using a microwave technique agree best with the results of Frost and Phelps (see Figure 4.43), but they are generally lower and indicate a deeper Ramsauer minimum.

It has become evident from the preceding discussion, that the comparison of the cross section data obtained by swarm and beam techniques relies heavily on the knowledge of the differential scattering cross section $I_s(\theta)$ through which the various transport cross sections are related to the total scattering cross section $\sigma_T$. For very low energies, $I_s(\theta)$ is difficult to obtain experimentally and one resorts to theoretical estimates. Often, effective range theory (O'Malley, 1963; Spruch, O'Malley and Rosenberg, 1960; O'Malley, Spruch and Rosenberg, 1961; O'Malley, Rosenberg and Spruch, 1962; and Berger, O'Malley and Spruch, 1965) has been employed to obtain low-energy electron–atom momentum transfer cross sections from data on $\sigma_T$ (O'Malley, 1963; Golden, 1966b; Frost and Phelps, 1964; Crompton, Elford and Jory, 1967).

---

*It should be borne in mind that $\sigma_T$ weights all scattering angles equally, whereas $\sigma_m$ weights the scattering by $(1 - \cos\theta)$, where $\theta$ is the scattering angle.

### 4.4.2 *The Ramsauer–Townsend Effect*

As it can be seen from Figure 4.43, the most remarkable feature of the electron–rare gas atom cross sections is the transparency of the heavier rare gas atoms toward low-energy electrons around $\sim 0.5$ eV. The explanation of this effect by quantum theory constituted a triumph for the theory itself (see discussion in Massey and Burhop (1952)). The first quantitative explanation of the effect required introduction of a polarization interaction necessary to account for the perturbation of the atomic field by the incident electron. In general, the interaction forces between the scattered electron and the scattering atom are of three types: static Coulomb, those due to exchange and those resulting from the distortion of the atomic system by the presence of the impacting electron. The third type of interaction is not only due to the effect of direct polarization but also due to higher order terms of the interaction as well. Most theoretical treatments assume that the velocity of the incident electron always remains substantially smaller than that of the atomic electrons (adiabatic approximation) and usually retain only the dipole term of the polarization interaction. However, at higher electron energies the atomic system fails to readjust instantaneously as the position of the incident electron changes, and thus the adiabatic approximation is no longer valid, i.e. velocity-dependent interaction terms should be considered.

Let us now see qualitatively why the Ramsauer–Townsend effect might be expected in the heavier rare gas atoms. Owing to the compact structure of a rare gas atom, the force it exerts on an approaching electron is a short-range one and it is strong for only small separation distances $r$. Let us, then, approximate the interaction potential by a deep narrow spherical potential well of depth $D$ and range $a$, so that $V(r) = -D$ for $r < a$ and $V(r) = 0$ for $r \geq a$. For a slow electron, the radius of the potential (effective range of electron–atom interaction) is small compared with the electron wavelength, and only $s$-wave scattering will make an important contribution to $\sigma_e$ (expressions (4.31) and (4.32)) since $\eta_l = 0$ for $l > 0$. Thus

$$\sigma_e = \frac{4\pi}{k^2} \sin^2 \eta_0 \qquad (4.73)$$

and is equal to zero for $\eta_0 = n\pi$; where $n$ is an integer. Now, for short-range and weak fields, such as those for helium (and neon), as $k \to 0$ (low impact velocities), $\sin \eta_0 \to 0$ at the same rate as does $k$, so that $\sigma_e$ remains finite at low impact velocities, i.e. no minimum in $\sigma_e$ or $\sigma_m$ is observed. When the short-range field is strong, such as for the higher-atomic-number rare gas atoms (Ar, Kr and Xe), the $s$-wave contribution to $\sigma_e$ may become zero for certain impact velocities (due to strong scattering introducing $\eta_0$ phase shifts equal to $n\pi$), and since the higher wave contributions are known to be small, $\sigma_e$ falls to a minimum value at these impact velocities. This effect may be regarded as a manifestation of the diffraction of the electron wave by the target atom. The effect is most pronounced for xenon, it is strong for krypton and argon, and is not observed for the lower-atomic-number rare gas atoms neon and helium (Figure 4.42). Since the Ramsauer–Townsend effect is expected in systems in which the attractive field is strong enough to introduce a phase shift of $n\pi$, the effect should not be significant in systems where the interaction is of sufficiently long range for higher-order phase shifts to be significant. In molecular systems, long-range dipole and other multipole-moment-interactions do not, as a rule, allow such an effect. Methane, owing to its high symmetry, appears to be an exception. A Ramsauer–Townsend effect has been observed (Ramsauer and Kollath, 1932) for this molecule indicating the short-range nature of the strong $e$–$CH_4$ interaction.

### 4.4.3 *Macroscopic Potentials Based on the Static Polarizability and the Permanent Electric Dipole Moment and the Scattering of Slow Electrons by Molecules*

The magnitude of the scattering cross section in an electron–atom (molecule) collision is a measure of the interaction potential, which in turn is determined by the atomic and molecular structure and is obtained, in principle, by solution of the Schrödinger equation. Quite generally, however, the scattering of low-energy electrons by even the simplest atoms is difficult to treat by solution of the Schrödinger equation. The development of a satisfactory theory which would enable one to work from the observed scattering to the charge distribution in the atom or the molecule is by no means easy. For atoms more complex than hydrogen the complexities of the interaction forces, especially at short separation distances, introduce serious difficulties. This is more so for molecules, where one not only has to be able to account for the scattering by fields which do not possess spherical symmetry, but he is hampered by the greater ignorance of molecular fields as compared to atomic. Molecules possessing a dominant long-range interaction term offer an advantage, since, owing to these terms, electrons are scattered at long distances from the molecule so that short-range effects do not appreciably affect the scattering. Recently, some calculations on elastic scattering have been performed which considered interaction potentials based on bulk or macroscopic molecular properties giving rise to a long-range interaction. Such molecular properties are the permanent electric dipole moment $\mu^*$ and the static polarizability $\alpha$. Analyses of electron transport data offer a test of the usefulness of such macroscopic potential functions in electron–molecule scattering.

#### 4.4.3.1 Scattering of slow electrons by non-polar molecules

When gaseous ions (or electrons) move through a gas of non-polar and

---

*Note that we use the same symbol as for the electron mobility.

relatively small molecules, the interaction potential may be represented by

$$V = -e^2 \alpha / 2r^4 \tag{4.74}$$

where $e$ is the ionic charge, $\alpha$ is the static polarizability, and $r$ is the distance between the ion (or electron) and the target molecule. It is important to recognize the severe limitations of this interaction potential – assumption of central force field, neglect of near-field corrections and polarization of individual molecular orbitals, and no inclusion of repulsive terms at the origin—which although they may be reasonable for ionic scattering (Vogt and Wannier, 1954), they are not expected to be so for electron scattering. Nevertheless, it is interesting to see how good a potential function expression (4.74) is in slow electron–molecule scattering. Prior to looking into this question, we may note that whenever theoretical calculations based on microscopic considerations are compared with macroscopic (static) quantities (e.g. $\alpha$ and $\mu$), the question always exists as to how well the static parameters hold on the microscopic level. Macroscopically, for example, the induced dipole $\mu_{\text{ind}}$, the static polarizability $\alpha$, and the field strength $E$, are connected by

$$\mu_{\text{ind}} = \alpha E \tag{4.75}$$

Equation (4.75), however, is limited to low field strengths and may well break down on the microscopic scale, in view of increased values of $E$ in the vicinity of a dipole (Coulson, Maccoll and Sutton, 1952).

Vogt and Wannier (1954) assumed that the interaction potential between ions and non-polar molecules is given by Equation (4.74) and in the limit of low velocities they gave a quantum mechanical cross section for spiralling collisions equal to

$$\sigma = 4\pi (e^2 \alpha / mv^2)^{\frac{1}{2}} \tag{4.76}$$

or

$$\sigma = A/v,$$

where

$$A = 4\pi e m^{-\frac{1}{2}} \alpha^{\frac{1}{2}}$$

which is twice its classical value; $m$ and $v$ are the reduced mass and the relative velocity, respectively; for electron scattering $m$ and $v$ can be taken as the electron mass and electron velocity, respectively. Although Equation (4.76) has been derived for heavy-ion collisions, it is of interest to inquire as to how well it holds for the case of slow electrons scattered by non-polar molecules.

Christophorou, Hurst and Hadjiantoniou (1966) determined $\langle \sigma_m(v) \rangle$ (Equations 4.38 and 4.70) from electron transport data at near thermal energies, assuming a Maxwell form for the distribution of electron energy and a momentum transfer cross section varying inversely with the electron velocity. Their results and some recent ones obtained by Christophorou and

Pittman (1969) are compared in Figure 4.44 with those obtained through Equation (4.76). There is a disagreement between experimental and calculated values by as much as a factor of 4 to 70, which strongly indicates that macroscopic potential functions based on the static polarizability cannot describe the scattering of low-energy electrons from non-polar (but

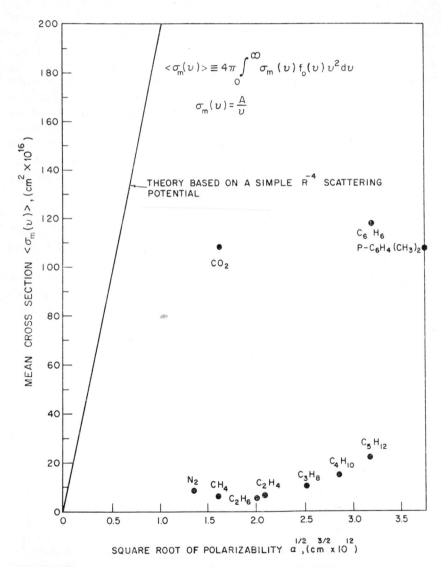

**Figure 4.44** Mean scattering cross sections for non-polar molecules at thermal energies as a function of the square root of the static polarizability

polarizable) molecules. It is interesting to note the enhanced cross sections for $CO_2$, $C_6H_6$ and $p$-$C_6H_4(CH_3)_2$. $CO_2$ has a large quadrupole moment and $C_6H_6$ and $p$-$C_6H_4(CH_3)_2$ possess $\pi$ electron structures. In the latter case (aromatic systems), it has been suggested (Coulson, Maccoll and Sutton, 1952) that superimposed on the classical polarizability is a non-classical one which involves the $\pi$ electron orbitals.

The situation is quite different for polar molecules, where long-range dipole forces provide a reasonable description of low-energy electron scattering from polar molecules.

### 4.4.3.2 Scattering of slow electrons by polar molecules

The potential a charged particle experiences near a polar molecule can be approximated at large distances by that of an extended dipole, and at very large distances it may be further approximated by a point-dipole potential. The great energy losses suffered by slow electrons travelling through polar gases is a result of such a long range dipole interaction between the colliding electron and the polar (target) molecule. In such a case (the same is true for molecules possessing a long-range quadrupole potential), analytical treatment of the scattering becomes possible since, if most of the scattering occurs at long distances from the target, higher multipoles as well as exchange effects may become relatively unimportant.

The scattering of low-energy electrons from polar molecules has first been considered by Altshuler (1957). He treated the target molecule as a fixed rotator which interacts with the incident electron by means of a point-dipole potential represented by

$$V = - \frac{\mu e}{r^2} \cos \theta \qquad (4.77)$$

where $\mu$ is the molecular electric dipole moment, $r$ is the distance of the electron from the centre of mass of the fixed rotator, and $\theta$ is the angle between $\mathbf{r}$ and $\boldsymbol{\mu}$. Altshuler evaluated the scattering amplitude in the first Born approximation and gave for the momentum transfer cross section the expression

$$\sigma_m(v) = \frac{8\pi}{3} \left( \frac{e\mu}{\hbar} \right)^2 \frac{1}{v^2} = \frac{A}{v^2} \quad \text{(c.g.s. units)} \qquad (4.78)$$

where

$$A = \frac{8\pi}{3} \left( \frac{e\mu}{\hbar} \right)^2 = 1.72\mu^2 \ (\mu \text{ in debye})$$

It is noted here again that the Altshuler calculation uses the Born approximation and extrapolates pure dipole interaction all the way to the origin. It also does not consider rotational excitation. Since no consideration was made of the short-range field and of the dynamical responses of molecules such as exchange and polarization the accuracy of the theory diminishes in the case of very weak dipole moments (Figure 4.45).

Following Altshuler, Mittleman and von Holdt (1965) gave exact solutions to the wave equation for electron scattering by a fixed point-dipole, while Shimizu (1963) has studied the scattering by a finite fixed dipole consisting of two opposite charges separated at a non-zero distance.

More recently there has been a strong interest in the quantum-mechanical bound states of an electron which moves in the field of a finite fixed dipole (Wallis, Herman and Milnes, 1960; Fox and Turner, 1966a,b; Mittleman and Myerscough, 1966; Turner and Fox, 1966; Lévy-Leblond, 1967; Brown and Roberts, 1967; Crawford and Dalgarno, 1967; Coulson and Walmsley, 1967; Crawford, 1967 and Fox, 1967). That there should exist bound states for an electron in the field of an extended dipole can be seen from the fact that the bound states for the electron–electric dipole system will be the states of the hydrogen atom perturbed by a distant charge. A number of independent determinations (Mittleman and Myerscough, 1966; Turner and Fox, 1966; Lévy-Leblond, 1967; Brown and Roberts, 1967; Crawford and Dalgarno, 1967; Coulson and Walmsley, 1967; and Crawford, 1967) gave for the minimum electric dipole moment, $\mu_{min}$, above which the electron can be bound temporarily to a stationary electric dipole the value of 0.639315 atomic units or 1.62487 debye for the ground state*. The values of $\mu_{min}$ required to bind an electron to several excited states of the dipole have also been calculated (Coulson and Walmsley, 1967; Crawford, 1967; and Lévy-Leblond and Provost, 1967), the value for the first excited state being 9.63 debye.

Lévy-Leblond and Provost (1967) discussed in some detail the behaviour of the scattering cross section for molecules with nearly critical moments (i.e. around $\mu_{min}$). Further, Takayanagi and Itikawa (1968) performed a partial wave analysis and found the appearance of a resonance with a peak at $\sim 2.5$ debye in the scattering cross section (both in $\sigma_T$ and $\sigma_m$) for thermal electrons when plotted as a function of the dipole moment when the magnitude of the dipole charges is varied at fixed separation. This resonance, which they referred to as a potential resonance, would give a large scattering cross section without temporary excitation of the target as suggested by Turner (1966) and further discussed by Itikawa (1967) in an effort to

---

*One should note that the value of $\mu$ above which an electron can be bound temporarily to a real rotating dipole exceeds this value. Garrett (1970), for example, calculated that for the case of a molecule with the moment of inertia of $H_2O$, $\mu_{min} = 1.95$ debye if molecular rotation is taken into consideration and the induced moments are neglected. The value of $\mu$ above which an electron can be bound temporarily to a polar molecule is a function of a number of parameters, including the moment of inertia of the molecule and the induced moments, and hence $\mu_{min}$ is actually varying from molecule to molecule.

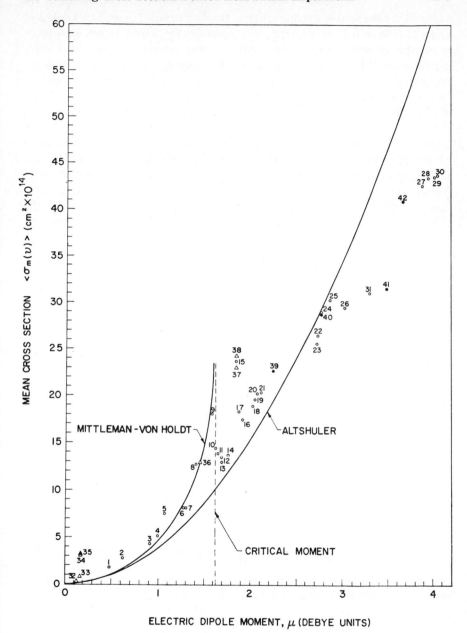

**Figure 4.45** Mean scattering cross sections for polar molecules at thermal energies as a function of the permanent electric dipole moment (Christophorou and Christo-doulides, 1969). The numbers identify polar molecules as shown in tables 1 and 2 of the paper by Christophorou and Christodoulides (1969)

explain the enhanced scattering cross sections observed (Hurst, Stockdale and O'Kelly (1963)) for certain triatomic molecules ($H_2S$ and $H_2O$). In a more recent paper where rotational excitation was considered (non-stationary dipole case), Itikawa (1969) has found that only $\sigma_m$ and not $\sigma_T$ showed the resonance behaviour. This implies that direct molecular rotation does not have an important effect on $\sigma_m$, but it does on $\sigma_T$. Turner and Fox (1968) also discussed the classical motion of an electron in an electric dipole field.

Prior to discussing the experimental information on the scattering of slow electrons by polar molecules, it might be appropriate to point out how some of the determinations of $\mu_{min}$ have been performed. We shall follow the treatment of Brown and Roberts (1967) which is similar to that of Lévy-Leblond (1967). Let $+q$ and $-q$ be the two dipole charges at a distance $d$ apart, so that $\mu = qd$. The energy eigenvalue $E(\mu,d)$ for the electron in the field of the dipole can be written as $E = \varepsilon(\mu)/d^2$ where $\varepsilon(\mu)$ is some function of $\mu$ alone, i.e. the values of $\mu$ for which $E = 0$ are independent of $d$. Let us then confine ourselves to $d = 0$ (point dipole) and search for the critical moment, $\mu_{min}$, for which the ground-state energy is zero (no binding). We note that for $d = 0$, $E = -\infty$ (infinitely large binding) for all $\mu > \mu_{min}$. Using atomic units, the Schrödinger equation for the system is

$$(-\tfrac{1}{2}\nabla^2 - \mu r^{-2} \cos\theta - E)\Psi = 0 \qquad (4.79)$$

which is separable in spherical polar coordinates.

Setting $\Psi(r,\theta,\phi) = R(r)\, Y(\theta,\phi)$, the radial equation for zero binding $(E = 0)$ is

$$\left[\frac{d^2}{dr^2} + \frac{2}{r}\frac{d}{dr} + \frac{C}{r^2}\right] R(r) = 0 \qquad (4.80)$$

where $C$ is a separation constant. We try a solution of the form $R(r) \sim r^s$ (Landau and Lifschitz, 1965) and from Equation (4.80) we have

$$s(s+1) + C = 0$$

A satisfactory critical solution $R(r) = A/r^{\frac{1}{2}}$, where $A$ is a constant, exists only when $s = -\tfrac{1}{2}$, i.e. $C = \tfrac{1}{4}$. Now, we may write the angular equation as

$$\frac{d}{dn}\left[(1-n^2)\left(\frac{dY}{dn}\right)\right] + (2\mu n - C)\, Y = 0 \qquad (4.81)$$

where $n = \cos\theta$, since the angular momentum about the $z$ axis is a constant (zero for the ground state). Seeking a solution of the form

$$Y = \sum_{l=0}^{\infty} b_l P_l(n)$$

where $P_l$ are the Legendre polynomials, we obtain (for $C = \tfrac{1}{4}$) the recursion relation

$$\left[\frac{l}{(2l-1)}\right]b_{l-1} - \left[\frac{(2l+1)^2}{8\mu}\right]b_l + \left[\frac{(l+1)}{(2l+3)}\right]b_{l+1} = 0 \qquad l = 0, 1, 2, \dots \quad (4.82)$$

For different $l$'s one gets an infinite set of homogeneous equations, their solution being equivalent to finding an eigenvalue of an infinite matrix from which $\mu_{min}$ is obtained. Restricting ourselves to the first three terms ($3 \times 3$ determinant), we obtain for $\mu_{min}$ the value 1.62496 debye. Accurate numerical evaluation gives 1.62487 debye.

Analyses of swarm data (Hurst, Stockdale and O'Kelly, 1963; Christophorou, Hurst and Hadjiantoniou, 1966; Christophorou, Hurst and Hendrick, 1966; Hamilton and Stockdale, 1966; Christophorou and Compton, 1967; Christophorou and Christodoulides, 1969; and Christophorou and Pittman, 1970) have confirmed the predictions of theory (Altshuler, 1957) that the scattering of slow electrons by polar molecules is predominantly controlled by the long-range electric dipole field of the molecule. Further analyses of swarm data on mixtures of polar molecules with ethylene (Stockdale and coworkers, 1967) and on pure polar gases (Christophorou and Christodoulides, 1969) have indicated the possible effect of the existence of $\mu_{min}$ on the scattering cross section. In Figure 4.45 the experimental data of Christophorou and Christodoulides (1969) on mean scattering cross sections $\langle\sigma_m(v)\rangle_{exp}$ (expression (4.70)) and for $\alpha = 2$ in Equation (4.65)) for thermal electrons are plotted against $\mu$ and are compared with the calculations of Altshuler (1957), $\langle\sigma(v)\rangle_A$, and Mittleman and von Holdt (1965). The electric molecular dipole moments range from 0 to $\sim 4.1$ debye (Christophorou and Christodoulides, 1969). It is seen that for $\mu > 0.6$ debye the Altshuler theory agrees with the experiment within less than a factor of two. For dipole moments in the range $0.6 \leqslant \mu \leqslant 4.1$ debye the agreement is within 50% in most cases. As expected, for $\mu \lesssim 0.6$ debye the theory cannot be applied; $\langle\sigma_m(v)\rangle_{exp}$ exceed $\langle\sigma(v)\rangle_A$ by a factor of 4, 8 and 25, respectively, for CO, $N_2O$ and $CO_2$ (Christophorou and Christodoulides, 1969). The overall experimental results favour the Born approximation calculation of Altshuler rather than the exact calculation of Mittleman and von Holdt.

The overall variation of $\langle\sigma_m(v)\rangle_{exp}$ with $\mu$ is striking; all molecules with $\mu \lesssim 2.4$ debye have $\langle\sigma_m(v)\rangle_{exp} > \langle\sigma(v)\rangle_A$ without exception, while all molecules with $\mu \gtrsim 2.4$ debye have $\langle\sigma_m(v)\rangle_{exp} < \langle\sigma(v)\rangle_A$. Further, there is evidence that the Born approximation calculation, which does not include the effect of the critical moment, is inadequate for analysing in detail the data, especially around the critical moment. The higher values of the experimentally determined cross sections close to $\mu_{min}$ and their lower values for $\mu \gtrsim 2.4$ debye may suggest a resonance type contribution to the cross section which is superimposed on the smooth variation of $\sigma_m(v)$ with $\mu$ in the pure $v^{-2}$ calculation. This would certainly be in accord with the calculation of Takayanagi and Itikawa (1968) (see further discussion in Christophorou and Christodoulides (1969) and Christophorou and Pittman (1970)). It should be pointed out that a potential resonance cannot occur for a point dipole. In the case of a point dipole all existing discrete levels have an infinitely large binding energy. Further experimental and theoretical work is necessary to elucidate this problem.

## 4.5 Elastic, Total and Differential Scattering Cross Sections Obtained by Electron-Beam Methods

At the beginning of this chapter, it was pointed out that in the scattering of low-energy electrons from atoms and molecules a compound negative ion intermediate may be formed which subsequently may decay elastically, inelastically, or through the process of dissociative attachment. Each of these decay channels offers a way of detecting the compound negative ion state. Compound negative ion states manifest themselves as resonances in electron scattering experiments. Such resonances appear to be a common feature of the early elastic and total scattering cross section data at low electron energies. This can be seen in Figure 4.46 where typical data of the 1930's are plotted * (Brode, 1933). A review of the early work on total elastic scattering cross sections for a number of atoms and molecules can be found in Brode (1933), Massey and Burhop (1952), and McDaniel (1964). The rapid variations of $p_c$ with electron velocity indicate the existence of resonances in the electron scattering cross section. The precise interpretation of these resonances was lacking until very recently. We will discuss the early data on the scattering cross sections later in this section (and in the next chapter) in connection with recent experimental and theoretical work. We may note at this point, however, that the early work has shown the existence of a close similarity in $\sigma_T$ within a chemically similar group of gases, and the large differences in $\sigma_T$ from one group to another. This is evident from a comparison of the data in Figures 4.46 and 4.43. At an energy of $\sim 2$ eV from Figure 4.43 we estimate a collision probability $p_c$ for Xe which is $\sim 40$ times smaller than that of Cs, which lies adjacent to Xe in the periodic table. The small values for the rare gases are due to their compact structure (closed outer electron shells), and the large values for the alkali metals is a manifestation of their diffused structure, each atom possessing a loosely bound valence electron. Since the major

---

*Note that in Figure 4.46 the collision probability $p_c$ rather than the total cross section $\sigma_T$ is plotted as a function of the electron velocity (in units of (volts)$^{1/2}$). The two quantities are related by $\sigma_T = 0.283 \times 10^{-16} p_c$ (in cm$^2$), where $p_c$ is defined as the average number of collisions experienced by the electron in 1 cm of path when the target gas pressure is 1 torr and the temperature 0 °C. Note also that $v$ (in cm/sec) $= 5.93 \times 10^7 \varepsilon^{1/2}$ ($\varepsilon$ in eV).

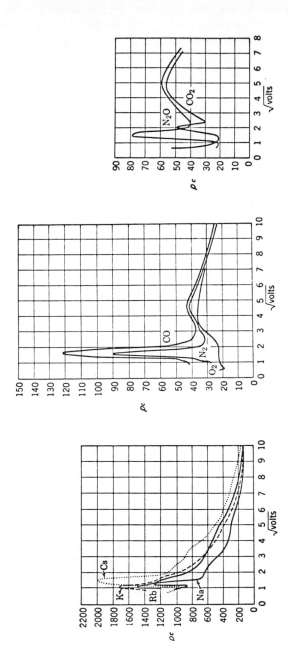

**Figure 4.46** Collision probability $p_c$ as a function of the electron velocity (in (volts)$^{\frac{1}{2}}$). The electron energy at any point along the abscissa is obtained (in eV) by squaring the value of the abscissa at that point (from Brode, 1933)

factor which determines the scattering cross section is the potential field of the outer electrons of the target atom or molecule as well as the distortion of this potential field by the incident particle, the scattering cross section should be a strong function of the atomic or molecular ionization potential and the polarizability. At higher energies the cross section falls off approximately as $v^{-1}$.

Recent advances in experimental techniques (especially in electron-beam experiments; Chapter 5) enabled many new resonances to be detected experimentally. In electron-beam experiments resonances have been observed both as enhancements in transmission currents and as dips in scattering currents. In this section we will restrict ourselves mainly to recent beam work on the elastic and total scattering of electrons from atoms and molecules. Resonances in inelastic electron scattering and in energy loss experiments will be elaborated upon in the next chapter. Autoionizing states observed in far ultraviolet absorption studies have been discussed in the previous chapter.

Resonances in electron–atom and electron–molecule scattering are helpful in elucidating problems in atomic and molecular physics as well as in other branches of physics and chemistry. Their understanding traces to nuclear resonance scattering theory. We refer the reader to general recent articles such as those by Burke (1965), Bardsley and Mandl (1968), Smith (1966) and to subsequent discussions in this chapter and Chapters 5, 6 and 7.

If one extends Fano's treatment of photoabsorption line-shapes to electron scattering the profile of an elastic resonance, if we neglect the effect of spin, can be represented by

$$\sigma(\varepsilon) = \sigma_0[(q+\varepsilon)^2/(1+\varepsilon^2)] + \sigma_A \qquad (4.83)$$

In Equation (4.83) $\varepsilon = (E - E_{res})/\frac{1}{2}\Gamma$, $E$ is the electron energy, $E_{res}$ is the idealized resonance energy, $\Gamma$ is the resonance width, $\sigma(\varepsilon)$ is the elastic scattering cross section for electrons of energy $E$, $\sigma_A$ is the nonresonant part of the cross section, $\sigma_0 + \sigma_A$ is the value of $\sigma$ away from the resonance, $q$ is a shape parameter equal to $-\cot \eta_l$ and $\eta_l$ is the scattering phase shift for the partial wave of angular momentum $l\hbar$ in which the resonance occurs. Formula (4.83) is due to Fano (1961) who developed a configuration interaction theory which he applied to the position, width, and shape of absorption lines in autoionization processes (see discussion in Chapter 3). We note that Equation (4.83) is still valid for electron energies exceeding the energy of the first excited state, but in this case $q$ is no longer simply related to the elastic phase shift. From Equation (4.83) it is seen that the resonance profile results in a maximum cross section ($= (q^2 + 1)\sigma_0 + \sigma_A$) at $\varepsilon = q^{-1}$ and a minimum cross section ($= \sigma_A$) at $\varepsilon = -q$. At the idealized resonance energy $E_{res}$ the cross section is $q^2\sigma_0 + \sigma_A$, which shows that neither the maximum nor the minimum of the cross section occurs at $E_{res}$. The idealized resonance energy $E_{res}$ is always somewhere between the minimum and maximum cross sections.

### 4.5.1 *Atoms*

#### 4.5.1.1 Hydrogen atom

Elastic scattering cross sections for atomic hydrogen are of great theoretical interest because they provide sensitive tests of the various approximations used in collision theory. The wave functions for atomic hydrogen are completely and exactly known, permitting accurate calculations. Many discussions exist on this topic (e.g. Moiseiwitsch, 1962; Burke and Smith, 1962). Some of the theoretical results are compared with experiment in Figure 4.47a. It is seen that there is good agreement between experiment and calculation. In the calculation by Temkin and Lamkin (1961), who used the method of polarized orbitals, higher order waves were taken into account. The contribution of the higher order waves, although small, should, generally, be considered.

Burke and Schey (1962) predicted the existence of a resonance in the elastic scattering of electrons from atomic hydrogen at 9.61 eV (with a width of 0.109 eV), i.e. just below the onset of electronic excitation. This result is shown in Figure 4.47a. Many other theoretical predictions on possible positions of a resonance in $e$–H scattering have been made which range from 9.559 to 10.198 eV. These have been discussed and cited in O'Malley and Geltman (1965).

The first experimental confirmation of the theoretical prediction of Burke and Schey (1962) that there are resonances in $e$–H scattering just below the $n = 2$ excitation threshold has been provided by Schulz (1964b) in a transmission experiment. He observed a rise in the transmitted current, centred at $9.7 \pm 0.15$ eV, indicating a dip in the elastic scattering cross section at this energy. Schulz's result has been confirmed by Kleinpoppen and Raible (1965) who observed the resonance in a scattering experiment (at scattering angles of 70°, 90° and 94°) and found it centred at $9.73 \pm 0.1$ eV, in agreement with Schulz (1964b). These observations are shown in Figures 4.47b and c. Schulz (1964b) pointed out that the hydrogen resonance was broader than that in helium at 19.3 eV (see later this section), indicating a shorter autoionization lifetime for $H^-$ than for $He^-$. It should be noted, however, that the width of this resonance as observed by Kleinpoppen and Raible (1965) is much smaller.

Experimental differential scattering cross sections for atomic hydrogen are shown in Figure 4.48 for four electron energies 3.8, 5.7, 7.1 and 9.4 eV. These cross sections were deduced by Gilbody, Stebbings and Fite (1961) by normalization of their data to the absolute data of Ramsauer and

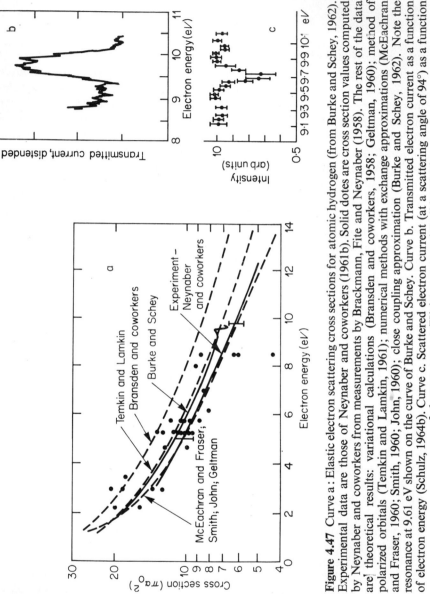

**Figure 4.47** Curve a : Elastic electron scattering cross sections for atomic hydrogen (from Burke and Schey, 1962). Experimental data are those of Neynaber and coworkers (1961b). Solid dotes are cross section values computed by Neynaber and coworkers from measurements by Brackmann, Fite and Neynaber (1958). The rest of the data are theoretical results: variational calculations (Bransden and coworkers, 1958; Geltman, 1960); method of polarized orbitals (Temkin and Lamkin, 1961); numerical methods with exchange approximations (McEachran and Fraser, 1960; Smith, 1960; John, 1960); close coupling approximation (Burke and Schey, 1962). Note the resonance at 9.61 eV shown on the curve of Burke and Schey. Curve b. Transmitted electron current as a function of electron energy (Schulz, 1964b). Curve c. Scattered electron current (at a scattering angle of 94°) as a function of electron energy (Kleinpoppen and Raible, 1965)

Kollath (1932) on molecular hydrogen. A comparison of these experimental results with theoretical calculations was given by Burke and Schey (1962). It is interesting to note the increase in large angle scattering with decreasing energy (Figure 4.48).

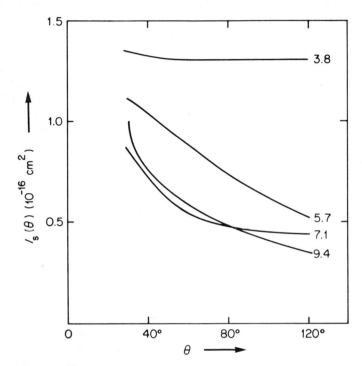

**Figure 4.48** Differential electron scattering cross section for atomic hydrogen (Gilbody, Stebbings, and Fite, 1961)

### 4.5.1.2 Rare gases

A similar resonance to that found in atomic hydrogen has been observed somewhat earlier by Schulz (1963) in the elastic scattering of electrons by He at an angle of 72°. This resonance appeared at $19.3 \pm 0.1$ eV, i.e. about 0.5 eV (as in H) below the first $n = 2$ excitation threshold. It appears (see data on other rare gases later this section) that 0.5 eV is typical of the binding energy produced by the polarization potential of an atom in an excited state. The discovery of the He resonance by Schulz (1963) had been preceded by theoretical discussions pointing out the possibility that such processes may exist in the rare gases (Baranger and Gerjuoy, 1957; Burke and Schey, 1962). The helium resonance has subsequently been confirmed by many other investigators (Simpson and Fano, 1963; Fleming and Higginson, 1963;

McFarland, 1964; Schulz, 1964a; Kuyatt, Simpson and Mielczarek, 1965a; Golden and Nakano, 1966). The sharpness of the resonance makes it ideally suited for use in energy scale calibrations in electron-beam experiments.

In Figure 4.49 the helium resonance is shown as observed by Schulz (1964a) in a scattering experiment (Figure 4.49b) and by Kuyatt, Simpson and Mielczarek (1965a) in a transmission experiment (Figure 4.49a). The dip in Figure 4.49b occurs at $19.30 \pm 0.05$ eV, and the peak in transmission (Figure 4.49a) at $19.31 \pm 0.03$ eV. In the transmission curve the points A and B are associated with the thresholds for exciting the $(1s2s)^3 S_1$ and $(1s2s)^1 S_0$ states of helium at 19.818 and 20.614 eV, respectively. Simpson and Fano (1963) interpreted the resonance at 19.3 eV in helium as being caused by interference between the formation of the $1s2s^2(^2S_{1/2})$ state of He$^-$ and the usual potential scattering. We may think of the $1s2s$ excited state configuration of helium as having a positive binding energy for a third electron in the unfilled $n = 2$ shell.

The observation of the resonance in helium was quickly followed by the discovery of similar resonances in neon (Schulz, 1963, 1964a,d; Simpson and Fano, 1963; Simpson, 1964; Kuyatt, Simpson and Mielczarek, 1965a), krypton and xenon (Schulz, 1964a; Kuyatt, Simpson and Mielczarek, 1965a), and argon as well as in mercury (Kuyatt, Simpson and Mielczarek, 1965a). For neon, Schulz (1964a) observed one resonance near 16 eV although Simpson (1964) reported two resonances at this energy. This apparent discrepancy is probably due to a difference in energy resolution. The two resonances observed by Simpson in neon have been seen clearly by Kuyatt, Simpson and Mielczarek (1965a) at $\sim 0.5$ to 0.6 eV below the lowest group of neon-excited levels $2p^5 3s(^2P_{3/2, 1/2})$, as a sharp decrease in transmission separated by $0.095 \pm 0.002$ eV. The compound states responsible for the two principal neon resonances were interpreted by Simpson and Fano (1963) as being due to highly excited Ne$^-$ ions formed by adding a $3s$ electron to the lowest excited configuration of neon, i.e. $(1s^2 2s^2 2p^5 3s^2)^2 P_{3/2, 1/2}$. The spacing of the resonances agrees very well with the $^2P_{3/2}$, $^2P_{1/2}$ splitting in the ground state of Ne$^+$ showing that the two $3s^2$ electrons have very little effect on the Ne$^+$ core.

The work of Schulz (1964a) and Kuyatt, Simpson and Mielczarek (1965a) on the rest of the rare gases has shown that elastic resonances are of widespread occurrence in the rare gases. In each of the rare gases the strongest resonance or resonances were related to the lowest excited atomic configuration and occurred around 0.5 eV below the first electronically excited state of the respective neutral atoms. Schulz (1964a) also noted that the magnitude of these resonances decreases with increasing atomic number. In analogy with neon, Kuyatt, Simpson and Mielczarek (1965a) proposed that the negative ion states in Ar$^-$, Kr$^-$ and Xe$^-$ are $(3p^5 4s^2)^2 P_{3/2, 1/2}$, $(4p^5 5s^2)^2 P_{3/2, 1/2}$ and $(5p^5 6s^2)^2 P_{3/2, 1/2}$, respectively. In support of this

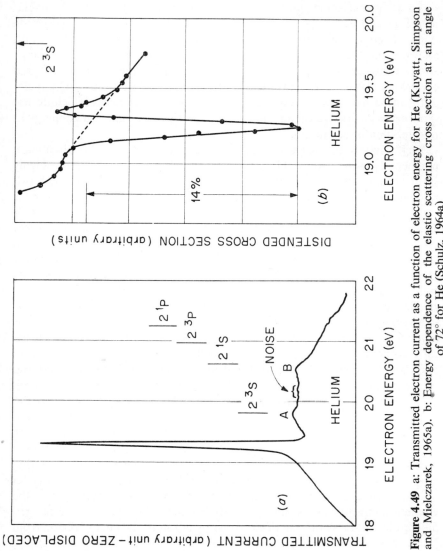

**Figure 4.49** a: Transmitted electron current as a function of electron energy for He (Kuyatt, Simpson and Mielczarek, 1965a). b: Energy dependence of the elastic scattering cross section at an angle of 72° for He (Schulz, 1964a).

assignment is the very interesting work of Andrick and Ehrhardt (1966) who measured the angular dependence of the elastic resonance scattering of low-energy electrons by He, Ne, Ar (and $N_2$) in the angular range from 8 to 110°. From their measurements these investigators derived the total angular momenta and hence the configuration of the quasi-stationary negative ion states. Their findings are in agreement with the assignments of Kuyatt, Simpson and Mielczarek (1965a). Angular distribution studies of low-energy electron scattering from $H_2$ and CO have been made by Ehrhardt and coworkers (1968). These investigators stressed the importance of angular distribution measurements and their use to detect resonances which do not show up as distinctive changes in the cross section as a function of energy in transmission or scattering.

Very importantly, Kuyatt, Simpson and Mielczarek (1965a) observed a number of resonances in He, Ne and Hg which are based on more highly excited configurations of these atoms. For He they have detected two resonances at $57.1 \pm 0.1$ and $58.2 \pm 0.1$ eV. The first of these resonances is 0.8 eV below the lowest doubly excited state of helium $(2s^2)^1S$ at 57.9 eV (Simpson, Mielczarek and Cooper, 1964), and the second is 0.3 eV below the second doubly excited state of He $(2s2p)^3P$ at 58.5 eV. Fano and Cooper (1965) attributed these resonances to temporary formation of highly excited states of $He^-$ in which all three electrons are in $n = 2$ quantum states and proposed the assignments $(2s^2 2p)^2P$ and $(2s2p^2)^2D$ for the two resonances.

Angular distributions of elastically scattered electrons from the rare gas atoms are of extreme interest. Massey and Burhop (1952) summarized the early work. In Figure 4.50 we reproduced some of their data on the rare gases and mercury. Some distinct features in these data are worth noting: (a) the maxima and minima in the differential scattering cross section which arise from diffraction of the electron waves by the target atom, (b) the increase in complexity of the angular pattern with increasing atomic number of the scattering atom, and (c) the variation of the differential scattering cross section with electron energy. At low energies large angle scattering becomes very probable, while at high electron energies forward scattering predominates. This, as we have pointed out earlier, has a profound effect on the relative magnitudes of the total scattering cross section $\sigma_T$ and the momentum transfer cross section $\sigma_m$. More recently Porteus (1964) measured differential elastic scattering of electrons by argon in the energy region from 75 to 500 eV and at angles from 4 to 160°. This data showed a strong backward scattering which is significant up to at least 250 eV, a feature which seems to be absent from the early data shown in Figure 4.50.

More recently, Vriens, Kuyatt and Mielczarek (1968) measured with a high-resolution electron spectrometer the angular dependence of elastic scattering of 100 to 400 eV electrons by helium at scattering angles of

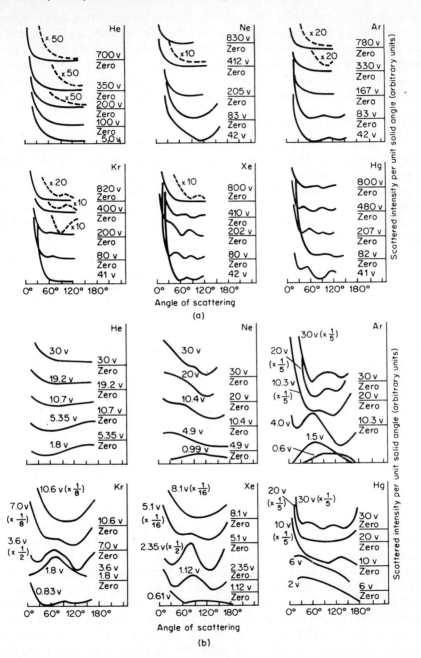

**Figure 4.50** Angular distribution of elastically scattered electrons for the rare gases and mercury (Massey and Burhop, 1952)

5 to 30°. They converted their relative differential scattering cross sections to absolute by measuring their ratio to known cross sections for excitation of the $2^1 P$ state. Integrating over angles, they obtained total scattering cross sections. They found large deviations from the Born approximation prediction particularly for small electron energies and small scattering angles which they have attributed to exchange and polarization effects.

### 4.5.1.3 Atomic nitrogen and oxygen

In Figure 4.51 the total collision cross sections for atomic nitrogen and atomic oxygen are shown. The experimental results in the case of nitrogen are those of Neynaber and coworkers (1963) who performed a crossed-beam experiment. For atomic oxygen, in addition to the data of Neynaber and coworkers (1961a), we have plotted the recent measurements of Sunshine, Aubrey and Bederson (1967) who employed an atomic-beam recoil method. The results of a number of theoretical calculations (see caption of Figure 4.51) are also shown in the figure. The recent calculations of Robinson and Geltman (1967) for the elastic scattering cross sections for atomic oxygen are in good agreement with the measurements of Sunshine, Aubrey and Bederson (1967). Robinson and Geltman (1967) calculated also elastic scattering cross sections for the elements C, F, Si, S, Cl, Br and I. The elastic scattering cross sections for Si, S, Cl, Br and I showed a pronounced Ramsauer–Townsend minimum, much as is observed in the corresponding neighbouring noble gas atoms. The elastic scattering cross section for fluorine did not show a Ramsauer–Townsend minimum which is consistent with it not being observed in experiments with neon.

### 4.5.2 *Molecules*

#### 4.5.2.1 $H_2$, HD and $D_2$

A series of sharp resonances in the transmission of electrons through molecular hydrogen in the energy range from 11.5 to $\sim 14$ eV has been observed by Kuyatt, Mielczarek and Simpson (1964). The observed spacings of the elastic resonances were very close to the vibrational spacings of the $1s\sigma 2p\pi\,^1 \Pi_u$ excited state of $H_2$, and appeared at electron energies $\sim 0.5$ eV below the energy required to excite the vibrational levels of this state. The resonance was ascribed to compound state formation with an excited state of $H_2$. The total scattering cross sections for the $H_2$ and $D_2$ resonances in the energy range from $\sim 10.7$ to 12.9 eV have been measured by Golden and Bandel (1965b) using a modified Ramsauer technique. From a further study of the $H_2$, HD and $D_2$ resonances in the energy region mentioned above, Kuyatt, Simpson and Mielczarek (1965b, 1966) found that there exist two electronic states of $H_2^-$ and $HD^-$, each with a well-developed vibrational structure. They have observed only one series of resonances for $D_2$, and this

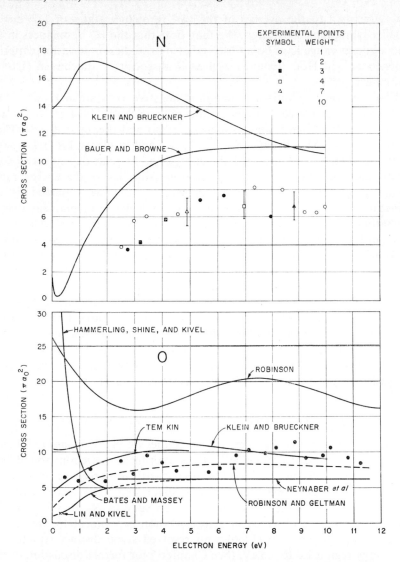

**Figure 4.51** Total electron scattering cross sections for atomic nitrogen and atomic oxygen. N: Experimental data are those of Neynaber and coworkers (1963). Theoretical calculations are those of Klein and Brueckner (1958) and Bauer and Browne (quoted by Neynaber and coworkers, 1963). O: Experimental data: Neynaber and coworkers (1961a); Lin and Kivel (1959) (obtained from a shock tube experiment); Sunshine, Aubrey, and Bederson (1967) (dark circles). Theoretical calculations: Robinson (1957); Temkin (1957); Klein and Brueckner (1958); Bates and Massey (1947); Robinson and Geltman (1967); Hammerling, Shine, and Kivel (1957); see also Henry (1967) and Garrett and Jackson (1967)

was attributed to the overlap of the two resonance states of $D_2^-$ due to smaller vibrational separation. We may note that the $H_2^-$ resonances in the above energy range have been observed also in studies on inelastic scattering (Menendez and Holt, 1966) as well as in dissociative attachment (Dowell and Sharp, 1968).

Evidence for an additional compound negative ion state in $H_2$ at $\sim 4$ eV has been accumulated both experimentally (Schulz, 1964c; Schulz and Asundi, 1967; Ehrhardt and coworkers, 1968) and theoretically (Bardsley, Herzenberg and Mandl, 1966; Eliezer ,Taylor and Williams, 1967). Ehrhardt and coworkers (1968) measured the energy and angular dependence of the elastic and inelastic scattering of electrons from $H_2$ in the energy range from 0.5 to 10 eV and for scattering angles 5 to 110°. They observed distinct structure with an elastic component and an inelastic component due to excitation of vibrational levels ($v = 1, 2, 3, 4$) of the neutral molecule via the short-lived negative ion state $^2\Sigma_u^+$ at $\sim 3.7$ eV (see also swarm data in Section 4.4 and further discussion in Chapter 6 on the involvement of this state in dissociative attachment). The elastic cross sections were found to be composed of potential and resonance scattering.

An enhancement in the total scattering cross section around 4 eV is seen in Figure 4.52a where beam and swarm data are compared with the results of some theoretical calculations. The effect of electron exchange is evident from the calculations of Massey and Ridley (1956) (compare curves III and IV in Figure 4.52a). Some experimental information on the angular dependence of the scattering of slow electrons from $H_2$ is given in Figure 4.52b.

### 4.5.2.2 $N_2$, $O_2$, CO and NO

As we have mentioned earlier, swarm studies on $N_2$ by Haas (1957) have shown the existence of a resonance in the scattering cross section around 2 eV. A resonance in $N_2$ at $\sim 2.3$ eV has been clearly observed in an inelastic scattering beam experiment by Schulz (1964c), and it has been attributed to the existence of a compound negative ion state in $N_2$ at $\sim 2.3$ eV. The resonance has been observed in a transmission experiment by Heideman, Kuyatt and Chamberlain (1966) who resolved some distinct structure in the energy region 1.8 to 3.5 eV. (See Chapter 5 for further discussion of this and of some additional apparent structure observed by Golden (1966a)). Andrick and Ehrhardt (1966) investigated the angular dependence of the elastic resonance scattering of low-energy electrons from $N_2$ and concluded that the increase in the elastic scattering from $N_2$ in the energy range 1.75 to 3.5 eV is due to the formation of $N_2^-$ ion in its ground electronic state. The $N_2$ resonances between 1.8 to 3.5 eV observed by Heideman, Kuyatt and Chamberlain (1966) in a transmission experiment are shown in Figure 4.53a. These investigators have observed also a very sharp and

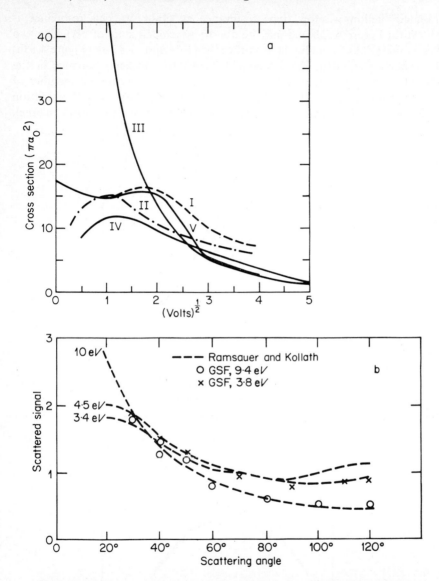

**Figure 4.52** Electron scattering cross sections for $H_2$ (from McDaniel, 1964). a: Scattering cross section as a function of electron energy: I. observed by Ramsauer method; II, observed by swarm diffusion methods; III, calculated by Massey and Ridley (1956), ignoring exchange; IV, calculated by Massey and Ridley (1956), including exchange; V, calculated by Fisk (1936), using an empirical molecular scattering potential with adjustable constants. b: Angular distribution of scattered electrons. Ramsauer and Kollath (1932): – – – –; Gilbody, Stebbings and Fite (1961): (○, ×)

isolated 'helium-window' type resonance at 11.48 eV. This resonance is
shown in Figure 4.53b and may be due to an excited state of $N_2^-$. The two
additional peaks found by these workers at 11.75 and 11.87 eV (Figure 4.53b)
have been attributed, the former (11.75 eV) to another resonance in the
elastic scattering cross section (perhaps the second vibrational member of
the large resonance at 11.48 eV), and the latter (11.87 eV) to a combination
of a resonance in the elastic scattering and the onset of an inelastic channel.

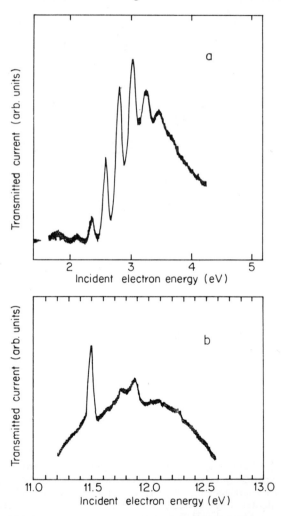

**Figure 4.53** Transmission of electrons by $N_2$ (Heideman, Kuyatt and Chamberlain, 1966). a: A series of resonances in the 1.8–3.5 eV energy region. b: A sharp
'helium-window' type resonance at $11.48 \pm 0.05$ eV. (See text for the other two
peaks at 11.75 and 11.87 eV)

Resonances in the transmission of low-energy electrons ($\lesssim 2$ eV) through $N_2$, $O_2$, NO and CO have been observed by Boness and Hasted (1966) and were attributed to compound negative ion states. In almost all cases a series of peaks has been observed below $\sim 2$ eV. A peak at $\sim 1.75$ eV in the inelastic scattering cross section of electrons by CO has been detected by Schulz (1964c) and has been attributed to the formation of a compound negative ion state at this energy. Angular studies on electron scattering from CO at low energies ($< 10$ eV) have been performed by Ehrhardt and coworkers (1968).

Finally, total cross sections for scattering of electrons by $O_2$ in the energy range 0.5 to 11.3 eV have been reported recently by Sunshine, Aubrey and Bederson (1967). Their data are shown in Figure 4.54 and are compared with the early beam work of Ramsauer and Kollath (1930) and Brüche (1927a).

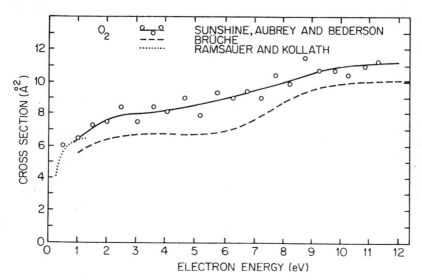

**Figure 4.54** Total electron scattering cross sections for $O_2$ in the 0.5 to 11.3 eV energy range (Sunshine, Aubrey, and Bederson, 1967). The solid line represents a best smooth-curve fit to the measurements of Sunshine and coworkers. The dashed and dotted lines refer to the early measurements of Brüche (1927a) and Ramsauer and Kollath (1930), respectively

### 4.5.2.3  Molecules with three or more atoms

Resonances in the elastic and inelastic scattering of slow electrons by polyatomic molecules have been found to be of common occurrence as is evidenced from the data collected in Table 4.4. Most of the resonances observed so far in polyatomic molecules occur at energies below the energy of

**Table 4.4** Resonances observed in the transmission and/or forward inelastic scattering of slow electrons through molecular gases which have been attributed to compound negative ion states

| Molecule | Resonance energy (eV) | Reference and method of observation[a] |
|---|---|---|
| $CO_2$ | ~3.3 | 1b |
| $N_2O$ | 2.2 | 2a |
| $CH_4$ | 1.9–3.0 | 3b |
| $C_2H_2$ | 2.0 | 4a |
| $C_2H_4$ | 1.7 | 4a |
|  | 0.2–1.8 | 3b |
|  | 1.8 | 5a |
| $C_2H_2Cl_2$ | ~0.6 | 6a |
| $(CH_3)_2C=CH_2$ | 2.3 | 5a |
| (cis-Butene-2) |  |  |
| $CH_3CH=CH_2$ | 1.8 | 4a |
| $CH_3C\equiv CH$ | 2.8 | 4a |
| $C_2H_5C\equiv CH$ | 2.4 | 4a |
| $C_6H_6$ | 0.8–2.2 | 7a |
|  | 0.8–1.8 | 3b |
|  | 1.3 | 5a |
| $C_6H_5CH_3$ | 1.45 | 7a, 9a |
| $C_6H_5F$ | 1.35 | 7a, 9a |
| $C_6H_5NO_2$ | 1.3, 3.64 | 8a, 9a |
| $o$-$C_6H_4CH_3Cl$ | 1.1 | 7a, 9a |
| $C_6H_5Cl$ | 0.86 | 7a, 9a |
| $C_6H_5Br$ | 0.84 | 7a, 9a |
| $o$-$C_6H_4Cl_2$ | 0.36 | 7a, 9a |
| $C_{10}H_8$ | 0.8 | 8a, 9a |
| (naphthalene) |  |  |
| $o$-$C_{10}H_7Cl$ | 0.55, 2.87 | 8a |
| ($o$-chloronaphthalene) |  |  |
| $C_5H_5N$ (pyridine) | 1.2, 1.6 | 10a |
| 1,3-$C_6H_4F_2$ | 0.6 | 11a |
| 1,3,5-$C_6H_3F_3$ | 0.3 | 11a |
| 1,2,3,4-$C_6H_2F_4$ | ~0 | 11a |

[a]   a denotes an inelastic scattering experiment, and b denotes a transmission experiment.

1. Boness and Hasted (1966).
2. Schulz (1961).
3. Boness and coworkers (1967).
4. Bowman and Miller (1965).
5. Brongersma, Hart, and Oosterhoff (1967).
6. Köllmann and Neuert (1968).
7. Compton, Christophorou, and Huebner (1966).
8. Huebner and coworkers (1968).
9. Compton and coworkers (1968).
10. Huebner, Compton, and Schweinler (1968).
11. Naff, Cooper and Compton (1968).

the first excited electronic state (see, however, Chapter 6). Figure 4.55 shows
the temporary negative ion resonances observed by Compton, Christophorou
and Huebner (1966) for a number of benzene derivatives. The benzene
resonance peaks at an energy of $\sim 1.5$ eV which corresponds closely to the
electron affinity of benzene as given by a number of theoretical calculations

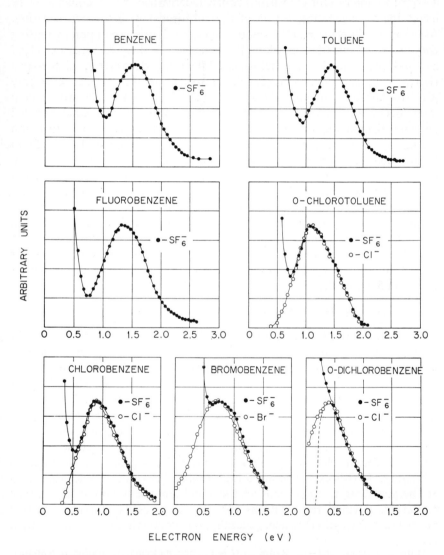

**Figure 4.55** Temporary negative ion resonances in benzene and some of its
derivatives (Compton, Christophorou and Huebner, 1966). The rising portion
of the $SF_6^-$ curves at low energies is due to the primary $SF_6^-$ current at $\sim 0.0$ eV

(Chapter 7). Although the benzene resonance in Figure 4.55 shows no evidence of any structure, Boness and coworkers (1967) observed vibrational structure in a transmission experiment for this resonance in the same energy region (see further discussion in Chapter 6). It is interesting to note the marked decrease in the peak energy of the resonance ('vertical attachment energy'; Chapter 6) for the various benzene derivatives which coincides with increasing resonance interaction of the substituent with the benzene ring. Also a marked decrease in the resonance energy is observed upon increasing the number of substitutions along the benzene periphery (compare $C_6H_5Cl$ and $o–C_6H_4Cl_2$). For the halogenated benzene derivatives, dissociative electron attachment is energetically possible at the resonance energy (Christophorou and coworkers (1966)). The dissociative attachment resonances observed by Christophorou and coworkers for some halogenated benzene derivatives are compared with the temporary negative ion resonances detected using the $SF_6$ scavenger technique (Chapter 5) in Figure 4.55. It is seen that the two resonances coincide in both shape and energy indicating that inelastic scattering and dissociative attachment proceed via the same compound negative ion state and are thus in competition. We note, finally, that the intense energy loss peak at 0.8 eV observed by Huebner and coworkers (1968) for the naphthalene molecule, and attributed to a temporary negative ion state, is about one-half as wide as the benzene resonance. This indicates a longer autoionization lifetime for the naphtalene compound negative ion state (See further discussion in Chapters 5, 6 and 7.)

## 4.6 Elastic Scattering of Positrons

High-energy positrons, as electrons, slow-down in matter losing their energy predominantly to the electronic system of the atoms or molecules making up the medium, while subexcitation positrons slow-down in matter, as electrons, losing energy in elastic collisions and in inelastic collisions involving vibrational and rotational excitation of molecules. There exist, however, notable differences between positron and electron scattering which arise from distinct differences between positron–molecule as compared to electron–molecule interactions. For example, the mean static field is attractive for an electron but it is repulsive for a positron; for electron scattering, electron exchange has a high probability of occurrence especially at energies close to the excitation thresholds, but it is absent in positron scattering. Further, for positrons the possibility exists of real or virtual positronium formation* which, generally, is in competition with moderation.

_____

*A threshold energy of $I-0.5$ Rydbergs ($I$ is the first molecular ionization potential) is required for the formation of a positronium. Positronium formation is considered to occur almost exclusively at positron kinetic energies $E$ in the range $I-0.5$ Ry $<E<E_1$ where $E_1$ is the energy of the first excited electronic state of the molecule.

Other differences may arise in positron–molecule scattering as compared to electron–molecule scattering. To illustrate this let us consider a long-range interaction such as the one we discussed in Section 4.4.3 for the scattering of slow electrons from molecules. Let us write for the potential energy $V(\mathbf{s},\mathbf{r})$ involving the interactions between an $e^-$ (or an $e^+$) and a diatomic molecule, where $\mathbf{r}$ is the distance from the centre of mass of the diatomic molecule with internuclear vector $\mathbf{s}$, the expression (Takayanagi and Inokuti, 1967)

$$V(\mathbf{s},\mathbf{r}) = -\frac{\alpha(\mathbf{s})e^2}{2r^4} \pm \frac{\mu(\mathbf{s})e}{r^2}P_1(\cos\theta) - \left[\frac{\alpha'(\mathbf{s})e^2}{2r^4} \pm \frac{Q(\mathbf{s})e}{r^3}\right]P_2(\cos\theta) + \dots$$
(4.84)

where the minus sign refers to the electron and the plus sign to the positron. In Equation (4.84) $P_n$ are the Legendre polynomials, $\theta$ is the angle between $\mathbf{r}$ and $\mathbf{s}$, $\mu$ and $Q$ are, respectively, the electric dipole and electric quadrupole moments, and $\alpha$ and $\alpha'$ are, respectively, the spherical and non-spherical parts of the molecular polarizability. From Equation (4.84) we see that although the polarizability terms are the same for both $e^-$ and $e^+$, the electrostatic terms change sign. Therefore, we expect differences in both the energy dependence and the magnitude of the cross sections for $e^-$ and $e^+$ when these dipole terms are important in the scattering. The cross sections for rotational excitation of a homonuclear molecule, for example, should be different for $e^+$ and $e^-$. The situation in bound to be different also for the formation of compound states which we have seen to occur abundantly in $e^-$–molecule scattering. Such states, we have emphasized earlier, represent properties of the combined electron–molecule system and thus have no direct bearing on positron–molecule scattering. Such effects specific to $e^+$ scattering seem to be less likely. Several investigations have been made of the differences between $e^-$ and $e^+$ scattering from simple atoms (see, for example, Mott and Massey (1965); Takayanagi and Inokuti (1967)).

There is no doubt as to the wealth of information which can be obtained from investigations of phenomena of positron interactions. However, experiments with low energy monoenergetic positron beams have, so far, not been feasible. In most experiments, positrons are introduced in matter with an appreciable initial kinetic energy and their ultimate mode of decay and interaction are investigated. An excellent reference on positron interactions with matter is the book by Green and Lee (1964).

## 4.7 References

W.P. Allis (1956). In S. Flügge (Ed.), *Handbuch der Physik*, vol. 21, Springer-Verlag, Berlin, p. 413.
S. Altshuler (1957). *Phys. Rev.*, **107**, 114.
D. Andrick and H. Ehrhardt (1966). *Z. Physik*, **192**, 99.

V. A. Bailey and W. E. Duncanson (1930). *Phil. Mag.*, **10**, 145.

V. A. Bailey and R. H. Healey (1935). *Phil. Mag.*, **19**, 725.

J. E. Bailey, R. E. B. Makinson and J. M. Somerville (1937). *Phil. Mag.*, **24**, 177.

V. A. Bailey and J. B. Rudd (1932). *Phil. Mag.*, **14**, 1033.

V. A. Bailey and J. M. Somerville (1934). *Phil. Mag.*, **17**, 1169.

J. B. Bannon and H. L. Brose (1928). *Phil. Mag.*, **6**, 817.

E. Baranger and E. Gerjuoy (1957). *Phys. Rev.*, **106**, 1182.

D. Barbiere (1951). *Phys. Rev.*, **84**, 653.

J. N. Bardsley, A. Herzenberg and F. Mandl (1966). *Proc. Phys. Soc. (London)*, **89**, 305, 321.

J. N. Bardsley and F. Mandl (1968). Rep. Prog. Phys. xxxi, Part 2, 471.

D. R. Bates (1962). *Atomic and Molecular Processes*, Academic Press, New York.

D. R. Bates and H. S. W. Massey (1947). *Proc. Roy. Soc. (London)*, **A192**, 1.

E. G. Bauer and H. N. Browne (1964). In M. R. C. McDowell (Ed.), *Atomic Collision Processes*, North Holland Publishing Co., Amsterdam, p. 16.

K. D. Bays, D. Kivelson and S. C. Wong (1962). *J. Chem. Phys.*, **37**, 1217.

G. Bekefi and S. C. Brown (1958). *Phys. Rev.*, **112**, 159.

P. R. Bell, quoted by Errett (1951).

R. O. Berger, T. F. O'Malley and L. Spruch (1965). *Phys. Rev.*, **137**, A1068.

M. A. Biondi (1963). In L. Marton (Ed.), *Advances in Electronic and Electron Physics*, Vol. 18, Academic Press, New York, pp. 67–165.

M. J. W. Boness and J. B. Hasted (1966). *Phys. Letters*, **21**, 526.

M. J. W. Boness and I. W. Larkin, J. B. Hasted and L. Moore (1967). *Chem. Phys. Letters*, **1**, 292.

M. J. W. Boness and G. J. Schulz (1968). *Phys. Rev. Letters*, **21**, 1031.

M. Born (1926). *Z. Physik*, **38**, 803.

T. E. Bortner, G. S. Hurst and W. G. Stone (1957). *Rev. Sci. Instr.*, **28**, 103.

J. C. Bowe (1957). *Drift Velocity of Electrons*, ANL-5829, pp. 14–29.

J. C. Bowe (1960). *Phys. Rev.*, **117**, 1411, 1416.

C. R. Bowman and D. E. Gordon (1967). *J. Chem. Phys.*, **46**, 1878.

C. R. Bowman and W. D. Miller (1965). *J. Chem. Phys.*, **42**, 681.

R. T. Brackmann, W. L. Fite and R. H. Neynaber (1958). *Phys. Rev.*, **112**, 1157.

N. E. Bradbury and R. A. Nielsen (1936). *Phys. Rev.*, **49**, 388.

J. Brambring (1964). *Z. Physik*. **179**, 539.

B. H. Bransden, A. Dalgarno, T. L. John and M. J. Seaton (1958). *Proc. Phys. Soc. (London)*, **A71**, 877.

R. B. Brode (1933). *Rev. Mod. Phys.*, **5**, 257.

H. H. Brongersma, J. A. v.d. Hart and L. J. Oosterhoff (1967). In *Proceedings of Nobel Symposium* 5, Interscience Publishers, New York, p. 211.

H. L. Brose (1925). *Phil. Mag.*, **50**, 536.

S. C. Brown (1959). *Basic Data of Plasma Physics*, John Wiley and Sons, New York.

W. B. Brown and R. E. Roberts (1967). *J. Chem. Phys.*, **46**, 2006.

E. Brüche (1927a). *Ann. Physik*, **83**, 1065.

E. Brüche (1927b). *Ann. Physik*, **84**, 280.

R. H. Bullis, T. L. Churchill, W. L. Wiegand and E. K. Schubert (1967). *Abstracts 5th International Conference on the Physics of Electronic and Atomic Collisions*, Nauka, Leningrad, p. 263.

P. G. Burke (1965). *Advan. Phys.*, **14**, 521.

P. G. Burke and H. M. Schey (1962). *Phys. Rev.*, **126**, 147.

P. G. Burke and K. Smith (1962). *Rev. Mod. Phys.*, **34**, 458.

N. P. Carleton and L. R. Megill (1962). *Phys. Rev.*, **126**, 2089; L. R. Megill (private communication).

G. Cavalleri, E. Gatti and P. Principi (1964). *Nuovo Cimento*, **31**, 302.

G. Cavalleri, E. Gatti and A. M. Interlenghi (1965). *Nuovo Cimento*, **40**B, 450.

L. M. Chanin and R. D. Steen (1964). *Phys. Rev.*, **136**, A138.

L. M. Chanin, A. V. Phelps and M. A. Biondi (1962). *Phys. Rev.*, **128**, 219.

S. Chapman and T. G. Cowling (1952). *The Mathematical Theory of Non-uniform Gases*, 2nd ed., Cambridge University Press, London, Chap. 18.

C. L. Chen (1963). *Phys. Rev.*, **131**, 2550.

C. L. Chen and M. Raether (1962). *Phys. Rev.*, **128**, 2679.

L. G. Christophorou and J. G. Carter (1968). *Chem. Phys. Letters*, **2**, 607.

L. G. Christophorou, E. L. Chaney and A. A. Christodoulides (1969). *Chem. Phys. Letters*, **3**, 363.

L. G. Christophorou and A. A. Christodoulides (1969). *J. Phys. B* (*Proc. Phys. Soc.* (*London*), **2**, 71.

L. G. Christophorou and R. N. Compton (1967). *Health Phys.*, **13**, 1277.

L. G. Christophorou, G. S. Hurst and A. Hadjiantoniou (1966). *J. Chem. Phys.*, **44**, 3506; see also Erratum, J. Chem. Phys. **47**, 1883 (1967).

L. G. Christophorou, R. N. Compton, G. S. Hurst and P. W. Reinhardt (1966). *J. Chem. Phys.*, **45**, 536.

L. G. Christophorou, G. S. Hurst and W. G. Hendrick (1966). *J. Chem. Phys.*, **45**, 1081.

L. G. Christophorou, R. N. Compton, G. S. Hurst and P. W. Reinhardt (1965) *J. Chem. Phys.*, **43**, 4273.

L. G. Christophorou and D. Pittman (1969). Unpublished results.

L.G. Christophorou and D. Pittman (1970). *J. Phys. B* (*Proc. Phys. Soc.* (*London*)), **3**, 1252.

L. G. Christophorou and J. A. Stockdale (1968). *J. Chem. Phys.*, **48**, 1956.

L. W. Cochran and D. W. Forester (1962). *Phys. Rev.*, **126**, 1785.

L. Colli and M. T. deLeonardis (1953). *J. Appl. Phys.*, **24**, 255.

L. Colli and U. Facchini (1952). *Rev. Sci. Instr.*, **23**, 39.

R. N. Compton, L. G. Christophorou and R. H. Huebner (1966). *Phys. Letters*, **23**, 656.

R. N. Compton, R. H. Huebner, P. W. Reinhardt and L. G. Christophorou (1968) *J. Chem. Phys.*, **48**, 901.

T. L. Cottrell and I. C. Walker (1965). *Trans. Faraday Soc.*, **61**, 1585.

T. L. Cottrell and I. C. Walker (1966). *Quart. Revs.*, **20**, 153.

T. L. Cottrell and I. C. Walker (1967). *Trans. Faraday Soc.*, **63**, 549.

C. A. Coulson, A. Maccoll and L. E. Sutton (1952). *Trans. Faraday Soc.*, **48**, 106.

C. A. Coulson and M. Walmsley (1967). *Proc. Phys. Soc.* (*London*), **91**, 31.

J. D. Craggs and H. S. W. Massey (1959). In S. Flügge (Ed.), *Handbuch der Physik*, Vol. XXXVII, Springer-Verlag, Berlin, pp. 314–415.

A. M. Cravath (1929). *Phys. Rev.*, **33**, 605.

O. H. Crawford (1967). *Proc. Phys. Soc.* (*London*), **91**, 279.

O. H. Crawford (1968). *Chem. Phys. Letters*, **2**, 461.

O. H. Crawford and A. Dalgarno (1967). *Chem. Phys. Letters*, **1**, 23.

O.H. Crawford, A. Dalgarno and P.B. Hays (1967). *Mol. Phys.*, **13**, 181.

R. W. Crompton and M. T. Elford (1963). Paper presented at the 6th International Conference on Ionization Phenomena in Gases, Paris.

R. W. Crompton, M. T. Elford and J. Gascoigne (1965). *Australian J. Phys.*, **18**, 409.

R. W. Crompton, M. T. Elford and R. L. Jory (1967). *Australian J. Phys.*, **20**, 369.

R. W. Crompton, M. T. Elford and A. I. McIntosh (1968). *Australian J. Phys.*, **21**, 43.

R.W. Crompton and R.L. Jory (1965). In *Proceedings of the IVth Inter. Conf.*

*on the Physics of Electronic and Atomic Collisions*, Quebec 1965 (Science Book-crafters; Hastings-on Hudson, N.Y., p. 118).

R. W. Crompton and R. L. Jory (1962). *Australian J. Phys.*, **15**, 451.

R. W. Crompton and A. I. McIntosh (1968). *Australian J. Phys.*, **21**, 637.

R. W. Crompton, J. A. Rees and R. L. Jory (1965). *Australian J. Phys.*, **18**, 541.

R. W. Crompton and D. J. Sutton (1952). *Proc. Roy. Soc. (London)*, **A215**, 467.

J. W. Daiber and H. F. Waldron (1966). *Phys. Rev.*, **151**, 51.

J. T. Dowell and T. E. Sharp (1968). *Phys. Rev.*, **167**, 124.

M. J. Druyvesteyn (1930). *Physica (Eindhoven)*, **10**, 69.

M. J. Druyvesteyn and F. M. Penning (1940). *Rev. Mod. Phys.*, **12**, 87.

H. Ehrhardt, L. Langhans, F. Linder and H. S. Taylor (1968). *Phys. Rev.*, **A173**, 222.

M. T. Elford (1966). *Australian J. Phys.*, **19**, 629.

K. G. Emeleus, R. W. Lunt and C. A. Meek (1936). *Proc. Roy. Soc. (London)*, **A156**, 394.

I. Eliezer, H. S. Taylor and J. K. Williams, Jr. (1967). *J. Chem. Phys.*, **47**, 2165.

A. G. Engelhardt and A. V. Phelps (1963). *Phys. Rev.*, **131**, 2115.

A. G. Engelhardt and A. V. Phelps (1964). *Phys. Rev.*, **133**, A375.

A. G. Engelhardt, A. V. Phelps and C. G. Risk (1964). *Phys. Rev.*, **135**, A1566.

D. D. Errett (1951). Ph.D. thesis, Purdue University.

U. Fano (1961). *Phys. Rev.*, **124**, 1866.

U. Fano and J. W. Cooper (1965). *Phys. Rev.*, **138**, A400.

J. B. Fisk (1936). *Phys. Rev.*, **49**, 167.

R. J. Fleming and G. S. Higginson (1963). *Proc. Phys. Soc. (London)*, **81**, 974.

K. Fox (1967). *Phys. Letters*, **25A**, 345.

K. Fox and J. E. Turner (1966a). *Am. J. Phys.*, **34**, 606.

K. Fox and J. E. Turner (1966b). *J. Chem. Phys.*, **45**, 1142.

L. Frommhold (1960). *Z. Physik*, **160**, 554.

L. Frommhold (1968). *Phys. Rev.*, **172**, 118.

L. S. Frost and A. V. Phelps (1962). *Phys. Rev.*, **127**, 1621.

L. S. Frost and A. V. Phelps (1964). *Phys. Rev.*, **136**, A1538.

A. P. Gagge (1933). *Phys. Rev.*, **44**, 808.

W. R. Garrett (1970). Private communication.

W. R. Garrett and H. T. Jackson, Jr. (1967). *Phys. Rev.*, **153**, 28.

S. Geltman (1960). *Phys. Rev.*, **119**, 1283.

E. Gerjuoy and S. Stein (1955). *Phys. Rev.*, **97**, 1671; **98**, 1848.

H. B. Gilbody, R. F. Stebbings and W. L. Fite (1961). *Phys. Rev.*, **121**, 794.

D. E. Golden (1966a). *Phys. Rev. Letters*, **17**, 847.

D. E. Golden (1966b). *Phys. Rev.*, **151**, 48.

D. E. Golden and H. W. Bandel (1965a). *Phys. Rev.*, **138**, A14.

D. E. Golden and H. W. Bandel (1965b). *Phys. Rev. Letters*, **14**, 1010.

D. E. Golden and H. W. Bandel (1966). *Phys. Rev.*, **149**, 58.

D. E. Golden and H. Nakano (1966). *Phys. Rev.*, **144**, 71.

J. Green and J. Lee (1964). *Positronium Chemistry*, Academic Press, New York.

R. Grünberg (1967). *Z. Physik*, **204**, 12.

R. Haas (1957). *Z. Physik*, **148**, 177.

R. D. Hake Jr. and A. V. Phelps (1967). *Phys. Rev.*, **158**, 70.

B. I. H. Hall (1955). *Australian J. Phys.*, **8**, 468.

P. Hammerling, W. W. Shine and B. Kivel (1957). *J. Appl. Phys.*, **28**, 760.

N. Hamilton and J. A. Stockdale (1966). *Australian J. Phys.*, **19**, 813.

J. B. Hasted (1964). *Physics of Atomic Collisions*, Butterworths, Washington.

R. H. Healey (1938). *Phil. Mag.*, **26**, 940.

R. H. Healey and J. W. Reed (1941). *The Behaviour of Slow Electrons in Gases*, Amalgamated Wireless of Australia, Sydney.

H. G. M. Heideman, C. E. Kuyatt and G. E. Chamberlain (1966). *J. Chem. Phys.*, **44**, 355.

R. J. W. Henry (1967). *Phys. Rev.*, **162**, 56.

P. Herreng (1942a). *Compt. Rend.*, **214**, 421.

P. Herreng (1942b). *Compt. Rend.*, **215**, 79.

P. Herreng (1943). *Compt. Rend.*, **217**, 75.

G. Herzberg (1950). *Molecular Spectra and Molecular Structure. I. Spectra of Diatomic Molecules*, 2nd ed., D. Van Nostrand, New York.

J. O. Hirschfelder, C. F. Curtiss and R. B. Bird (1954). *Molecular Theory of Gases and Liquids*, John Wiley and Sons, New York.

C. R. Hoffmann and H. M. Skarsgard (1969). *Phys. Rev.*, **178**, 168.

T. Holstein (1946). *Phys. Rev.*, **70**, 367.

B. Huber (1968). *Z. Naturforsch*, **23a**, 1228.

R. H. Huebner, R. N. Compton, L. G. Christophorou and P. W. Reinhardt (1968). ORNL-TM-2156.

R. H. Huebner, R. N. Compton and H. C. Schweinler (1968). *Chem. Phys. Letters*, **2**, 407.

C. A. Hurst and B. S. Liley (1965). *Australian J. Phys.*, **18**, 521.

G. S. Hurst, L. B. O'Kelly and T. E. Bortner (1961). *Phys. Rev.*, **123**, 1715.

G. S. Hurst, L. B. O'Kelly, E. B. Wagner and J. A. Stockdale (1963). *J. Chem. Phys.*, **39**, 1341.

G. S. Hurst and J. E. Parks (1966). *J. Chem. Phys.*, **45**, 282.

G. S. Hurst, J. A. Stockdale and L. B. O'Kelly (1963). *J. Chem. Phys.*, **38**, 2572.

L. G. H. Huxley (1940). *Phil. Mag.*, **30**, 396.

L. G. H. Huxley (1956). *Australian J. Phys.*, **9**, 44.

L. G. H. Huxley (1959). *J. Atmospheric and Terrest. Phys.*, **16**, 46.

L. G. H. Huxley and R. W. Crompton (1955). *Proc. Phys. Soc.* (*London*), **B68**, 381.

L. G. H. Huxley and R. W. Crompton (1962). In D. R. Bates (Ed.), *Atomic and Molecular Processes*, Academic Press, New York, p. 335.

L. G. H. Huxley, R. W. Crompton and C. H. Bagot (1959). *Australian J. Phys.*, **12**, 303.

L. G. H. Huxley, R. W. Crompton and M. T. Elford (1966). *Brit. J. Appl. Phys.*, **17**, 1237.

L. G. H. Huxley and A. A. Zaazou (1949). *Proc. Roy. Soc.* (*London*), **A196**, 402.

Y. Itikawa (1967). *Phys. Letters*, **24A**, 495.

Y. Itikawa (1969). *J. Phys. Soc. Japan*, **27**, 443.

G. Jäger and W. Otto (1962). *Z. Physik*, **169**, 517.

T. L. John (1960). *Proc. Phys. Soc.* (*London*), **76**, 532.

J. M. Kirschner and D. S. Toffolo (1952). *J. Appl. Phys.*, **23**, 594.

O. Klein (1932). *Z. Physik*, **76**, 226.

M. M. Klein and K. A. Brueckner (1958). *Phys. Rev.*, **111**, 1115.

H. Kleinpoppen and V. Raible (1965). *Phys. Letters*, **18**, 24.

E. D. Klema and J. S. Allen (1950). *Phys. Rev.*, **77**, 661.

von K. Köllmann and H. Neuert (1968). Private communication.

C. E. Kuyatt, S. R. Mielczarek and J. A. Simpson (1964). *Phys. Rev. Letters*, **12**, 293.

C. E. Kuyatt, J. A. Simpson and S. R. Mielczarek (1965a). *Phys. Rev.*, **138**, A385.

C. E. Kuyatt, J. A. Simpson and S. R. Mielczarek (1965b). In *Proceedings IVth International Conference on the Physics of Electronic and Atomic Collisions*, Quebec, 1965. (Science Bookcrafters, New York, p. 113).

C. E. Kuyatt, J. A. Simpson and S. R. Mielczarek (1966). *J. Chem. Phys.*, **44**, 437.

R. W. LaBahn and J. Callaway (1964). *Phys. Rev.*, **135**, A1539.
R. W. LaBahn and J. Callaway (1966). *Phys. Rev.*, **147**, 28.
L. D. Landau and E. M. Lifschitz (1965). *Quantum Mechanics*, 2nd ed., Pergamon Press, London.
H. Lehning (1968). *Phys. Letters*, **28A**, 103.
N. E. Levine and M. A. Uman (1964). *J. Appl. Phys.*, **35**, 2618.
J-M. Lévy-Leblond (1967). *Phys. Rev.*, **153**, 1.
J-M. Lévy-Leblond and J-P. Provost (1967). *Phys. Letters*, **26B**, 104.
B. S. Liley (1967). *Australian J. Phys.*, **20**, 527.
S. C. Lin and B. Kivel (1959). *Phys. Rev.*, **114**, 1026.
J. J. Lloyd (1960). *Proc. Phys. Soc.*, **75**, 387.
L. B. Loeb (1935). *Phys. Rev.*, **48**, 684.
L. B. Loeb (1955). *Basic Processes of Gaseous Electronics*, University of California Press, Berkeley and Los Angeles.
L. B. Loeb (1956). In S. Flügge (Ed.), *Handbuch der Physik*, Vol. XXI, Springer-Verlag, Berlin, pp. 445–503.
J. J. Lowke (1963). *Australian J. Phys.*, **16**, 115.
J. J. Lowke and J. H. Parker, Jr. (1969). *Phys. Rev.*, **181**, 302.
J. J. Lowke and J. A. Rees (1963a). 63-1 Ion Diffusion Unit, The Australian National University, Cambera, May.
J. J. Lowke and J. A. Rees (1963b). *Australian J. Phys.*, **16**, 447.
H. Margenau (1946). *Phys. Rev.*, **69**, 508.
H. S. W. Massey (1956). In S. Flügge (Ed.), *Handbuch der Physik*, Vol. XXXVI, Springer-Verlag, Berlin, pp. 232–408.
H. S. W. Massey and E. H. S. Burhop (1952). *Electronic and Ionic Impact Phenomena*, Clarendon Press, Oxford.
H. S. W. Massey and R. O. Ridley (1956). *Proc. Phys. Soc.* (*London*), **A69**, 659.
E. W. McDaniel (1964). *Collision Phenomena in Ionized Gases*, John Wiley and Sons, New York.
R. P. McEachran and P. A. Fraser (1960). *Can. J. Phys.*, **38**, 317.
R. H. McFarland (1964). *Phys. Rev.*, **136**, A1240.
J. D. McGee and J. C. Jaeger (1928). *Phil. Mag.*, **6**, 1107.
A. I. McIntosh (1966). *Australian J. Phys.*, **19**, 805.
M. G. Menendez and H. K. Holt (1966). *J. Chem. Phys.*, **45**, 2743.
M. H. Mittleman and V. P. Myerscough (1966). *Phys. Letters*, **23**, 545.
M. H. Mittleman and R. E. von Holdt (1965). *Phys. Rev.*, **140A**, 726.
B. L. Moiseiwitsch (1962). In D. R. Bates (Ed.), *Atomic and Molecular Processes*, Academic Press, New York, Chap. 9.
P. M. Morse (1929). *Phys. Rev.*, **34**, 57.
N. F. Mott and H. S. W. Massey (1965). *The Theory of Atomic Collisions*, Clarendon Press, Oxford.
W. T. Naff, C. D. Cooper and R. N. Compton (1968). *J. Chem. Phys.*, **49**, 2784.
T. Nagy, L. Nagy and S. Dési (1960). *Nucl. Instr. Methods*, **8**, 327.
D. R. Nelson (1968). Ph. D. Thesis, University of Tenn.
D. R. Nelson and G. E. Whitesides (1968). Private communication.
R. H. Neynaber, L. L. Marino, E. W. Rothe and S. M. Trujillo (1961a). *Phys. Rev.*, **123**, 148.
R. H. Neynaber, L. L. Marino, E. W. Rothe and S. M. Trujillo (1961b). *Phys. Rev.*, **124**, 135.
R. H. Neynaber, L. L. Marino, E. W. Rothe and S. M. Trujillo (1963). *Phys. Rev.*, **129**, 2069.
R. A. Nielsen (1936). *Phys. Rev.*, **50**, 950.

R. A. Nielsen and N. E. Bradbury (1937). *Phys. Rev.*, **51**, 69.

C. E. Normand (1930). *Phys. Rev.*, **35**, 1217.

T. F. O'Malley (1963). *Phys. Rev.*, **130**, 1020.

T. F. O'Malley and S. Geltman (1965). *Phys. Rev.*, **137**, A1344.

T. F. O'Malley, L. Spruch and L. Rosenberg (1961). *J. Math. Phys.*, **2**, 491.

T. F. O'Malley, L. Rosenberg and L. Spruch (1962). *Phys. Rev.*, **125**, 1300.

J. L. Pack and A. V. Phelps (1961). *Phys. Rev.*, **121**, 798.

J. L. Pack, R. E. Voshall and A. V. Phelps (1962). *Phys. Rev.*, **127**, 2084; Westinghouse Research Labs. Scientific Paper: 62-928-113-P1.

J. H. Parker, Jr. and J. J. Lowke (1968). *Bull. Am. Phys. Soc.* **13**, 201.

J. H. Parker, Jr. and J. J. Lowke (1969). *Phys. Rev.* **181**, 290.

J. E. Parks and G. S. Hurst (1965). ORNL-TM-1287.

J. Perel, P. Englander and B. Bederson (1962). *Phys. Rev.*, **128**, 1148.

A. V. Phelps (1968). Private communication.

A. V. Phelps (1968). *Rev. Mod. Phys.*, **40**, 399.

A. V. Phelps, O. T. Fundingsland and S. C. Brown (1951). *Phys. Rev.*, **84**, 559.

W. J. Pollock (1968). *Trans. Faraday Soc.*, **64**, 2919.

J. O. Porteus (1964). In M. R. C. McDowell (Ed.), *Proceedings Third International Conference on the Physics of Electronic and Atomic Collisions*, University College London, July 1963, North-Holland Publishing Co., Amsterdam, p. 72.

A. N. Prasad and G. P. Smeaton (1967). *Brit. J. Appl. Phys.*, **18**, 371.

H. Raether (1961). *Ergel. der Naturw.*, **33**, 175.

C. Ramsauer (1921a). *Ann. Physik.*, **64**, 513.

C. Ramsauer (1921b). *Ann. Physik*, **66**, 546.

C. Ramsauer and R. Kollath (1929). *Ann. Physik*, **3**, 536.

C. Ramsauer and R. Kollath (1930). *Ann. Physik*, **4**, 91.

C. Ramsauer and R. Kollath (1932). *Ann. Physik*, **12**, 529, 837.

J. A. Rees (1964). Measurements of Townsend's energy factor $k_1$ for electrons in carbon dioxide, Ion Diffusion Unit of the Australian National University, Canberra.

J. A. Rees (1965). *Australian J. Phys.*, **18**, 41.

R. H. Ritchie and G. E. Whitesides (1961). ORNL-3081. Revised data (Private communication, 1968).

E. J. Robinson and S. Geltman (1967). *Phys. Rev.*, **153**, 4.

L. B. Robinson (1957). *Phys. Rev.*, **105**, 922.

R. Rydberg (1932). *Z. Physik*, **73**, 376.

R. Rydberg (1933). *Z. Physik*, **80**, 514.

H. Ryzko (1966). *Arkiv för Fysik*, **32**, 1.

L. I. Schiff (1955). *Quantum Mechanics*, 2nd ed., McGraw-Hill Book Co., New York.

G. J. Schulz (1959). *Phys. Rev.*, **116**, 1141.

G. J. Schulz (1961). *J. Chem. Phys.*, **34**, 1778.

G. J. Schulz (1963). *Phys. Rev. Letters*, **10**, 104.

G. J. Schulz (1964a). *Phys. Rev.*, **136**, A650.

G. J. Schulz (1964b). *Phys. Rev. Letters*, **13**, 583.

G. J. Schulz (1964c). *Phys. Rev.*, **135**, A988.

G. J. Schulz (1964d). In M. R. C. McDowell (Ed.), *Proceedings of the Third International Conference on the Physics of Electronic and Atomic Collisions*, London July 1963, North-Holland Publishing Co., Amsterdam, p. 124.

G. J. Schulz (1965). In *Abstracts of the 4th International Conference on the Physics of Electronic and Atomic Collisions*, Quebec, Science Bookcrafters, New York, p. 117.

G. J. Schulz and R. K. Asundi (1967). *Phys. Rev.*, **158**, 25.

H. C. Schweinler (1967). ORNL-4168 p. 133.

M. Shimizu (1963). *J. Phys. Soc. Japan*, **18**, 811.

J. A. Simpson (1964). In M. R. C. McDowell (Ed.), *Proceedings of the Third International Conference on the Physics of Electronic and Atomic Collisions*, London, July 1963, North-Holland Publishing Co., Amsterdam, p. 128.

J. A. Simpson and U. Fano (1963). *Phys. Rev. Letters*, **11**, 158.

J. A. Simpson and S. R. Mielczarek (1963). *J. Chem. Phys.*, **39**, 1606.

J. A. Simpson, S. R. Mielczarek and J. Cooper (1964). *J. Opt. Soc. Am.*, **54**, 269.

M. F. Skinker (1922). *Phil. Mag.*, **44**, 994.

M. F. Skinker and J. V. White (1923). *Phil. Mag.*, **46**, 630.

K. Smith (1960). *Phys. Rev.*, **120**, 845.

K. Smith (1966). *Rep. Progr. Phys. XXIX*, 373, part 2.

V. N. Smith and J. F. Fidiam (1964). *J. Chromatography*, **36**, 1739.

L. Spruch, T. F. O'Malley and L. Rosenberg (1960). *Phys. Rev. Letters*, **5**, 374.

J. A. Stockdale, L. G. Christophorou, J. E. Turner and V. E. Anderson (1967). *Phys. Letters*, **25A**, 510.

J. A. Stockdale and G. S. Hurst (1964). *J. Chem. Phys.*, **41**, 255.

G. Sunshine, B. B. Aubrey and B. Bederson (1967). *Phys. Rev.*, **154**, 1.

K. Takayanagi and M. Inokuti (1967). *J. Phys. Soc. Japan*, **23**, 1412.

K. Takayanagi and Y. Itikawa (1968). *J. Phys. Soc. Japan*, **24**, 160.

H. S. Taylor and F. E. Harris (1963). *J. Chem. Phys.*, **39**, 1012.

A. Temkin (1957). *Phys. Rev.*, **107**, 1004.

A. Temkin and J. C. Lamkin (1961). *Phys. Rev.*, **121**, 788.

J. S. Townsend (1900). *Phil. Trans. Roy. Soc. London*, **A193**, 129.

J. S. Townsend (1901). *Phil. Trans. Roy. Soc. London*, **195**, 259.

J. S. Townsend (1908a). *Proc. Roy. Soc. (London)*, **A80**, 207.

J. S. Townsend (1908b). *Proc. Roy. Soc. (London)*, **A81**, 464.

J. S. Townsend (1928). *Proc. Roy. Soc. (London)*, **A120**, 511.

J. S. Townsend (1947). *Electrons in Gases*, Hutchinson's Scientific and Technical Publications, London.

J. S. Townsend and V. A. Bailey (1921). *Phil. Mag.*, **42**, 873.

J. S. Townsend and V. A. Bailey (1922a). *Phil. Mag.*, **43**, 593.

J. S. Townsend and V. A. Bailey (1922b). *Phil. Mag.*, **44**, 1033.

J. S. Townsend and V. A. Bailey (1923). *Phil. Mag.*, **46**, 657.

J. S. Townsend and H. T. Tizard (1912). *Proc. Roy. Soc. (London)*, **A87**, 357.

J. E. Turner (1966). *Phys. Rev.*, **141**, 21.

J. E. Turner and K. Fox (1966). *Phys. Letters*, **23**, 547.

J. E. Turner and K. Fox (1968). *J. Phys. A (Proc. Phys. Soc.)* **1**, 118.

M. A. Uman (1964). *Phys. Rev.*, **133**, A1266.

E. Vogt and G. H. Wannier (1954). *Phys. Rev.*, **95**, 1190.

L. Vriens, C. E. Kuyatt and S. R. Mielczarek (1968). *Phys. Rev.*, **170**, 163.

E. B. Wagner, F. J. Davis and G. S. Hurst (1967). *J. Chem. Phys.*, **47**, 3138.

K. H. Wagner (1964). *Z. Physik*, **178**, 64.

R. F. Wallis, R. Herman and H. W. Milnes (1960). *J. Mol. Spectry*, **4**, 51.

R. W. Warren and J. H. Parker, Jr. (1962a). *Phys. Rev.* **128**, 2661.

R. W. Warren and J. H. Parker Jr. (1962b). Westinghouse Research Laboratories Scientific Paper 62-908-113-P6.

J. H. Williamson and M. R. C. McDowell (1965). *Proc. Phys. Soc. (London)* **85**, 719.

# 5 *Inelastic electron scattering*

## 5.1 Basic Background

### 5.1.1 *Introduction*

Studies of inelastic electron scattering by atoms and molecules are mainly aimed at the measurement of the inelastic scattering cross sections and their energy dependence; the location of, and the excitation function for, rotational, vibrational, electronic, and compound negative ion states, as well as the excitation functions and appearance potentials of assorted ions are of primary interest. In general, low-energy electron impact spectroscopy provides a unique tool to investigate the structure of atoms and molecules. For sufficiently low-energy electrons optical selection rules involving change in multiplicity, angular momentum, and symmetry are greatly relaxed, and a study of both optically allowed and optically forbidden transitions, becomes possible. A large portion of the work on electronic excitation by electron impact has dealt with electrons of energy sufficiently high ($\gtrsim 200\,\mathrm{eV}$, depending on the state involved) for the Born approximation to be valid (the electronic excitation spectrum so obtained closely resembles the optical-dipole excitation spectrum). A wealth of information, however, has been accumulating recently using low-energy electron beams. For electron impact energies below $\sim 30$ to $40\,\mathrm{eV}$ the excitation spectrum frequently differs from the optical one, and thus low-energy electron impact studies complement optical experiments which are often difficult to perform in the vacuum ultraviolet region (Chapter 3).

Most of the low-energy electron impact work has been on atoms and simple molecules (diatomic and triatomic); the study of polyatomic molecules has been limited, although the importance of such investigations has been increasingly recognized (see, for example, Christophorou and Compton (1967)).

In this chapter we shall elaborate on the various inelastic electron scattering processes and present typical data. Various aspects of the subject have been reviewed by Massey and Burhop (1952), Massey (1956), Craggs and Massey (1959), Fite (1962), McDaniel (1964) and Hasted (1964). A review of experimental cross sections for electron impact ionization of atoms, atomic ions, and diatomic molecules was given by Kieffer and Dunn (1966). Moiseiwitsch and Smith (1968) reviewed electron impact excitation of atoms, while Burke (1965) and Smith (1966) reviewed resonances in electron–atom scattering, and Bardsley and Mandl (1968) resonances in electron–molecule scattering.

### 5.1.2 *Electron Impact Excitation and Ionization Cross Sections*

A discussion of collision cross sections has been given in Section 4.1.2. It has been emphasized there that in addition to elastic electron scattering which is particularly important at subexcitation energies, a number of other processes contribute to the total collision cross section, especially for electron energies exceeding the lowest electronic excitation threshold of the target atom or molecule. We shall now elaborate briefly on the excitation and ionization cross sections.

#### 5.1.2.1 Electron impact excitation cross section

Consider a parallel beam of $N_e$ electrons per cm$^2$ per second striking the target gas containing $N_t$ atoms or molecules, in the volume from which scattered electrons are detected, in the state 0. Let $dN_s$ be the number of incident electrons scattered per second at an angle $\theta$ into the solid angle $d\Omega$, having excited the target from the initial state 0 to a final state $n$. Under single collision conditions $dN_s$ is given by

$$dN_s = I_{0n}(\theta) N_e N_t \, d\Omega \tag{5.1}$$

where $I_{0n}(\theta)$ is called the differential excitation cross section and has the dimensions cm$^2$/atom (or molecule) steradian. It defines the angular distribution of electrons which are scattered having excited the target to the $n$th state. Since we shall be concerned only with atoms and molecules initially in the ground state we shall abbreviate $I_{0n}(\theta)$ to $I_n(\theta)$. The total cross section (all scattering angles) to the $n$th state is

$$\sigma_n = 2\pi \int_0^\pi I_n(\theta) \sin \theta \, d\theta \tag{5.2}$$

### 5.1.2.2 Electron impact ionization cross section

If the inelastic collision process leads to ionization, viz.

$$A + e \rightarrow A^+ + 2e \tag{5.3}$$

we may write for the ionization cross section $\sigma_i$ for (5.3)

$$\sigma_i = \int_\varepsilon \int_\Omega I_\varepsilon(\theta) \, d\Omega \, d\varepsilon \tag{5.4}$$

$I_\varepsilon(\theta) \, d\varepsilon$ is the *differential ionization cross section*, i.e. the differential cross section for an ionizing collision in which the energy of the ejected electron lies between $\varepsilon$ and $\varepsilon + d\varepsilon$. The integrations in (5.4) are carried out over all scattering angles and all possible continuum states. For collisions leading to multiple ionization, viz.

$$A + e \rightarrow A^{n+} + (n+1)e \quad (n \geqslant 1) \tag{5.5}$$

we shall use $\sigma_{ni}$ rather than $\sigma_i (\sigma_i \equiv \sigma_{1i})$, and we shall refer to the cross sections for single, double and multiple ionization ($\sigma_i, \sigma_{2i}, \ldots \sigma_{ni}$) as the partial cross sections for $n$-fold ionization. When the produced ions are measured by recording the positive current due to a mixture of all singly and multiply charged ions (and for molecules also for fragment ions), the cross section

$$\sigma_{gi} = \sum_n n\sigma_{ni} \tag{5.6}$$

is called the 'gross' or total ionization cross section.

### 5.1.3 *Molecular Dissociation by Electron Impact*

Electronic transitions in diatomic and polyatomic molecules ending at upper bound or repulsive states have been discussed in Chapter 3 within the framework of the Franck–Condon principle. The process of predissociation has been discussed in Chapter 3, also. Molecular dissociation upon electron impact $(AX + e \rightarrow A + X + e)$, although a simple inelastic process, has been poorly studied due, probably, to the difficulty in detecting and identifying slow neutral species. The dissociation collision may be understood as a molecular transition to an antibonding state, similar to that of dissociative ionization $(AX + e \rightarrow A^+ + X + 2e)$. The cross section functions for the two processes may have a similar form.

Transitions in diatomic molecules resulting in positive and/or negative ion formation, namely ion-pair formation $(AX + e \rightarrow A^+ + X^- + e)$, and dissociative attachment $(AX + e \rightarrow A + X^-)$ are discussed in Chapter 6. The process of dissociative ionization and ion-pair formation may occur for any incident electron energy above threshold; the process of dissociative attachment, however, is a resonance one occurring in a restricted energy range (Chapter 6).

For all processes mentioned above, the kinetic energy of the products is fixed by the electronic transitions induced in the molecule by the electron impact. Vertical Franck–Condon transitions to bonding and antibonding states are shown in Figure 5.1a, b for the case of molecular ionization and molecular dissociative ionization producing positive ions only (Craggs and McDowell, 1955). A number of pertinent parameters are indicated on the figure.* Transitions ending at the repulsive upper state (Figure 5.1b) will lead to products with appreciable kinetic energy.

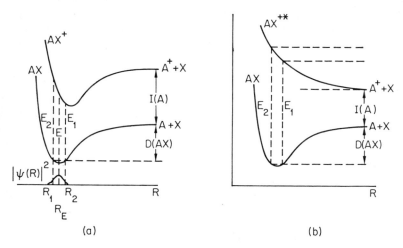

**Figure 5.1** Diagram illustrating molecular ionization and dissociative ionization energetics

The appearance potential $A_{kin}(A^+)$ of the ion $A^+$ with kinetic energy $E_{kin}(A^+)$ is

$$A_{kin}(A^+) = D(AX) + I(A) + E_{ex}(A^+, X) + E_{kin}(A^+, X) \tag{5.7}$$

where $D(AX)$ is the dissociation energy of $AX$, $I(A)$ is the ionization potential of $A$, $E_{ex}(A^+, X)$ is the sum of the excitation energies of the products, and $E_{kin}(A^+, X)$ is the sum of their kinetic energies; the ejected electron is

---

*The so-called vertical ionization energy corresponds to the energy of a vertical transition between the molecule and the molecular-ion potential energy curve, according to the Franck–Condon principle. When the equilibrium distances are different in the molecule and molecular ion, the vertical ionization energy will exceed the adiabatic ionization energy, i. e. the energy between the zero-point vibrational levels of the molecule and molecular ion.

assumed to carry a negligible amount of the available kinetic energy. The division of the total kinetic energy between the fragments is obtained by considering the conservation of momentum. For a polyatomic molecule the situation is complicated since the kinetic energies of all fragments have to be determined. For further discussion the reader is referred to the review by Craggs and Massey (1959) and to the books by Reed (1962), Cottrell (1958), and Field and Franklin (1957); see these references and Vedeneyev and coworkers (1966) for data on dissociation energies, appearance and ionization potentials.

### 5.1.4 Theoretical Considerations

#### 5.1.4.1 Cross sections for electronic excitation and ionization by electron impact

In most theoretical studies on the excitation and ionization of atoms and molecules by electron impact, quantum-mechanical calculations have been performed and use was made, by necessity, of a number of approximations. Many theoretical approaches were also restricted to conditions where the Born approximation is valid. The dependence of the excitation and ionization cross section on the energy of the incident electron has been thoroughly discussed elsewhere (e.g. Massey, 1956; Bates, 1962; Mott and Massey, 1965). A summary of quantal and other approximations can be found, also, in a recent review by Rudge (1968).

The approximations made in theoretical treatments of the interaction between an electron and a target atom or molecule relate to the time allowed for the impinging electron to interact with the target. Hence, the relative speeds (and thus the energies) of the impinging particles will be of utmost importance. Impinging particles of the same kinetic energy but of different masses may induce different transitions owing to their different velocities.

Most theoretical calculations employed the Born approximation. A general discussion of this approximation as it applies to inelastic collisions can be found in Massey and Burhop (1952) (see also a discussion in Schram (1966a)). Owing to the wide usage of the Born approximation and the many efforts to test its validity at low electron energies, a summary of the expressions for the differential and total excitation and ionization cross sections in this approximation will now be given.

Following Massey and Burhop (1952) the differential cross section for excitation of an atom from its ground state 0 to an excited electronic state $n$ upon collision with an electron is

$$I_n = \frac{4\pi^2 m^2}{h^4} \cdot \frac{k_n}{k} \left| \int V_n(\mathbf{r}') \exp\left[i(k\mathbf{n}_0 - k_n\mathbf{n})\cdot\mathbf{r}'\right] d\tau' \right|^2 \tag{5.8}$$

In Equation (5.8) $2\pi/k$ and $2\pi/k_n$ are, respectively, the de Broglie wave-lengths of the impinging and scattered electrons, and $\mathbf{n}_0$ and $\mathbf{n}$ are the unit vectors in the direction of incidence and scattering; $V_n$ is given by

$$V_n(\mathbf{r}) = e^2 \int \sum_1^N \frac{1}{r_{0,s}} \psi_0(\mathbf{r}_1, \mathbf{r}_2, ..., \mathbf{r}_N)\psi_n^*(\mathbf{r}_1, ..., \mathbf{r}_N)\, d\tau_1 \, ... \, d\tau_N \qquad (5.9)$$

where $r_{0,s}$ is the distance between the incident electron and the $s$th atomic electron, and $\psi_0$ and $\psi_n$ are the wave functions for the ground and the $n$th excited atomic states.

For single electron excitation one obtains for $\sigma_n$

$$\sigma_n = \frac{2\pi}{kk_n} \int_{K_{min}}^{K_{max}} I_n(K)K \, dK \qquad (5.10)$$

where $K_{min} = k - k_n$ and $K_{max} = k + k_n$. It may be further shown (Massey and Burhop, 1952) that a good approximation to (5.10) is

$$\sigma_n \simeq \frac{4\pi m^2 e^4}{k^2 \hbar^4} |z_{0,n}|^2 \ln \frac{2mv^2}{E_n - E_0} \qquad (5.11)$$

where $z_{0,n}$ is the dipole moment matrix element for the transition, and $E_n - E_0$ is the energy of the excited state $n$ referred to the ground state. From expression (5.11), which is valid for optically allowed transitions ($|z_{0,n}| \neq 0$), it is seen that the excitation cross section for large velocities falls off as $\varepsilon^{-1} \ln \varepsilon (\varepsilon = k^2 \hbar^2 / 2m)$.

If the $0 \rightarrow n$ transition is optically forbidden, $|z_{0,n}|$ is zero and the transition is associated with the quadrupole moment. In this case of a dipole-forbidden transition an approximate expression for $\sigma_n$ is

$$\sigma_n \simeq \frac{2\pi m^3 e^4}{k^2 \hbar^6} |(z^2)_{0,n}|^2 |E_0| \qquad (5.12)$$

where $(z^2)_{0,n}$ is the quadrupole moment matrix element. It is seen that for optically forbidden transitions the cross section decreases for large velocities faster (as $1/\varepsilon$) than for optically allowed transitions. In general, $\sigma_n$ for optically forbidden transitions are much smaller than $\sigma_n$ for optically allowed transitions at large energies.

The excitation probabilities of both allowed and forbidden transitions are influenced by electron exchange. The total electron spin (electron's plus target's) should be conserved. Electron exchange makes possible a change in multiplicity in electron–target collisions. For a target of electron spin $s\hbar$, the multiplicity is $2s + 1$, and the total spin (electron plus target) $(s \pm \frac{1}{2})\hbar$. If $s'\hbar$

is the spin of the target in the state $n$ reached by excitation, the total spin is $(s' \pm \frac{1}{2})\hbar$, and for spin-allowed transitions $s' \pm \frac{1}{2}$ must lie within $s \pm \frac{1}{2}$, i.e. $s' = s - 1$, $s$, or $s + 1$. When exchange is involved the multiplicity can change by $\pm 2$. This argument neglects spin–orbit coupling so that the spin of the free electron does not change during the collision. Such a situation may not be true for heavy atoms. Electron exchange is expected to be more probable for incident electron energies low enough to allow for an exchange during the collision. See results later this chapter as to the effect of electron exchange on the excitation cross sections, especially close to threshold.

Equations (5.11) and (5.12) are for transitions between the ground state and an excited state $n$. A summation of such expressions over all possible transitions in the energy range considered should yield the overall excitation cross section.

An expression analogous to (5.11) for optical excitation of an atom by a fast positive ion of charge $z'e$ is (Mott and Massey, 1965)

$$\sigma_n \simeq \frac{16\pi^4 z'^2 e^4}{mv^2} |z_{0,n}|^2 \; \ln \; \frac{2mv^2}{E_n - E_0} \tag{5.13}$$

Expressions (5.11) and (5.12) for $\sigma_n$ may be extended to the case of ionization of outer-shell electrons (for inner-shell ionization, see Mott and Massey (1965)) by taking into account the kinetic energy of the ejected electron. The electron-impact ionization cross section for the ejection of an electron with quantum numbers $n$ and $l$ is (Bethe, 1930)

$$\sigma_i(n, l) \simeq \frac{2\pi e^4}{mv^2} \cdot \frac{c_{nl}}{|I_{nl}|} \cdot Z_{nl} \; \ln \; \frac{2mv^2}{C_{nl}} \tag{5.14}$$

where

$$c_{nl} = \frac{Z_{eff}^2}{n^2 a_0^2} \int |z_{nl,k}|^2 \; dk \tag{5.15}$$

$Z_{eff}$ is the effective nuclear charge; $Z_{nl}$ is the number of electrons in the $n,l$ shell; $I_{nl}$ is the ionization energy for the $n,l$ electron; $C_{nl}$ is an energy of the order of $I_{nl}$; and $k$ is the wave number of the ejected electron. It is seen that at high incident velocities, $\sigma_i(n,l)$ decreases with increasing impact velocity in the same manner that $\sigma_n$ does for optically allowed excitations.

Similarly, in the Bethe–Born approximation the cross section for ionization by a fast positive ion of charge $z'e$ is (Mott and Massey, 1965)

$$\sigma_i(n,l) \simeq \frac{2\pi z'^2 e^4 c_{nl} Z_{nl}}{mv^2 |I_{nl}|} \; \ln \; \frac{2mv^2}{C_{nl}} \tag{5.16}$$

Here $v$ is the relative velocity of approach between the projectile and the target. Note that the mass appearing explicitly in (5.16) is that of the electron.

Expression (5.16) applies strictly to completely stripped projectile nuclei (e.g. $H^+$, $He^{2+}$). For singly charged heavy bombarding ions an effective ion charge must be used. The expressions given above for $\sigma_i$ apply to direct ionization and do not include the effects of autoionization and preionization.

As will be discussed in Section 5.2 in energy-loss experiments a measurement is made of the number of electrons which are scattered at an angle after suffering a given energy loss due to excitation and/or ionization. In analyses of such energy-loss measurements use is often made of the "generalized oscillator strength" (Bethe, 1930) which is analogous to the optical oscillator strength discussed in Chapter 3 (see also Chapter 1). For an atom* the generalized oscillator strength for a transition from the ground state 0 to the state $n$ can be written as

$$f_n(K) = (E_n/R)(Ka_0)^{-2} \left| \sum_j (n|\exp(i\mathbf{K} \cdot \mathbf{r}_j)|0) \right|^2 \qquad (5.17)$$

where $a_0$ is the Bohr radius, $R$ is the Rydberg energy, $\mathbf{r}_j$ is the coordinate vector of the $j$th atomic electron and $(n| \,0)$ is a matrix element; $f_n(K)$ is averaged over degenerate substates. A Born-approximation expression for the differential cross section in which the electron is scattered with a momentum change $K\hbar$ and concomitant excitation of the target from 0 to $n$ has been given in Chapter 1 (Equation (1.16)).

### 5.1.4.2 Excitation of molecular rotation and vibration by electron impact

In spite of the fact that direct excitation of molecular rotation and vibration by electron impact is not expected to be large, it is well established that subexcitation electrons lose energy to excitation of molecular vibrations and rotations (see Section 5.3). Early investigators (see Massey and Burhop (1952) and Craggs and Massey (1959)) realized that the energy lost by slow electrons in collisions with molecules is much greater than can be accounted for by elastic electron scattering (see Figure 4.37).

The vibrational excitation in an electron–molecule collision will be small unless the duration of the collision, $\tau_{coll}$, is approximately equal to the natural vibrational period of the molecule, $\tau_{vib}$, i.e.

$$\tau_{coll}/\tau_{vib} \simeq 1 \qquad (5.18)$$

For energies in the electron volt range the electron passes by the molecule much too fast to allow sufficient energy transfer, and thus the electron must be temporarily bound to the molecule in order to satisfy Equation (5.18). The large vibrational excitation cross sections observed recently in electron swarm and electron beam experiments for a number of molecules (see

---

*For a molecule appropriate averaging must be made over molecular orientation and internal degrees of freedom.

Section 5.3) were interpreted as proceeding through such temporary negative ion intermediates. It may be possible that rotational excitation is enhanced through temporary negative ion states in a similar fashion. Direct rotational excitation may be enhanced through the long-range interaction of low-energy electrons with polar and quadrupolar molecules. Several theoretical studies have been made of the excitation of molecular rotations and vibrations by electron impact (see, for example, Massey and Burhop (1952); Mott and Massey (1965); Carson (1954); Morse (1953); Breig and Lin (1965); Takayanagi (1965, 1966); Takayanagi and Geltman (1965)). Theoretical investigations of the vibrational excitation of molecules by electron impact via compound negative ion intermediates have been made, also, (see, for example, Herzenberg and Mandl (1962); Chen (1964a, 1966); Bardsley, Herzenberg and Mandl (1966); and Taylor, Nazaroff and Golebiewski (1966)) and have generally been successful in explaining the experimental data (see Section 5.3). The excitation of compound negative ion intermediates which subsequently decay into a free electron plus the molecule in various vibrational states of excitation may lead to unusually large vibrational excitation cross sections and high-energy (apparent) onsets for vibrational excitation (see Section 5.3).

The long-range interaction of low-energy electrons with polar and quadrupolar molecules has been shown (Stein, Gerjuoy and Holstein (1954); Gerjuoy and Stein (1955a, b); Lane and Geltman (1967) and references cited therein) to be primarily responsible for the large rotational excitation cross sections consistent with observation (see Section 5.3). Neglect of these long-range interactions is largely responsible for the small cross section values obtained by earlier investigators (Morse, 1953; Carson, 1954). Gerjuoy and Stein (1955a, b), using the Born approximation, gave for molecules possessing a quadrupole moment the following expressions for the rotational excitation cross sections $\sigma_{J,J+2}$ in which the incident electron loses energy to the molecule, and $\sigma_{J \to J-2}$ in which the incident electron gains energy:

$$\sigma_{J \to J+2}(\varepsilon) = \frac{(J+2)(J+1)}{(2J+3)(2J+1)} \sigma_0 \left(1 - \frac{(4J+6)B_0}{\varepsilon}\right)^{\frac{1}{2}} \qquad (5.19)$$

$$\sigma_{J \to J-2}(\varepsilon) = \frac{J(J-1)}{(2J-1)(2J+1)} \sigma_0 \left(1 + \frac{(4J-2)B_0}{\varepsilon}\right)^{\frac{1}{2}} \qquad (5.20)$$

where $\sigma_0 = (8/15)\pi Q^2 a_0^2$, $Q$ is the electric quadrupole moment in units of $ea_0^2$, $\varepsilon$ is the incident energy, $J$ is the rotational quantum number, and $B_0$ is the rotational constant (Herzberg, 1951) which is related to the energy levels of the rotating molecule approximately by

$$E_J = J(J+1)B_0 \qquad (5.21)$$

We note that for a rotational quantum number $J$ there is a selection rule $\Delta J = \pm 2$, so that the cross sections $\sigma_{J \to J + J}{}'$ are taken equal to zero for all $J'$ except $\pm 2$. The cross sections are zero at the appropriate onset energy and rise to a saturation at sufficiently large electron impact energy $\varepsilon$ (see Section 5.3).

For a dipolar molecule, Takayanagi (1966) derived the following expressions for the Born cross section $\sigma_{J \to J + 1}(\varepsilon)$ for excitation and $\sigma_{J \to J - 1}(\varepsilon)$ for de-excitation of the $J$th rotational level:

$$\sigma_{J, J+1}(\varepsilon) = \frac{(J+1) \mathrm{Ry}\, \sigma_r}{(2J+1)\varepsilon} \ln \frac{[\varepsilon^{\frac{1}{2}} + (\varepsilon - \varepsilon_J)^{\frac{1}{2}}]}{[\varepsilon^{\frac{1}{2}} - (\varepsilon - \varepsilon_J)^{\frac{1}{2}}]} \tag{5.22}$$

$$\sigma_{J, J-1}(\varepsilon) = \frac{J \mathrm{Ry}\, \sigma_r}{(2J+1)\varepsilon} \ln \frac{[(\varepsilon + \varepsilon_{-J})^{\frac{1}{2}} + \varepsilon^{\frac{1}{2}}]}{[(\varepsilon + \varepsilon_{-J})^{\frac{1}{2}} - \varepsilon^{\frac{1}{2}}]} \tag{5.23}$$

In the above expressions $\sigma_r = (8\pi/3)\mu^2 a_0^2$, $\mu$ is the electric dipole moment in units of $ea_0$, $a_0$ is the Bohr radius, $J$ is the rotational quantum number of the initial state, and Ry is the Rydberg energy. The energy, $\varepsilon_J$, lost by the electron (Equation 5.22) is

$$\varepsilon_J = 2(J+1)B_0 \tag{5.24}$$

while the energy, $\varepsilon_{-J}$, gained by the electron (Equation 5.23) is

$$\varepsilon_{-J} = 2JB_0 \tag{5.25}$$

$B_0$ is the rotational constant of the molecule (Herzberg, 1951). The expressions given above for homonuclear and heteronuclear molecules have been employed by Phelps and coworkers in analyses of electron transport data (see Section 5.3) to obtain cross sections for rotational excitation.

### 5.1.4.3 Resonant scattering of electrons by atoms and molecules

Resonance effects in electron scattering from atoms and molecules have been investigated for many years, but it is only recently that they have become the subject of intense experimental and theoretical scrutiny. As stated in Chapter 4, a resonance can be produced when the energy of the incident particle is sufficient to excite the target to a state in which the electron is temporarily retained by the target. Such a resonance is not a true bound (stationary) state but is a temporary one which is capable of decaying with a characteristic lifetime (and thus width). If the lifetime of the resonance is long compared with the time the electron takes to traverse the target, the formation of the resonance from the target and the electron will show up in electron collision experiments. There are different mechanisms for retaining a projectile electron in the target, each producing different characteristic features in scattering experiments. Many of these features have been observed (see reviews by Burke (1965); Smith (1966); Bardsley and Mandl (1968); Chapters 3, 4 and 6 and Sections 5.3 and 5.4 of this chapter).

Various types of resonances in electron–atom (molecule) scattering have been distinguished such as *shape resonances, electron-excited Feshbach resonances,* and for molecules, *nuclear-excited Feshbach resonances.* In a *shape resonance* the electron is retained by a potential barrier in either the ground or an excited electronic target state. Thus the trapping occurs when the incident particle experiences a region of attractive potential surrounded by a repulsive potential barrier. When in an excited state the incident electron excites the electronic state whose energy is less than the resonance energy, the extra electron has thus sufficient energy to escape, leaving the target electronically excited. If the potential experienced by the extra electron in the field of the excited target contains a barrier, its escape may be hindered and the electron may be temporarily bound to the target. In this case the resonance is called *core-excited shape resonance.* In *electron-excited Feshbach* resonances the electron loses energy to electronic excitation of the target and is left with insufficient energy to escape, while the target remains in its excited state. Before the electron is re-emitted it must absorb energy from the target.

In molecules, nuclear motion allows new features to arise, as compared to atoms; exchange of energy between electronic and nuclear motion may show up in elastic scattering and in inelastic scattering (e.g. vibrational excitation). Inelastic scattering cross sections may be greatly enhanced by the comparatively long lifetime of a resonance. For molecules, the simplest case of an *electron-excited Feshbach resonance* would be that in which the incident electron excites only a single electronic state of the target for which the vibrational levels lie above the resonance energy. In molecules, further, another type of resonance is possible; namely, *nuclear-excited Feshbach resonances.* In these the kinetic energy of the incident electron is absorbed solely into nuclear motion of the molecule. The collision, in this case, does not involve excitation of the electronic motion in the target.

All types of resonances mentioned above occur as intermediate states in collision processes. Nuclear-excited Feshbach resonances show up beautifully in electron attachment studies (Chapter 6). Most of the information about resonances has been derived from analyses of electron scattering experiments as has been discussed previously and will be further elaborated upon in this and the next chapter.

Theoretical studies of angular distributions of resonant scattering of electrons by molecules have also been made (Read, 1968; Bardsley and Read, 1968; O'Malley and Taylor, 1968). The angular distribution of inelastically scattered electrons is, in certain cases, determined uniquely by a comparison of the symmetry of the resonant state with the symmetry of the initial and final states of the target molecule (see Sections 5.3 and 5.4 for results).

## 5.2 Experimental Methods

Some of the techniques discussed in Chapter 4 in connection with elastic electron scattering, and some to be described in Chapter 6 in connection with electron attachment studies can be similarly employed for the study of inelastic electron scattering. In this section certain aspects of electron impact techniques will be discussed, and a description will be given of special arrangements relevant to the excitation and ionization of atoms and molecules by electron impact.

### 5.2.1 *Techniques for the Study of Excitation of Atoms and Molecules by Electrons*

#### 5.2.1.1 Electron beam techniques

Two classes of such techniques can be distinguished: energy-loss and threshold-electron excitation. In energy-loss experiments an electron beam of fixed energy (usually well above the excitation thresholds for valence electrons) strikes the scattering target, and the energy spectrum of the scattered electrons, referred to as the *energy-loss spectrum*, is measured either in the forward direction or at an angle (s). The basic components of an apparatus of this kind are an electron source, an energy selector, a collision chamber, a scattered-electron energy analyser, and an electron detector. In threshold-electron excitation, on the other hand, a quasi-monoenergetic electron beam of varying (mean) energy strikes the target gas, and those electrons which have lost essentially all of their energy at collision are detected, independently of the scattering angle. In this method, the incident electron energy matches the excitation thresholds of the target, and the scattered-electron current (due to the complete collection of 'zero-energy' electrons) is plotted as a function of the incident electron energy, reflecting the excitation spectrum of the target, to be referred to as the *threshold-electron excitation spectrum*.

A. *Energy-loss techniques.* The techniques for the production of mono-energetic incident electron beams, and the energy analysis of the scattered electrons have had an impressive progress recently (see, for example, Marnet and Kerwin (1960); Schulz (1962, 1964); Simpson (1964); Lassettre and coworkers (1964); Boersch, Geiger and Stickel (1964); Andrick and Ehrhardt (1966); Golden and Bandel (1965a); see also a review by

Klemperer (1965)). The resolution achieved ranges from ~0.15 to ~0.05 eV, although a better resolution has at times been reported, e.g. 0.017 eV Boersch, Geiger and Stickel, 1964). The reader is referred to Hasted (1964) for a discussion of electron sources and velocity selection of electrons.

In Figure 5.2 a schematic illustration of an energy-loss experiment is shown, and in Figures 5.3a and 5.3b, respectively, schematic diagrams of Simpson's transmission experiment and Schulz's scattering experiment are given. In Simpson's (Simpson, 1964) S-shape spectrometer an intense and well-collimated beam leaves the electron gun and is subsequently decelerated, made monochromatic by energy selection in a spherical electrostatic deflector, and accelerated into the collision chamber. The transmitted electron beam is decelerated, energy analysed, and accelerated into a

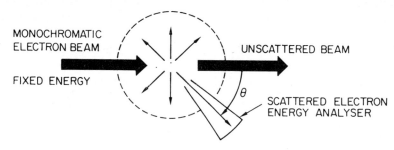

**Figure 5.2** Schematic illustration of energy-loss experiments

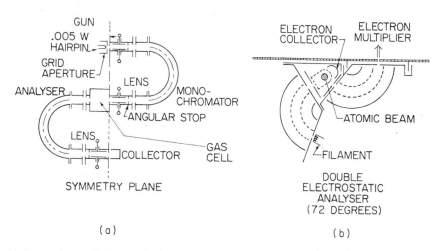

**Figure 5.3** (a) Schematic illustration of Simpson's S-shape analyser. (b) Schulz's 72° double electrostatic analyser

Faraday-cup collector, the electron current being measured with a vibrating-reed electrometer. The use of differential pumping enables low pressures to be maintained in all parts of the apparatus except the scattering chamber. In these experiments the gas-filled scattering chamber is a separate unit not interfering with energy selection parts of the apparatus. This situation does not always prevail in other beam experiments (e.g. Ramsauer-type apparatus). For small electron energy losses—excitation of low-vibrational quantum states, for example—transmission experiments suffer from interference from elastically scattered electrons (Hasted and Awan, 1969).

In the double electrostatic analyser of Schulz (1962) electrons emitted from a filament traverse the first electrostatic analyser, are accelerated and focussed at the entrance electrode near the collision chamber, and are crossed in the collision chamber by an atomic or molecular beam in an equipotential region. The unscattered portion of the primary electron beam is collected by an electron collector (see Figure 5.3b), while those electrons scattered at an angle of $\sim 72°$ are accepted by a second electrostatic analyser (identical in dimensions to the first). Electrons passing the exit slit of the second electrostatic analyser impinge on an electron multiplier, and the output current is measured by a vibrating-reed electrometer.

In single-collision scattering experiments care must be taken that the pressures are kept sufficiently low for single-collision conditions to prevail. Serious errors may arise, especially for differential cross section measurements at large angles, due to plural scattering (Chamberlain and coworkers, 1967; Kieffer and Dunn, 1966; Lassettre, 1959).

B. *Threshold-electron excitation techniques.* As stated earlier in this section, in these techniques the energy of the incident electron beam is varied, kept close to the excitation thresholds of the target. The scattered electrons, which have lost essentially all of their energy upon collision with the target, are trapped by either a small potential well, or a suitable scavenger gas which surrounds the scattering centre. The principle of such experiments, where electrons are collected at all scattering angles ($4\pi$-geometry), is shown schematically in Figure 5.4. Figure 5.5 shows schematically the trapped electron apparatus of Schulz (1958, 1959). A quasi-monoenergetic electron beam is obtained by application of the retarding potential difference method (RPD) (Fox and coworkers, 1955). The electrons are accelerated to a potential $V_A$, and in passing along the axis of the collision chamber some undergo scattering and are trapped, the unscattered beam being collected at the collector E. The entrance and exit slits, and the grid G are held at the same potential, while the cylindrical collector M, which surrounds the grid, is maintained at a positive potential $V_M$ with respect to the grid. The penetration of the potential $V_M$ into the collision region produces a well-depth $W$, between the centre of the chamber and the collision chamber

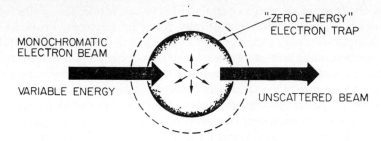

MONOCHROMATIC
ELECTRON BEAM

"ZERO-ENERGY"
ELECTRON TRAP

VARIABLE ENERGY

UNSCATTERED BEAM

**Figure 5.4** Schematic illustration of threshold-electron excitation
experiments

electrodes. (The collision chamber electrodes are plated with platinum black
to reduce electron scattering from surfaces.) The potential distribution along
the axis is shown on the lower portion of the figure. Electrons losing energy
between $V_A$ and $V_A + W$ will be collected; the well-depth, $W$, is generally
$\sim 0.1$ eV although, often, larger well-depths (0.2–0.3 eV) have been used. In
the arrangement of Figure 5.5 electrons scattered elastically at any direction

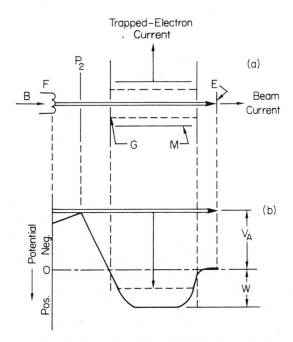

**Figure 5.5** (a) Schematic diagram of the trapped-
electron apparatus (Schulz, 1958, 1959). (b) Potential distribution along the axis

except the forward, as well as negative ions (when formed) are collected, in addition to the zero-energy inelastically scattered electrons. Provided that such complications are properly accounted for and that $W$ is kept small so that only essentially zero energy electrons are collected, a plot of the trapped-electron current versus incident electron energy yields the threshold-electron excitation spectrum of the target (see experimental results in Section 5.4). An advantage of the trapped-electron method is its higher sensitivity compared to the electrostatic analysers. However, its energy resolution is poor.

As will be discussed in Chapter 6 there are a number of compounds such as $SF_6$ and $CCl_4$ which possess a very sharp and intense electron attachment resonance at $\sim 0.0$ eV. For $SF_6$, for example, the width of the zero-peaking electron attachment resonance was found to be instrumental even with a resolution of 0.017 eV (Schulz, 1969). Electrons, then, which have lost essentially all of their energy upon collision can be detected with such 'indicator compounds', when these are added simultaneously with the target in the collision chamber (Curran, 1963; Compton, Christophorou and Huebner, 1966). The sequential processes may be written for the case of $SF_6$ as

$$e_{fast} + AX \rightarrow AX^* + e_{thermal} \tag{5.26a}$$

$$e_{thermal} + SF_6 \rightarrow SF_6^- \tag{5.26b}$$

Thus a plot of the $SF_6^-$ current versus the incident electron energy ($e_{fast}$) yields the threshold-electron excitation spectrum of AX.

The threshold-electron techniques which are otherwise well-suited for the detection of optically forbidden transitions and compound negative ion states—provided the latter yield near-zero energy electrons upon decay—do not allow information as to the variation of the peak intensity with incident electron energy over a wide energy range. Also, the sensitivity of the electron scavenger technique is low, and the presence of a scavenger compound in the collision chamber simultaneously with the species under study, may complicate the experiment. The scavenger negative ion current, however, is not ordinarily affected by other negative ions which may form simultaneously in the collision chamber or by elastically scattered electrons, both of which may affect the trapped-electron current. Also avoided are complications arising from electrostatic potentials which are applied in the trapped-electron method to the collision chamber for scattered electron collection. Such potentials can affect the mean electron energy and resolution. The zero-energy electron attachment resonance of the scavenger may be used to obtain an approximate energy scale calibration; an energy scale calibration based on the zero-energy electron attachment process is, however, usually underestimated (Schulz, 1960a; Christophorou and coworkers, 1965).

Energy scale calibrations are essential, but rather difficult to make accurately. A number of energy calibration techniques are in use: retarding analysis before or after the collision—but not in the collision chamber itself—and calibrations utilizing well-defined positions of resonances in transmission, (e.g. the 19.31 eV 'window' resonance in He (Kuyatt, Simpson and Mielczarek, 1965), the $11.48 \pm 0.05$ eV resonance in $N_2$ (Heideman, Kuyatt and Chamberlain, 1966b), or inelastic scattering (e.g. the 19.81 eV $(2^3S)$ state in He (Compton and coworkers, 1968), the $11.82 \pm 0.05$ eV $(E^3\Sigma_g^+)$ state in $N_2$ (Brongersma and Oosterhoff, 1967)), or dissociative electron attachment established by the swarm-beam technique (Christophorou and coworkers, 1965; Chapter 6). In the latter cases the calibrating gas is mixed with the gas under study in the collision chamber.

C. *Electron-beam techniques for optical measurement of the cross section for excitation.* In these experiments a monoenergetic electron beam of sufficient energy to excite electronically the target passes through the target gas, producing excited atomic species which undergo radiative de-excitation. A measurement of the absolute intensity of the emitted radiation from a given line, in connection with all higher levels that combine with the level under consideration can, in principle, be used to determine the excitation cross section function. Because in these experiments the light intensity is weak, a high detection sensitivity is necessary to enable one to work with low gas pressures and thus to avoid complications arising from re-absorption and collisions of the second kind. True excitation cross sections are difficult to determine with this technique and often one measures only the variation of the atomic emission with incident energy and normalizes this measurement at high energies to a Born-approximation calculation of the excitation cross section. It should be noted that metastable atomic levels usually have lifetimes long enough to allow diffusion of the metastable species out of the collision region to the surrounding surfaces prior to radiative decay. In this case detection and measurement of a known part of the metastable atom flux is required (see further discussion in Bates (1962) and Hasted (1964)).

### 5.2.1.2 Electron swarm techniques

The electron swarm method is not suitable for studies of electronic excitation functions but, in certain cases, it can be successfully applied to studies of rotational and vibrational excitation functions (see Section 5.3). The study of the latter by electron beam techniques presents special problems in view of the low energies involved, although recently such problems are being overcome* (e.g. Ehrhardt and Linder, 1969). In particular, the small energy losses characteristic of rotational excitation in heavy gases lend

---

*See Maier-Leibnitz (1955) for a modification of the electron beam technique which has been used in few studies of vibrational excitation. See, also, relevant papers by Ramien (1931), Haas (1957) and Chantry, Phelps and Schulz (1966).

themselves easily to a study by electron swarm techniques (and electron energy relaxation techniques). A characteristic energy (Chapter 4) of 0.007 eV can be reached for a gas temperature of 77°K; the energy resolution is of the same order of magnitude ($\sim$0.01 eV). It has to be borne in mind, however, that quantitative cross section data derived from swarm experiments are indirect and often difficult to extract from the measured electron transport coefficients since rotational excitation is masked by other energy-loss processes.

The determination of inelastic scattering cross sections from analyses of measured electron transport coefficients which—to the advantage of the swarm technique—can be measured accurately (see Chapter 4) is due to Phelps and coworkers (Frost and Phelps, 1962; Engelhardt and Phelps, 1963; Engelhardt, Phelps and Risk, 1964; Hake and Phelps, 1967). As stated in Chapter 4, these investigators first assume a set of cross sections for elastic and inelastic scattering—based on any available information—which they substitute into Equation (4.53) which is solved with the aid of a computer for the electron energy distribution function. This distribution function is then used with the momentum transfer cross section to calculate the transport coefficients $w$, $D/\mu$, $v_m/N$ and $v_u/N$ (Equations (4.71) and (4.72)). The computed and experimental values for $v_m/N$ and $v_u/N$ are compared and, on the basis of the discrepancies found, the initial elastic and inelastic cross sections are adjusted and a recalculation is made of the electron energy distribution function and the electron transport coefficients. The new values of $v_m/N$ and $v_u/N$ are compared again with the experimental values. The process is repeated until the agreement between the calculated and the experimental transport coefficients is comparable to the accuracy of the experimental data. In this analysis, then, one obtains a set of elastic and inelastic cross sections and a distribution function which are self-consistent but not unique. Results obtained this way are presented and discussed in Section 5.3.

Up until recently information on vibrational and rotational excitation has not been abundant, partly because of the fairly small cross sections and the low-energy thresholds for the processes considered; investigations of such excitations by emission of radiation following excitation by electron impact are not practical because of weak radiative transitions and thus difficulties in detection. The measurement of the relaxation of electron energy using high frequency techniques may yield information about rotational and vibrational excitation cross sections (Golant, 1961; Mentzoni and Row, 1963).

### 5.2.2 Techniques for the Study of Ionization of Atoms and Molecules by Electron Impact

The many types of apparatus employed for the study of ionization of atoms and molecules by electrons fall into four groups: (i) total ionization apparatus, (ii) total ionization tubes with charge-to-mass ($q/M$) discrimination, (iii) mass spectrometers, (iv) crossed-beam apparatus.

In the first group a collimated and nearly monoenergetic electron beam passes through a 'thin' target, and a collection is made of essentially all of the positive ions formed in ionizing events. The target gas is kept 'thin' (low-pressure) for three main reasons: firstly, that a small fraction of the incident beam will undergo collisions; secondly, that an insignificant fraction of the projectiles will experience more than one collision; and thirdly, that an insignificant amount of ionization is caused by the ionization electrons. The collector positive ion current $i_i(\varepsilon)$ for an incident electron beam of energy $\varepsilon$, can be expressed as

$$i_i(\varepsilon) = i_e(\varepsilon) N \sigma_{gi}(\varepsilon) x \qquad (5.27)$$

where $i_e(\varepsilon)$ is the incident electron beam current of energy $\varepsilon$, $N$ is the number density, $x$ is the collision path length over which the measured current is collected, and $\sigma_{gi}(\varepsilon)$ (Equation 5.6) is the gross or total ionization cross section. Assuming an ideal-gas behaviour, we may write for $\sigma_{gi}(\varepsilon)$

$$\sigma_{gi}(\varepsilon) = \frac{i_i(\varepsilon)}{i_e(\varepsilon)} \frac{T}{273} \frac{1}{3.535 \times 10^{16} Px} \quad (\text{cm}^2) \qquad (5.28)$$

where $P$ and $T$ are the gas pressure and absolute temperature expressed in torr and °K, respectively, and $3.535 \times 10^{16}$ is the number of particles per cm$^3$ at 1 torr and 273°K.

Total (gross) cross section data appear to be more abundant and accurate (see Section 5.5) than data on partial ionization cross sections. In spite of the fact that in such measurements no information is obtained as to the charge, mass, and state of the various ions formed, gross ionization cross sections are important for a number of reasons. Firstly, they are intimately connected with the measurement of number density of atomic scatterers at low pressure ($< 10^{-3}$ torr); secondly, they are more accurately known than cross sections for creating specific ions; and thirdly, in many experiments $\sigma_{gi}$ are used to normalize relative ionization cross sections for other (often unstable) species. In such comparisons the ion currents $i_x$ of the sample under study and $i_s$ of the standard are measured at the respective energies and at normalized electron current and gas pressure, and the cross section $\sigma_i$ of the sample is obtained from

$$\sigma_i = \sigma_i(s) \frac{i_x}{i_s} \qquad (5.29)$$

where $\sigma_i(s)$ is the cross section for the standard. No discussion will be made of ionization techniques other than of the new arrangement by Ehrhardt and coworkers (1969). The reader is referred to Kieffer and Dunn (1966)

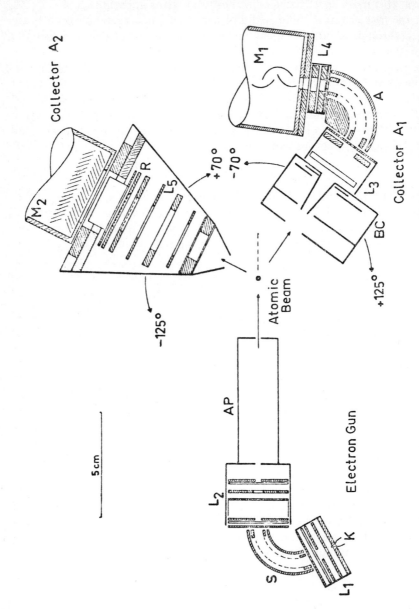

**Figure 5.6** Schematic diagram of the experimental arrangement of Ehrhardt and coworkers (1969)

for a complete discussion and literature. We note, however, that total ionization arrangements may incorporate a $q/M$ analysis while, although mass spectrometers yield easily a $q/M$ separation, they present serious problems for obtaining absolute cross section values. Unstable species are studied by crossed-beam arrangements, the modulated crossed-beam technique being mostly suited. Partial cross sections for $n$-fold atomic ionization are expressed as $\sigma_{ni}/\sigma_{gi}$, the positive current $i_{ni}$ due to the $n$-fold ionized species being assumed proportional to $n\sigma_{ni}$. Much of the uncertainty in these studies arises from pressure measurements.

Recently, Ehrhardt and coworkers (1969) described an arrangement whereby the incident electron energy $\varepsilon_p$, energy $\varepsilon_s$ and angle $\theta_s$ of the scattered electron, and the energy $\varepsilon_e$ and angle $\theta_e$ of the ejected electron, for single ionization by slow electrons can be chosen and measured (the azimuthal angle $\varphi_e$ was fixed in their experiments with helium). Their arrangement is shown schematically in Figure 5.6. Electrons from the filament K are collimated by the lens system $L_1$ and are injected into the $127°$ analyser S. After energy selection the electrons are accelerated by the lens system $L_2$ and are focused onto the atomic beam. The aperture box AP reduces the number of background electrons from the gun. The two outgoing electrons are collected in collectors $A_1$ and $A_2$; the former collector $(A_1)$ can be rotated in the angular range from $-70$ to $+125°$ and the latter $(A_2)$ from $+70$ to $-125°$. The primary electron beam is collected for small values of $\theta_s$ in the beam collector BC of $A_1$. Ionizing collisions are identified by the coincidence signal from $M_1$ and $M_2$ (the time resolution was 10 nsec). The coincidence counts were registered in a multichannel analyser, the channel number of which was set proportional to the angular position of $A_2$. The parameters $\varepsilon_p$, $\varepsilon_s$, $\varepsilon_e$, $\theta_s$ and $\varphi_e$ in one experiment were fixed and $\theta_e$ was varied.

## 5.3 Inelastic Electron Scattering Resulting in Excitation of Molecular Rotation and Vibration

In spite of the experimental difficulties outlined in Sections 5.2.1.1 and 5.2.1.2, considerable progress has been made on this subject since the review by Craggs and Massey (1959). Many beam techniques are now in use which yield fundamental information on rotational and vibrational excitation of molecules. A recent review of experimental and theoretical studies of rotational and vibrational excitation of simple molecules by low-energy electrons was given by Phelps (1968).

### 5.3.1 *Rotational Excitation by Electron Impact*

#### 5.3.1.1 $H_2$

A number of investigators studied rotational excitation of $H_2$ by slow electrons. With the exception of some recent work, rotational excitation cross sections for $H_2$ have been summarized by Phelps (1968). The cross sections for the $J = 0 \rightarrow J' = 2$ transition are shown in Figure 5.7. The solid curves are the swarm data of Engelhardt and Phelps (1963). The lowest broken curve is the result of a calculation by Gerjuoy and Stein (1955b), who considered only the quadrupole interaction, and the results of Dalgarno and Henry (1965), who included short-range interaction terms but no polarization in their calculation. Dalgarno and Moffett (1963) included asymmetric polarization terms in their calculation and obtained the next higher broken curve shown on the figure. The upper two curves shown on

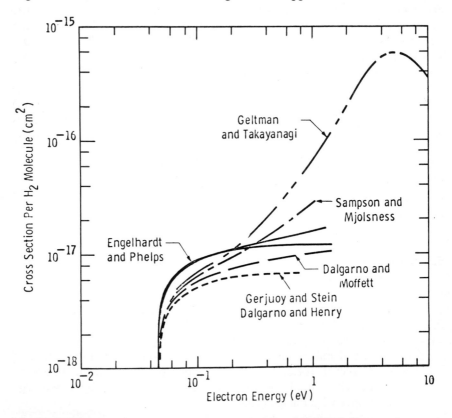

**Figure 5.7** Cross section for rotational excitation of $H_2$ from the $J = 0$ to the $J = 2$ level (Phelps, 1968). The solid curves are the result of swarm analyses and the broken curves that of various calculations (see text)

the figure are the results of calculations by Sampson and Mjolsness (1965) and Geltman and Takayanagi (1966). Recent close coupling calculations by Lane and Geltman (1967) (see this reference for a literature survey) yielded similar results to those of Sampson and Mjolsness (1965). There seem to exist significant differences between all theoretical values and the experiment below $\sim 0.1$ eV.

Rotational excitation of $H_2$ has been successfully studied by Ehrhardt and Linder (1969), using monoenergetic electrons (with an energy resolution of 30 mV) in the energy range 1 to 10 eV. Figure 5.8 shows the energy dependence of the absolute total cross section for rotational excitation of the $J = 1 \rightarrow J' = 3$ transition of $H_2$ in its ground electronic and vibrational

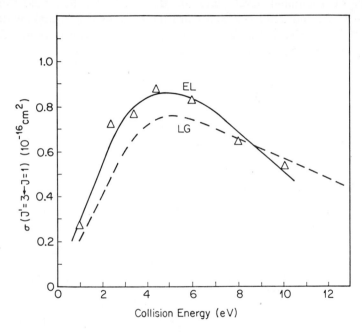

**Figure 5.8** Comparison of the total cross section for the rotational transition $J = 1 \rightarrow J' = 3$ ($\Delta v = 0$) measured by Ehrhardt and Linder (1969), with the theoretically predicted cross section of Lane and Geltman (1967)

states. Ehrhardt and Linder normalized their data to the absolute total cross section measurements of Golden, Bandel and Salerno (1966). Their results are seen to agree well (particularly in shape) with the calculated values of Lane and Geltman (1967). Measurements of the differential cross section for the $J = 1 \rightarrow J' = 3$ transition using 4.42 eV electrons seem to show that the cross section contains large contributions from $s$, $p$ and $d$ waves (Ehrhardt and Linder, 1969). Recent calculations by Abram and Herzenberg (1969) are in fair agreement with the data of Ehrhardt and Linder on the excitation of certain rotational transitions of $H_2$ with the simultaneous excitation of a vibrational quantum state.

The importance of rotational excitation has been further shown by recent swarm data on ortho and para hydrogen and on hydrogen and deuterium. The observed differences in the values of the transport coefficients ($w$ and $D/\mu$) for ortho and para hydrogen at near thermal energies have been attributed to differences in the statistical weights of the rotational levels for the two molecules (Crompton and McIntosh, 1967). Also, differences in the statistical weights of rotational levels and in the excitation threshold for the $J = 0 \rightarrow J' = 2$ and $J = 1 \rightarrow J' = 3$ transitions for hydrogen and deuterium can explain the observed differences in $D_L/\mu$ and $w$ shown in Figure 5.9 for these gases (Crompton, Elford and McIntosh, 1968). Crompton, Elford and McIntosh calculated that at 77°K about 57.2% of the $D_2$ molecules are in the $J = 0$ state and 33.1% in the $J = 1$ state; for $H_2$ these percentages are, respectively, 25% and 75%. The $J = 0 \rightarrow J' = 2$ and $J = 1 \rightarrow J' = 3$ energy

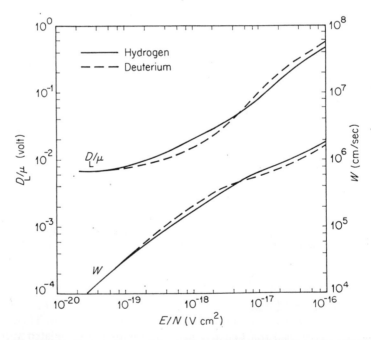

**Figure 5.9** Comparison of the variation of $w$ and $D_L/\mu$ with $E/N$ for electrons in hydrogen and deuterium at 77°K (Crompton, Elford and McIntosh, 1968)

thresholds are 0.023 and 0.038 eV for $D_2$ and 0.045 and 0.075 eV for $H_2$. The lower excitation thresholds for $D_2$, coupled with the higher percentage of $D_2$ molecules in the $J = 0$ state, result in a slower initial rise of the mean electron energy with increasing $E/N$ compared with $H_2$. Hence, initially $D_L/\mu$ rises faster with increasing $E/N$ for $H_2$ than for $D_2$ until $E/N$ is large enough for a significant number of electrons in the swarm to be capable of exciting the $J = 1 \rightarrow J' = 3$ transition of $H_2$. When this situation is reached and, hence, rotational transitions play a significant role in determining the energy distribution in both $D_2$ and $H_2$, the power loss due to collisions resulting in rotational excitation may be greater for $H_2$ than for $D_2$. This can account for the larger values of $D_L/\mu$ for $D_2$ compared to those for $H_2$ above an $E/N$ value of $\sim 6 \times 10^{-18}$ V cm$^2$. Of course, at higher values of $E/N$ vibrational excitation becomes important and strongly affects the $D_L/\mu$ values. A similar qualitative explanation can be applied to the drift velocity data shown in Figure 5.9.

### 5.3.1.2 $N_2$, $O_2$ and CO

Studies of rotational excitation of these molecules in collisions with slow electrons are more difficult than for $H_2$ and the analysis of swarm transport data is rather less straightforward. Experimental and theoretical cross section data for $N_2$, $O_2$ and CO have been summarized by Phelps (1968). For gases such as $O_2$ which have large attachment cross sections for electrons of thermal and near thermal energies, swarm determinations of the transport coefficients are not possible at very low energies. Measurements of the electron energy relaxation times in $O_2$ have indicated rather large rotational excitation cross sections (Phelps, 1968). For the heteronuclear molecule CO the dominant interaction is due to the molecular electric dipole moment and the selection rule for $J$ is $\Delta J = \pm 1$. The rotational excitation cross section function for this molecule shown in Figure 5.10 has a behaviour typical of an allowed dipole transition, decreasing as $\varepsilon^{-1} \ln \varepsilon$ at high electron energies $\varepsilon$. In Figure 5.10a $\sigma_{4 \rightarrow 5}$ is the cross section for rotational excitation from the $J = 4$ to the $J = 5$ level and $\sigma_{5 \rightarrow 4}$ is the cross section for rotational de-excitation from the $J = 5$ to the $J = 4$ level.

### 5.3.2 *Vibrational Excitation by Electron Impact*

### 5.3.2.1 Diatomic molecules

A. $H_2$. The energy required for the $v = 0 \rightarrow v = 1$ transition in $H_2$ is 0.52 eV. Figure 5.11 (Phelps, 1968) compares the experimental cross sections of Engelhardt and Phelps (1963) obtained from an analysis of swarm transport coefficients and those of Schulz (1964) obtained from electron scattering at 72° with the results of some theoretical calculations. Schulz's cross

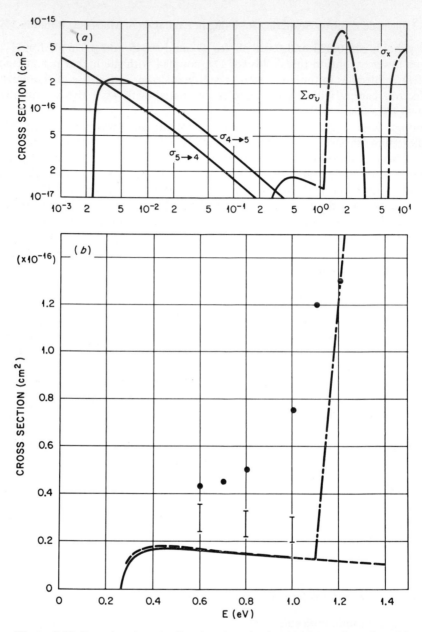

**Figure 5.10** Rotational and vibrational excitation cross sections for CO (Phelps, 1968). (a) Solid curves are the result of swarm analyses and the broken curve is the sum of the vibrational excitation cross sections of Schulz (see text). (b) Vibrational excitation cross sections for CO near threshold. Solid curve is again the result of swarm analyses, points are the beam results of Schulz, and the vertical error bars and the short dashed curve are the results of theoretical calculations (see text)

sections for both the excitation of the $v = 1$ and $v = 2$ levels are included in the figure. Schulz (1964) obtained his cross section data by assuming isotropic scattering. According to Bardsley, Herzenberg and Mandl (1966), Schulz's cross sections should be increased by a factor of 1.4 to correct for non-isotropic electron scattering. This correction would bring the swarm and beam cross sections at higher energies in comfortable agreement, although

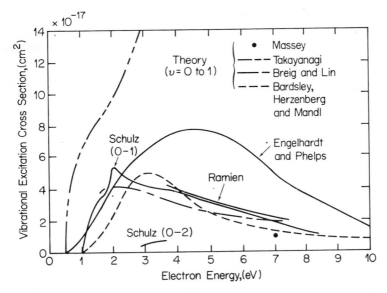

**Figure 5.11** Vibrational excitation cross sections for $H_2$ (Phelps, 1968). Solid curves are experimental results and broken curves are the results of theoretical calculations

since Engelhardt and Phelps (1963) assumed that only the cross section for the $v = 1$ level is significant, their cross section may be somewhat overestimated. The curve marked Ramien (1931) is the result of a multiple scattering experiment for the sum of the cross sections for excitation of the $v = 1$ and $v = 2$ levels. There seems to be a disagreement between the electron swarm and electron beam cross sections in the energy region 0.5 to 1.0 eV. However, both techniques clearly show that vibrational excitation cross sections for $H_2$ are large. We may note that Menendez and Holt (1966) in an energy-loss experiment observed energy losses corresponding to excitation of the $v = 1$ and $v = 2$ vibrational levels of the $H_2(X^1\Sigma_g^+)$ ground state, and for very small scattering angles they found that the $v = 1$ vibrational excitation cross section was rising linearly from 0.5 eV. This finding is consistent with the swarm data of Engelhardt and Phelps (1963).

With the exception of the result by Takayanagi (1965), there is reasonable agreement between the experimental and theoretical (Massey, 1935; Breig and Lin, 1965; Bardsley, Herzenberg and Mandl, 1966) cross sections shown in Figure 5.11. Bardsley, Herzenberg and Mandl employed resonance scattering theory, the resonance for $H_2$ being rather broad with a peak at $\sim 3$ eV. The resonance scattering theory also yielded a ratio for the excitation cross sections for the $v = 1$ and $v = 2$ levels in reasonable agreement with Schulz's experimental data.

**B. $N_2$.** Vibrational excitation of molecular nitrogen via the decay of a compound negative ion state at 2.3 eV is a unique example which demon-

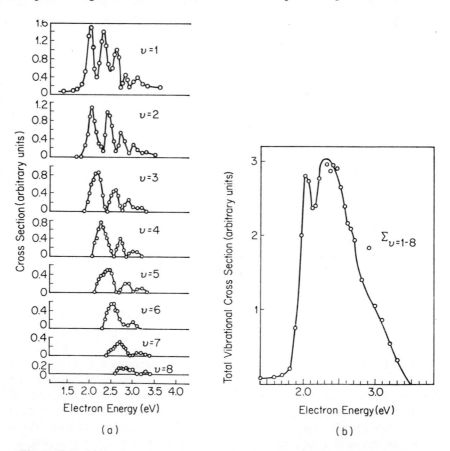

**Figure 5.12** (a) Energy dependence of the vibrational cross section of $N_2$ by electron impact (Schulz, 1964). When the ordinate numbers are multiplied by $10^{-16}$ a cross-section scale (in cm$^2$) is obtained such as to give a total vibrational cross section in agreement with Haas (1957). (b) Sum of the vibrational cross sections to individual states as a function of energy in $N_2$ (Schulz, 1964)

strates the importance of the resonance character of vibrational excitation. This mode of vibrational excitation was first discovered by Haas (1957) and has subsequently been studied by many workers in some detail (e.g. Schulz, 1959, 1962, 1964; Heideman, Kuyatt and Chamberlain, 1966b; Boness, Hasted and Larkin, 1968). Figure 5.12a shows the cross sections for excitation of the first eight vibrational levels of $N_2$ obtained by Schulz (1964). The sum over the vibrational levels $v = 1$ to 8 is shown in Figure 5.12b. Schulz estimated a cross section of $\sim 3 \times 10^{-16}$ cm$^2$ at $\sim 2.2$ eV, a value consistent with the swarm work of Engelhardt, Phelps and Risk (1964). It is noted that whereas the cross sections for $v = 2$ to 8 have a threshold greater than 1.7 eV, the cross section for $v = 1$ extends to lower energies. The low-energy tail for the $v = 1$ cross section has been attributed by Schulz to direct excitation of the first vibrational level by electron impact. Engelhardt, Phelps and Risk (1964) found that the 'tail' extends down to the threshold of the $v = 1$ state at 0.3 eV. The swarm-determined cross sections near threshold agree with Chen's calculations (1964a, b; 1966) in this energy region but they appear not to agree with the beam cross sections (Phelps, 1968). Figure 5.13 illustrates the formation of the compound $N_2^-(^2\Pi_g)$ state which subsequently decays to the various vibrational states of the ground electronic state $(X^1\Sigma_g^+)$ of $N_2$.

The resonance theory has been successful in providing a rather good fit of the experimental cross sections for $N_2$ (Herzenberg and Mandl, 1962; Chen, 1964a, b, 1966). The models which have been used to fit the $N_2$ cross sections did not include a series of lower magnitude resonances persisting down to $\sim 0.3$ eV reported by Golden (1966) who performed a total Ramsauer-type scattering experiment. However, Boness, Hasted and Larkin (1968) were unable to find such resonances down to 0.1 eV and concluded that Golden's peaks are false resonances.

The angular distribution of electrons inelastically scattered in the vibrational excitation of $N_2$ has been measured by Ehrhardt and Willmann (1967) over the angular range 0 to 110° for several incident energies (1.9 to 3.1 eV) and vibrational quantum numbers $v = 1$, 3 and 5. They have found the angular distribution to be characteristic of $d$ wave scattering. From earlier measurements of the angular dependence of elastically scattered electrons from $N_2$, Andrick and Ehrhardt (1966) assigned the configuration $^2\Pi_g$ to the $N_2^-$ resonance at 2.3 eV as it has been proposed earlier by Gilmore (1965). Bardsley, Mandl and Wood (1967) also discussed this assignment and concluded that a $^2\Pi_g$ configuration is consistent with both the experimental data and their theoretical interpretation.

Studies of vibrational excitation of $N_2$ using higher incident electron energies have also been made. Geiger and Wittmaack (1965a) used very high energies (33 keV) and reported for a number of molecules excitation of only electric-dipole allowed vibrations, in good agreement with the Born approximation. Recent experiments by Skerbele, Dillon and Lassettre (1968a) with electron energies between 40 and 50 eV have shown that $v = 1$ levels can easily be detected for $N_2$ for such incident energies, although excitation of ground-state vibrations is electric-dipole forbidden for homonuclear molecules. Since at these incident energies resonance excitation mechanisms are unlikely, the observed transitions show the importance of polarization and quadrupole contributions to the vibrational cross section.

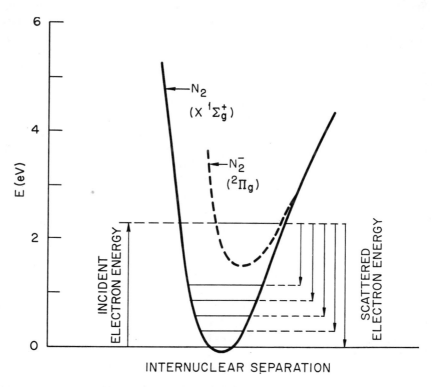

**Figure 5.13** Schematic diagram illustrating the formation of transient negative ions in $N_2$

C. $O_2$. Schulz and Dowell (1962) postulated a resonance behaviour in $O_2$. They interpreted the observed peaks in their trapped-electron current as being due to vibrational excitation of the $v = 1$ to $v = 8$ states of $O_2$. Although Schulz and Dowell's results indicated a small vibrational excitation cross section for $O_2$ near threshold, the transport coefficient data of Hake and Phelps (1967) require larger cross sections—much larger than for $N_2$ in the characteristic energy range 0.1 to 0.8 eV. Boness and Hasted (1966) observed a series of resonance-like peaks in a transmission experiment at energies below $\sim 1$ eV (see Boness, Hasted and Larkin (1968) for a possible configuration of compound states of $O_2$). Excitation of the $v = 1$ level of $O_2$ has been observed by Skerbele, Dillon and Lassettre (1968a), using 40 to 50 eV electrons. These workers detected, also, for these incident energies excitation of the forbidden electronic transitions*

$$a^1 \Delta_g \leftarrow X^3 \Sigma_g^- \quad (0.98\,\text{eV})$$

$$b^1 \Sigma_g^+ \leftarrow X^3 \Sigma_g^- \quad (1.63\,\text{eV})$$

D. CO. In the energy range above 1 eV Schulz (1959, 1962, 1964) obtained for CO a very large resonance similar to that in $N_2$. His overall vibrational excitation cross section is shown in Figure 5.10. The magnitude of the cross section is of the same order as in the isoelectronic $N_2$ molecule. In addition to the strong resonance at $\sim 1.75$ eV (a peak cross section value of $8 \times 10^{-16}$ cm$^2$ has been obtained for this process by Schulz assuming isotropic scattering), there is evidence for another inelastic process below this energy. The value of the cross section near threshold is much smaller than at the peak of the resonance at 1.75 eV, but it is very much larger than for $N_2$ (compare Figures 5.10 and 5.12). In Figure 5.10b the solid line are the swarm data of Hake and Phelps (1967) which agree well with the cross section for direct vibrational excitation calculated by Hake and Phelps (1967) (short dashed line) using the Born approximation expression for the cross section for vibrational excitation and for the infrared transition probability the value given by Penner (1959). The vertical error bars show the range of calculated cross sections by Breig and Lin (1965) and the closed circles are the beam data of Schulz (1959, 1962, 1964). The almost vertical line is, to a first approximation, the resonance structure which would be necessary to fit the electron swarm results of Hake and Phelps (1967).

Vibrational excitation in CO, then, appears to be taking place: (i) directly through dipole transitions near threshold and (ii) indirectly through a resonance at higher energies, leading to very large cross sections. Boness, Hasted and Larkin (1968) assigned the configuration $^2\Pi$ to the CO$^-$ resonance at $\sim 1.75$ eV. Additionally, the 1.75 eV CO$^-$ resonance has been studied in transmission by Boness and Hasted (1966) and by Ehrhardt and coworkers (1968) in angular distribution measurements. The inelastic angular distributions of the last group of investigators had a $p$ wave character.

E. NO. Here as in CO one has the possibility of vibrational excitation by a dipole transition. Boness and Hasted (1966) observed in a transmission experiment a series of resonances in the range 0.4 to $\sim 1.6$ eV, while Boness,

---

*These transitions were also detected by Schulz (1962) but with much lower incident energies. The radiative lifetime of $O_2^*$ in $^1\Delta_g$ is extraordinarily long (45 minutes (Badger, Wright and Whitlock, 1965); 60 minutes (Jones and Harrison, 1958)).

Hasted and Larkin (1968) assigned the configuration $^1\Sigma^+$ to the $NO^-$ resonance around 1 eV. See also swarm analyses by Phelps (1968) and Christophorou and Carter (1968).

## 5.3.2.2 Polyatomic molecules

A. $CO_2$. Hake and Phelps (1967), using electron transport coefficient data, obtained cross sections for a set of four resonances for the linear triatomic molecule $CO_2$, and pointed out that the resonances in each case occurred very close to the threshold for the onset of the particular energy loss processes, a situation rather different from that in $H_2$, $N_2$ and CO where resonances occur well above the vibrational excitation threshold. They also argued that there must be an additional resonance very close to the onset for vibrational excitation at 0.08 eV (see also Phelps (1968) and Christophorou and Carter (1968)). Recent beam work is in agreement with the general conclusions of Hake and Phelps (1967).

Boness and Schulz (1968) employed a double electrostatic analyser and measured the energy-loss spectrum at fixed incident electron energies ranging from 0.3 to 4.5 eV at 45, 72 and 90-degree scattering angles. For incident energies in the range 0.3 to 3 eV the predominant features of the energy-loss spectrum were peaks at 0.29 and 0.58 eV which they associated with the asymmetric* stretch vibrational modes of $CO_2$. For incident energies 3 to 4.5 eV they observed six equally spaced peaks separated by 0.17 eV which they associated with excitations of the symmetric stretch mode. Figure 5.14 shows the energy dependence of the 0.29, 0.58 and 0.17 eV energy-loss peaks. The energy dependence of the elastic electron scattering cross section is also shown. The peak in the elastic cross section was first observed by Ramsauer and Kollath (1930) and has recently been studied by Boness and Hasted (1966). Boness and Schulz (1968) interpreted their data in terms of two short-lived compound states of $CO_2^-$, one at 0.9 eV exhibiting a $p$ wave character and another at 3.8 eV exhibiting an $s$ wave character. The resonance at 0.9 eV leads to excitation of asymmetric and that at 3.8 eV leads predominantly to symmetric stretch excitations upon decay. Boness and Schulz (1968) reported an approximate value of $3 \times 10^{-16}$ cm$^2$ for the 0.29 eV cross section which they obtained by normalization to the elastic scattering from the 3.8 eV compound state.

The 0.08 eV energy-loss process suggested by the swarm data was not observed by Boness and Schulz (1968) probably due to the low threshold of this process and the large cross section for elastic scattering. However, Stamatović and Schulz (1969) observed energy losses corresponding to all three fundamental vibrational modes at their respective thresholds.

---

*$CO_2$ has three fundamental vibrational modes: asymmetric stretch (0.29 eV spacing), symmetric stretch (0.17 eV spacing) and bending (0.08 eV spacing).

Additionally, Skerbele, Dillon and Lassettre (1968b) observed excitation of all fundamental vibrations and combination bands in $CO_2$ using 30 and 60 eV electrons (at a scattering angle of 2°).

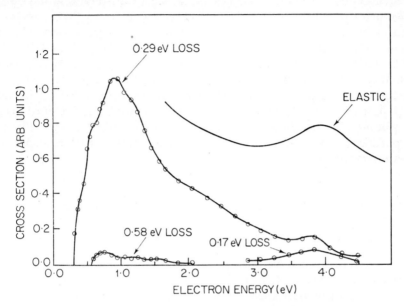

**Figure 5.14** Energy dependence of the 0.17-, 0.29- and 0.58-eV (100, 001, and 002 vibrational modes) loss cross sections and the elastic cross section in $CO_2$ (Boness and Schulz, 1968)

B. $H_2O$, $N_2O$ and $NO_2$. Skerbele, Dillon and Lassettre (1968b) using 30 and 60 eV electrons observed vibrational excitation of fundamental vibrations and combination bands in $H_2O$ at a scattering angle of 2°. Schulz (1961) found a large inelastic process in $N_2O$ at 2.3 eV which coincided in energy with the main $O^-/N_2O$ dissociative attachment peak (see Chapter 6). He invoked a compound negative ion resonance at this energy which either decays elastically or inelastically, leaving $N_2O$ vibrationally excited, or by a dissociative attachment process giving $O^-$ and $N_2$ in various states of vibrational excitation. The $N_2O^-$ resonance at 2.3 eV has recently been studied by Boness, Hasted and Larkin (1968) who also found no well-defined structure. This lack of structure may be either due to a very short lifetime of the compound state or due to such a superposition of vibrational modes as to obliviate structure. See also a discussion of this resonance by Christophorou and Carter (1968).

Boness, Hasted and Larkin (1968) reported a series of bumps in the transmission current through $NO_2$ in the range ~0.2 to 1.6 eV.

C. *Organic molecules.* Evidence for vibrational excitation of molecules directly or indirectly through the decay of compound negative ion resonance states has been found for a large number of polyatomic molecules from analyses of electron swarm data (see Figure 4.37, Christophorou and Carter, 1968 and Cottrell and Walker, 1965). Large energy losses for a number of organic molecules have also been observed at low energies using electron beam techniques: $C_2H_2(2.0)$;* $C_2H_4(1.7)$; $CH_3CH = CH_2(1.8)$; $CH_3C \equiv CH$ (2.8); $C_2H_5C \equiv CH(2.4)$ (Bowman and Miller, 1965); $CH_4(1.9–3.0$, with a vibrational spacing of $\sim 0.25$ eV); $C_2H_4(1.2–1.8)$; $C_6H_6(0.8–1.8$, with vibrational spacing of $\sim 0.12$ eV) (Boness and coworkers, 1967); $C_6H_6(1.55)$; $C_6H_5CH_3(1.45)$; $C_6H_5F(1.35)$; $o–C_6H_4CH_3Cl(1.10)$; $C_6H_5Cl(0.86)$; $C_6H_5Br$ (0.85); $o–C_6H_4Cl_2(0.36)$ (Compton, Christophorou and Huebner, 1966);** naphthalene (0.8); 1-chloronaphthalene (0.55, 1.21, 2.87) (Huebner and coworkers, 1968).

## 5.4 Inelastic Electron Scattering Resulting in Electronic Excitation

### 5.4.1 *Electronic Excitation of Atoms by Electron Impact*

There is a considerable amount of experimental data on the electronic excitation of atoms by electron impact. Moiseiwitsch and Smith (1968) have summarized theoretical approaches and experimental results for a number of atoms (H, O, N, Hg, rare gases, alkali metals) and certain positive ions. The reader is referred to this review for literature and detailed discussion. In this section we will discuss typical data on H, He and Hg. No detailed account of the field will be given.

Excitation functions for allowed transitions in atoms rise sharply from a zero value at the onset, to a maximum at an incident electron energy which is well-above the excitation threshold, and then decrease again with increasing impact energy (see, for example, Figure 5.15). On the other hand, the excitation of spin-forbidden transitions by electron impact, is dominated by electron exchange close to threshold, except for heavy elements where spin–orbit coupling is important. Exchange dominated excitation functions show much sharper maxima than those for optically allowed transitions. Forbidden transitions which do not involve a change in multiplicity have sharp cross section function maxima also.

---

*The number in parenthesis indicates the peak energy for the energy range of the observed resonance in eV.

**Although no structure was observed in the $SF_6^-$ current, the shape of the dissociative attachment resonance for the halogenated benzene derivatives (see Chapter 6), which coincides in position with the energy-loss peaks, evidenced some unresolved structure. This, coupled with the data of Boness and coworkers (1967) on benzene, may indicate that the energy-loss resonances for the benzene derivatives exhibit fine structure.

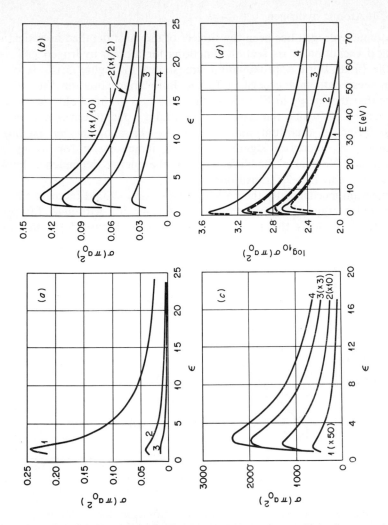

**Figure 5.15** Total cross sections for the excitation of atomic hydrogen using the first Born approximation (Moiseiwitsch and Smith, 1968). Note that the electron energy is in threshold units defined by $\varepsilon = E/(E_n - E_{n'}) = k_{n'}^2/\Delta E_{nn'}$, where $\Delta E_{nn'}$ is the excitation energy in Rydberg units and $k_{n'}$ is the wave number of the incident electron. (a) Curve 1: $1s \rightarrow 2s$; curve 2: $1s \rightarrow 3s$; curve 3: $1s \rightarrow 4s$ (Vainshtein, 1965; Veldre and Rabik, 1965). (b) Curve 1: $1s \rightarrow 2p$; curve 2: $1s \rightarrow 3p$; curve 3: $1s \rightarrow 4p$; curve 4: $1s \rightarrow 5p$ (Vainshtein, 1965; Veldre and Rabik, 1965). (c) Curve 1: $2s \rightarrow 3s$; curve 2: $3s \rightarrow 4s$; curve 3: $4s \rightarrow 5s$; curve 4: $5s \rightarrow 6s$ (Vainshtein, 1965; Veldre and Rabik, 1965). (d) Solid curves denote first Born approximation, while dashed curves denote Bethe approximation. Curve 1: $4s \rightarrow 5p$; curve 2: $4p \rightarrow 5d$; curve 3: $4d \rightarrow 5f$; curve 4: $4f \rightarrow 5g$ (Fisher, Milford and Pomilla, 1960)

### 5.4.1.1 Atomic hydrogen

Atomic hydrogen has been the subject of the most detailed investigations. Typical excitation cross section functions for atomic hydrogen calculated in the (first) Born approximation are shown in Figure 5.15. It is seen from these data that according to the Born approximation the value $\varepsilon_M$ (note that the energy in Figures 5.15a, b and c, is in threshold-energy units) of energy $\varepsilon$ for which the cross section attains a maximum is not sensitive to $n$ (principal quantum number) for $1s \to ns$ and $1s \to np$ excitations. However, $\varepsilon_M$ increases significantly with increasing $n$ for $ns \to (n+1)s$ transitions. Note also the differences in the energy dependence between the $1s \to 2s$ and $1s \to 2p$ transitions.

Quantum mechanically the $s$ wave cross section should increase from the threshold energy, $\varepsilon_{th}$, proportionally with $(\varepsilon - \varepsilon_{th})^{\frac{1}{2}}$. Although, for certain transitions, such as the $1s \to 2p$ in H shown in Figure 5.16, this initial

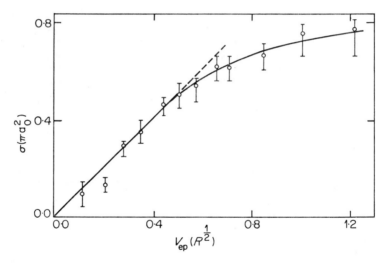

**Figure 5.16** Absolute electron excitation function for the $2p$ level of the hydrogen atom from the ground state $1s$ (Fite and Brackmann, 1958b); Fite, Stebbings and Brackmann, 1959). The abscissa is the post-collision velocity of the impinging electron in (Rydberg)$^{\frac{1}{2}}$

behaviour is observed, it is generally difficult to determine the energy range above $\varepsilon_{th}$ (usually small) to wherein the square root behaviour holds. In Figure 5.17 experimental and theoretical excitation cross section functions for the $1s \to 2p$ transition in H are shown. The experimental results were normalized to the Born approximation value of the cross section at energies $\geqslant 200$ eV. The three sets of experimental data are seen to be very consistent

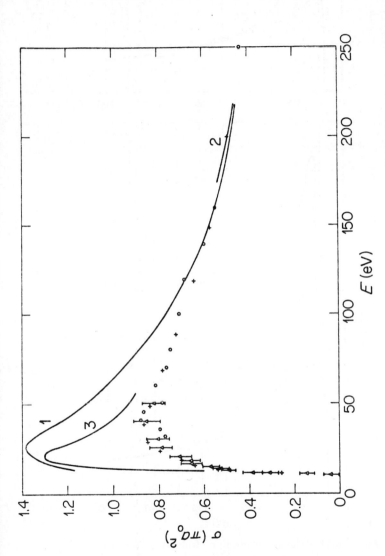

**Figure 5.17** Relative measurements of the excitation cross section for the $1s \rightarrow 2p$ transition in atomic hydrogen (Long, Cox and Smith, 1968). ($+$) Long, Cox and Smith (1968). ($\circ$) Fite and Brackmann (1958b); authors gave confidence limits of $\pm 12\%$. ($\triangle$) Fite, Stebbings and Brackmann (1959); error bars shown are confidence limits. Curves 1 and 2 are the Born approximation result without and with cascading, respectively (Moiseiwitsch and Smith, 1968). Curve 3 is a $1s$-$2s$-$2p$ close coupling calculation (Burke, Schey and Smith, 1963). The experimental results are normalized to curve 2

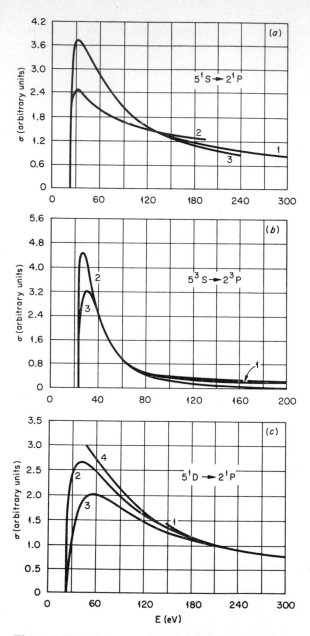

**Figure 5.18** Relative excitation functions for the (a) $5^1S \to 2^1P$, (b) $5^3S \to 2^3P$, and (c) $5^1D \to 2^1P$ helium transition (taken from Moiseiwitsch and Smith (1968) to which the reader is referred for references to individual calculations). Curves 1, 2 and 4 are experimental results and curves 3 are the result of calculations

and clearly show the existence of a large discrepancy between the experimental and theoretical results below $\sim 50$ eV. The Born approximation would not be expected to yield accurate cross sections apart from the high-energy limit. From analyses of early data Massey and Burhop (1952) have shown that for allowed transitions the Born approximation holds reasonably well down to about seven times the excitation energy. Below this energy the calculated values in this approximation are higher than those derived from experiments.

### 5.4.1.2 Rare gases

With the exception of some very recent work the available theoretical and experimental information on the rare gases has been summarized and discussed by Moisewitsch and Smith (1968). Helium has been especially studied, both theoretically and experimentally, with a variety of methods. Experimental studies employed monoenergetic electron beams with energies close to the excitation thresholds, a few Rydbergs of the ionization energy, and high enough ($> 200$ eV) for the Born approximation to be valid. Relative excitation cross sections for the $5^1S \rightarrow 2^1P$, $5^3S \rightarrow 2^3P$ and $5^1D \rightarrow 2^1P$ transitions are shown in Figure 5.18, and clearly show that the shapes of the excitation cross section functions for the various transitions differ considerably.

Low-energy electron impact total excitation cross section functions for helium metastables are shown in Figure 5.19. The measurements are those of Dorrestein (1942), Schulz and Fox (1957), and Holt and Krotkov (1966) who used basically the same method, namely detection of metastables by measuring the secondary electron current emitted from a metal surface due to the energy provided by the de-excitation of the metastables. The shapes of all three sets of experimental cross section functions compare well; the slight differences can probably be attributed to differences in energy resolution. The first peak in Figure 5.19 can be unambiguously ascribed to the $2^3S$ metastable. Schulz and Fox (1957) reported for the peak of the $2^3S$ excitation a value of $(4 \pm 1.2) \times 10^{-18}$ cm$^2$, which compares well with that, $(2.6 \pm 0.4) \times 10^{-18}$ cm$^2$, obtained by Fleming and Higginson (1964). Above 20.6 eV the metastable current includes contributions from $2^1S$ and $2^3P$, the latter state cascading into $2^3S$, although the interpretation of each of the humps and bumps in Figure 5.19 in terms of excitation of different states is difficult. The inelastic electron scattering studies of Schulz and Philbrick (1964) for He at $72°$ scattering angle have shown distinct structure in the $2^3S$ excitation cross section function at 21 and 22.4 eV (see curve 4, Figure 5.19). This finding strongly suggests that the maximum near 21 eV in the total metastable cross section curve is due to the second peak in the $2^3S$ excitation function rather than due to $2^1S$ or $2^3P$ production. The bump near 22.5 eV has been attributed to resonance structure. Similarly,

Chamberlain and Heideman (1965) and Chamberlain (1967) have observed for zero angle scattering resonances in all four $(2^3S, 2^1S, 2^1P, 2^3P)$ helium $n = 2$ inelastic channels.

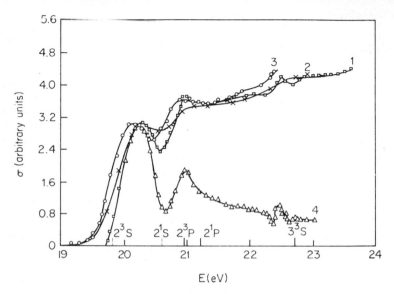

**Figure 5.19** Total metastable excitation functions for helium. Curve 1: Schulz and Fox (1957); Schulz and Philbrick (1964). Curve 2: Holt and Krotkov (1966). Curve 3: Dorrestein (1942). Curve 4: $2^3S$ excitation cross section measured at 72° by Schulz and Philbrick (1964); see discussion in text

Recent studies of the excitation functions of helium have been made by Dugan, Richards and Muschliz (1967), Olmsted, Newton and Street (1965) and Pichanick and Simpson (1968). All three groups employed the metastable detection technique; the second group of workers reported excitation cross section functions for metastable states in Ne, Ar and Kr, while Pichanick and Simpson investigated He, Ne, Ar, Kr and Xe and have attributed several narrow resonances to compound negative ion states. See also Zapesochnyi and Shpenik (1966) and Čermák (1966).

Additionally, the threshold-electron excitation spectrum of He has been studied using the $SF_6$ scavenger technique (Compton and coworkers, 1968). Some apparent differences between the threshold-electron excitation spectrum of Compton and coworkers and that obtained earlier (Schulz, 1958) by the trapped-electron method, as to the relative heights and slopes of the peaks associated with the $2^3S$ and $2^1S$ states may be reconciled in view of the fact that $SF_6$ accepts electrons with energies closer to zero ($< 0.05$ eV) as compared with the trapped-electron method where electrons of a broader range of energies are accepted.

Optically forbidden—by symmetry and/or spin selection rules—transitions for He have been studied using electron impact methods with varying incident energy and also with varying the scattering angle. Simpson, Menendez and Mielczarek (1966) have reported on the angular dependence of several transitions in He at 56.5 eV for scattering angles up to 50°, and have investigated a number of inelastic scattering resonances. Low-energy large-angle electron impact spectra were reported for He by Doering and Williams (1967) and Rice, Kuppermann and Trajmar (1968). These studies have shown that not only the differential scattering cross sections for forbidden excitations are enhanced relative to those for allowed transitions at low incident energies but also at large scattering angles. Doering and Williams pointed out that large-angle scattering experiments at low energies are a convenient method to study forbidden transitions since under such conditions the optical selection rules do not apply. At large scattering angles singlet–triplet transitions are favoured compared to optically allowed transitions which predominate in energy-loss spectra at small angles. The low-energy large-angle scattering technique may be employed to supplement ordinary optical as well as small-angle scattering experiments.

We may finally note the beam work of Lassettre and collaborators with higher energy electrons ($> 200$ eV). In Figure 5.20a the experimental data of Silverman and Lassettre (1964) on the differential cross section for the $1^1S \rightarrow 2^1P$ excitation of He using 500 eV are compared with the Born approximation calculations of Lassettre and Jones (1964) (see this reference for a discussion of the wave functions used) and Silverman and Lassettre (1964). It is seen that the agreement between the shape of the theoretical and experimental data is remarkably good. A satisfactory agreement between the theoretical (Fox (1965); Born approximation) and experimental (Lassettre, Krasnow and Silverman (1964); 500 eV incident electron energy) values for the differential cross section for the $1^1S \rightarrow 2^1S$ excitation is also seen from the data presented in Figure 5.20b. The reader is referred to the work of Lassettre and coworkers for further discussion and results on excitation cross section functions and generalized oscillator strength calculations.

### 5.4.1.3 Other atoms

Excitation cross section functions for a number of atoms have been reported by Zapesochnyi and collaborators. In Figure 5.21 the results of Zapesochnyi and Shpenik (1966)—who used a 127° cylindrical electrostatic analyser of 0.05 to 0.1 eV energy resolution—for the $6^3P_1$ and $6^3D_3$ states in mercury are shown. The pronounced maxima (the width of some of these may be instrumental) observed in the energy dependence of the excitation

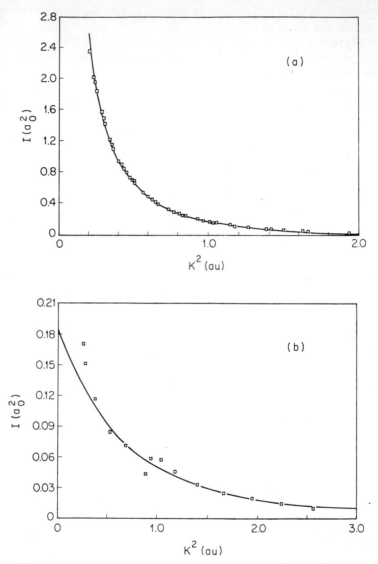

**Figure 5.20** (a) Differential cross section for the $1^1S \to 2^1P$ excitation of helium at $\sim 500$ eV. The points are experimental data for $\sim 500$ eV (Silverman and Lassettre, 1964), and the solid curve is the first Born approximation result (Lassettre and Jones, 1964; Silverman and Lassettre, 1964) (taken from Moiseiwitsch and Smith, 1968). (b) Differential cross section for the $1^1S \to 2^1S$ excitation of helium at $\sim 500$ eV. The points are experimental data for $\sim 500$ eV (Lassettre, Krasnow and Silverman, 1964), and the solid curve is the first Born approximation result (Fox, 1965)

cross section function are typical of a number of other transitions in a variety of atoms (see Zapesochnyi and Shpenik (1966) for excitation cross section functions for a number of transitions in Zn, Cd, Hg, N, K and He). The distinct maxima shown in Figure 5.21 have been related by Zapesochnyi and Shpenik to excitation of higher levels with subsequent cascading. Electron impact excitation functions for mercury have been reported by Anderson, Lee and Lin (1967); see also literature cited in Moiseiwitsch and Smith (1968), and a recent study of mercury vapour by Skerbele, Ross and Lassettre (1969) with 50 and 300 eV incident electron energies.

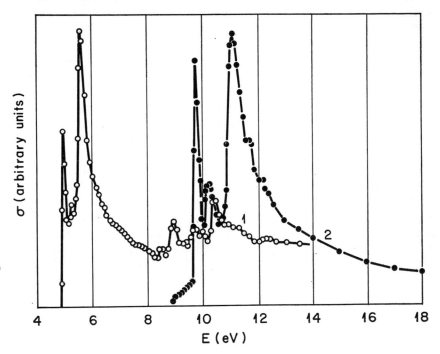

**Figure 5.21** Excitation cross sections of mercury versus electron energy (Zapesochnyi and Shpenik, 1966). Curve 1: $\lambda = 2537$ Å ($6^3P_1$ state). Curve 2: $\lambda = 3650$ Å($6^3D_3$ state)

### 5.4.2 *Electronic Excitation of Molecules by Electron Impact*

As for the case of atoms, the excitation of molecules by electron impact has been studied by either looking at the scattered electron or the radiation emitted following de-excitation of the excited molecule, or—for metastable state excitation—by employing detection of metastable molecules. The excited molecule may also dissociate. From the data to be discussed in this section it will be apparent that at energies in excess of $\sim 200$ eV the Born approximation may satisfactorily account for the observed energy-loss spectra. At primary electron energies less than $\sim 50$ to 60 eV forbidden transitions are easily detectable in diatomic molecules in forward scattering, but even more easily at large scattering angles. For polyatomic molecules primary electron energies closer to the excitation thresholds are required to violate optical selection rules in forward scattering. Certain forbidden transitions not detectable in forward scattering can be studied with ease at large scattering angles. In this respect low-energy large-angle electron scattering experiments and threshold-electron excitation studies are very valuable.

### 5.4.2.1 Diatomic molecules

A. $H_2$. Recent electron impact excitation studies of $H_2$ employed detection of inelastically scattered electrons. Schulz (1958) using the trapped-electron method found a broad peak extending from 6.5 to 10.5 eV and peaking at 9.5 eV, which he attributed to the $b^3\Sigma_u^+$ state and another and sharper peak extending from 10.5 to 13 eV and peaking at 11.8 eV which he identified as $B^1\Sigma_u^+$. Using 33 eV incident energy electrons and performing an energy analysis of the scattered electrons, Kuyatt, Mielczarek and Simpson (1964) showed well-resolved vibrational levels of several electronic states of $H_2$. Several resonances have also been reported in the total electron scattering from $H_2$ between 11 and 13 eV (e.g. Golden and Bandel (1965b)). For incident electron energies 50.7 and 30.7 eV Heideman, Kuyatt and Chamberlain (1966a) found that the observed main energy-loss peaks were due to vibrational series of two optically allowed excitations: $B^1\Sigma_u^+$ starting at 11.19 eV and $C^1\Pi_u$ starting at 12.27 eV. For 15.7 and 13.7 eV incident energies two loss peaks, one at 11.7 eV and another at 12.06 eV, became very prominent, compared with the $B^1\Sigma_u^+$. These were attributed to the triplet states $c^3\Pi_u$ starting at 11.76 eV and $a^3\Sigma_g^+$ starting at 11.79 eV. No distinction could be made between the energy-loss peaks of the two series since their vibrational spacings are too nearly equal ($\sim 0.3$ eV).

A re-examination of the threshold-electron excitation of $H_2$ has been undertaken by Dowell and Sharp (1967), using the trapped-electron method. A series of eight peaks were resolved for $H_2$ in the trapped-electron current in the range 11.7–13.5 eV, in agreement with the findings of Heideman, Kuyatt and Chamberlain (1966a). These peaks were identified as vibrational levels of the $c^3\Pi_u$ state. Ten peaks were resolved for $D_2$ beginning at 11.78 eV. Dowell and Sharp found that in the quoted energy range threshold-electron excitation of other singlet or triplet states was weaker compared to that of the $c^3\Pi_u$ state. Several peaks were observed by these workers in the range

13.8 to 15.4 eV for both $H_2$ and $D_2$ but could not be assigned to vibrational excitation of specific electronic states.

B. $N_2$. Four distinct energy-loss peaks are evident in the threshold-electron excitation spectrum of $N_2$ below 12 eV (Schulz, 1959; Curran, 1963; Brongersma and Oosterhoff, 1967; Compton and coworkers, 1968). The first peak has a maximum at 2.3 eV and is due to the $N_2^-(^2\Pi_g)$ compound negative ion state (Chapter 4). The observation of this resonance with the $SF_6$ scavenger technique (Curran, 1963; Compton and coworkers, 1968) requires that a vibrational level of $N_2^-(^2\Pi_g)$ matches a vibrational level of $N_2(^1\Sigma_g^+)$ so that the ejected electron has essentially zero energy ($< 0.05$ eV). The intensity of the $SF_6^-$ signal at 2.3 eV was much weaker, however, compared to the $SF_5^-$ signal. The amplification of the $SF_5^-$ signal is due to the fact that the primary $SF_5^-$ ion current extends from 0 to $\sim 1$ eV (Chapter 6); consequently, electrons which have suffered partial energy losses resulting in scattered electron energies from $\sim 0$ to $\sim 1$ eV can be detected. The second energy-loss peak extends from 7 to 9 eV. Although this energy-loss has been attributed initially by Schulz to the $A^3\Sigma_u^+$ state, subsequently Brongersma and Oosterhoff (1967) observed, with better resolution, that the vibrational spacing for this band agreed better with that of the $B^3\Pi_g$ state. Also, the observed relative intensities corresponded closer to the Franck–Condon factors for the $X^1\Sigma_g^+ \to B^3\Pi_g$ transition. No definite assignment for the third energy-loss peak(s) in the range $\sim 9.5$–11 eV has been made although certain authors (Schulz, 1959; Compton and coworkers, 1968) suggested excitation of the $a^1\Pi_g$ state (see, however, Brongersma and Oosterhoff (1967)). Assignment of the fourth energy-loss peak has been controversial. Schulz (1959) and Compton and coworkers (1968) have found this sharp energy-loss process peaking at 11.5 eV and have associated it with excitation of the $C^3\Pi_u$ state. However, Brongersma and Oosterhoff (1967) and Brongersma, Hart and Oosterhoff (1967) observed the peak at 11.82 eV and have associated it with the $X^1\Sigma_g^+ \to E^3\Sigma_g^+$ transition, suggesting that this peak is the same as that observed by Schulz (1959) at $\sim 0.3$ eV lower energies due to that much error in his energy scale. This conclusion is in accord with an earlier one by Heideman, Kuyatt and Chamberlain (1966b) who observed a very sharp peak due to the $E^3\Sigma_g^+$ state at 11.87 eV. Meyer and Lassettre (1966) also observed the energy-loss process at 11.87 eV with 400 eV electrons and have similarly associated it with the $E^3\Sigma_g^+$ state. Further, Olmsted (1967) discussed complications which arise if the 11.5 eV in Schulz's spectrum is ascribed to the $C^3\Pi_u$ state rather than to the $E^3\Sigma_g^+$ state (see also Winters (1965)). The sharpness of this energy-loss peak has led to the suggestion that the $E^3\Sigma_g^+$ state is resonantly excited by electron impact.

Heideman, Kuyatt and Chamberlain (1966b) were able to detect singlet–

triplet transitions in forward scattering at incident energies 15.7 eV, but not at 30 eV. Skerbele, Dillon and Lassettre (1967a), however, have observed three singlet–triplet transitions both at 35 and 50 eV at scattering angles from 0 to 12°, while an enormous enhancement in the intensity of the singlet–triplet energy-loss peaks with increasing scattering angle has been found by Doering and Williams (1967) at primary energies 16.1 and 36.9 eV.

Geiger and Stickel (1965) measured the energy-loss spectra in $N_2$ for 33 keV primary electrons and compared their results with the optical absorption data of Huffman, Tanaka and Larrabee (1963). Although the two spectra, electron-impact and optical, exhibited considerable similarities, marked differences were observed in the energy range 12.8 to 14.5 eV. This was in accordance with the anomaly in the $N_2$ energy-loss spectrum reported by Lassettre and coworkers (1965) using 200 to 500 eV electrons at zero scattering angle, which they have attributed to a failure of the Born approximation for this portion of the spectrum. Lassettre and coworkers found that the most intense transition in $N_2$ occurred at $12.91 \pm 0.03$ eV rather than at 12.74 eV as in the optical spectrum (Huffman, Tanaka and Larrabee, 1963) and interpreted this as a strong indication that the distribution of intensity among the various vibrational levels is not the same in optical and in electron-impact spectra. For the remainder of the $N_2$ spectrum optical and electron-impact spectra agreed well. However, more completely resolved spectra at 200 eV incident energies (see review by Lassettre (1969)) have shown that the observed anomaly is fortuitous and the implication of the results by Lassettre and coworkers (1965) that the Born approximation does not apply equally accurately in every region of a given spectrum is incorrect, i.e. *no* failure of the Born approximation is involved.

C. $O_2$.   The ground electronic state of $O_2$ is the triplet $X^3\Sigma_g^-$. Transitions from this state to the singlets $a^1\Delta_g$ and $b^1\Sigma_g^+$, and the triplets $^3\Sigma_u^+$ and $B^3\Sigma_u^-$ have been observed by Schulz and Dowell (1962) in a threshold-electron excitation study of $O_2$. Schulz and Dowell covered the energy range $\sim 0$–12 eV and estimated for the $a^1\Delta_g$ and $b^1\Sigma_g^+$ states a cross section less than $3 \times 10^{-20}$ and $6 \times 10^{-21}$ cm$^2$, respectively, at 0.16 eV above threshold. The $X^3\Sigma_g^- \rightarrow a^1\Delta_g$ and $X^3\Sigma_g^- \rightarrow b^1\Sigma_g^+$ transitions have been observed, also, with 40 and 50 eV primary electrons (Skerbele, Dillon and Lassettre, 1968a). The energy-loss spectrum of $O_2$ in the energy range 6.8–21 eV has been shown (Geiger and Schröder, 1968) to be in fair agreement with the optical spectrum for 25 keV primary electrons.

D. HCl.   The threshold-electron excitation spectrum of HCl obtained with the $SF_6$ scavenger technique is shown in Figure 5.22. It very closely resembles the atomic-like spectra observed for HCl by Price (1938). A well-defined Rydberg series is observed above $\sim 10$ eV which converges to the ionization potential. The break at $12.5 \pm 0.1$ eV agrees well with the

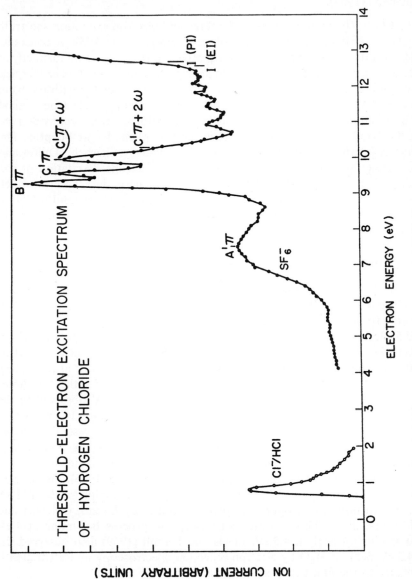

**Figure 5.22** Threshold-electron excitation spectrum of HCl (Compton and coworkers, 1968)

electron impact ionization potential at $12.5 \pm 0.1$ eV (Fox, 1960a) but it is $\sim 0.2$ eV lower than the photoionization potential determined by Watanabe (1957).

E. CO.   The main features of the threshold-electron excitation spectrum of CO are similar to those for the isoelectronic $N_2$ (Schulz, 1959; Brongersma and Oosterhoff, 1967); a compound negative ion state shows up distinctly at $\sim 1.75$ eV (see Section 5.3.2.1.D) and so do triplet excitations at higher energies. Excitation of the lowest triplet state $a^3\Pi$ has been associated with the band between 5.5 and 7 eV observed by Schulz using the trapped-electron method with a well-depth of 0.7 eV. Six distinct vibrational levels were observed for this band by Brongersma and Oosterhoff with smaller well-depths. The structure of the $a^3\Pi$ state has been resolved also by Skerbele, Dillon and Lassettre (1967b) using 30 and 60 eV primary electrons. Excitation of CO with higher energy ($> 200$ eV) electrons gave energy-loss spectra which compared well with optical data except in the region around 12.79 eV (Meyer, Skerbele and Lassettre, 1965). See detailed discussion in Lassettre (1969).

## 5.4.2.2 Polyatomic molecules

A. $H_2O$.   The threshold-electron excitation spectrum of water vapour obtained using the $SF_6$ scavenger technique (Compton and coworkers, 1968), and the trapped-electron method (Schulz, 1960b) is shown in Figures 5.23b and 5.23c, respectively, and is compared with the energy-loss spectrum taken (Skerbele, Dillon and Lassettre, 1968b) with 53 eV primary electrons at 5° scattering angle (Figure 5.23a). All three spectra show a peak at the energy of the first singlet state ($^1B_1$) of $H_2O$. At higher energies the threshold-electron excitation spectra do not show the peak at 11 eV observed in the energy-loss spectrum with 53 eV electrons. With the exception of the energy-loss peak below $\sim 6$ eV the inelastic energy-loss processes essentially agree with optical data. Agreement between energy-loss spectra and optical absorption spectra for $H_2O$ has been found also for 200 eV incident electrons and zero scattering angle (Skerbele, Meyer and Lassettre, 1965). The energy-loss process below 6 eV has been associated with the first triplet state of $H_2O$. It is worth noting that optical absorption studies by Larzul, Gélébart and Johannin-Gilles (1965) indicated an absorption process beginning at 4 eV and reaching a peak at 4.5 eV. Lewis and Hamill (1969) also observed the $\sim 4.5$ eV peak in $H_2O$ by electron-reflection spectroscopy and attributed it similarly to the first triplet state of water.*

---

*Very recently Kuppermann (1970) has presented experimental data on the angular dependence of the cross section for the $\sim 4.5$ eV energy-loss process which confirm the assignment of this energy-loss peak to the excitation of a spin-forbidden transition.

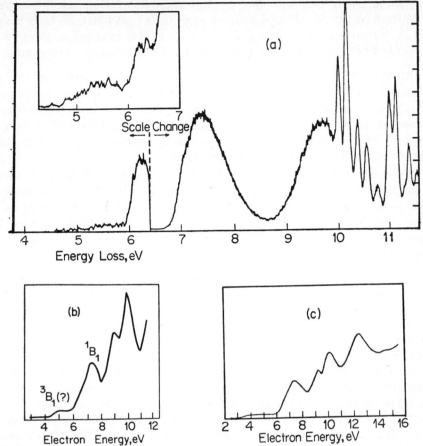

**Figure 5.23** (a) Electron-impact spectrum of $H_2O$ at 53 eV primary electrons and 5° scattering angle (Skerbele, Dillon and Lassettre, 1968b). The inset shows the 4.5–7 eV energy-loss region with no change in the energy scale. (b) Threshold-electron excitation spectrum of $H_2O$ using the $SF_6$ scavenger technique (Compton and coworkers, 1968). (c) Threshold-electron excitation spectrum of $H_2O$ using the trapped-electron method (Schulz, 1960b)

The observation of the first triplet state of $H_2O$ by Skerbele, Dillon and Lassettre (1968b) is interesting since these workers (Lassettre and coworkers, 1968; Skerbele, Dillon and Lassettre, 1967a, b) failed to observe singlet–triplet transitions in forward scattering in polyatomic molecules, although they did observe some such transitions for atoms and diatomic molecules under similar experimental conditions. Singlet–triplet transitions in polyatomic molecules at energies 30–60 eV have been observed at large scattering angles (see later this section). Skerbele, Dillon and Lassettre (1968b) have attributed the excitation of the first triplet state in $H_2O$ with 53 eV primary electrons principally to a spin–orbit coupling mechanism, since at 53 eV electron exchange is expected to be relatively unimportant.

B. $N_2O$; $CO_2$. The threshold-electron excitation spectrum of $N_2O$ (Schulz, 1961) showed a reasonable agreement with its optical absorption spectrum except for an energy loss peak at ~5.6 eV which has not been observed in optical absorption. Lassettre and coworkers (1968) using ~50 eV primary electrons found no low-energy losses in $N_2O$ which might have resulted from excitation of triplet states.

The energy-loss spectrum of $CO_2$ has been measured by Skerbele, Dillon and Lassettre (1968b) using 30 to 60 eV primary electrons. Energy-loss peaks due to vibrational excitation including excitation of a forbidden vibrational transition have been observed (see Section 5.3.2.2.A). Electron impact spectra with 150 to 400 eV primary electrons were in good agreement with optical ones (Meyer and Lassettre, 1965).

C. $C_2H_2$ (*acetylene*). The main features of the threshold-electron excitation spectrum of acetylene reported by Bowman and Miller (1965) using the trapped-electron method are similar to those of the optical absorption spectrum of this compound for energies in excess of ~4 eV. Below this energy Bowman and Miller have detected an energy-loss process peaking at 2 eV which they tentatively attributed to excitation of a triplet state. This energy-loss process may be due to compound negative ion state excitation. It should be noted, however, that Trajmar and coworkers (1968) failed to observe the 2 eV energy-loss process at 10 and 50° scattering angle with 25 eV primary electrons. In the same experiment Trajmar and coworkers observed energy-loss maxima at 5.2 eV (onset at 4.5 eV) and 6.1 eV which they have attributed to excitation of low-lying triplet states.

D. $C_2H_4$ (*ethylene*). The threshold-electron excitation spectrum of $C_2H_4$ reported by Bowman and Miller (1965) (see also Brongersma, Hart and Oosterhoff (1967)) showed energy-loss peaks at 1.7, 4.4, 7.7 and 9.2 eV. The peak at 1.7 eV has been ascribed as excitation of a compound negative ion state while that at 4.4 eV as excitation of the first triplet state at 4.6 eV. The peaks at 7.7 and 9.2 eV correlated well with optical absorption maxima and with energy-loss peaks observed by Lassettre and Francis (1964) at 7.66 and 9.03 eV in forward scattering with 390 eV primary electrons. Kuppermann and Raff (1962, 1963) observed energy-losses peaking at 4.6, 6.5, 7.7, 8.8 and 10.5 eV at large scattering angles and 40 eV primary electrons. Although the energy-loss peaks at 4.6, 7.7, 8.8 and 10.5 eV are in accord with other optical and energy-loss data the peak at 6.5 eV seems to be a feature of only this spectrum. This peak ascribed by Kuppermann and Raff to triplet-state excitation, has not been observed in the threshold-electron excitation spectra. Neither did Simpson and Mielczarek (1963) detect the 6.5 eV (and 4.6 eV) energy-loss peak in forward scattering at 50 eV nor did Doering (1967) at 90° and 50 eV, nor Doering and Williams (1967) at 90° and 27.1, 18.1 and 10.9 eV primary electron energies.*

Ross and Lassettre (1966) reported energy-loss spectra for $C_2H_4$ at 200 eV in forward scattering. The intensity distribution of the energy-loss spectrum agreed well with optical absorption data and measurements made at 33 keV (Geiger and Wittmaack, 1965b), except in the region of the 7.45eV transition. This discrepancy has been attributed by Ross and Lassettre to a forbidden quadrupole transition which occurs in this energy region. Such a transition is probably hidden by allowed transitions at 33 keV but not at lower energies where the intensity of the allowed transition is appreciably reduced.

E. $CH_4$ (*methane*); $C_2H_6$ (*ethane*); $C_3H_8$ (*propane*); $C_4H_{10}$ (*butane*). The threshold-electron excitation spectrum of $CH_4$ showed a slightly resolved peak at 10.2 eV and a broad band at 11.8 eV while that of $C_2H_6$ showed a single maximum at 9.87 eV (Bowman and Miller, 1965), in agreement with the energy-loss spectra of Lassettre and Francis (1964) at 390 eV and forward scattering. The reader is referred to Lassettre, Skerbele and Dillon (1968) for energy-loss spectra for $CH_4$, $C_2H_6$, $C_3H_8$ and $C_4H_{10}$ at scattering angles 0° and 6°, 0° to 9°, 0° and 0°, respectively, and at 50 to 180 eV primary energies. See this reference for energy-loss spectra of perdeuteroethane ($C_2D_6$) in forward scattering at 50 and 100 eV.

F. $C_6H_6$ (*benzene*); $C_{10}H_8$ (*naphthalene*). Considerable emphasis has been placed on benzene ($C_6H_6$), the simplest of the aromatics. The threshold-electron excitation spectrum of Compton and coworkers (1968) using the $SF_6$ scavenger technique, and that of Brongersma, Hart and Oosterhoff (1967) using the trapped-electron method are compared in Figure 5.24 with the energy-loss spectrum of Skerbele and Lassettre (1965) obtained with 300 eV primary electrons in forward scattering. The spectrum of Skerbele and Lassettre is in general agreement with the optical absorption spectrum of benzene and it shows distinct peaks at 4.93, 6.21 and 6.96 eV. Higher-energy peaks appear at 8.45, 8.70, 9.27, 10.58, 11.02, 11.32 and 11.96 eV. The two threshold-electron excitation spectra although in qualitative overall agreement show some differences in the energy range 4 to 6 eV which are not understood.

The most intense peak in all three energy-loss spectra shown in Figure 5.24 is that due to excitation of the electric dipole transition $^1A_{1g}^-(^1A) \rightarrow {}^1E_{1u}^+(^1B)$

---

*Based on their recent experimental results Trajmar, Rice and Kuppermann (1970) concluded that the early observations of Kuppermann and Raff (1962, 1963) on the 6.5 eV band were spurious and that the band in question is either too weak to detect or, more likely, does not lie in this energy region of the spectrum.

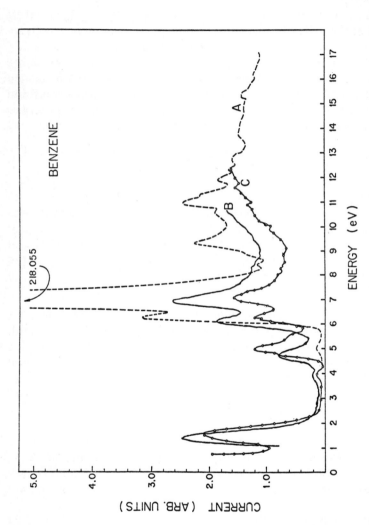

**Figure 5.24** (A) Energy-loss spectrum of $C_6H_6$ obtained with 300 eV primary electrons in forward scattering (Skerbele and Lassettre, 1965). (B) Threshold-electron excitation spectrum of $C_6H_6$ using the trapped-electron method (Brongersma, Hart and Oosterhoff, 1967). (C) Threshold-electron excitation spectrum of $C_6H_6$ using the $SF_6$ scavenger technique (Compton and coworkers, 1968)

at 6.96 eV, which is symmetry allowed.* The intensity of this energy-loss peak for 300 eV primary electrons far exceeds those ascribed to the symmetry forbidden transitions $^1B_{1u}^+(^1L_a)$ and $^1B_{2u}^-(^1L_b)$ at, respectively, 6.21 and 4.93 eV. All three energy-loss peaks show comparable intensities in the threshold-electron excitation spectra. Doering and Williams (1967) found the intensity of energy loss peaks due to the forbidden transitions $^1A_{1g}^-(^1A) \rightarrow {}^1B_{1u}^+(^1L_a)$ and $^1A_{1g}^-(^1A) \rightarrow {}^1B_{2u}^-(^1L_b)$ to increase relative to that for the $^1A_{1g}^-(^1A) \rightarrow {}^1E_{1u}^+(B)$ transition in lowering the primary electron energy from 19.2 to 12.6 eV at 90° scattering angle. The energy-loss spectrum reported by Doering and Williams agreed well with that of Skerbele and Lassettre with the exception of an additional peak at 3.95 eV which Doering and Williams have ascribed to excitation of the first $\pi$-triplet state $^3B_{1u}^+(^3L_a)$. This energy-loss process with an onset at 3.65 eV is clearly observable in the threshold-electron excitation spectra shown in Figure 5.24. Additionally, the shoulder at $\sim$4.7 eV on the low-energy side of the $^1B_{2u}^-(^1L_b)$ (see curve C in Figure 5.24) has been attributed by Birks, Christophorou and Huebner (1968) and Compton and coworkers (1968) to excitation of the second $\pi$-triplet state $^3E_{1u}^+(^3B)$. The threshold for this peak at 4.4 eV agrees closely with the optical absorption onset of 4.56 eV reported by Colson and Bernstein (1965, 1966) using $O_2$ and NO perturbation techniques. Similarly the shoulder at 5.4 eV on the high-energy side of the $^1B_{2u}^-$ energy-loss peak has been tentatively ascribed to excitation of the third triplet state $^3B_{2u}^-(^3L_b)$ of benzene. The energy-loss resonance at 1.5 eV (see curves B and C in Figure 5.24) has been associated with excitation of a temporary negative ion state at this energy (Chapter 4) and as such it is not expected to show-up in spectrum A (Figure 5.24).

We finally note the calculation of differential and total excitation cross sections for the first three $\pi$-singlet excited states of benzene by Read and Whiterod (1965) in the Born approximation. These workers used simple LCAO $\pi$-electron wave functions and neglected all but nearest neighbour overlap. Their conclusion was similar to that reached earlier by Inokuti (1958) and Matsuzawa (1963), namely that $\pi$-electron states are more easily excited than $\sigma$-electron states by high-energy incident electrons.

The threshold-electron excitation spectrum of naphthalene reported by Compton and coworkers (1968) using the $SF_6$ scavenger technique has been discussed by Birks, Christophorou and Huebner (1968) and by Compton and coworkers (1968), in connection with optical absorption and theoretical data. A broad energy-loss peak at 3.2 eV (onset 2.6 eV) was identified as due to the first $\pi$-triplet state $^3B_{2u}^+(^3L_a)$, with possible contribution from excitation of the second $\pi$-triplet state $^3B_{3u}^+(^3B_b)$. The asymmetric energy-loss peak a-round 4.8 eV was ascribed to excitation of the two lowest $\pi$-singlet states

---

*Notation in parenthesis is that of Platt (Chapter 3).

$^1B_{3u}^-(^1L_b)$ and $^1B_{2u}^+(^1L_a)$; another energy-loss peak at 6.0 eV was associated with excitation of the third $\pi$-singlet state $^1B_{3u}^+(^1B_b)$. Two intense energy-loss maxima at 0.8 and 5.4 eV were new features in the energy-loss spectrum in that they have not been observed earlier in optical absorption or other studies. The peak at 0.8 eV was attributed to excitation of a compound negative ion state while the peak at 5.4 eV was tentatively assigned by Birks, Christophorou and Huebner (1968) to excitation at the second $^3B_{2u}^+(^3B_a)$ state.

G. *Other organic molecules.* Essentially all features in the threshold-electron excitation spectrum of ethylene are evident in the respective spectrum of propylene ($C_3H_6$) which showed (Bowman and Miller, 1965) a compound state at $\sim 1.8$ eV and energy-loss peaks at 4.4, 7.8 and 8.8 eV (see discussion on $C_2H_4$). The threshold-electron excitation spectra of propyne ($CH_3C \equiv CH$) and 1-butyne ($C_2H_5 \equiv CH$) showed (Bowman and Miller, 1965), respectively, energy-loss peaks at 2.8 and 2.4 eV which may be attributed to excitation of compound negative ion states. Energy-loss peaks in the range 6.1 to 8.3 eV are evident for both molecules in their respective threshold-electron excitation spectra.

Brongersma, Hart and Oosterhoff (1967) reported threshold-electron excitation spectra for *cis*-butene-2, butadiene and 1,1-biphenyl ethylene. An energy loss at 2.3 eV observed in the *cis*-butene-2 spectrum has been associated with excitation of a compound negative ion state, while a vague shoulder at 3.3 eV and a broad peak at 3.8 eV in the butadiene spectrum have been attributed to excitation of the first and second triplet states of this molecule; a shoulder at 3.4 eV in the 1,1-biphenyl ethylene spectrum, not observed in optical absorption studies, has been explained as excitation of a triplet state.

## 5.5 Inelastic Electron Scattering Resulting in Single and Multiple Ionization

The study of electron ejection from atoms and molecules by electron impact is one of great intrinsic value and of practical interest. It provides a test of theoretical assumptions and approximations and hence it aids the development of the theory of atomic collisions. Knowledge of electron-impact ionization cross sections for atoms and molecules is necessary for the interpretation of physical phenomena in a variety of fields such as astrophysics, upper atmosphere physics, plasma physics, radiation sciences, gaseous electronics, vacuum technology, radiation detection and dosimetry. Electron-impact determinations of the first and higher ionization potentials of atoms and molecules remain an important task for the experimentalist in the field, in spite of the fact that ionization potentials of atoms and molecules are sufficiently well known from optical spectroscopic data. The ionization potential of the most loosely bound electron is a quantity required for the understanding of many physical phenomena, atomic and molecular structure, molecular reactions and chemical reactivity. Inner-shell atomic ionization potentials and higher-order ionization potentials of molecules are required for the development of an understanding of matter and its inter-action with radiation. Ionization of atoms and molecules is an important dissipative process via which ionizing radiation slows-down and is absorbed in matter. Electron-impact ionization cross sections as a function of electron energy constitute the corner stone of stopping-power calculations.

Single ionization can occur when the energy of the incident electron $\varepsilon$ exceeds the threshold energy, namely the ionization potential $I$. The process can be studied (Section 5.2) either by measuring the positive ions formed, the detached electrons, or the kinetic energy loss of the impacting electron. Here, the ejected electron may possess kinetic energy from zero to $(\varepsilon - I)$. In our definition of the ionization cross section in Section 5.1.2 care was taken so that the cross section would refer to the total electron ejection, integrated over the entire energy spectrum. In general, most of the impacting electrons lose energies little greater than $I$, i.e. the bulk of the electrons are ejected with small energies.

For both atoms and molecules the cross section for single ionization (see later this section) dominates over those for $n$-fold (multiple) ionization so that in many instances it is sufficient to measure the total number of ions produced without mass spectroscopic identification. At threshold, the ionization cross section is zero but it rises rapidly with increasing incident energy above threshold. The initial rise is more or less linear (Section 5.5.4) becoming slower at higher energies. The cross section reaches a broad maximum usually in the region of $\sim 100$ eV and it falls to one-half its maximum value around 500 eV but the energy where the cross section maximizes is a strong function of the species involved. For alkali vapours, for example, the cross section maximizes at energies below $\sim 20$ eV, while for the rare gas atoms in the range 80 to 180 eV (Table 5.1). Calculations based on the Born approximation (Equation (5.14)) show that the single-ionization cross section $\sigma_i$ varies with the incident electron energy $\varepsilon$ as

$$\sigma_i \propto \varepsilon^{-1} \ln B\varepsilon \qquad (5.30)$$

where $B^{-1}$ is a quantity of the order of $I$. This energy dependence is similar to that for the excitation cross section for optically allowed transitions and has been verified experimentally for incident energies largely exceeding that which is transferred to the target in the ionization process (e.g. Schram and coworkers, 1965, 1966a; Fiquet-Fayard and coworkers, 1968; Section 5.5.1). The cross section for direct multiple ionization (process 5.5) behaves in a somewhat similar manner, but is has a different threshold dependence and it is significant at higher impact energies (Section 5.5.1); the higher the

**Table 5.1** Peak values of the total ionization cross section and peak energies for He, Ne, Ar, Kr and Xe[a]

| Atom | Peak cross section $(\pi a_0^2)$ | Peak energy[b] (eV) |
|------|------|------|
| He | 0.425 | 125 |
| Ne | 0.89 | 180 |
| Ar | 3.25 | 92 |
| Kr | 4.84 | 80 |
| Xe | 6.21 | 115 |

[a] From Rapp and Englander-Golden (1965).
[b] Note that for some atoms the total ionization cross section $(\sigma_i)_T$ peaks at lower energies; for the alkali metals, in particular, the cross section for single ionization by electron impact peaks at energies $\lesssim 20$ eV (Brink, 1962; Tate and Smith, 1934).

degree of ionization, the higher the cross section function onset. At high incident energies the cross section $\sigma_{2i}$ for direct double ionization varies with incident energy as (Section 5.5.1)

$$\sigma_{2i} \propto \varepsilon^{-1} \tag{5.31}$$

that is, in the same way as does the cross section for an optically forbidden transition.

The early work has been summarized by many authors (e.g. Massey and Burhop, 1952; Massey, 1956; Craggs and Massey, 1959; McDaniel, 1964; Kieffer, 1965; Kieffer and Dunn, 1966). The available experimental data are far from exhaustive and often disagree considerably. Much of the uncertainty in experimental determinations of ionization cross sections comes from pressure measurements and 'pumping errors'. Other disturbing effects exist such as (at high energies) ionization by secondary electrons which may have higher cross sections for ionization compared to the primary ones. A discussion of such disturbing effects can be found in Schram and coworkers (1965) and Kieffer and Dunn (1966). We, finally, note that very little experimental information is available as regards the ionization of excited atoms and molecules by electron impact. Angular distribution measurements of the ejected electrons are very limited also.

Theoretically, the basic formulation of the problem is different from that of excitation by electron impact. Rudge (1968) gave an excellent review of the quantum theory of ionizing collisions and concluded that although theoretical treatments of ionizing collisions have improved in recent years, a really adequate theoretical approach to the calculation of single ionization cross sections is still lacking. The situation as regards multiple ionization is a good deal worse, virtually the only available estimates being those of Gryzinski (1959, 1965a,b,c,d). A treatment which gives correct threshold behaviour and correct angular distribution of the reaction products is, also, still lacking. The Born approximation and its variations which take into account exchange effects, predict with reasonable success cross-section values at large incident energies $(\gtrsim 200 \text{ eV})$. The Born approximation and its variations are, however, inadequate to explain the cross section behaviour at low energies, especially near threshold. The existing theoretical treatments become less successful when extended from atomic hydrogen to polyelectronic atoms.

### 5.5.1 *Electron Impact Ionization Cross Section Functions for Atoms*

#### 5.5.1.1 Total ionization cross section functions

Total ionization cross section functions are shown in Figure 5.25 for the atoms H, N and O, in Figure 5.26 for the rare gases He, Ne, Ar, Kr and Xe and in Figure 5.27 for the alkali metals Li, Na, K, Rb and Cs and Hg. Detailed references are given in the figure captions and the reader is directed to these sources as well as to Kieffer and Dunn (1966) for details. For atomic hydrogen the experimental results are seen to be in a rather good agreement considering the difficulties in the techniques employed. This, however, is not seen to be the case for atomic N and O, where the experimental data differ appreciably. The Born approximation result for H is seen to lie above the experimental values, the discrepancy, as expected, increasing with decreasing energy. Seaton's theoretical values for the $\sigma_i$ of N and O are in fair agreement with the experimental data (Figure 5.25).

From the data shown in Figure 5.26 on the rare gases it is apparent that the total cross sections for these inert atoms are considerably uncertain in spite of the fact that they are generally regarded as well known. In particular, Smith's early work (Smith, 1930) has been considered classic and his data have been used as a standard by many workers for comparison or normalization purposes. However, his results profoundly differ from the recent work of Rapp and Englander-Golden (1965). In view of the recent work of Srinivasan and Rees (1967) on Ar, and Srinivasan, Rees and Craggs (1967) on Kr, and their suggested correction of the Asundi and Kurepa (1963) data (see caption of Figure 5.26), one would tend to conclude that the recent experimental data are consistent and more accurate than the early data of Smith. The results of Schram and coworkers (1965) in the range 0.6 to 20 keV are low compared to the other data in Figure 5.26. Schram and coworkers, however, argued that in spite of the uncertainty in their measurements due to multiple ionization, the energy dependence of their cross sections is in agreement with theoretical predictions. They have used the Bethe–Born expression for $\sigma_i$ (Equations (2.22) and (5.14)) and having corrected $\varepsilon$ for relativistic effects they have plotted $\sigma_i \varepsilon / 4\pi a_0^2 R$ as a function of $\ln \varepsilon$. Their result for He is shown in Figure 5.28. It is seen that the energy dependence of the cross section of Schram and coworkers is in excellent

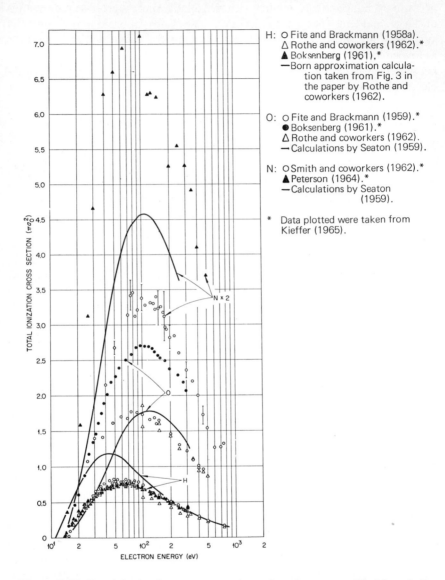

Figure 5.25 Total ionization cross sections for the atoms H, N and O. The data for atomic nitrogen have been multiplied by a factor of two for convenience of display.

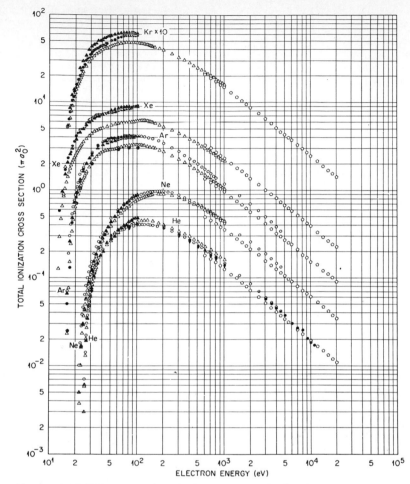

He: O Smith (1930).* ● Liska (1934).* Δ Harrison (1956).* ▲ Asundi and Kurepa (1963).
◆ Rapp and Englander—Golden (1965). ◌ Schram and coworkers (1965).
Ne: O Smith (1930).* ▲ Asundi and Kurepa (1963). Δ Rapp and Englander—Golden (1965).
◌ Schram and coworkers (1965).
Ar: O Smith (1930).* ● Tozer and Craggs (1960). ▲ Asundi and Kurepa (1963).†
Δ Rapp and Englander—Golden (1965). ◌ Schram and coworkers (1965).
◆ Srinivasan and Rees (1967).
Kr and Xe: ● Tozer and Craggs (1960). ▲ Asundi and Kurepa (1963).‡
Δ Rapp and Englander—Golden (1965). ◡ Schram and coworkers (1965).

**Figure 5.26** Total ionization cross sections for He, Ne, Ar, Kr, and Xe.

* Data plodded were take from Kieffer (1965).
† According to Srinivasan and Rees (1967), these data have to be multiplied by a factor of 0.83 to correct for 'pumping errors' whiéh will bring them into better agreement with the data of Rapp and Englander-Golden.
‡ For Kr, Srinivasan, Rees, and Craggs (1967) confirmed the data of Rapp and Englander-Golden (1965) and pointed out that the data of Asundi and Kurepa (1963) should be multiplied by a factor of 0.77 to correct for pressure errors. This would bring the data of Asundi and Kurepa (1963) into agreement with those of Rapp and Englander-Golden.

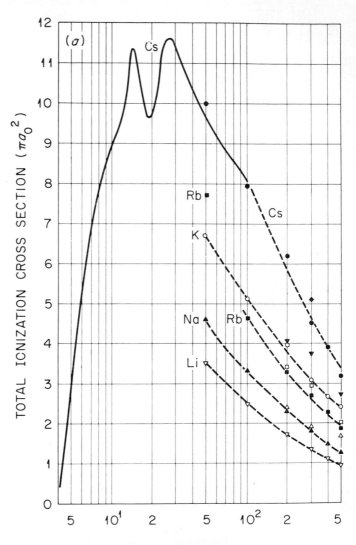

**Electron Energy (eV)**

(a) Li ▽ , Na ▲ , K ○ , Rb ■ , Cs ● :
McFarland and Kinney (1965).
Na △ , K ▼ , Rb′□ , Cs ◆ : Brink (1964).
Cs (solid line): Nygaard (1968).

**Figure 5.27** Total ionizat

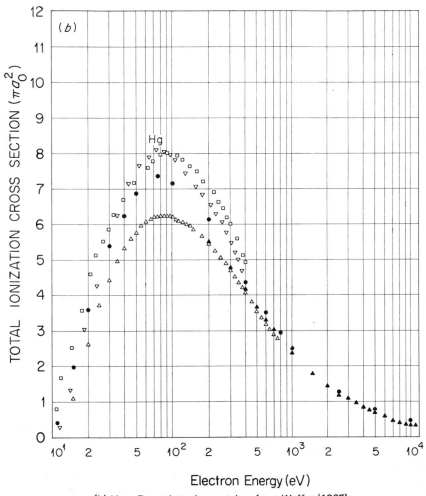

(b) Hg:   Data plotted were taken from Kieffer (1965).
□ Jones (1927). ▽ Bleakney (1930a). Δ Smith (1931).
▲ Liska (1934). ● Harrison (1956).

ss sections for Li, Na, Rb, K, Rb, Cs and Hg

agreement with the theory, as is to be expected for energies where the Bethe–Born approximation is valid. Smith's data are seen not to lie on a straight line and Liska's data have the wrong energy dependence. Schram and coworkers obtained similar plots for other atomic and molecular species (see also Schram and coworkers (1966b)) and determined through Equations (2.22) and (2.23) values of $M_i^2$ and $c_i$. In Table 5.2 we list their estimates of $M_i^2$ for He, Ne, Ar, Kr and Xe. The values of $M_i^2$ of Schram and

**Table 5.2** Values of $M_i^2$ for He, Ne, Ar, Kr and Xe

| Atom[a] | Electron impact[b] | Photoionization | Proton ionization[f] |
|---|---|---|---|
| He | 0.489 | 0.494[c] | 0.471 |
| Ne | 1.87 | 1.70[d] | 1.68 |
| Ar | 4.50 | 3.15[e] | 4.12 |
| Kr | 7.51 | — | — |
| Xe | 11.75 | — | — |

[a] For theoretical and other values of $M_i^2$ see Table V in Schram and coworkers(1965) from which the data listed in this table were taken.
[b] Schram and coworkers (1965).
[c] Baker, Bedo and Tomboulian (1961).
[d] Ederer and Tomboulian (1964).
[e] Samson (1964).
[f] Hooper and coworkers (1961, 1962).

coworkers are seen to compare reasonably well with photoionization data showing the relation of $M_i^2$ to the optical oscillator strength (Chapter 3).

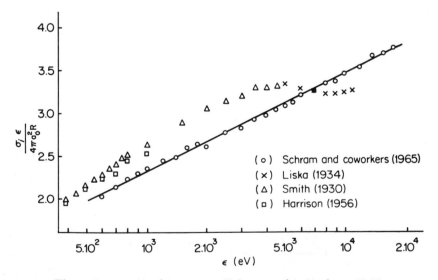

**Figure 5.28** $\sigma_i \varepsilon / 4\pi a_0^2 R$ versus $\varepsilon$ (Schram and coworkers, 1965)

In Table 5.2 values of $M_i^2$ for high-energy protons are also listed and are seen to agree well with the electron-impact ionization and photoionization results. From expressions (5.14) and (5.16) it is clear that at sufficiently high velocities the ionization cross section $\sigma_i$ for electrons and protons of the same velocity are equal. Hooper and coworkers (1961, 1962) have shown that this prediction holds for the rare gases at electron energies greater than $\sim 300$ eV. The corresponding proton energy is 552 keV. In Figure 5.29

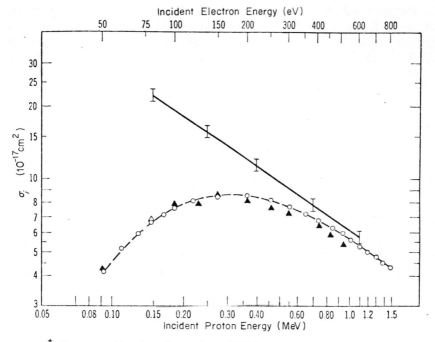

H$^+$ + Ne:  ——  Hooper and coworkers (1962).

e + Ne:  -o-o-  Smith (1930).

△  Tozer and Craggs (1960).

▲  Bleakney (1930 b).

**Figure 5.29** Comparison of experimental apparent ionization cross sections for protons and electrons of equal velocity incident on Ne (McDaniel, 1964)

we present an example of the results of Hooper and coworkers for the case of Ne. It is seen that at high energy the cross section curves appear essentially linear on a log–log plot in the manner predicted by the Bethe–Born equation.

In addition to the data plotted in Figure 5.27 for the alkali metals, relative ionization cross sections for these systems have been reported by a number of other workers. Brink (1962), for example, has measured the relative ionization cross section of Li, Na and K and has found that for Li (and Na) the cross-section values lay on a straight line on a log–log plot from 500 eV to ~30 eV, in accord with the Born approximation. The relative cross sections obtained by Brink (1962) for $K^+$ and $K^{2+}$ are shown in Figure 5.30. The break in the curve for $K^+$ at ~19 eV has been associated by Brink with autoionization, namely ionization through the $3p^6 4s \rightarrow 3p^5 4s^2$ excited state of the atom, which has a threshold at 19 eV. He pointed out that similar structure found earlier by Tate and Smith (1934) for K, Rb and Cs has a similar origin. The structure in the Nygaard (1968) cross section curve for Cs (Figure 5.27) is in agreement with that observed by Tate and Smith (1934).

Single and double ionization cross section functions for silver have been reported by Crawford and Wang (1967).

The ionization cross section functions for excited atoms begin at a lower energy and can be greater than those of the unexcited species. Very little work has been done on the ionization of excited atoms.

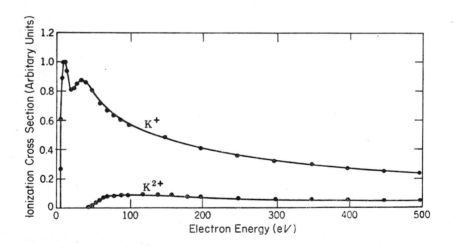

**Figure 5.30** Relative ionization cross section of potassium (Brink, 1962)

### 5.5.1.2 Multiple ionization cross section functions

Typical results on the probability of production of specific ionic species are shown in Figure 5.31 for Ar and Hg. The data have been plotted as $\sigma_{ni}/(\sigma_i)_T$ on a log–log plot. To obtain $\sigma_{ni}$ at a given energy one multiplies the ratio $\sigma_{ni}/(\sigma_i)_T$ by the corresponding value of $(\sigma_i)_T$. Plots similar to that in Figure 5.31 for other species can be found in Kieffer and Dunn (1966). Generally, single ionization takes place in the great majority of ionizing collisions, but at high energies a significant number of multiply charged ions is produced. The higher the degree of ionization, the higher the energy onset and the smaller the magnitude of the cross section. The preponderance of ionizing collisions, then, is seen to involve ejection of outer-shell electrons.

Multiply ionized species can usually be produced by two different basic mechanisms: (i) direct multiple ionization where two or more electrons are ejected by the impinging electron, and (ii) inner-shell ionization followed by one or more Auger processes. At large $\varepsilon$ the cross section $\sigma_{ni}(n \geqslant 2)$ has an energy dependence determined by either mechanism (i) or (ii) or by their superposition. It has been found experimentally (Monahan and Stanton, 1962, Fiquet-Fayard, 1965; Fiquet-Fayard and coworkers, 1968) that for direct double ionization the cross section $\sigma_{2i}$ varies as $\varepsilon^{-1}$. The formation of multiply charged ions by the Auger process (ii) yields $\sigma_{ni} \propto \varepsilon^{-1} \ln B\varepsilon$, since such ions are formed by the initial removal of only one electron (Schram, 1966b; Schram, Boerboom and Kistemaker, 1966; Fiquet-Fayard and coworkers, 1968). When both mechanisms (i) and (ii) contribute to the measured $\sigma_{ni}$, the energy dependence of $\sigma_{ni}$ will be a superposition of the energy dependences of (i) and (ii). Alternatively, in certain cases departure of $\sigma_{ni}$ from its $\varepsilon^{-1}$ dependence may be considered as evidencing significant Auger-effect contribution. Additionally, Fiquet-Fayard and coworkers (1968) presented experimental evidence for formation of doubly-charged molecular ions by an Auger process (e.g. $COS^{2+}$, $H_2S^{2+}$, $CS_2^{2+}$, $NH_3^{2+}$). This is an interesting observation since inner-shell ionization raises the energy of the molecule by more than it is necessary for its disruption and one would have expected the molecule to 'explode' under the action of this strong perturbation. Actually, extensive multiple ionization resulting in a highly charged molecular ion can lead to molecular decomposition. Such Coulombic molecular 'explosions' have been reported for heavy-atom containing molecules (Carlson and White, 1966; Carlson, Hunt and Krause, 1966). These latter observations in addition to their physical value are of practical interest, especially in the radiation damage of systems whose fate depends on a particular heavy atom which they contain (e.g. enzymes or chromosomes containing a heavy atom).

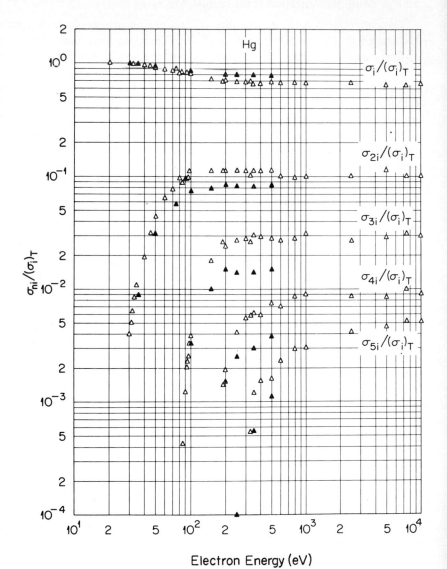

**Electron Energy (eV)**

Hg:  ▲ Bleakney (1930a). △ Harrison (1956). Data plotted
were taken from Kieffer (1965).

**Figure 5.31** Relative cross sections for the production

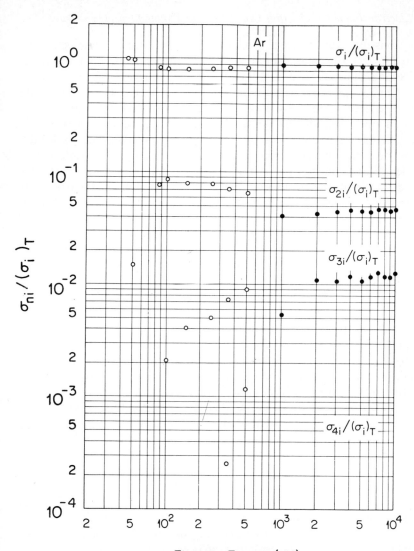

**Electron Energy (eV)**

Ar: O Bleakney (1930b). ● Schram and coworkers (quoted in Kieffer (1965)). Data plotted were taken from Kieffer (1965).

$^+$ ($n$ = 1 to 5) from Hg and $Ar^{n+}$ ($n=1$ to 4) from Ar.

### 5.5.2 *Electron Impact Ionization Cross Section Functions for Molecules*

Many possible transitions can arise in the ionization of molecules by electron impact. The energy of a vertical transition between the neutral molecule and the molecular ion potential energy curves according to the Franck–Condon principle (Section 5.1.3) yields the onset or appearance potential, referred to as the vertical 'ionization potential'. This quantity in many molecules exceeds the 'adiabatic ionization potential', namely the energy difference between the $v = 0$ states of the molecule and molecular ion, implying that the equilibrium internuclear distances are different in the molecule and the molecular ion (see Appendix II). Often, breaks are observed in the molecular ionization cross section functions particularly near threshold. Some of these discontinuities arise from distinct transitions to the residual positive ion, and are discussed in Section 5.5.4.

Mass-spectrometric analyses of polyatomic molecules, especially organic, have been made in abundance using commercially available instruments. Much of this work can be found in tables and books on mass spectrometry. Provided that certain complicating factors such as metastable ion production, ionization of free radicals, and ion–atom interchange processes are properly accounted for, mass spectrometric studies of ionic products arising from collisions of electrons with polyatomic molecules can reveal a great deal about molecular structure and the distribution of the excess energy among the many molecular degrees of freedom (see, for example, Reed (1962)). To this end, statistical approaches to the fragmentation (see, for example, Rosenstock and coworkers (1952) and Rosenstock and Krauss (1963)) of a complex molecule have been formulated. In such treatments the excess energy is assumed to be distributed radiationlessly among the many vibrational degrees of freedom and methods of statistical mechanics are applied to find an expression for the rate constant of the decomposition of a particular state. In many cases such treatments explained satisfactorily the experimental results. However much is still desirable.

Total ionization cross section functions are shown in Figure 5.32 for diatomic molecules and in Figure 5.33 for polyatomic molecules. The absolute ionization cross sections (including multiple ionization and ionic fragments) of Schram and coworkers (1965) are seen to be lower than the other results shown in Figure 5.32b (see Schram and coworkers (1966b) for similar data on hydrocarbons). The cross section values of Harrison (1956) for $H_2$ are considerably higher than the other data plotted. Molecular nitrogen and carbon monoxide are seen to have very close cross section functions; the two molecules have the same number of electrons. The shapes of the cross-section curves in Figures 5.32 and 5.33 are similar to one another especially at incident energies above $\sim 50$ eV. The same is seen to be true for the dissociative ionization cross section functions shown in Figure 5.34. The latter, however, are generally of lower magnitude. Absolute cross sections for simultaneous ionization and excitation of specific ionic states seem to be having similar shapes also. Recently McConkey, Burns and Woolsey (1968) using a crossed-beam apparatus found that the cross-section functions for the $^2\Sigma_u$ and $^2\Pi_u$ states of $CO_2^+$ are similar in shape to the total ionization cross section functions of Rapp and Englander–Golden (1965) for this molecule. The two cross section functions had a broad maximum at $\sim 150$ eV. Additional determinations of the cross sections for ionization into excited ionic states by measuring the radiation which accompanies radiative de-excitation of the ion have been made (see St. John and Lin (1964) for He and Stewart (1956), Sheridan, Oldenberg and Carleton (1961) and Hayakawa and Nishimura (1964) for $N_2$). Such studies do not actually yield the cross sections for ionization into a single excited ionic state but rather those for the production of radiation of a certain wavelength. Cascading and branching rations often complicate cross section function determinations this way.

Multiple ionization cross section functions for molecules are rare, although a number of doubly-charged molecular ions have been observed in electron-impact studies (e.g. Dorman and Morrison, 1961b; Herron and Dibeler, 1960). We may note that Daly and Powell (1966) obtained the cross-section function for $N_2^{2+}$ up to 240 eV, and reported a cross section function maximum of $3.9 \times 10^{-18}$ cm$^2$ at $\sim 120$ eV.

### 5.5.3 *Ionization of Positive Ions by Electron Impact*

Ionization cross section functions for the positive ions $He^+$, $Ne^+$ and $N^+$, obtained with crossed-beam techniques, are shown in Figure 5.35 Additional data can be found in Kieffer and Dunn (1965). The cross sections plotted are for the process

$$A^+ + e \rightarrow A^{2+} + 2e \qquad (5.32)$$

since mass analysis has been incorporated in these experiments (see references in figure caption). The cross section for process (5.32) is lower and the maximum of the cross section is shifted to higher energies compared to the analogous quantities for the neutral atom.*

The basic differences between ionization of a neutral atom and of a positive ion by electron impact arise from the Coulomb attractive field of the ion. The influence of the unscreened field of the ion nucleus, the ionic

---

*Positive ions may be studied by the trapped-ion method whereby positive ions are trapped in an electric field for times that are long enough for the trapped ion to be repeatedly struck by electrons in successive impacts.

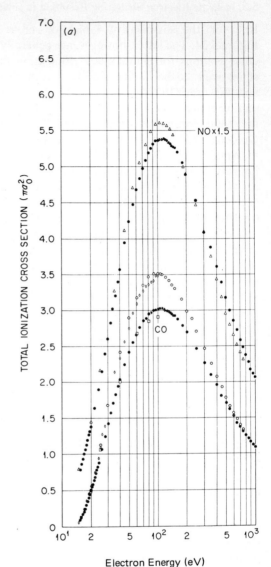

Figure 5.32 Total ionization cr◀

NO:[†]   △ Tate and Smith (1932).*
         ● Rapp and Englander—Golden (1965).
CO:      ○ Tate and Smith (1932).*
         ◇ Asundi, Craggs and Kurepa (1963).[‡]
         ● Rapp and Englander—Golden (1965).
         □ Srinivasan and Rees (1967).

* Data plotted were taken from Kieffer (1965).
† Note that the NO data have been multiplied by 1.5 for convenience of display.
‡ These data, according to Srinivasan and Rees (1967), should be multiplied by 0.95 to for correct for 'pumping errors'.

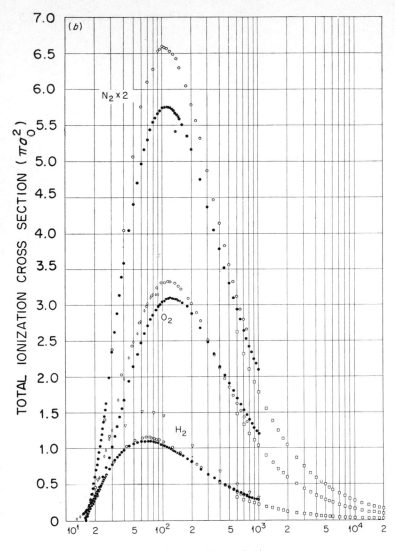

Electron Energy (eV)

H₂:   O Tate and Smith (1932).* ∇ Harrison (1965).*
      ● Rapp and Englander—Golden (1965). ☐ Schram and coworkers (1965).
N₂‡:   O Tate and Smith (1932).* ● Rapp and Englander—Golden (1965).
      See also Srinivasan, Rees, and Craggs (1967).
      ☐ Schram and coworkers (1965).
      O Tate and Smith (1932).* ◇ Asundi, Craggs, and Kurepa (1963).
      ● Rapp and Englander—Golden (1965). ☐ Schram and coworkers (1965).

tions for dia tomic molecules.

    * Data plotted were taken from Kieffer (1965).
    ‡ Note that data for N₂ have been multiplied by two for convenience of plotting.

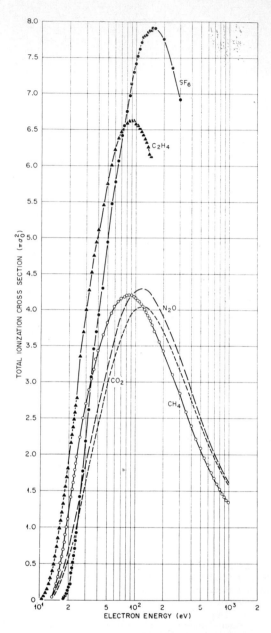

**Figure 5.33** Total ionization cross sections for polyatomic molecules (taken from Table VII of Rapp and Englander-Golden (1965))

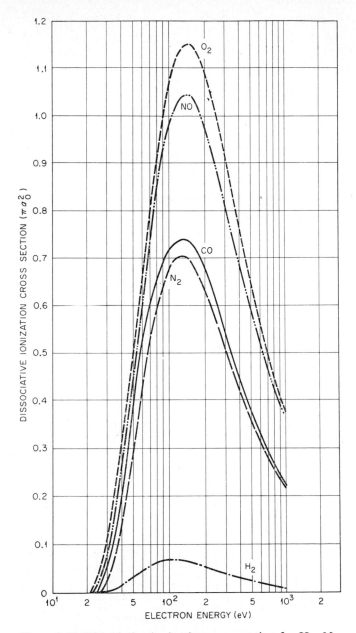

**Figure 5.34** Dissociative ionization cross section for $H_2$, $N_2$, $O_2$, NO and CO yielding product ions with kinetic energies greater than 0.25 eV (2.5 eV for $H_2$). These data were taken from tables given by Kieffer (1965) and are the results of Englander-Golden and Rapp (1964)

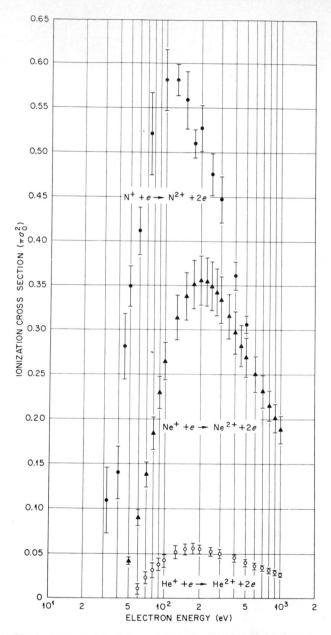

**Figure 5.35** Cross sections for ionization of singly charged positive atomic ions. He⁺: Dolder, Harrison and Thonemann (1961). Ne⁺: Dolder, Harrison and Thonemann (1963). N⁺: Harrison, Dolder and Thonemann (1963); the data plotted for N⁺ were taken from Kieffer (1965)

field, can clearly be seen by comparison of the cross sections for electron-impact ionization of H and $He^+$. Dolder, Harrison and Thonemann (1961) compared their experimental cross sections for removal of an electron by electron impact from the hydrogenic $He^+$ ion (Figure 5.35) with the electron impact ionization cross section for H measured by Fite and Brackmann (1958a) (Figure 5.25). They have found that when the $He^+$ cross sections are multiplied by the classical scaling factor $(I_2/I_1)^2$, where $I_1$ and $I_2$ are, respectively, the ionization potentials of H and $He^+$, the cross-section functions for $He^+$ and H are remarkably similar with the exception of the energy region around threshold where the $He^+$ cross section was found to increase more rapidly. Classically, this difference can be reconciled from the fact that the long range ionic field deflects and accelerates slow incident electrons towards $He^+$, thereby increasing the ionization cross section, while at high incident energies the electron trajectory is not appreciably perturbed. Born approximation calculations using Coulomb distorted waves (Burgess, 1960) gave cross-section functions in reasonable agreement with experiment at high energies, but they were rather high at lower impact energies.

### 5.5.4 Threshold Behaviour of and Structure in the Ionization Efficiency Curves of Atoms and Molecules

A great deal of research effort has been directed in recent years at discovering the threshold behaviour and identifying the structure in the ionization efficiency curves of atoms and molecules. However, in spite of recent progress in experimental techniques—especially as regards energy resolution—much of the existing information is still conflicting and much is to be wished.

Many threshold laws have been predicted for single ionization by electron impact (see review by Rudge, 1968). In the vicinity of the ionization threshold most of the theoretical approximations predict a power law energy dependence of the ionization cross section $\sigma_i$, viz.

$$\sigma_i \propto (\varepsilon - \varepsilon_i)^n \tag{5.33}$$

where $\varepsilon$ is the incident electron energy and $\varepsilon_i$ is the ionization threshold energy. Depending upon the assumptions chosen, the value of the exponent $n$ is found to be either one (Rudge and Seaton, 1964, 1965; Rudge and Schwartz, 1966; Geltman, 1956; Peterkop, 1961) or 1.127 (Wannier, 1953), or 1.5 (Omidvar, 1965; Temkin, 1966). A more complex threshold law has been reported by Omidvar (1967). A first power threshold ionization efficiency law has been given earlier by Bates and coworkers (1950) in the first Born approximation. For single ionization of ions Wannier (1953) predicts a value for the exponent $n$ between 1.127 and 1. The formation of an ion with $n$ charges is supposed to follow an $n$th power law. None of the theoretical treatments can tell us over what energy range a threshold law strictly applies. Such information will have to come from experiment. Let us now summarize very briefly the experimental information on the subject.

For atomic hydrogen Fite and Brackmann (1958a) reported a linear variation of the cross section function up to ~5 eV. McGowan and Clarke (1968) found, with better energy resolution (~0.06 eV), that the apparent linearity of the cross section for H extends down to ~0.4 eV, but below this energy it is non-linear. Their results below ~0.4 eV could well be described by a $1.13 \pm 0.03$ power law although within ~0.05 eV of the ionization threshold a higher power law would be consistent with their measurements. These results clearly show that the threshold behaviour of $\sigma_i$ is a complicated function of energy. Similar recent work on single ionization of He by electron impact by Brion and Thomas (1968) shown in Figure 5.36a indicates a non-linear dependence for the He ionization cross section as a function of electron energy above threshold. One of the two curves shown in Figure 5.36a gives the experimental results of Brion and Thomas and the other represents the energy dependence of the ionization cross section for a 1.127 power law. The two curves have been normalized at 10 eV above threshold and fit each other closely at higher energies but a power law in excess of 1.127 is indicated nearer to threshold. Additionally,

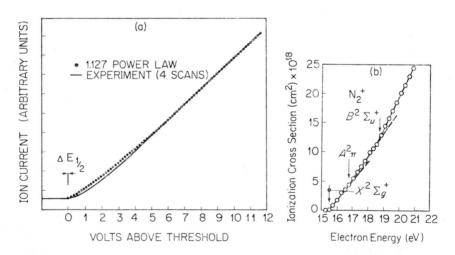

**Figure 5.36** (a) Ionization efficiency curve for $He^+$ (Brion and Thomas, 1968). The energy resolution of Brion and Thomas is seen from $\Delta E_{\frac{1}{2}}$ shown. (b) Ionization efficiency curve for $N_2^+$ (Fox, 1961). The arrows indicate the designated energy states of $N_2^+$ determined from spectroscopic data. Absolute cross sections shown were obtained by Fox by comparison of his relative results with the absolute data of Tate and Smith (1932)

Marchand, Paquet and Marmet (1969) found that the best fitting power law for the He ionization cross section from threshold to several energies between 0.4 and 12 eV varied from 1.17 to 1 as the energy range of fitting was increased. A power law of 1.17 close to the threshold is consistent with their measurements while power fitting to within 12 eV of threshold yielded a power law of $1.02 \pm 2\%$. These results are consistent with those of Brion and Thomas and contradict earlier studies (e.g. Fox and coworkers, 1955; Morrison, 1953) which showed a first power law for $He^+$. In the early studies, however, the energy resolution was inferior. Marchand, Paquet and Marmet reported that their results for Ar could be represented by two power laws with powers 1.3 and 1.34, respectively, joining at the energy of the $^2P_{\frac{1}{2}}$ level of $Ar^+$.

Multiple ionization of the rare gases has been shown to be consistent with the $n$th power threshold law (see, for example, Dorman, Morrison and Nicholson (1959); Dorman and Morrison (1961b); Morrison and Nicholson (1959); Krauss, Reese and Dibeler (1959); Kiser (1962)). Thus the double ionization of He has been found to vary in accordance with the square of the excess energy above threshold (Fox, 1959; Krauss, Reese and Dibeler 1959). Linear and square power laws for single and double ionization, respectively, of Na and K have also been demonstrated (Kaneko, 1961). Clarke (1954) has reported a threshold law for $Xe^{2+}$ consistent with a square-law dependence on excess energy above threshold, although Fox (1959, 1960b) and Hickam, Fox and Kjeldaas (1953) found that a series of straight-line segments fit their data most satisfactorily. Similarly, Blais and Mann (1960) fitted straight-line segments to double ionization of Au, but Dorman and Morrison (1961b) indicated that a plot of the square root of ionization efficiency versus energy is also a straight line over at least 10 eV, within experimental error. Double ionization of NO, CO, $N_2$, and of some polyatomic molecules has been found by Dorman and Morrison (1961a) to increase in a fashion which is consistent with a second power law. For a discussion of other data see this reference and Kiser (1962). One should, of course, keep in mind that higher ionization cross section functions may be complicated by effects arising from excited ions and superexcitation. Such complications are absent for $H^+$ and $He^{2+}$. Although more work is needed in this area one may conclude that there may be cases for which the $n$th power law holds and others for which it is masked.

Let us now turn our attention to the structure in the ionization efficiency curves for atoms and molecules. Here as in the case of threshold laws the experimental results are often contradictory and inconclusive, but the existence of ionization mechanisms other than direct transitions of electrons to the first continuum has been well established. The relative importance of these additional mechanisms is still not fully known but the subject is of current interest. In general three basic mechanisms can contribute to the structure in the ionization efficiency curve. Firstly, excitation of high lying (excited) ionic states; the cross-section function for an excited singly-charged ion most probably rises linearly from the appropriate threshold as that of the ground-state ion. Secondly, excitation of superexcited atomic and molecular states. Such states are subject to autoionization and the energy dependence of their cross section is characteristic of autoionizing levels (Chapters 3 and 4). Thirdly, Auger processes can contribute to the structure in the ionization efficiency curve. For molecules dissociative transitions also contribute to the structure in the ionization efficiency curve. All three mechanisms mentioned above have been invoked at one instance or another to interpret breaks in the ionization efficiency curves for a variety of systems both atomic and molecular. Structure in the rare gas cross section functions, for example, has been interpreted as due to excited ionic states (see, for example, Hickam, Fox and Kjeldaas, 1954; Fox, 1961; Morrison, 1964), and autoionizing states (e.g. Kr and Xe; Burns, 1964; Morrison, 1964; Fox, Hickam and Kjeldaas, 1953). Also, Auger transitions have been conclusively found to be responsible for anomalies in the shapes of multiple ionization curves (e.g. Fox, 1959; Fiquet-Fayard and Lahmani, 1962; Fiquet-Fayard and coworkers, 1968). The structure in the ionization efficiency curves for a number of molecules has in many cases been correlated with spectroscopically identified levels. An example is shown in Figure 5.36b for $N_2^+$. The first linear portion of the cross section function curve is attributed to ionization arising from excitation to the $X^2\Sigma$ ground ionic state, while the first and second breaks are ascribed, respectively, to the ionization thresholds of the $A^2\Pi$ and $B^2\Sigma$ excited ionic states. There is no doubt that in many molecules anomalies in the ionization efficiency curves are caused by superexcitation. A recent study of direct and auto-ionization processes near threshold for $H_2$ has been made by McGowan and coworkers (1968). Also recently Brion (1969) correlated the fine structure he observed in the threshold region of the ionization efficiency curve for acetylene with known vibrational levels of this molecule.

### 5.5.5 *Angular and Energy Distribution of Electrons after Ionization*

The presence of two outgoing electrons in an ionizing electron–atom (molecule) collision complicates the analyses of such measurements. The excess energy $(\varepsilon - I)$ is distributed between the two outgoing electrons unequally. One of the outgoing electrons carries away most of the energy, while the other is slow. Little information exists on this matter outside the early work summarized and discussed in Massey and Burhop (1952). A recent experimental set-up by Ehrhardt and coworkers promises valuable information, in this respect. In Figure 5.37 we reproduce some of the data collected by Massey and Burhop which show the principal features observed

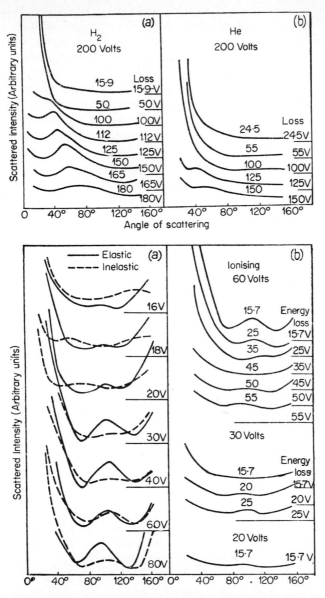

**Figure 5.37** *Upper portion.* Angular distribution in scattering of fast electrons in ionizing collisions (a) in $H_2$ and (b) in He. *Lower portion.* Observed angular distribution of electrons scattered in argon. (a) Elastic (———) and inelastic (– – – –) collisions involving excitation of the $^3P_1$ level in which the energy loss is 11.6 eV. (b) Ionizing collisions with different energy losses

in the angular distribution of ionizing collisions in which the electrons have suffered various amounts of energy loss. The scattered intensity is seen to peak in the forward direction for small energy losses, but it becomes more uniform as the energy loss increases. Note also the similarities in structure at large angles between the elastically scattered electrons and the inelastically scattered ones involving small energy losses. A resemblance between the distribution curves for electrons which have suffered different energy losses in ionizing collisions with those for elastic scattering is evident for low-energy losses, but this similarity becomes less apparent as the energy loss in the ionizing collision increases.

Experimental information as to the energy spectrum of the electrons after an ionizing collision is lacking since it is impossible to distinguish between the two outgoing electrons. According to McDaniel (1964) if the energy of the incident electron is less than a few times the threshold energy $I$, all of the ejected electrons have energies no more than a few eV, whereas at incident energies of several keV about half of the ejected electrons have energies greater than $I$. These conclusions are supported by measurements of kinetic energies of electrons ejected in ionizing proton–atom collisions.

It is, finally, noted that in an ionizing collision only a small amount of momentum is transferred to the target, and the ions produced in simple ionizing collisions have essentially the same kinetic energy as the original targets. However, in dissociative ionization internal energy is carried away as kinetic energy of the fragments (Section 5.1.3). The considerable work devoted for the measurements of the energy distribution of ionic fragments in dissociative ionization has been discussed by Massey and Burhop (1952), Craggs and Massey (1959) and Field and Franklin (1957). Dunn (1962) argued that the angular distributions of the dissociation products are anisotropic in a majority of cases (see also a discussion on dissociative ionization of $H_2$ by Dunn and Kieffer (1963)).

## 5.6 References

R. A. Abram and A. Herzenberg (1969). *Chem. Phys. Letters*, **3**, 187.

R. J. Anderson, E. T. P. Lee and C. C. Lin (1967). *Phys. Rev.*, **157**, 31.

D. Andrick and H. Ehrhardt (1966). *Z. Physik*, **192**, 99.

R. K. Asundi, J. D. Craggs and M. V. Kurepa (1963). *Proc. Phys. Soc. (London)*, **82**, 967.

R. K. Asundi and M. V. Kurepa (1963). *J. Electronics Control*, **15**, 41.

R. M. Badger, A. C. Wright and R. F. Whitlock (1965). *J. Chem. Phys.*, **43**, 4345.

D. J. Baker, D. E. Bedo and D. H. Tomboulian (1961). *Phys. Rev.*, **124**, 1471.

J. N. Bardsley, A. Herzenberg and F. Mandl (1966). *Proc. Phys. Soc. (London)*, **89**, 321.

J. N. Bardsley and F. Mandl (1968). *Rep. Progr. Phys.*, **31**, 471.

J. N. Bardsley, F. Mandl and A. R. Wood (1967). *Chem. Phys. Letters*, **1**, 359.

J. N. Bardsley and F. H. Read (1968). *Chem. Phys. Letters*, **2**, 333.

D. R. Bates (1962). *Atomic and Molecular Processes*, Academic Press, New York.

D. R. Bates, A. Fundaminsky, J. W. Leech and H. S. W. Massey (1950). *Phil. Trans. Roy. Soc.*, **243**, 93.

H. Bethe (1930). *Ann. Physik*, **5**, 325.

J. B. Birks, L. G. Christophorou and R. H. Huebner (1968). *Nature*, **217**, 809.

N. C. Blais and J. B. Mann (1960). *J. Chem. Phys.*, **33**, 100.

W. Bleakney (1930a). *Phys. Rev.*, **35**, 139.

W. Bleakney (1930b). *Phys. Rev.*, **36**, 1303.

H. Boersch, J. Geiger and W. Stickel (1964). *Z. Physik*, **180**, 415.

A. Boksenberg, (1961). Ph.D. Thesis, University of London ; in Kieffer (1965).

M. J. W. Boness and J. B. Hasted (1966). *Phys. Letters*, **21**, 526.

M. J. W. Boness, J. B. Hasted and I. W. Larkin (1968). *Proc. Roy. Soc. (London)*, **A305**, 493.

M. J. W. Boness, I. W. Larkin, J. B. Hasted and L. Moore (1967). *Chem. Phys. Letters*, **1**, 292.

M. J. W. Boness and G. J. Schulz (1968). *Phys. Rev. Letters*, **21**, 1031.

C. R. Bowman and W. D. Miller (1965). *J. Chem. Phys.*, **42**, 681.

E. L. Breig and C. C. Lin (1965). *J. Chem. Phys.*, **43**, 3839.

G. O. Brink (1962). *Phys. Rev.*, **127**, 1204.

G. O. Brink (1964). *Phys. Rev.*, **134**, A345.

C. E. Brion (1969). *Chem. Phys. Letters*, **3**, 9.

C. E. Brion and G. E. Thomas (1968). *Phys. Rev. Letters*, **20**, 241.

H. H. Brongersma, J. A. v.d. Hart and L. J. Oosterhoff (1967). *Nobel Symposium*, **5**, 211.

H. H. Brongersma and L. J. Oosterhoff (1967). *Chem. Phys. Letters*, **1**, 169.

A. Burgess (1960). *Astrophys. J.*, **132**, 503.

P. G. Burke (1965). *Advan. Phys.*, **14**, 521.

P. G. Burke, H. M. Schey and K. Smith (1963). *Phys. Rev.*, **129**, 1258.

J. F. Burns (1964). In M. R. C. McDowell (Ed.), *Atomic Collision Processes*, North-Holland Publishing Co., Amsterdam, p. 451.

T. A. Carlson, W. E. Hunt and M. O. Krause (1966). *Phys. Rev.*, **151**, 41.

T. A. Carlson and R. M. White (1966). *J. Chem. Phys.*, **44**, 4510.

T. R. Carson (1954). *Proc. Phys. Soc. (London)*, **A67**, 909.

V. Čermák (1966). *J. Chem. Phys.*, **44**, 3774.

G. E. Chamberlain (1967). *Phys. Rev.*, **155**, 46.

G. E. Chamberlain and H. G. M. Heideman (1965). *Phys. Rev. Letters*, **15**, 337.

G. E. Chamberlain, J. A. Simpson, S. R. Mielczarek and C. E. Kuyatt (1967). *J. Chem. Phys.*, **47**, 4266.

P. J. Chantry, A. V. Phelps and G. J. Schulz (1966). *Phys. Rev.*, **152**, 81.

J. C. Y. Chen (1964a). *J. Chem. Phys.*, **40**, 3507, 3513.

J. C. Y. Chen (1964b). *Phys. Letters*, **8**, 183.

J. C. Y. Chen (1966). *J. Chem. Phys.*, **45**, 2710.

L. G. Christophorou and J. G. Carter (1968). *Chem. Phys. Letters*, **2**, 607.

L. G. Christophorou and R. N. Compton (1967). *Health Phys.*, **13**, 1277.

L. G. Christophorou, R. N. Compton, G. S. Hurst and P. W. Reinhardt (1965). *J. Chem. Phys.*, **43**, 4273.

E. M. Clarke (1954). *Can. J. Phys.*, **32**, 764.

S. D. Colson and E. R. Bernstein (1965). *J. Chem. Phys.*, **43**, 2661.

S. D. Colson and E. R. Bernstein (1966). *J. Chem. Phys.*, **45**, 3873.

R. N. Compton, L. G. Christophorou and R. H. Huebner (1966). *Phys. Letters*, **23**, 656.

R. N. Compton, R. H. Huebner, P. W. Reinhardt and L. G. Christophorou (1968). *J. Chem. Phys.*, **48**, 901.

T. L. Cottrell (1958). *The Strengths of Chemical Bonds*, 2nd ed., Butterworths, London.

T. L. Cottrell and I. C. Walker (1965). *Trans. Faraday Soc.*, **61**, 1585.

J. D. Craggs and H. S. W. Massey (1959). In S. Flügge (Ed.), *Handbuch der Physik*, Vol. XXXVII, Springer-Verlag, Berlin, pp. 314–415.

J. D. Craggs and C. A. McDowell (1955). *Rept. Prog. Phys.*, **18**, 374.

C. K. Crawford and K. I. Wang (1967). *J. Chem. Phys.*, **47**, 4667.

R. W. Crompton and A. I. McIntosh (1967). *Phys. Rev. Letters*, **18**, 527.

R. W. Crompton, M. T. Elford and A. I. McIntosh (1968). *Australian J. Phys.*, **21**, 43.

R. K. Curran (1963). *J. Chem. Phys.*, **38**, 780.

A. Dalgarno and R. J. W. Henry (1965). *Proc. Phys. Soc. (London)*, **85**, 679.

A. Dalgarno and R. J. Moffett (1963). *Proc. Nat. Acad. Sci. India*, **A33**, 511.

N. R. Daly and R. E. Powell (1966). *Proc. Phys. Soc. (London)*, **89**, 273.

J. P. Doering (1967). *J. Chem. Phys.*, **46**, 1197.

J. P. Doering and A. J. Williams III (1967). *J. Chem. Phys.*, **47**, 4180.

K. T. Dolder, M. F. A. Harrison and P. C. Thonemann (1961). *Proc. Roy. Soc. (London)*, **A264**, 367.

K. T. Dolder, M. F. A. Harrison and P. C. Thonemann (1963). *Proc. Roy. Soc. (London)*, **A274**, 546.

F. H. Dorman and J. D. Morrison (1961a). *J. Chem. Phys.*, **35**, 575.

F. H. Dorman and J. D. Morrison (1961b). *J. Chem. Phys.*, **34**, 1407.

F. H. Dorman, J. D. Morrison and A. J. C. Nicholson (1959). *J. Chem. Phys.*, **31**, 1335.

R. Dorrestein (1942). *Physica*, **9**, 447.

J. T. Dowell and T. E. Sharp (1967). *J. Chem. Phys.*, **47**, 5068.

J. L. G. Dugan, H. L. Richards and E. E. Muschliz, Jr. (1967). *J. Chem. Phys.*, **46**, 346.

G. H. Dunn (1962). *Phys. Rev. Letters*, **8**, 62.

G. H. Dunn and L. J. Kieffer (1963). *Phys. Rev.*, **132**, 2109.

D. L. Ederer and D. H. Tomboulian (1964). *Phys. Rev.*, **133**, A1525.

H. Ehrhardt and F. Linder (1969). *Phys. Rev. Letters*, **21**, 419.

H. Ehrhardt, M. Schulz, T. Tekaat and K. Willmann (1969). *Phys. Rev. Letters*, **22**, 89.

H. Ehrhardt, L. Langhans, F. Linder and H. S. Taylor (1968). *Phys. Rev.*, **173**, 222.

H. Ehrhardt and K. Willmann (1967). *Z. Physik*, **204**, 462.

A. G. Engelhardt and A. V. Phelps (1963). *Phys. Rev.*, **131**, 2115.

A. G. Engelhardt, A. V. Phelps and C. G. Risk (1964). *Phys. Rev.*, **135**, A1566.

P. Englander-Golden and D. Rapp (1964). Lockheed Missiles and Space Company Report No. LMSC 6-74-64-12, Palo Alto, California.

F. H. Field and J. L. Franklin (1957). *Electron Impact Phenomena*, Academic Press, New York.

F. Fiquet-Fayard (1965). *J. Chim. Phys.*, **62**, 1065.

F. Fiquet-Fayard, J. Chiari, F. Muller and J-P. Ziezel (1968). *J. Chem. Phys.*, **48**, 478.

F. Fiquet-Fayard and M. Lahmani (1962). *J. Chim. Phys.*, **59**, 1050.

L. Fisher, S. N. Milford and F. R. Pomilla (1960). *Phys. Rev.*, **119**, 153.

W. L. Fite (1962). In D. R. Bates (Ed.), *Atomic and Molecular Processes*, Academic Press, New York, pp. 421–492.

W. L. Fite and R. T. Brackmann (1958a). *Phys. Rev.*, **112**, 1141.

W. L. Fite and R. T. Brackmann (1958b). *Phys. Rev.*, **112**, 1151.

W. L. Fite and R. T. Brackmann (1959). *Phys. Rev.*, **113**, 815.

W. L. Fite, R. F. Stebbings and R. T. Brackmann (1959). *Phys. Rev.*, **116**, 356.

R. J. Fleming and G. S. Higginson (1964). *Proc. Phys. Soc. (London)*, **84**, 531.

M. A. Fox (1965). *Proc. Phys. Soc. (London)*, **86**, 789.

R. E. Fox (1959). In J. D. Waldron (Ed.), *Advances in Mass Spectrometry*, Pergamon Press, London, pp. 397–412.

R. E. Fox (1960a). *J. Chem. Phys.*, **32**, 385.

R. E. Fox (1960b). *J. Chem. Phys.*, **33**, 200.

R. E. Fox (1961). *J. Chem. Phys.*, **35**, 1379.

R. E. Fox and W. M. Hickam (1954). *J. Chem. Phys.*, **22**, 2059.

R. E. Fox, W. M. Hickam, D. J. Grove and T. Kjeldaas, Jr. (1955). *Rev. Sci. Instr.*, **26**, 1101.

R. E. Fox, W. M. Hickam and T. Kjeldaas, Jr. (1953). *Phys. Rev.*, **89**, 555.

L. S. Frost and A. V. Phelps (1962). *Phys. Rev.*, **127**, 1621.

J. Geiger and B. Schröder (1968). *J. Chem. Phys.*, **49**, 740.

J. Geiger and W. Stickel (1965). *J. Chem. Phys.*, **43**, 4535.

J. Geiger and K. Wittmaack (1965a). *Z. Physik*, **187**, 433.

J. Geiger and K. Wittmaack (1965b). *Z. Naturforsch.*, **20a**, 628.

S. Geltman (1956). *Phys. Rev.*, **102**, 171.

S. Geltman and K. Takayanagi (1966). *Phys. Rev.*, **143**, 25.

E. Gerjuoy and S. Stein (1955a). *Phys. Rev.*, **97**, 1671.

E. Gerjuoy and S. Stein (1955b). *Phys. Rev.*, **98**, 1848.

F. R. Gilmore (1965). *J. Quant. Spec. Rad. Trans.*, **5**, 369.

V. E. Golant (1961). *Soviet Phys.—Tech. Phys.*, **5**, 1197.

D. E. Golden (1966). *Phys. Rev. Letters*, **17**, 847.

D. E. Golden and H. W. Bandel (1965a). *Phys. Rev.*, **138**, A14.

D. E. Golden and H. W. Bandel (1965b). *Phys. Rev. Letters*, **14**, 1010.

D. E. Golden, H. W. Bandel and J. A. Salerno (1966). *Phys. Rev.*, **146**, 40.

M. Gryzinski (1959). *Phys. Rev.*, **115**, 374.

M. Gryzinski (1965a). *Phys. Rev.*, **A138**, 305.

M. Gryzinski (1965b). *Phys. Rev.*, **A138**, 322.

M. Gryzinski (1965c). *Phys. Rev.*, **A138**, 336.

M. Gryzinski (1965d). *Phys. Rev. Letters*, **14**, 1059.

R. Haas (1957). *Z. Physik*, **148**, 177.

R. D. Hake and A. V. Phelps (1967). *Phys. Rev.*, **158**, 70.

H. Harrison (1956). Thesis, Catholic University Press, Washington, D. C.

M. F. A. Harrison, K. T. Dolder and P. C. Thonemann (1963). *Proc. Phys. Soc. (London)*, **82**, 368.

J. B. Hasted (1964). *Physics of Atomic Collisions*, Butterworths, London.

J. B. Hasted and A. M. Awan (1969). *J. Phys. B (Proc. Phys. Soc.)*, **2**, 367.

S. Hayakawa and N. Nishimura (1964). *J. Geomagnetism and Geoelectricity (Japan)*, **16**, 72.

H. G. M. Heideman, C. E. Kuyatt and G. E. Chamberlain (1966a). *J. Chem. Phys.*, **44**, 440.

H. G. M. Heideman, C. E. Kuyatt and G. E. Chamberlain (1966b). *J. Chem. Phys.*, **44**, 355.

J. T. Herron and V. H. Dibeler (1960). *J. Chem. Phys.*, **32**, 1884.

G. Herzberg (1951). *Molecular Spectra and Molecular Structure*. I. *Spectra of Diatomic Molecules*, 2nd ed., Van Nostrand, New York.

A. Herzenberg and F. Mandl (1962). *Proc. Roy. Soc. (London)*, **A270**, 48.

W. M. Hickam, R. E. Fox and T. Kjeldaas, Jr. (1954). *Phys. Rev.*, **96**, 63.

H. K. Holt and R. Krotkov (1966). *Phys. Rev.*, **144**, 82.
J. W. Hooper, D. S. Harmer, D. W. Martin and E. W. McDaniel (1962). *Phys. Rev.*, **125**, 2000.
J. W. Hooper, E. W. McDaniel, D. W. Martin and D. S. Harmer (1961). *Phys. Rev.*, **121**, 1123.
R. H. Huebner, R. N. Compton, L. G. Christophorou and P. W. Reinhardt (1968). Oak Ridge National Laboratory Report No. ORNL-TM-2156, Oak Ridge, Tennessee.
R. E. Huffman, Y. Tanaka and J. C. Larrabee (1963). *J. Chem. Phys.*, **39**, 910.
M. Inokuti (1958). *J. Phys. Soc. Japan*, **13**, 537.
T. J. Jones (1927). *Phys. Rev.*, **29**, 822.
A. V. Jones and A. W. Harrison (1958). *J. Atm. Terr. Phys.*, **13**, 45.
Y. Kaneko (1961). *Proc. Phys. Soc. Japan*, **16**, 2288.
L. J. Kieffer (1965). Joint Institute for Laboratory Astrophysics Report No. JILA 30, Boulder, Colorado.
L. J. Kieffer and G. H. Dunn (1966). *Rev. Mod. Phys.*, **38**, 1.
R. W. Kiser (1962). *J. Chem. Phys.*, **36**, 2964.
O. Klemperer (1965). *Rep. Prog. Phys.*, **28**, 77.
M. Krauss, R. M. Reese and V. H. Dibeler (1959). *J. Res. Natl. Bur. Std. (U. S.)*, **63A**, 201.
G. J. Krige, S. M. Gordon and P. C. Haarhoff (1968). *Z. Naturforsch.*, **23a**, 1383.
A. Kuppermann (1970). Paper presented at the *IVth International Congress of Radiation Research*, Evian, France, June 29–July 4, 1970.
A. Kuppermann and L. M. Raff (1962). *J. Chem. Phys.*, **37**, 2497.
A. Kuppermann and L. M. Raff (1963). *Discussions Faraday Soc.*, **35**, 30.
C. E. Kuyatt, S. R. Mielczarek and J. A. Simpson (1964). *Phys. Rev. Letters*, **12**, 293.
C. E. Kuyatt, J. A. Simpson and S. R. Mielczarek (1965). *Phys. Rev.*, **138**, A385.
N. F. Lane and S. Geltman (1967). *Phys. Rev.*, **160**, 53.
H. Larzul, F. Gélébart and A. Johannin-Gilles (1965). *Compt. Rend. Paris*, **261**, 4701.
E. N. Lassettre (1959). *Rad. Res. Suppl.*, **1**, 530.
E. N. Lassettre (1969). *Can. J. Chem.*, **47**, 1733.
E. N. Lassettre, A. S. Berman, S. M. Silverman and M. E. Krasnow (1964). *J. Chem. Phys.*, **40**, 1232.
E. N. Lassettre and S. A. Francis (1964). *J. Chem. Phys.*, **40**, 1208.
E. N. Lassettre, F. M. Glaser, V. D. Meyer and A. Skerbele (1965). *J. Chem. Phys.*, **42**, 3429.
E. N. Lassettre and E. A. Jones (1964). *J. Chem. Phys.*, **40**, 1218.
E. N. Lassettre, M. E. Krasnow and S. M. Silverman (1964). *J. Chem. Phys.*, **40**, 1242.
E. N. Lassettre and S. M. Silverman (1964). *J. Chem. Phys.*, **40**, 1256.
E. N. Lassettre, S. M. Silverman and M. E. Krasnow (1964). *J. Chem. Phys.*, **40**, 261.
E. N. Lassettre, A. Skerbele and M. A. Dillon (1968). *J. Chem. Phys.*, **49**, 2382.
E. N. Lassettre, A. Skerbele, M. A. Dillon and K. J. Ross (1968). *J. Chem. Phys.*, **48**, 5066.
D. Lewis and W. H. Hamill (1969). *J. Chem. Phys.*, **51**, 456.
J. W. Liska (1934). *Phys. Rev.*, **46**, 169.
R. L. Long, Jr., D. M. Cox and S. J. Smith (1968). *J. Res. Natl. Bur. Std. (U. S.)*, **72A**, 521.
H. Maier-Leibnitz (1935). *Z. Physik*, **95**, 499.

P. Marchand, C. Paquet and P. Marmet (1969). *Phys. Rev.*, **180**, 123.

P. Marmet and L. Kerwin (1960). *Can. J. Phys.*, **38**, 787.

H. S. W. Massey (1935). *Trans. Faraday Soc.*, **31**, 556.

H. S. W. Massey (1956). In S. Flügge (Ed.), *Handbuch der Physik*, Springer-Verlag, Berlin, pp. 307–408.

H. S. W. Massey and E. H. S. Burhop (1952). *Electronic and Ionic Impact Phenomena*, Clarendon Press, Oxford, see also 1969 edition.

M. Matsuzawa (1963). *J. Phys. Soc. Japan*, **18**, 1473.

J. W. McConkey, D. J. Burns and J. M. Woolsey (1968). *J. Phys. B. (Proc. Phys. Soc.)*, **1**, 71.

E. W. McDaniel (1964). *Collision Phenomena in Ionized Gases*, John Wiley and Sons, New York.

R. H. McFarland and J. D. Kinney (1965). *Phys. Rev.*, **137**, A1058.

J. W. McGowan and E. M. Clarke (1968). *Phys. Rev.*, **167**, 43.

J. W. McGowan, M. A. Fineman, E. M. Clarke and H. P. Hanson (1968). *Phys. Rev.*, **167**, 52.

M. G. Menendez and H. K. Holt (1966). *J. Chem. Phys.*, **45**, 2743.

M. H. Mentzoni and R. V. Row (1963). *Phys. Rev.*, **130**, 2312.

V. D. Meyer and E. N. Lassettre (1965). *J. Chem. Phys.*, **42**, 3436.

V. D. Meyer and E. N. Lassettre (1966). *J. Chem. Phys.*, **44**, 2535.

V. D. Meyer, A. Skerbele and E. N. Lassettre (1965). *J. Chem. Phys.*, **43**, 805.

B. L. Moiseiwitsch and S. J. Smith (1968). *Rev. Mod. Phys.*, **40**, 238.

J. E. Monahan and H. E. Stanton (1962). *J. Chem. Phys.*, **37**, 2654.

J. D. Morrison (1953). *J. Chem. Phys.*, **21**, 1767.

J. D. Morrison (1964). *J. Chem. Phys.*, **40**, 2488.

J. D. Morrison and A. J. C. Nicholson (1959). *J. Chem. Phys.*, **31**, 1320.

P. M. Morse (1953). *Phys. Rev.*, **90**, 51.

N. F. Mott and H. S. W. Massey (1965). *The Theory of Atomic Collisions*, 3rd ed., Clarendon Press, Oxford.

K. J. Nygaard (1968). *J. Chem. Phys.*, **49**, 1995.

J. III, Olmsted (1967). *Rad. Res.*, **31**, 191.

J. III, Olmsted, A. S. Newton and K. Street, Jr. (1965). *J. Chem. Phys.*, **42**, 2321.

T. O'Malley and H. S. Taylor (1968). *Phys. Rev.*, **176**, 207.

K. Omidvar (1965). *Phys. Rev.*, **140**, A26.

K. Omidvar (1967). *Phys. Rev. Letters*, **18**, 153.

S. S. Penner (1959). *Quantitative Molecular Spectroscopy and Gas Emissitivities*, Addison-Wesley Co., Inc., Reading, Massachusetts, p. 53.

R. K. Peterkop (1961). *Proc. Phys. Soc. (London)* A**77**, 1220.

J. R. Peterson (1964). In M. R. C. McDowell (Ed.), *Atomic Collision Processes*, North-Holland Publishing Co., Amsterdam, p. 465.

A. V. Phelps (1968). *Rev. Mod. Phys.*, **40**, 399.

F. M. J. Pichanick and J. A. Simpson (1968). *Phys. Rev.*, **168**, 64.

W. C. Price (1938). *Proc. Roy. Soc. (London)*, **167**, 216.

H. Ramien (1931). *Z. Physik*, **70**, 353.

C. Ramsauer and R. Kollath (1930). *Ann. Physik*, **4**, 91.

D. Rapp and P. Englander-Golden (1965). *J. Chem. Phys.*, **43**, 1464.

F. H. Read (1968). *J. Phys. B (Proc. Phys. Soc.) (ser. 2)* **1**, 893.

F. H. Read and G. L. Whiterod (1965). *Proc. Phys. Soc. (London)*, **85**, 71.

R. I. Reed (1962). *Ion Production by Electron Impact*, Academic Press, New York.

J. K. Rice, A. Kuppermann and S. Trajmar (1968). *J. Chem. Phys.*, **48**, 945.

H. M. Rosenstock, M. B. Wallenstein, A. L. Wahrhaftig and E. H. Eyring (1952). *Proc. Natl. Acad. Sci. U. S.*, **38**, 667.

H. M. Rosenstock and M. Krauss (1963). In R. M. Elliot (Ed.), *Advances in Mass Spectrometry*, Vol. II, Pergamon Press Inc., New York, pp. 251–284.

K. J. Ross and E. N. Lassettre (1966). *J. Chem. Phys.*, **44**, 4633.

E. W. Rothe, L. L. Marino, R. H. Neynaber and S. M. Trujillo (1962). *Phys. Rev.*, **125**, 582.

M. R. H. Rudge (1968). *Rev. Mod. Phys.*, **40**, 564.

M. R. H. Rudge and S. B. Schwartz (1966). *Proc. Phys. Soc. (London)*, **88**, 563, 579.

M. R. H. Rudge and M. J. Seaton (1964). *Proc. Phys. Soc. (London)*, **83**, 680.

M. R. H. Rudge and M. J. Seaton (1965). *Proc. Roy. Soc. (London)*, **A283**, 262.

D. H. Sampson and R. C. Mjolsness (1965). *Phys. Rev.*, **140**, A1466.

J. A. R. Samson (1964). *J. Opt. Soc. Am.*, **54**, 421.

B. L. Schram (1966a). Ph. D. Thesis, University of Amsterdam.

B. L. Schram (1966b). *Physica*, **32**, 197.

B. L. Schram, A. J. H. Boerboom and J. Kistemaker (1966). *Physica*, **32**, 185.

B. L. Schram, A. J. H. Boerboom, M. J. van der Wiel, F. J. de Heer and J. Kistemaker (1966a). In W. L. Mead (Ed.), *Advances in Mass Spectrometry*, Vol. 3, The Institute of Petroleum, London, p. 273.

B. L. Schram, F. J. de Heer, M. J. van der Wiel and J. Kistemaker (1965). *Physica*, **31**, 94.

B. L. Schram, M. J. van der Wiel, F. J. de Heer and H. R. Moustafa (1966b). *J. Chem. Phys.*, **44**, 49.

G. J. Schulz (1958). *Phys. Rev.*, **112**, 150.

G. J. Schulz (1959). *Phys. Rev.*, **116**, 1141.

G. J. Schulz (1960a). *J. Appl. Phys.*, **31**, 1134.

G. J. Schulz (1960b). *J. Chem. Phys.*, **33**, 1661.

G. J. Schulz (1961). *J. Chem. Phys.*, **34**, 1778.

G. J. Schulz (1962). *Phys. Rev.*, **125**, 229.

G. J. Schulz (1964). *Phys. Rev.*, **135**, A988.

G. J. Schulz (1969). Paper presented at the Atomic and Molecular Physics Congress, Manchester, England, March 31-April 3, 1969.

G. J. Schulz and J. T. Dowell (1962). *Phys. Rev.*, **128**, 174.

G. J. Schulz and R. E. Fox (1957). *Phys. Rev.*, **106**, 1179.

G. J. Schulz and J. W. Philbrick (1964). *Phys. Rev. Letters*, **13**, 477.

M. J. Seaton (1959). *Phys. Rev.*, **113**, 814.

W. F. Sheridan, O. Oldenberg and N. P. Carleton (1961). In *Second International Conference on the Physics of Electronic and Atomic Collisions*, W. A. Benjamin, Inc., New York, p. 159.

S. M. Silverman and E. N. Lassettre (1964). *J. Chem. Phys.*, **40**, 1265.

J. A. Simpson (1964). *Rev. Sci. Instr.*, **35**, 1698.

J. A. Simpson, M. G. Menendez and S. R. Mielczarek (1966). *Phys. Rev.*, **150**, 76.

J. A. Simpson and S. R. Mielczarek (1963). *J. Chem. Phys.*, **39**, 1606.

A. Skerbele, M. A. Dillon and E. N. Lassettre (1967a). *J. Chem. Phys.*, **46**, 4161.

A. Skerbele, M. A. Dillon and E. N. Lassettre (1967b). *J. Chem. Phys.*, **46**, 4162.

A. Skerbele, M. A. Dillon and E. N. Lassettre (1968a). *J. Chem. Phys.*, **49**, 3543.

A. Skerbele, M. A. Dillon and E. N. Lassettre (1968b). *J. Chem. Phys.*, **49**, 5042.

A. Skerbele and E. N. Lassettre (1964). *J. Chem. Phys.*, **40**, 1271.

A. Skerbele and E. N. Lassettre (1965). *J. Chem. Phys.*, **42**, 395.

A. Skerbele, V. D. Meyer and E. N. Lassettre (1965). *J. Chem. Phys.*, **43**, 817.

A. Skerbele, K. J. Ross and E. N. Lassettre (1969). *J. Chem. Phys.*, **50**, 4486.

A. C. H. Smith, E. Caplinger, R. H. Neynaber, E. W. Rothe and S. M. Trujillo (1962). *Phys. Rev.*, **127**, 1647.

K. Smith (1966). *Rep. Progr. Phys.*, **29**, 373.

P. T. Smith (1930). *Phys. Rev.*, **36**, 1293.
P. T. Smith (1931). *Phys. Rev.*, **37**, 808.
V. Srinivasan and J. A. Rees (1967). *Brit. J. Appl. Phys.*, **18**, 59.
V. Srinivasan, J. A. Rees and J. D. Craggs (1967). *Proc. VIth International Conference on the Physics of Electronic and Atomic Collisions*, Publishing House 'Nauka', Leningrad, p. 56.
R. M. St. John and C. C. Lin (1964). *J. Chem. Phys.*, **41**, 195.
A. Stamatović and G. J. Schulz (1969). *Phys. Rev.*, **188** (No. 1), 213.
S. Stein, E. Gerjuoy and I. Holstein (1954). *Phys. Rev.*, **93**, 934.
D. T. Stewart (1956). *Proc. Phys. Soc. (London)*, A69, 437.
K. Takayanagi (1965). *J. Phys. Soc. Japan*, **20**, 562.
K. Takayanagi (1966). *J. Phys. Soc. Japan*, **21**, 507.
K. Takayanagi and S. Geltman (1965). *Phys. Rev.*, **138**, A1003.
J. T. Tate and P. T. Smith (1932). *Phys. Rev.*, **39**, 270.
J. T. Tate and P. T. Smith (1934). *Phys. Rev.*, **46**, 773.
H. S. Taylor, G. V. Nazaroff and A. Golebiewski (1966). *J. Chem. Phys.*, **45**, 2872.
A. Temkin (1966). *Phys. Rev., Letters*, **16**, 835.
B. A. Tozer and J. D. Craggs (1960). *J. Electronics Controls*, **8**, 103.
S. Trajmar, J. K. Rice and A. Kuppermann (1970). In I. Prigogine and S. A. Rice (Eds.), *Advances in Chemical Physics*, Vol. XVIII, Interscience, New York, pp. 15–90.
S. Trajmar, J. K. Rice, P. S. P. Wei and A. Kuppermann (1968). *Chem. Phys. Letters*, **1**, 703.
L. A. Vainshtein (1965). *Opt. and Spectry*, **18**, 538.
V. I. Vedeneyev, L. V. Gurvich, V. N. Kondrat'yev, V. A. Medvedev and Ye. L. Frankevich (1966). *Bond Energies, Ionization Potentials and Electron Affinities*, St. Martin's Press, New York.
V. Ya. Veldre and L. L. Rabik (1965). *Opt. and Spectry*, **19**, 265.
G. H. Wannier (1953). *Phys. Rev.*, **90**, 817.
K. Watanabe (1957). *J. Chem. Phys.*, **26**, 542.
H. F. Winters (1965). *J. Chem. Phys.*, **43**, 926.
I. P. Zapesochnyi and O. B. Shpenik (1966). *Soviet Physics—JETP*, **23**, 592.

# 6 Negative ions

## 6.1 Introduction

The result of a collision between a low-energy electron (e.g. subionization or subexcitation) and an atom or a molecule may be the formation of a negative ion. A negative ion is an atom or a molecule (or a molecular complex) with a net negative charge. Such ions exist in the gaseous phase and in the condensed phase (liquid or solid). Our discussion will be confined to gaseous ions. Negative ions formed in the liquid phase are expected to differ in certain respects from those formed in the gaseous phase due to environmental effects.

Negative-ion formation has been a field of research since essentially the beginning of the century. J. J. Thomson performed pioneering work on negative and positive ions around 1910. The major effort of the early work concentrated on atoms and simple molecules, and progress was handicapped by inadequate experimental methods. The relatively great fragility of negative ions, the apparent absence of any discrete optical negative ion spectra, the often lower production efficiency (compared to positive ions), and the fact that negative ions are often produced with excess kinetic energy, which greatly reduces the collection efficiency in mass spectrometers, explain partly the relatively slow progress in the field for the first half of the century. Nevertheless, by the 1950's a wealth of information had been accumulated which has been summed up in the excellent work of Healey and Reed (1941), Massey (1950), Massey and Burhop (1952) and Loeb (1955, 1956). Especially the work of Loeb (1955, 1956) gives an exhaustive coverage of the historical development of the field. Recent improved methods and refined techniques allowed the renewed interest—pressed by many problems in several fields—in free negative ion physics to expand drastically. The more recent advances, mostly on atoms and small molecules, have been reviewed in a number of articles and texts (Massey, 1956; Branscomb, 1957; Bučel'nikova, 1960;

Prasad and Craggs, 1962; Branscomb, 1962; McDaniel, 1964; Hasted, 1964; Fiquet-Fayard, 1965). Some of the most important developments appeared in the last few years. Furthermore, the growing interest of radiation and life sciences in the field pressed for an expansion of the gaseous negative ion studies to more complex molecular systems.

Negative ions are of intrinsic physical interest in that they allow basic information on atoms and molecules such as molecular structure, electron affinities, and bond dissociation energies to be obtained. They are also of direct practical importance. In chemistry, negative ions have been shown to be important intermediates in radiation-induced reactions. Mass spectroscopists often find negative-ion spectra of complex molecules simpler and more revealing than their positive-ion spectra, while nuclear physicists find negative ions interesting for accelerating purposes. Electron attachment and detachment processes occur in the upper atmosphere and in space. This explains the large interest of the astrophysicist in the field. Radio and television communications which rely upon the existence of free electrons in the upper atmosphere, determined by negative ion intermediates, take special interest in negative-ion studies of atmospheric gases. Further, the relevance of electron attachment to the breakdown strength of electron-attaching gases and gas discharges is evident. The high dielectric strengths of many gases can be explained by the ability of the molecules to remove slow electrons before they attain enough energy to start an avalanche initiating electrical breakdown of the dielectric. In radiation detection and radiation dosimetry, the operation of ionization chambers and gas counters is influenced by the production of negative ions which produce delayed and spurious counts, complicating the determinations of absolute electron yields. The property of certain molecules to capture strongly low-energy electrons (dissociatively or non-dissociatively) makes them suitable for detection (scavenging) of low-energy electrons in gases or liquids.

But, perhaps, the most important role of negative ions is (yet) to be found in some of the basic and challenging problems of radiation and life sciences. It has been stated earlier that in the action of ionizing radiation with matter large numbers of low-energy electrons are ultimately produced. The sheer number of low-energy events could thus make the low-energy electron interactions biologically more significant than the primary interaction. Further, no fundamental understanding of matter and its interaction with ionizing radiation can be satisfactory without information concerning the interaction of low-energy electrons with individual atoms and molecules. For example, the development of a complete picture of the slowing-down of radiation in matter requires knowledge of the various processes through which low-energy electrons interact with matter and lose energy. In this respect, dissociative electron attachment and excitation of compound negative ion states provide effective ways of removing or slowing-down subexcitation electrons which in the past have been thought of as having prolonged lifetimes in the absence of dissipative processes (other than elastic). Also, as will be discussed in Chapter 9, the ability of molecules to capture low-energy electrons and to form stable or temporary negative ions is believed to play an important role in the biological action of molecules, such as the cancer inducing ability of chemical carcinogens. The toxicity of certain halogenated aliphatic hydrocarbons has been related to electron attachment. The belief is also held that electron attachment studies may provide insight into the action of radiation upon the cell where the molecule in question is normally involved in subcellular biosynthetic activity or structural function. Although application of the properties of gaseous ions to explain biological actions of molecules has to be done with extreme caution—the behaviour of free electrons in gases may be different from that in a biological system—knowledge of electron transport properties of biologically important molecules in the gas phase is considered fundamental and basic.

## 6.2 Modes of Production of Negative Ions

### 6.2.1 Atomic Negative Ions

A stable atomic negative ion can be formed if its binding energy exceeds that of the neutral atom. For an atom of $Z$ electrons for which the binding energy for the $i$th atomic electron is $E_i^0$ prior to attachment and $E_i^-$ following attachment, the criterion for atomic negative ion stability is (Massey, 1950)

$$E_1 - \sum_{i=1}^{Z} (E_i^0 - E_i^-) > 0 \tag{6.1}$$

where $E_1$ is the binding energy for the attached (extra) electron. The quantity on the left-hand side of (6.1) represents the difference in the total energy of the ground states of the neutral atom and the ion and is called the electron affinity ($EA$) of the atom. Even for a positive electron affinity (binding energy for the extra electron greater than zero) the atomic negative ion immediately after attachment has to lose its surplus energy in order to become stabilized. The excess energy, which is the sum of the incident (attached) electron's kinetic energy, $\varepsilon$, plus $EA$, can either be radiated in a so-called radiative attachment process, viz.

$$e + A \rightarrow A^- + hv \tag{6.2}$$

or, alternatively, it can be transferred through collision with a third body (electron, atom, ion, or molecule) to kinetic energy. The stabilization cross section is a function of the stabilizing agent (Section 6.5). Radiative attachment of a free electron to a neutral atom is a very important process at low pressures where stabilization by collision has a small probability, although it has been customary to regard the probability of stabilization by radiation as small. This stems from the fact that radiative atomic and molecular lifetimes are of the order of $10^{-8}$ to $10^{-9}$ sec (Chapter 3) while, say, a 10 eV electron spends $\sim 10^{-15}$ sec in the field of the atom. Thus, the probability of radiative stabilization is $\sim 10^{-7}$ per collision. For a lower energy electron, which remains in the atomic field longer, stabilization by radiation may become more probable.

The radiative capture of electrons with kinetic energy $\varepsilon$ by a neutral atom gives rise to a continuous emission spectrum, which in the absence of excited ionic states extends over wavelengths given by $\lambda = hc/(\varepsilon + EA)$. Such a spectrum is known as the *radiative attachment continuum or as the affinity spectrum* (Branscomb, 1962). Wildt (1939) identified as the dominant source in the visible range of the solar continuum the radiative attachment spectrum of $H^-$. The radiative attachment spectrum of $H^-$ was observed years later in the laboratory by Lochte-Holtgreven (1951) and Weber (1958). The affinity spectra of $N^-$ and $O^-$ were investigated by Boldt (1959). The absolute intensity of the continuum emission which results from electron attachment to Cl, Br and I was measured recently by Rothe (1969). These intensity measurements were used to determine the cross sections for the reverse process of photodetachment (Chapter 7).

In Equation (6.1), $E_i^0$ and $E_i^-$ are not experimentally measurable quantities. We may, then, simply define $EA$ as the difference in the total energies of the ground state of the neutral atom and the atomic negative ion. This energy difference equals the minimum energy (say, photon energy $h\nu$) required to detach the least tightly bound electron from the atomic negative ion referred to as the *electron detachment energy*. If detachment takes place by photon impact, viz.

$$h\nu + A^- \rightarrow A + e \qquad (6.3)$$

a measurement of the photodetachment cross section allows determination of the cross section for the reverse process, namely the radiative attachment process (6.2). The cross sections for the two processes are related by the principle of detailed balancing. Measurement of the photodetachment threshold yields directly $EA$ (Chapter 7).

A negative atomic ion may be considered in terms of the stationary states of an electron in an attractive field which falls off very rapidly with distance. The attractive force exerted by the neutral atom on the extra electron has a relatively short range; it falls off with approximately the inverse fourth power of distance (Massey, 1950). Due to this short-range attractive force, the negative ion has only a finite number of stationary states which is in contrast to the case of a positive ion for which an infinite number of stationary states exist because of the Coulomb field seen by the valence electron. This, together with the Pauli exclusion principle, severely restricts the elements which can form negative ions. Furthermore, the number of excited atomic negative ion states are expected to be rare, metastable, lie near the continuum, and be finite in number (at most one to two) because the nuclear Coulomb force is strongly screened so that the electron is no longer stably bound. An atomic negative ion for which there exists experimental evidence of a stable excited state is $C^-$. An excited negative ion in $Si^-$ may also exist (Hasted, 1964) and, further, possibly in $B^-$, $Al^-$ and $P^-$ (Chapter 7). A list of stable ground-state atomic negative ions is given in Chapter 7. The atomic hydrogen negative ion, $H^-$, has been identified experimentally. The hydrogen atom with a vacancy in the $n = 1$ level permits a second electron to be captured into the $1s$ shell. Hylleraas (1950, 1951) has shown theoretically that the $2s2p^3P$ state of $H^-$ is stable with respect to electron ejection. The existence of such states leads to the possibility of the radiationless formation of a negative ion by dielectronic attachment, the inverse process to autoionization.* Thus, although the nuclear field in the He atom is strongly attenuated by the atomic electrons at distances corresponding to shells with principal quantum number $n > 1$ and the $He^-$ ion is thought not to exist in the $(1s^2 2s)(^2S)$ ground state, $He^-$ has been observed experimentally (Hiby, 1939; Dukel'skii, Afrosimov and Fedorenko, 1956; Windham, Joseph and Weinman, 1958; Riviere and Sweetman, 1960; Smirnov and Chibisov, 1966), and variation calculations by Holøien and Midtdal (1955) showed that it is in the doubly excited metastable state $(1s2s2p)^4 P_{5/2}$. The state has a measurably long lifetime ($> 10 \mu sec$, Sweetman (1960); $18.2 \pm 2.7 \mu sec$, Nicholas, Trowbridge and Allen (1968)), although the extra electron is very weakly bound (Chapter 7, Table 7.2). The nitrogen negative ion in a doubly excited $^1D$ state is also known (Fogel', Kozlov and Kalmykov, 1959). Dielectronic attachment can be described as the process in which the incoming, extra, electron excites the target atom (or molecule) and is simultaneously captured in a doubly excited state of the negative ion. It is quite common in atoms with several electrons to have states that correspond to double excitation, i.e. two electrons being in states other than the ground state.

The doubly excited negative ion subsequent to its formation may either autoionize or become stabilized by the emission of radiation with the atom returning to the ground state in either case. If $\tau_a$ and $\tau_r$ are the lifetimes against autoionization and radiation, respectively, then $\tau_a/(\tau_a + \tau_r)$ gives the probability that the doubly excited ion will be stabilized by radiation. For

---

*Often, for the case of metastable negative ions the term autodetachment is used instead of autoionization. However, autoionization is a more general term and we will use both indiscriminately. The same will apply for the terms autoionization lifetime and autodetachment lifetime.

atoms (and small molecules) autoionization lifetimes are generally much shorter than radiative lifetimes so that in most cases $\tau_a/(\tau_a + \tau_r)$ is expected to be small. Dielectronic attachment, as a rule, is probably much less likely to occur than direct radiative attachment. Calculations by Bates and Massey (1943) for atomic oxygen indicated that this is the case, at least for this atom. It should be stated, however, that autodetachment* lifetimes of doubly excited negative ions can, under certain conditions, be quite long, when, for example certain transitions are forbidden by the Pauli exclusion principle.

The additional electron which is attached to a neutral atom (to form a negative ion) would have to be in a state of higher total quantum number than the outermost electron of the neutral atom. For these states the attractive force is too weak. Thus, in general, atoms with completely filled outer-shells or subshells such as the rare gases, are unlikely to form stable negative ions, while atoms with single vacancies in their outer-shells or subshells such as the halogens, should be the most likely to attach electrons. In the latter case the outer atomic electrons have very little effect in shielding the added electron from the nucleus. Atomic negative ions and electron affinities of atoms will be discussed further in Chapter 7. It has to be noted at this point, however, that detailed calculations are necessary for a quantitative understanding of the behaviour of individual atomic negative ions. Further, although doubly (and multiply) charged negative ions form in the liquid phase, no such ions are known[†] in the gas phase. It is very unlikely that a second electron could be attached to an atom in the gaseous phase for two reasons: (i) because the binding energy for a single electron is small, and (ii) a second electron would suffer strong Coulomb repulsion.

### 6.2.2 *Types of Electron Attachment Processes in Molecules*

Molecular negative ions differ from atomic negative ions. In molecules, nuclear and electronic motion are separable in the Born–Oppenheimer approximation, and application of the Franck–Condon principle permits electron attachment to molecules to proceed by vertical transitions between potential energy curves (surfaces) from the equilibrium position of the atoms. The attachment of an electron to a neutral molecule can be regarded as occurring between two electronic states of the molecular ion: (i) the initial state in which one of the electrons occupies an unbound orbital and the potential energy curve is that of the neutral molecule, and (ii) the potential

---

*See note on page 413.
†Evidence for doubly charged negative ions has been reported for oxygen and the three lighter halogens (Stuckey and Kiser, 1966). However, there are alternative explanations for these observations, the meaning of which is still in doubt (Fremlin, 1966). For a polyatomic molecule, however, the situation may be different. Evidence for a doubly charged polyatomic negative ion in the gas phase has been reported by Dougherty (1969).

energy curve of the upper state. The position of the upper potential energy curve of the negative ion relative to the lower determines the processes which follow electron attachment.

Molecules have larger numbers of low-lying electronic states than atoms, and molecular negative ions may have a number of excited states. Molecular negative ions in an excited state may not rapidly autoionize. Each of the molecular negative-ion states has the same set of rotational and vibrational levels found in the neutral molecule (Massey and Burhop, 1952). The electron affinity of the molecule may be defined as the difference in energy between the neutral molecule plus an electron at rest at infinity and the molecular negative ion, when both neutral molecule and negative ion are in their ground electronic, vibrational and rotational states. If the energy of the unexcited negative ion lies below that of the unexcited neutral molecule, the electron affinity is positive and the ion is stable with respect to electron ejection. For a negative electron affinity, the negative ion state lies above that of the neutral and the negative ion is metastable. Whether the electron affinity of a molecule is positive or negative has an important consequence on the interaction of low-energy electrons with molecules. For a positive electron affinity the electron can be captured into a long-lived negative-ion state (Section 6.5), while for a negative electron affinity the negative-ion state can manifest itself by acting as an intermediate for elastic and inelastic electron scattering (Section 6.5 and Chapters 4 and 5). Although the distinction between the two processes is the time in which the electron is bound to the molecule, the observed physical effects are quite different.

The molecular electron affinity is a difficult quantity to obtain accurately, in contrast to some other molecular physical parameters such as the ionization potential. Due to the complexity of molecular negative ions and to the inherent difficulty in determining electron affinities, there exist very few accurate determinations of this physical quantity.

Another parameter, analogous to the electron detachment energy discussed in the previous section, is the *vertical detachment energy* defined as the minimum energy required to eject the electron from the negative ion (in its ground electronic and nuclear state) without changing the internuclear separation. The vertical detachment energy is easily obtained, but since the vertical transition may leave the neutral molecule in an excited vibrational state, the vertical detachment energy although the same as the electron affinity for atoms is, in general, different from the electron affinity for molecules. The electron affinity of molecules is discussed in Chapter 7.

In view of the many temporary negative-ion resonances discovered recently in the electron scattering from atoms and molecules, a term analogous to the vertical detachment energy was introduced (Christophorou and Compton, 1967), namely the *vertical attachment energy*. This quantity is defined as the difference in energy between the neutral molecule in its ground electronic, vibrational and rotational states plus an electron at rest at infinity, and the molecular negative ion formed by addition of an electron to the neutral molecule without allowing a change in the internuclear separation of the constituent nuclei. The vertical attachment energy is expected to be less than or equal to the true electron affinity of the molecule. For the nitrogen molecule the vertical attachment energy is $-2.3$ eV (Schulz, 1959b) and for benzene $-1.5$ eV (Compton, Christophorou and Huebner, 1966).

Free electrons attach to molecules in the gas phase essentially in two different ways: (i) dissociatively and (ii) non-dissociatively. Various modes of negative-ion formation exist within these two broad categories and will be distinguished further, as required, in subsequent discussions. The non-resonance process of ion-pair formation (iii) differs from (i) and (ii) in that the electron is not captured, but merely provides the necessary energy to excite the molecule to an unstable state which spontaneously dissociates into a positive and a negative ion.

(i) *Resonance dissociative electron attachment.* Dissociative electron attachment being a resonance process occurs over a narrow range of energy (from $\sim 0$ to 15 eV). The energy for the process is essentially equal to the energy difference between the ground state $((AX+e)$; ground state molecule and electron at rest at infinite separation) and the excited states of the compound negative ion, viz.

$$e+AX \rightarrow AX^-{}^* \rightarrow A(\text{or } A^*)+X^-(\text{or } X^-{}^*) \qquad (6.4)$$

Thus, we think of (6.4) as proceeding in two stages and according to the Franck–Condon principle. Electrons within a restricted energy range (between $E_1$ and $E_2$; Figure 6.1a) are first captured by the neutral molecule without alteration of the position and velocities of the nuclei (the speed of the incident electron is large compared to that of the nuclei of the molecule). The resulting compound (complex) system $AX^-{}^*$ dissociates into the final products (Figure 6.1). For polyatomic molecules A and/or $X^-$ may be an atom or a radical with or without excess energy. Figure 6.1 shows schematic potential energy diagrams for various modes of dissociative electron attachment to diatomic (or 'diatomic-like') molecules. In Figure 6.1a a general type of dissociative electron attachment process is illustrated where the potential energy curve for $AX^-{}^*$ is purely repulsive. The asymptote of the $AX^-{}^*$ curve, $A+X^-$, lies below the asymptote of the AX curve, $A+X$, by an amount equal to the electron affinity, $EA(X)$, of X. The difference in energy between the asymptote of the AX curve and the 'zero-point energy' (energy of the lowest vibrational level of AX) is equal to the dissociation energy, $D(AX)$, of AX. In the region from $R=R_1$ to $R=R_c$, where the potential energy curve of $AX^-{}^*$ lies above that of AX, the molecular negative ion is unstable toward electron ejection. This region will be referred

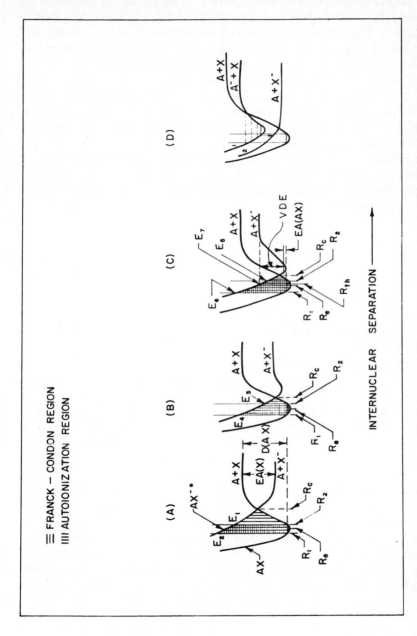

**Figure 6.1** Schematic potential energy diagrams for various modes of dissociative electron attachment to diatomic molecules discussed in text. Note that the ground state indicated is for the neutral molecule AX plus the electron at rest at infinity

to as 'autoionization region' and is shown in some of the diagrams of Figure 6.1 by vertical lines. Thus process (6.4) may not necessarily lead to dissociation, but alternative channels to dissociation such as elastic and inelastic scattering may take place. Dissociation of $AX^-*$ is in competition with autoionization for all interatomic distances $R$ between $R_1$ and $R_c$. Studies of process (6.4) may yield information on the nature and position of the potential energy curves (surfaces) of molecules of the type $AX^-*$. Dissociative attachment does not require stabilization. The quantity of importance here is the dissociative attachment cross section, $\sigma_{da}(\varepsilon)$, as a function of electron energy $\varepsilon$.

The negative ions resulting from dissociative attachment through $AX^-*$ of Figure 6.1a will have considerable distribution of kinetic energy. Assuming that the dissociation products are in their ground state (unexcited) the kinetic energies of $X^-$ will vary from

$$[E_{kin}(X^-)]_{min} = \frac{M_A}{M_A + M_X}[EA(X) - D(AX) + E_1]$$

to

$$[E_{kin}(X^-)]_{max} = \frac{M_A}{M_A + M_X}[EA(X) - D(AX) + E_2]$$

where $M_A$ is the mass of the neutral fragment A and $M_X$ the mass of the negative ion $X^-$.

The energy balance for (6.4) leads to

$$AP(X^-) = D(AX) - EA(X) + E_{kin} + E_{ex} \tag{6.5}$$

where $AP(X^-)$ is the 'appearance potential' for $X^-$ and $E_{kin}$ and $E_{ex}$ represent the kinetic and excitation energies of the fragments formed. If $E_{kin} + E_{ex} = 0$, the sum of the quantities of the right-hand side represents the minimum energy required for the reaction. When this is ensured, $D(AX)$ or $EA(X)$ can be evaluated if either is known (see further discussion in Section 7.2.1.2.D).

Similarly, appropriate kinetic energy expressions can be written for the case shown in Figure 6.1b. Here the $AX^-*$ potential energy curve has an attractive region outside the Franck–Condon and autoionization regions at $R > R_c$, and the minimum electron energy for dissociative attachment is $E_3 (> D(AX) - EA(X))$. As far as dissociative attachment is concerned, this case is similar to that shown in Figure 6.1a. If, however, a situation like that in Figure 6.1c exists, where $E_5 < D(AX) - EA(X)$, then both dissociative and non-dissociative electron attachment can occur. However, the dissociative attachment cross section, $\sigma_{da}(\varepsilon)$, will exhibit a vertical onset behaviour, i.e. the negative ion $X^-$ will appear at $E_7$ and will show a very sharp rise at or very close to this energy. Because dissociation is energetically possible for only a limited range of interatomic distances, $R_1 \leqslant R \leqslant R_{th}$, in the

Franck–Condon region, ($R_{th}$ is the internuclear distance corresponding to the threshold energy, $E_{th} = E_7$) and because the products at the threshold, $E_7$, are formed with essentially zero kinetic energy, $\sigma_{da}(\varepsilon)$ is smaller (in certain cases very much smaller) than when all $R$ were allowed. On the other hand, some of the final possible states fall at energies less than $E_7$. Transitions induced by electron attachment in the energy range from $E_5$ to $E_7$ (Figure 6.1c) will produce vibrationally excited molecular ions, $AX^-*$. Since in the case of Figure 6.1c $EA(AX) < 0$, the negative ion is unstable with respect to electron ejection (see also Figure 6.2). In Figure 6.1c the *vertical detachment energy* (*VDE*) is shown. Here the molecular-ion-potential energy curve has a minimum at an internuclear separation greater than the minimum of the neutral molecule potential energy curve, and it is clearly seen that the *VDE* required to remove the electron is quite different from the electron affinity of the molecule $EA(AX)$, also shown in Figure 6.1c. Figure 6.1d shows a dissociative attachment process taking place through bound states of the negative ion which undergo radiationless intramolecular transitions to repulsive states of the negative ion as suggested by Chen (1963). Electrons are first captured into the bound state (curve 1) and prior to being auto-detached an intramolecular radiationless transition to curve 2 occurs leading to dissociation into $A + X^-$. Such a process results from the overlapping of discrete states with a continuum of states of $AX^-*$, and could be important in polyatomic molecules due to the large overlapping of electronic states provided by the vibrational molecular levels. This process is similar to that of predissociation discussed in Chapter 3.

(ii) *Resonance non-dissociative electron attachment.* This is a resonance process occurring over a limited (narrow) range of electron energies usually less than a few eV. In actual fact, in most cases of parent negative ion formation by electron capture in the field of the molecular ground state, the ions have a maximum probability of formation at $\sim 0.0$ eV. In general, the process can result in the formation of a 'permanent' parent negative ion, a long-lived temporary negative ion, or it may manifest itself as a short-lived temporary negative ion intermediate with as short a lifetime as $10^{-15}$ sec in electron-molecule scattering. Electron capture can take place in both the field of the ground and excited electronic states of the molecule (Section 6.5). In the case of non-dissociative electron attachment the cross section, $\sigma_0$, for capture (formation of $AX^-*$), the average lifetime for autoionization, $\tau_a$, and the cross section for stabilization, $\sigma_{st}$, are important. We may, then, write*

$$e + AX \underset{\tau_a^{-1}}{\overset{\sigma_0}{\rightleftarrows}} AX^-* \overset{\sigma_{st}}{\rightarrow} AX^- + energy \tag{6.6}$$

---

*Note that in Equation (6.6) $\tau_a^{-1}$ and $\sigma_0$ are in different units; $\tau_a^{-1}$ has the same units as $N_{AX} v \sigma_0$.

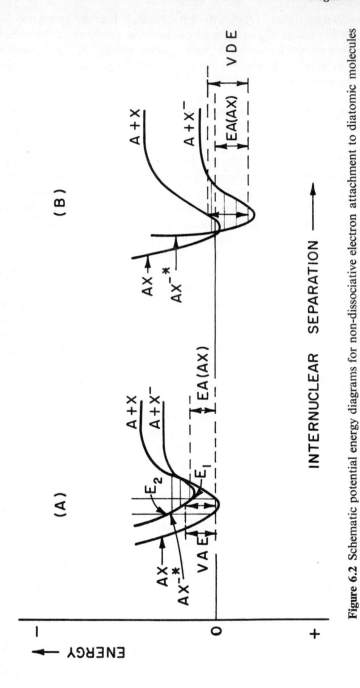

**Figure 6.2** Schematic potential energy diagrams for non-dissociative electron attachment to diatomic molecules

If we assume that stabilization occurs through collision and denote by $\tau_{st}$ the average time taken for $AX^{-*}$ to transfer its excess energy to a gas molecule in a collision ($\tau_{st}$ is inversely proportional to pressure; $\tau_a$ is pressure-independent), then the pressure dependence of the attachment probability for process (6.6) can be found as follows (Bloch and Bradbury, 1935): Consider an excited ion at $t = 0$. The probability that it will survive destruction (autoionization) in time $t$ is $e^{-t/\tau_a}$. The probability of energy transfer in the time interval between $t$ and $t + dt$ is $e^{-t/\tau_{st}}\, dt/\tau_{st}$. Then the total probability, $p_t$, that the ion will transfer its excess energy in a collision before destruction becomes

$$p_t = \int_0^\infty \frac{\exp-\left(\dfrac{t}{\tau_a} + \dfrac{t}{\tau_{st}}\right)}{\tau_{st}}\, dt = \frac{\tau_a}{\tau_a + \tau_{st}} = \frac{P}{P + P_0} \tag{6.7}$$

where $P$ is the gas pressure and $P_0$ is a critical pressure for which $\tau_a = \tau_{st}$. At high $P(\gg P_0)$ the attachment probability will be independent of $P$ and at low $P$ it will be proportional to it.

Figures 6.1c and 6.2a, b show schematically certain types of non-dissociative electron attachment. Electron attachment into discrete states of $AX^-$ will occur between $E_5$ and $E_7$ in Figure 6.1c and between $E_1$ and $E_2$ in Figure 6.2a, giving a vibrationally excited molecular ion. Figures 6.1c and 6.2a are examples of molecular negative ions for which $EA(AX) < 0$ and the potential energy curve of $AX^{-*}$ crosses the Franck–Condon region. Because of the negative electron affinity of $AX$, $AX^{-*}$ is extremely short-lived. The vertical attachment energy is shown in Figure 6.2a. Figure 6.2b shows an example where $EA(AX) > 0$. This is distinctly a different mode of attachment in that the $AX^{-*}$ potential energy curve does not cross the Franck–Condon region. The *VDE* is shown in this diagram. Electron attachment through this mechanism has been proposed to occur via the vibrational excitation of the neutral molecule and subsequent capture of the incident electron. The process can alternatively be thought of as a non-vertical transition since the energy of the incoming electron is sufficiently small to allow long enough electron–molecule-interaction times for the nuclei to relax to a new position on the negative-ion curve. As will be discussed in Section 6.5, a number of molecules attach electrons by uni-molecular electron capture for times greater than a microsecond. These molecules are large and generally symmetric so that the excess energy of the captured electron is shared with the many degrees of freedom of the molecule for a time which is long enough for the ion to be detected in a conventional time-of-flight mass spectrometer.

We may summarize the various electron attachment processes to molecules through (6.8):

$$AX+e \rightarrow AX^{-}* \overset{\underset{p_a \nearrow}{\sigma_0}}{\underset{p_{st} \searrow}{\overset{p_{da}}{\rightarrow}}} \begin{array}{ll} \text{Pa} & AX(\text{or } AX^*)+e \quad (a) \\ & A(\text{or } A^*)+X^- \quad (b) \\ \text{Pst} & AX^-+\text{energy} \quad (c) \end{array} \qquad (6.8)$$

where $p_a$, $p_{da}$ and $p_{st}$ are, respectively, the probability for autoionization, dissociative attachment and stabilization.

For a complete understanding of (6.8), the cross section $\sigma_0$ for the formation of the compound negative ion $AX^-*$ as well as the cross sections for the subsequent decay channels (a), (b) and (c), have to be determined. The cross section for the formation of $AX^-*$, $\sigma_0$, is a function of $AX$ and the energy of the incoming electron and can be measured at high pressures where complete stabilization of $AX^-*$ by collision can take place (for an indirect calculation of $\sigma_0$, see Section 6.4). The autoionization lifetime of $AX^-*$, $\tau_a$, (decay through channel (a)) is determined by the molecular structure and the internal energy of the system and can be measured at present in low-pressure beam experiments when $\tau_a \gtrsim 10^{-6}$ sec. The dissociative attachment cross section $\sigma_{da}$ is primarily determined by $\sigma_0$ and the competition between dissociation and autoionization of $AX^-*$, and can be measured by various techniques such as total ionization techniques, Lozier tubes, mass spectrometers, and the swarm-beam combination (see following section). Finally, the cross section for permanent parent-negative-ion formation is a function of $\sigma_0$, the relative magnitudes of the other possible channels of decay, and the cross section for stabilization $\sigma_{st}$. The most effective way of stabilizing $AX^-*$ is that through collisions with molecules of the same or of different kind (Section 6.5).

From discussions in Chapters 4 and 5 and from further discussions in Sections 6.4 and 6.5 of this chapter, it is evident that dissociative and non-dissociative electron attachment to molecules proceeds via a compound negative ion intermediate which can be formed in either of two ways: (i) by electron capture in the field of the ground electronic state, or (ii) by electron capture in the field of an electronic excited state. Electron capture in the field of the ground electronic state for very short times ($\sim 10^{-15}$ sec) has been observed as resonances in elastic and inelastic electron scattering experiments (e.g. $H_2$—Schulz(1964a), Schulz and Asundi(1967), Ehrhardt and coworkers (1968); $N_2$—Haas (1957), Schulz (1964a); $N_2O$—Schulz (1961)) and in dissociative electron attachment studies (e.g. Christophorou and Stockdale (1968); Section 6.4). For polyatomic molecules which can bind an extra electron, long-lived (lifetimes $\gtrsim 10^{-6}$ sec) compound negative ions have been detected (Christophorou and Compton, 1967; Christophorou and

Blaunstein, 1968; Edelson, Griffiths and McAfee, 1962; Christophorou and coworkers, 1970a). Electron capture in the field of an excited electronic state for very short times ($\lesssim 10^{-14}$ sec) has also been observed as resonances in electron scattering experiments (e.g. H—Schulz (1964c), Kleinpoppen and Raible (1965); rare gases—Schulz (1963, 1964b), Simpson and Fano (1963), Fleming and Higginson (1963), McFarland (1964), Kuyatt, Simpson and Mielczarek (1965); $H_2$, IID, $D_2$—Kuyatt, Mielczarek and Simpson (1964, 1966); $N_2$—Heideman, Kuyatt and Chamberlain (1966)) and in dissociative electron attachment studies (group II in Christophorou and Stockdale (1968); Christophorou, Carter and Christodoulides (1969); Sections 6.4 and 6.5). The compound negative ion states which have been associated with excited electronic states appeared slightly below ($\sim 0.5$ eV in some cases) the energy of the corresponding excited state of the neutral species (e.g. Christophorou and Stockdale, 1968; Simpson and Fano, 1963; Kuyatt, Mielczarek and Simpson, 1964). Long-lived polyatomic negative ions formed by electron capture in the field of an excited electronic state have also been observed* (Christophorou, Carter and Christodoulides, 1969; Chaney and coworkers, 1970; Section 6.5).

(iii) *Ion-pair formation.* This is a non-resonance electron attachment process occurring upon collision of an electron with a molecule when the electron energy is adequate to excite the molecule to an unstable state which dissociates into a positive–negative ion pair, viz.

$$e + AX \rightarrow A^+ + X^- + e \qquad (6.9)$$

The process will set in at some definite electron energy and persist to quite high energies, as the free electron carries away the excess energy. The variation of the cross section with electron energy above threshold is distinctly different from that for the resonance processes (i) and (ii). The cross section increases (from a zero value at the threshold) with energy above threshold approximately linearly close to the threshold up to an electron energy of about three to four times of the threshold energy, after which it declines steadily approximately as the inverse of energy or more slowly. The process is illustrated in Figure 6.3. The shaded area refers again to the Franck–Condon region. It is seen from the figure that the maximum energy shared by the fragment ions $A^+$ and $X^-$ ranges from

$$E_{min} = E_1 - D(AX) - I(A) + EA(X) \quad \text{to} \quad E_{max} = E_2 - D(AX) - I(A) + EA(X)$$

where $I(A)$ refers to the ionization potential of the fragment (atom or radical) A. Measurement of $E_{min}$ or $E_{max}$ and $E_1$ or $E_2$ combined with knowledge of $I(A)$ and $D(AX)$ provides a method of determining $EA(X)$.

---

*See Section 6.2.1 for long-lived metastable He$^-$ ions formed in the field of an excited electronic state.

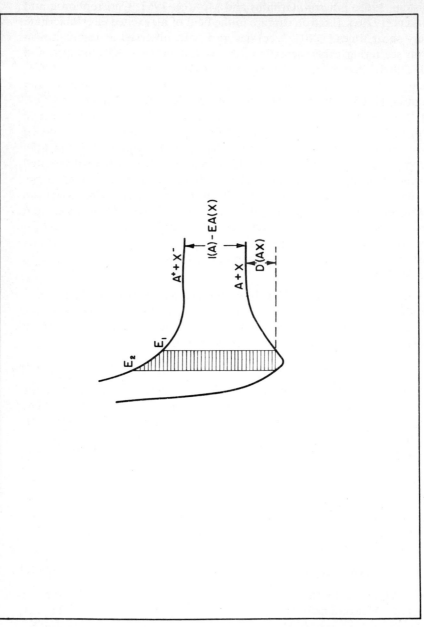

Figure 6.3 Schematic potential energy diagram illustrating the process of ion-pair formation

Negative-ion production by process (iii) is not restricted in energy range (as for processes (i) and (ii)) and is a strong dissipative process in the action of ionizing radiation with matter especially at electron energies $\varepsilon$ in the range $I < \varepsilon \lesssim 100$ to 200 eV.

(iv) *Production of negative ions by other processes.* Gaseous negative ions are also produced in charge changing (or charge exchange, or charge transfer, or charge permutation) collisions (or reactions). One or more electrons are transferred in a neutral–neutral $(A + B \rightarrow A^- + B^+)$ or a positive ion–neutral $(C^+ + D \rightarrow C^- + D^{2+})$ collision. This mechanism is significant in the production of negative ions (such as $He^-$ )which are not produced directly by electron impact (see also Chapter 8).

Additional mechanisms yielding negative ions involve surfaces. These will not be discussed. The topic is discussed elsewhere (e.g. Massey, 1950; Loeb, 1955; McDaniel, 1964).

## 6.3 Techniques for Negative Ion Studies

Electron attachment processes in the gas phase are frequently studied by two experimental methods: the electron swarm and the electron beam. Both methods trace back to the beginning of the century and have undergone large changes and improvements since then. They have also been successfully combined and continue to supplement each other. We shall elaborate on the physical parameters obtained by both methods and discuss briefly some of the most important. The reader is referred to the following texts and review articles for more details: Massey (1950); Massey and Burhop (1952); Loeb (1955, 1956); Branscomb (1957); Field and Franklin (1957); Craggs and Massey (1959); Prasad and Craggs (1962); McDaniel (1964); Fiquet-Fayard (1965). The work of Loeb (1955, 1956) and Prasad and Craggs (1962) is especially recommended.

### 6.3.1 *Electron Swarm Methods*

#### 6.3.1.1 Rate of electron attachment

In swarm experiments electrons are produced at the beginning of the drift space by various means such as photon absorption, particle ionization, and thermionic emission. They make many collisions with the gaseous medium through which they travel under the influence of a uniform electric field and hence their energies are spread over a wide range. The product $P \times d$, where $P$ is the gas pressure and $d$ is the overall drift distance, is large enough that the electrons attain an equilibrium energy distribution within a distance $x(\ll d)$ from the point or the plane of formation regardless of their initial energy of liberation. The distribution of electron energies is characterized by a function $f(\varepsilon, E/P)$ which depends on the gaseous medium and $E/P$; $\varepsilon$ is the electron energy and $E/P$ is the 'pressure-reduced electric field', usually expressed in units of volts/cm torr at a specified gas temperature. The distribution function $f(\varepsilon, E/P)$ is defined as $f(\varepsilon, E/P)d\varepsilon \equiv$ fraction of electrons in an energy range $d\varepsilon$ about $\varepsilon$. Electron swarm experiments measure quantities which are averaged over the energy distribution $f(\varepsilon, E/P)$. Most of the early work on electron attachment obtained by the electron swarm method is complicated because of unknown electron energy distributions.

The swarm data are usually compared by plotting the attachment coefficient, $\alpha$, or the probability of attachment per collision, $h$, versus $E/P$ or versus the mean electron energy, $\langle \varepsilon \rangle$, obtained from independent electron diffusion measurements (Chapter 4). In swarm experiments $\alpha$ and $w$ (the electron swarm drift velocity) are directly obtainable quantities. The product $\alpha w$ is the absolute rate of electron attachment expressed in $\sec^{-1} \text{torr}^{-1}$. The absolute rate of electron attachment is related to $f(\varepsilon, E/P)$ and the electron attachment cross section, $\sigma_a(v)$* (in $\text{cm}^2$), by

$$\alpha(E/P) \times w(E/P) = N \int_0^\infty v\sigma_a(v) f(v, E/P)\, dv$$

or, in terms of energy, by

$$\alpha(E/P) \times w(E/P) = N(2/m)^{\frac{1}{2}} \int_0^\infty \varepsilon^{\frac{1}{2}} \sigma_a(\varepsilon) f(\varepsilon, E/P)\, d\varepsilon \tag{6.10}$$

where $N$ is the number of attaching gas molecules per cubic centimeter per unit torr (at specified temperature). A knowledge of $f(\varepsilon, E/P)$ is necessary to determine $\sigma_a(\varepsilon)$. The energy distribution function $f(\varepsilon, E/P)$ is not easy to obtain for each and every gas studied. In fact, $f(\varepsilon, E/P)$ has been calculated for only some of the rare gases and very few simple molecules such as $N_2$, $H_2$ and $D_2$ (see discussion in Chapter 4). Due to this difficulty a number of electron attachment studies have been performed in binary gaseous mixtures in which the predominant gas (carrier gas) does not attach electrons in the energy region of interest and its role is to allow a knowledge of $f(\varepsilon, E/P)$ at each $E/P$.

### 6.3.1.2 Electron swarm energy distribution functions for carrier gases

As discussed in Chapter 4, electron energy distribution functions have become known for a number of non-electron attaching gases such as Ar,

---

*We use the notation $\sigma_a(v)$ to denote the electron attachment cross section indiscriminately of the specific products formed.

$N_2$ and $C_2H_4$. This allowed accurate analyses of electron swarm data through Equation (6.10), especially when the functional form of $\sigma_a(\varepsilon)$ is known (Section 6.4). In this respect the calculations by Carleton and Megill (1962) for $N_2$ are of special interest since for this gas $f(\varepsilon, E/P)$ peaks at energies $< 2$ eV for convenient values of $E/P$, in which energy region most interesting electron attachment processes occur. In Figure 6.4 the normalized distribution functions calculated by Carleton and Megill (1962) are plotted for three values of $E/P$: 0.11, 0.33 and 0.99 volts/cm torr. It is seen that for $N_2$ a change in $E/P$ from 0.1 to 1 shifts the energy distribution from near thermal energies to $\sim 1$ eV. Electron attachment processes in this energy range can be studied by mixing the gas to be investigated in small proportions with nitrogen which does not attach electrons.* Many polyatomic non-electron attaching gases yield at low $E/P (E/P \to 0)$ a Maxwell form of the energy distribution. Ethylene has been used quite extensively (Stockdale and Hurst, 1964; Christophorou and coworkers, 1965, 1966) to provide a Maxwell distribution of electron energies for $E/P < 0.1$ volts/cm torr for both thermal electron attachment and thermal electron scattering studies. Other gases employed to yield thermal ($T \simeq 300°$K) energy distribution functions at low $E/P$ include $C_2H_5OH$ and $CO_2$ (Bouby, Fiquet-Fayard and Abgrall, 1965). In Figure 6.4 a normalized Maxwell distribution function has also been plotted. It is characteristic of electron energy distributions in polyatomic gases at low $E/P$. The distribution functions calculated by Ritchie and Whitesides (1961, 1968) for Ar are also shown in the figure. It becomes thus clear from Figure 6.4 that by changing carrier gas and $E/P$, the electron energy distribution function $f(\varepsilon, E/P)$ can be set to peak at any energy from thermal to $\sim 10$ eV. Some representative dissociative attachment resonances ($I^-/HI$; $Cl^-/HCl$; $O^-/N_2O$; $O^-/O_2$ (Section 6.4)) are included in the figure to show the overlapping of the energy distribution functions with these resonances and to indicate how electron attachment can be studied in mixtures with appropriate non-electron attaching gases in suitable $E/P$ regions (Section 6.4). Thus, it is profitable to perform electron swarm attachment experiments by mixing the attaching gas to be studied in small proportions (often in less than one part in $10^5$) with a gas which does not attach electrons in the energy region of interest. This is advantageous in two ways: (i) a non-electron attaching gas can be used for which $f(\varepsilon, E/P)$ is known for certain $E/P$ values and (ii) a study can be made of highly electron attaching gases which are often very difficult to study alone. It is noted, however, that the number of gases which can be used as carrier gases is quite limited. And even for these limited cases there has been some question as to the correctness of the electron energy distribution functions used (see also discussion in Chapter 4).

---

*Nitrogen forms a temporary negative ion (Chapters 4 and 5) at $\sim 2.3$ eV, but this is very short-lived and does not show up as attachment in electron attachment studies.

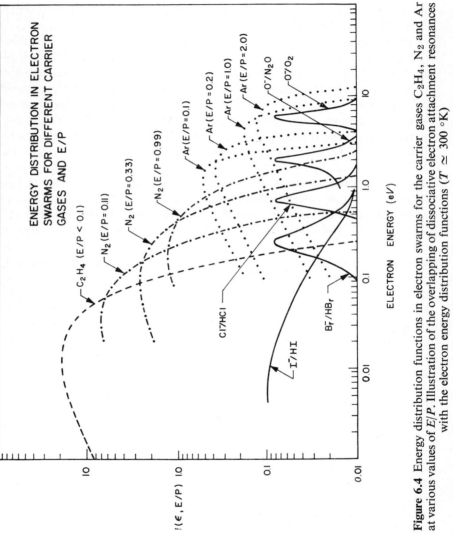

**Figure 6.4** Energy distribution functions in electron swarms for the carrier gases $C_2H_4$, $N_2$ and Ar at various values of $E/P$. Illustration of the overlapping of dissociative electron attachment resonances with the electron energy distribution functions ($T \simeq 300\ ^\circ K$).

Experimental evidence as to the correctness of the electron energy distribution functions for three carrier gases Ar, $N_2$ and $C_2H_4$ has been provided by Christophorou, Chaney and Christodoulides (1969) and Christodoulides and Christophorou (1970). These workers have measured electron attachment rates for a number of molecules in each of the carrier gases $C_2H_4$, $N_2$ and Ar independently, and have plotted them as a function of the mean electron energy $\langle \varepsilon \rangle$ for the respective carrier gas as shown in Figure 6.5 (see also Figures 6.25 and 6.26). The data provide an accurate basis for comparison of the electron energy distribution functions for the three carrier gases. The measured attachment rates in the three gases are in excellent agreement when plotted versus $\langle \varepsilon \rangle$, calculated from the revised distribution functions of Ritchie and Whitesides (1961, 1968) for Ar, from the distribution functions of Phelps and coworkers (Engelhardt and Phelps, 1963; Engelhardt, Phelps and Risk, 1964; Phelps, 1968) for $N_2$, and from $D_l/\mu$ data (Wagner, Davis and Hurst, 1967; Davis, 1969), using a Maxwell form for the distribution function for $C_2H_4$. These three carrier gases with the respective distribution functions just mentioned are then recommended as standard carrier gases for electron attachment work. The attachment rates, plotted as a function of $\langle \varepsilon \rangle$ obtained from Carleton and Megill (1962) are seen (Figure 6.5) to be in less satisfactory agreement with the Ar data for which the distribution functions are well accepted.

The variation of $\langle \varepsilon \rangle$ with $E/P$ (volts/cm torr) for each of the three carrier gases $C_2H_4$, $N_2$ and Ar is shown in Figure 6.6. This graph helps to demonstrate the useful energy range of each of the three carrier gases, as well as the $E/P$ range for which the three gases overlap energy wise. Table 6.1 lists $\langle \varepsilon \rangle$ for a number of $E/P$ for the three gases.

The comparisons discussed above are valid insofar as the distribution functions have the same shapes in the respective regions of overlap. This has been shown to be the case for the Ar distributions of Ritchie and Whitesides and the $N_2$ distributions of Phelps and coworkers (Christophorou, Chaney and Christodoulides, 1969). At low values of $E/P$ the distribution functions in $N_2$ approach a Maxwellian shape, whereas at high $E/P$ values they approach a shape characteristic of those in Ar.

## 6.3.1.3 Mean electron attachment cross sections

The absolute rate of electron attachment given by Equation (6.10) can be combined with the negative ion yield, $I(\varepsilon)$, of a particular ion as a function of electron energy, obtained independently by the electron beam method (Section 6.3.2) to determine $\sigma_a(\varepsilon)$ through the swarm-beam technique (Section 6.3.3). Prior to the establishment of the swarm-beam combination (Christophorou and coworkers, 1965) or in the absence of beam information, the swarm data alone can be used to determine mean attachment cross sections as a function of the mean electron energy $\langle \varepsilon \rangle$ in the swarm.

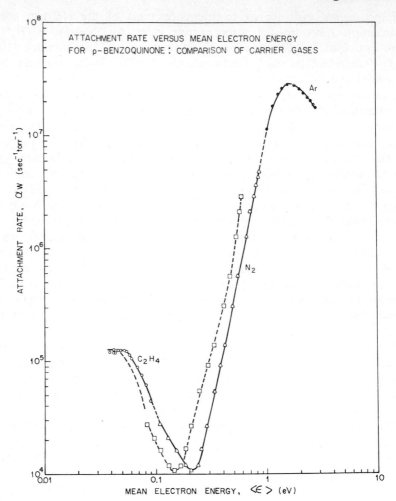

(i) From Carleton and Megill (1962) (curve □---□---□)
and (ii) from Phelps and coworkers (1968) (curve △—△—△).
The mean electron energies for $C_2H_4$ were calculated from $D_l/\mu$
(ratio of longitudinal electron diffusion coefficient to electron mobility)
data (see Table 6.1) by assuming (i) a Maxwell (curve ○—○—○)
and (ii) a Druyvesteyn (curve ------) form of the electron energy
distribution function.

**Figure 6.5** Electron attachment rates in *p*-benzoquinone as a function
of mean electron energy; comparison of carrier gas energy distribution
functions. The mean electron energies for Ar were calculated from the
revised distribution functions of Ritchie and Whitesides (1961, 1968).
Those for $N_2$ were calculated from two sources as indicated above.

**Table 6.1** Mean electron energies as a function of $E/P$ for the carrier gases Ar, $N_2$, and $C_2H_4$

| Ar[a] | | $N_2$[b] | | $C_2H_4$[c] | |
|---|---|---|---|---|---|
| $E/P_{298}$ (V/cm torr) | $\langle \varepsilon \rangle$ (eV) | $E/P_{293}$ (V/cm torr) | $\langle \varepsilon \rangle$ (eV) | $E/P_{298}$ (V/cm torr) | $\langle \varepsilon \rangle$ (eV) |
| | | | | 0.01 | 0.040 |
| 0.001 | 0.064 | 0.033 | 0.057 | 0.05 | 0.043 |
| 0.05 | 0.935 | 0.066 | 0.093 | 0.10 | 0.048 |
| 0.10 | 1.285 | 0.132 | 0.162 | 0.15 | 0.051 |
| 0.15 | 1.559 | 0.198 | 0.226 | 0.20 | 0.054 |
| 0.20 | 1.797 | 0.330 | 0.349 | 0.25 | 0.055 |
| 0.25 | 2.010 | 0.412 | 0.421 | 0.30 | 0.057 |
| 0.30 | 2.197 | 0.495 | 0.490 | 0.40 | 0.060 |
| 0.40 | 2.526 | 0.577 | 0.538 | 0.50 | 0.064 |
| 0.50 | 2.808 | 0.660 | 0.583 | 0.60 | 0.068 |
| 0.70 | 3.283 | 0.824 | 0.656 | 0.8 | 0.076 |
| 1.00 | 3.871 | 0.989 | 0.710 | 1.0 | 0.083 |
| 1.20 | 4.213 | 1.154 | 0.754 | 1.2 | 0.090 |
| 1.50 | 4.677 | 1.320 | 0.787 | 1.5 | 0.099 |
| 1.80 | 5.107 | 1.648 | 0.841 | 1.8 | 0.108 |
| 2.00 | 5.383 | 1.980 | 0.878 | 2.0 | 0.113 |
| | | | | 2.5 | 0.125 |
| | | | | 3.0 | 0.136 |

[a] Calculated from the revised distribution functions of Ritchie and Whitesides (1961, 1968).
[b] Calculated from distribution functions obtained by Phelps and coworkers (1968).
[c] Calculated from the $D_1/\mu$ data of Wagner, Davis and Hurst (1967) ($E/P \leqslant 1$) and of Davis (1969) ($1 \leqslant E/P \leqslant 3.0$ V/cm torr) assuming a Maxwell form for the electron energy distribution. Wagner, Davis and Hurst (1967) and Davis (1969) measured longitudinal diffusion coefficients. Their values are in reasonable agreement with the lateral diffusion coefficient data of Bannon and Brose (1928), but differ appreciably from the lateral diffusion data of Cochran and Forester (1962). The data of the latter group of workers seem to disagree substantially with other well-accepted values of electron diffusion coefficients for other systems.

Let us define a mean attachment cross section as

$$\langle \sigma_a(\varepsilon) \rangle_v \equiv \frac{\int_0^\infty \sigma_a(\varepsilon) \varepsilon^{\frac{1}{2}} f(\varepsilon, E/P) \, d\varepsilon}{\int_0^\infty \varepsilon^{\frac{1}{2}} f(\varepsilon, E/P) \, d\varepsilon} \tag{6.11}$$

which can appropriately be called the 'mean velocity-weighted attachment

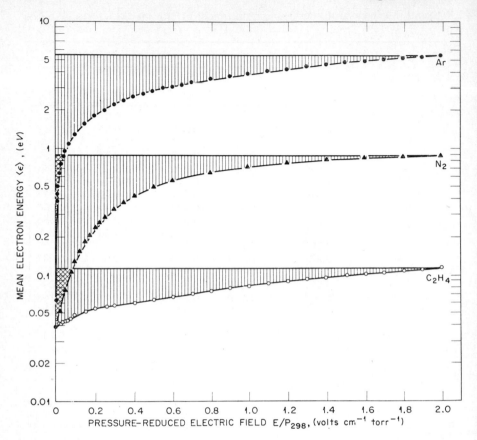

**Figure 6.6** Mean electron energy as a function of $E/P_{298}$ for the carrier gases Ar, $N_2$, and $C_2H_4$. The energy range of usefulness of each carrier gas and the degree of energy overlap is shown in the figure. The mean electron energies were calculated as discussed in text and in Table 6.1

cross section'. Combining Equation (6.11) with (6.10), we have

$$\langle \sigma_a(\varepsilon)\rangle_v = \frac{\alpha w}{N(2/m)^{\frac{1}{2}}} \frac{1}{\displaystyle\int_0^\infty \varepsilon^{\frac{1}{2}} f(\varepsilon, E/P)\, d\varepsilon} = \frac{\alpha w}{N(2/m)^{\frac{1}{2}} \langle \varepsilon^{\frac{1}{2}}\rangle} \qquad (6.12)$$

If $f(\varepsilon, E/P)$ is known as a function of $E/P$, $\langle \varepsilon^{\frac{1}{2}}\rangle$ can be determined for each $E/P$ for which $\alpha w$ is measured and thus $\langle \sigma_a(\varepsilon)\rangle_v$ can be obtained through Equation (6.12). For most cases, however, $f(\varepsilon, E/P)$ is unknown. In the absence of precise information on $f(\varepsilon, E/P)$, typical energy distributions such as Maxwell or Druyvesteyn have been used in the determination of $\langle \sigma_a(\varepsilon) \rangle_v$. Such simple forms of $f(\varepsilon, E/P)$, however, may not resemble the true energy distribution. In cases where the attaching gas is studied in mixtures with a carrier gas, as has been done by Christophorou and coworkers, $f(\varepsilon, E/P)$ is characteristic of the carrier gas and such distributions are known for a number of carrier gases as has already been discussed earlier in this section.

Independent measurements of three swarm parameters are needed for the determination of $\langle \sigma_a(\varepsilon) \rangle_v$ through Equation (6.12), namely the parameter $k_1$ —related to the Townsend energy factor by Equation (4.36)—, the attachment coefficient $\alpha$, and the drift velocity $w$. The values of $\langle \sigma_a(\varepsilon) \rangle_v$ depend markedly on the accuracy of the values used for $k_1$. Calculations of $\langle \sigma_a(\varepsilon) \rangle_v$ have been made* in connection with comparisons of high-pressure swarm with low-pressure beam data. Thus the mean attachment cross sections derived from swarm data have been compared (e.g. Prasad and Craggs, 1960a, b; Crompton, Rees and Jory, 1965; Craggs, Thorburn and Tozer, 1957; Chanin, Phelps and Biondi, 1962) with values of the same quantity calculated by averaging the monoenergetic attachment cross sections, $\sigma_a(\varepsilon)$, over $f(\varepsilon, E/P)$. Such comparisons can also be made by using the energy-dependent values of the attachment cross section found in beam experiments to compute $\alpha$ for a series of $E/P$. The accuracy of such calculations, obviously, depends on that of $f(\varepsilon, E/P)$.

A detailed comparison of various forms of mean electron attachment cross sections obtained from swarm studies with beam data has been given by Blaunstein and Christophorou (1968). These authors concluded that although the $\langle \sigma_a(\varepsilon) \rangle_v$ derived from swarm and beam experiments are in reasonable agreement (which depends on the accuracy of $f(\varepsilon, E/P)$), the cross section $\sigma_a(\varepsilon)$ obtained by the beam method differs markedly from $\langle \sigma_a(\varepsilon) \rangle_v$ especially when the dissociative attachment resonances peak at energies $\gtrsim 1$ eV. The difference, however, between $\sigma_a(\varepsilon)$ and $\langle \sigma_a(\varepsilon) \rangle_v$ becomes smaller with decreasing peak energy of the dissociative attachment resonance (Blaunstein and Christophorou, 1968), and for low enough energies ($\lesssim 0.5$ eV) and precise forms of $f(\varepsilon, E/P)$, as in experiments with mixtures with carrier gases, $\sigma_a(\varepsilon) \simeq \langle \sigma_a(\varepsilon) \rangle_v$. Mean electron attachment cross sections for assumed Maxwell and/or Druyvesteyn forms of the energy distribution function have been reported for a number of systems such as $O_2$ (Craggs, Thorburn and Tozer, 1957; Tozer, Thorburn and Craggs, 1958; Thompson, 1959) $H_2O$ (Crompton, Rees and Jory, 1965) and dry air (Rees and Jory, 1964; Prasad 1959).

---

*An equation similar to (6.12) was employed for the calculation of $\langle \sigma_a(\varepsilon) \rangle_v$ by a number of workers. However, the square root of the mean energy ($\langle \varepsilon \rangle^{\frac{1}{2}}$) and not the mean-square root $\langle \varepsilon^{\frac{1}{2}} \rangle$ was used. This introduces an error which in the case of a Maxwell distribution function is $\sim 8\%$.

The mean cross sections obtained from swarm data alone, as discussed in this section, are total mean cross sections, i.e. for the total removal of electrons. No identification of the particular ions formed is made, and multiple fragment negative ions are known to form in certain polyatomic molecules. Total mean attachment cross sections at low energies are very useful in efforts to relate electron attachment to biological action of certain molecules (Chapter 9).

### 6.3.1.4 Cross sections for subthermal-peaking electron attachment resonances

A large number of polyatomic molecules have been found to capture electrons non-dissociatively for long periods of time ($> 10^{-6}$ sec) (e.g., Compton and coworkers, 1966; Christophorou and coworkers, 1970a; Section 6.5.3). In most cases (e.g., Christophorou and Compton, 1967; Christophorou and Blaunstein, 1969; Section 6.5.3) the cross section for such processes increased sharply at thermal electron energies, and the electron beam techniques were incapable of giving accurately the magnitude of the cross section and its energy dependence. The shapes of the negative ion currents as a function of electron energy as obtained in electron beam studies are, quite generally, instrumental in such cases (see Figures 6.38 and 6.39).

A simple procedure to obtain both the magnitude and the energy dependence of thermal (or subthermal)-peaking electron attachment cross-section functions from swarm experiments alone has been given recently by Christophorou, McCorkle, and Carter (1970). Let us assume that $\alpha w$ has been measured for a number of $E/P$ values in a carrier gas where $f(\varepsilon, E/P)$ are known down to very low $E/P$ (Section 6.3.1.2; Table 6.1). Let us further make the assumption that for an initial energy region $\sigma_a(\varepsilon)$ can be represented by

$$\sigma_a(\varepsilon) = \frac{A_\gamma}{\varepsilon^\gamma} \tag{6.13}$$

Introducing Equation (6.13) into Equation (6.10) and considering the values of the constants involved, we have

$$\alpha w = 1.52 \times 10^{30} \int_0^\infty \varepsilon^{\frac{1}{2}} f(\varepsilon, E/P) A_\gamma \varepsilon^{-\gamma} \, d\varepsilon \tag{6.14}$$

Hence, when $\alpha w$ is measured in a carrier gas for which $f(\varepsilon, E/P)$ are known down to very low $E/P$ values, the $\gamma$ and $A_\gamma$ can be determined by a least-squares procedure whereby for each set of values of $\gamma$ and $A_\gamma$, $\alpha w$ is calculated constants at each $i$ th value of $E/P$ and compared with the experimental results until the residuals

$$\sum_i \left[(\alpha_i w_i)_{exp} - (\alpha_i w_i)_{cal}\right]^2 \qquad (6.15)$$

are minimized. The optimum values of $\gamma$ and $A_\gamma$ yield both the energy dependence and the magnitude of the attachment cross section for the assumed cross-section form (Equation (6.13)). Christophorou, McCorkle, and Carter (1970) successfully applied this method to a number of systems and their results on four molecules are given in Table 6.2 (see also Figure 6.40 in Section 6.5.3). In the last column of Table 6.2 the energy range (above $3/2\ kT$) over which $\sigma_a(\varepsilon)$ can be approximated by Equation (6.13) with the values of $A_\gamma$ and $\gamma$ given in the table is listed.

**Table 6.2.** Values of $\gamma$, $A_\gamma$, $\sigma_a$ (0.05 eV), $\delta$, and $(\frac{1}{2}+\delta)$

| Compound | $\gamma$ | $A_\gamma$ [(eV)$^\gamma$ cm$^2$] | $\sigma_a$(0.05 eV) (cm$^2$) | $\delta$ | $(\frac{1}{2}+\delta)$ | Energy Range[a,b] (eV) |
|---|---|---|---|---|---|---|
| Sulphurhexafluoride | 1.12 | $4.06 \times 10^{-16}$ | $1.17 \times 10^{-14}$ | 0.64 | 1.14 | $\leqslant 0.11$ |
| 1,4-Naphthoquinone | 1.412 | $3.16 \times 10^{-17}$ | $2.17 \times 10^{-15}$ | 0.863 | 1.363 | $\leqslant 0.5$ |
| Anthracene | 1.148 | $3.16 \times 10^{-18}$ | $1 \times 10^{-16}$ | 0.622 | 1.122 | $\leqslant 0.3$ |
| 1,2-Benzanthracene | 1.031 | $3.16 \times 10^{-18}$ | $5.2 \times 10^{-16}$ | 0.521 | 1.021 | $\leqslant 0.35$ |

[a]This is the energy range over which ln $\alpha w$ varies linearly with ln $\langle\varepsilon\rangle$ and in which region $\sigma_a$ can be approximated by Equation (6.13) with the values of $A_\gamma$ and $\gamma$ listed in this table.
[b]The attachment rates used in this analysis were measured at 298 °K for sulphurhexafluoride (Christophorou, McCorkle and Carter, 1970), at 348°K for 1,4-naphthoquinone (Collins and coworkers, 1970), at 383 °K for anthracene (Christophorou and Blaunstein, 1969), and at $\sim$423°K for 1,2-benzanthracene (Christophorou and Blaunstein, 1969).

Additionally, a straightforward way to check the validity of the assumed simple exponential behaviour of $\sigma_a(\varepsilon)$ (Equation (6.13)) and to find with less labour the power $\gamma$ of $\varepsilon$ has been suggested by Christophorou, McCorkle, and Carter (1970). They have plotted $\alpha w$ versus $\langle\varepsilon\rangle$ on a log–log graph and they have found that below a certain value of $\langle\varepsilon\rangle$ the data could be fitted to a straight line, showing that the attachment rate can be represented by

$$\alpha w = \text{const} \langle\varepsilon\rangle^{-\delta} \qquad (6.16)$$

The values of $\delta$ they obtained this way by a least-squares fitting to their data for the molecules listed in Table 6.2 are given in column 5 ot Table 6.2. If we now take

$$\alpha w \propto \sigma_a \langle\varepsilon\rangle^{\frac{1}{2}} \qquad (6.17)$$

we have

$$\sigma_a \propto \langle\varepsilon\rangle^{-(\frac{1}{2}+\delta)} \qquad (6.18)$$

The values of $(\frac{1}{2}+\delta)$ are listed in column 6 of Table 6.2 and are seen to be in good agreement with the values of $\gamma$ listed in column 2 of Table 6.2. Hence, from measurements of $\alpha w$ versus $E/P$ in $N_2$ and the $\langle\varepsilon\rangle$ versus $E/P$ data for $N_2$ listed in Table 6.1, one can obtain from a log–log plot of $\alpha w$ versus $\langle\varepsilon\rangle$ a good estimate of both the velocity dependence and the magnitude of $\sigma_a(\varepsilon)$ at near-thermal energies quickly and without either the need of a computer or a detailed knowledge of $f(\varepsilon, E/P)$.

### 6.3.1.5 Measurement of the electron attachment coefficient and the rate of electron attachment by the swarm method

Numerous swarm experiments have been designed to study electron attachment processes in gases. The choice of a particular method depends to some extent on the order of magnitude of $\alpha$ to be expected and the nature of the gas under investigation. Basically, in swarm experiments the quantity measured is the rate of electron removal from the swarm which is drifting in a gas (at total pressures of several torr) under the influence of a uniform electric field $E$. The swarm methods differ among themselves mainly in the mode of production of electrons or the way in which the removal of electrons from the swarm is measured. Let $n$ be the number of electrons at a distance $x$ from a positively charged electrode (collector). Then, because of attachment a number

$$\mathrm{d}n = -n\alpha P\,\mathrm{d}x \tag{6.19}$$

where $P$ is the attaching gas pressure in torr, is removed from the swarm when it drifts a distance $\mathrm{d}x$ from $x$ in the direction of the field. The ratio of electron numbers or electron currents $i_2$ and $i_1$ at two distances $x_1$ and $x_2$ is given by

$$i_2/i_1 = e^{-\alpha P(x_2-x_1)} \tag{6.20}$$

Equations (6.19) and (6.20) form the basis of all swarm methods used to determine $\alpha$ or $h$. Expression (6.20) does not include the increase in electron population due to electron impact ionization which takes place at high $E/P$.

Six main electron swarm methods have been employed for the study of electron attachment and these can be distinguished as follows:

(i) *The steady-state diffusion method.* This method was introduced by Bailey (1925). The work of Bailey and his coworkers and subsequent improvements of Bailey's method are discussed by Healey and Reed(1941). The original method has been superseded by the similar and highly improved method of Huxley and coworkers (Huxley, Crompton and Bagot, 1959; Hurst and Huxley, 1960; Huxley and Crompton, 1962; Crompton and Jory, 1962). In this method the coefficient of electron attachment per centimeter of travel in the field direction $\alpha'(=\alpha P)$ and the lateral diffusion coefficient to mobility ratio $D_L/\mu$ can be measured simultaneously. The experimental equipment is similar to that described in Chapter 4 for the measurement of $D_L/\mu$. A schematic diagram is given in Figure 6.7. A stream

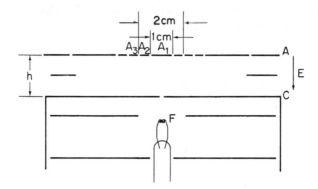

**Figure 6.7** Schematic diagram of the steady-state diffusion method for measurement of the coefficient of electron attachment (Crompton, Rees and Jory, 1965)

of electrons from the filament F enters the diffusion chamber through a small hole in the cathode C and is collected at the anode A having moved the distance CA through the gas under the influence of the uniform electric field $E$. In the presence of an electron attaching gas some of the electrons become attached to the molecules. The collecting electrode (anode) consists of separately insulated sections and the experimental parameters are chosen so that all of the ions entering from the source are collected by the centre disc (earthed throughout the experiment) and so an inappreciable current of electrons and ions falls outside the larger annulus. For a given gas pressure $P$ and $E/P$, the ratio $R$ of the current falling on the annular section $A_2$ to that falling on sections $A_2$ and $A_3$ of the collecting electrode is determined for two (or more) values of the length $h$ of the diffusion chamber. The currents to the sections of the collecting electrode result from both free electrons and negative ions formed by electron attachment in the diffusion chamber. The attachment coefficient $\alpha$ and the ratio $D_L/\mu$ are obtained for each $E/P$ from a pair of measurements of $R$ taken at two values of $h$ (see details in Huxley, Crompton and Bagot, 1959). The method has been applied to the study of electron attachment in a number of molecules such as $O_2$ (Rees, 1965) and $H_2O$ (Crompton, Rees and Jory, 1965).

(ii) *Steady-state electron filter method.* The method was introduced by Loeb in 1926 (see Loeb (1955, 1956)). It consists of a scheme to remove electrons from a mixed swarm of electrons and negative ions drifting through a gas in an electric field. The electron filter method has been used by Bradbury (1933) in his early electron attachment work. Recent electron filter measurements are discussed by Prasad and Craggs (1962). The principle of the method is shown in Figure 6.8. A high-frequency oscillator is fed to the alternate wires of the grids $G_1$ and $G_2$ so that the mobile electrons are collected at the wires, whilst the negative ions may pass through the grid. Either of the two grids $G_1$ or $G_2$ can be moved in the path of the swarm.

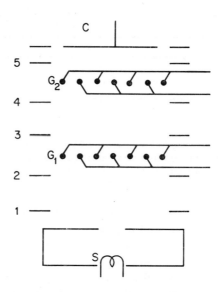

**Figure 6.8** Schematic illustration of the principle of the steady-state electron filter method

This allows the distance between the electron source S and either of the grids to be changed and $\alpha$ to be obtained through Equation (6.20). The total negative current (without the electrons) that survived the distances $SG_1$, $SG_2$ is collected at C. A uniform electric field is maintained by the guard rings 1 through 5. The diffusion method partly replaced the steady-state electron filter method.

(iii) *Microwave method.* Biondi and Brown (1949) have shown that the resonance frequency of a microwave cavity depends on the number density of electrons present. From a measurement of the resonance frequency with and without an attaching gas in the microwave cavity the attachment coefficient can be calculated. The method has been used to study dissociative electron attachment in $I_2$ (Biondi, 1958; Fox, 1958; Biondi and Fox, 1958) and is useful for thermal electron attachment studies. Mulcahy, Sexton and Lennon (1962) and Chantry, Wharmby and Hasted (1962) have applied the microwave method to the study of electron attachment in oxygen.

(iv) *Pulsed drift tube methods.* In these techniques free electrons are produced either photoelectrically or by direct ionization of the gas. A measurement is made of the time dependence of the currents of electrons which survive attachment and the currents of negative ions produced by electron attachment. The method of Chanin, Phelps and Biondi (1959) which is an extension of Doehring's (1952) method, merits attention as it provided (in its various forms of operation) recent information on a number of molecules, especially at very low $E/P$. The principle of the method of Chanin and coworkers is shown in Figure 6.9a. Electrons subsequently to their release at the photocathode by a short-duration ultraviolet pulse drift through the gas-filled tube under the action of the applied uniform electric field. The unattached electrons pass through the control grid and are collected at the collector. The negative ions being much slower are drifting across the tube and from the time dependence of the negative ion current arriving at the control grid the attachment coefficient is obtained. The control grid is used to sample the ion current at some time $t$ delayed relative to the short light pulse. In the time-sampling method of Chanin and coworkers the grid is normally kept closed (no ions or electrons are transmitted) by equal and opposite bias voltages applied to alternative wires. During selected evenly spaced time intervals the grid is made transmitting. This is done by application of rectangular voltage pulses which reduce the field between wires to zero so that most of the ions and electrons pass through to the collector. The grid is set to span periodically at a certain delay time with respect to the photopulse and the currents transmitted to the collector during the transmitting intervals are integrated over a large number of cycles by the electrometer which is connected to the collector. This is done for each delay-time interval $t$ so that the ion current reaching the grid is accurately determined as a function of delay time $t$ without actually measuring the grid current directly. In Figure 6.9b the change of collector current as a function of time delay between the occurrence of the light pulse and the application of the pulse voltage to the grid is shown. The dashed line is an idealized wave form observed at room temperature and at low pressures such that not all of the electrons are attached before they reach the grid. The sharp spike at very early times is due to collection of unattached electrons. It is followed by a slow linear rise (note the semilog plot) during which negative ions are collected. Chanin and coworkers showed that the ion current during this rise is

$$I(t) = I_0 \, e^{\alpha' w_1 t} \qquad (6.21)$$

where $I_0$ is the initial current reaching the control grid, $\alpha'(=\alpha P)$ is the attachment coefficient per unit distance, and $w_i$ is the ionic drift velocity. The product $\alpha' w_i (= v_a w_i / w$; where $v_a$ is the attachment frequency and $w$ the electron drift velocity) is determined from the slope of the ln $I$ versus $t$ plot.

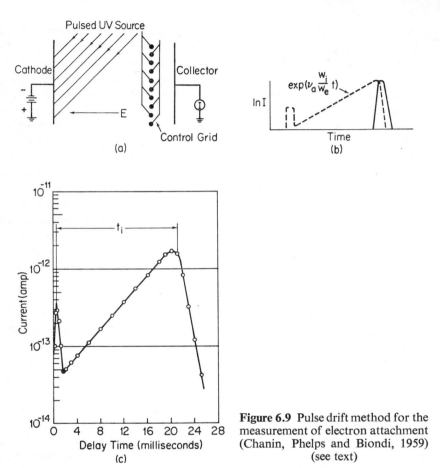

Figure 6.9 Pulse drift method for the measurement of electron attachment (Chanin, Phelps and Biondi, 1959) (see text)

The quantity $w_i$ is evaluated from the observed transit time of the ions formed adjacent to the photocathode. A typical observed current wave form is shown in Figure 6.9c for the case of $O_2$ for $E/P = 0.25$ volts/cm torr. It is to be noted that the mean electron energy is not measured simultaneously in this method. If the attachment coefficient is to be expressed as a function of the mean electron energy $\langle \varepsilon \rangle$, it is necessary to use an independent measurement of $\langle \varepsilon \rangle$. Results obtained by this method are discussed in Section 6.5.

(v) *The pulse-shape method.* The method has been developed by Bortner and Hurst (1958) and is based on the finding that electron attachment appreciably modifies the shape of pulses obtained in a parallel plate ionization chamber. A somewhat similar technique was developed earlier by Herreng (1952). Electrons are produced by α-particle ionization from an O-ring type $^{239}$Pu α-source in a plane normal to the applied uniform electric field at a fixed distance from the collector. To achieve full absorption of the α-particle energy, high total pressures ($\geqslant 200$ torr) are needed and thus all attachment studies by this method have been restricted to gas mixtures. This, as discussed in the previous section, has the advantage of allowing knowledge of the energy distribution functions, characteristic of the particular carrier gas used. Further, in swarm experiments at high $E/P$ an appreciable number of electrons attain enough energy to cause ionization which complicates electron attachment studies. In this method this complication does not arise since the charge collected when the chamber is filled with a mixture of the attaching and non-attaching gases is normalized to that collected for the pure non-attaching gas alone. However, one has to work with as low attaching gas pressures as possible to avoid affecting the carrier gas energy distribution function and the drift velocity, although changes in the latter can be properly accounted for. Low attaching gas pressures are also required to avoid complications due to the Jesse effect (Chapter 2), especially when Ar is used as a carrier gas. The main part of the apparatus used in this method is shown in Figure 6.10 (Christophorou and coworkers, 1965). The apparatus has been modified for high temperature electron scattering (Christophorou, Hurst and Hendrick, 1966) and high temperature electron attachment studies (Christophorou and Blaunstein, 1969). The large amount of information on electron attachment obtained by this method is discussed in subsequent sections of this chapter. The information obtained on elastic electron scattering has been discussed in Chapter 4.

The physical basis of measuring α is as follows: Suppose $n_0$ electrons are formed at a distance $d(= 7$ cm, see Figure 6.10) from the collecting electrode. At a distance $x$ from the point of formation ($x = 0$), only $n$ electrons remain, the rest having been attached. The number of electrons, $dn$, attached in travelling a distance $dx$ in the field direction is given by Equation (6.19). The time variation of the positive electrode (collector) potential due to the arrival of the unattached portion of the electron swarm is

$$V_c(t) = (A/f)\,[1 - e^{-(ft/\tau_0)}] \tag{6.22}$$

where $A = n_0 V_0$, $f = \alpha P_1 d$, $\tau_0$ is the collection time, $P_1$ is the attaching gas pressure, and $V_0$ is the change in the potential of the collector due to a single electron moving the entire drift distance $d$. If the pulse given by Equation (6.22) is observed with a linear pulse amplifier having a response to a step function given by

$$V_s(t) = (t/t_1)\,e^{-t/t_1} \tag{6.23}$$

where $t_1$ is the amplifier differentiating and integrating time constant (assumed equal), the output of the amplifier will be given by

$$V(\tau) = \int_0^{\tau_0} \frac{dV_c(t)}{dt}\, V_s(\tau - t)\, dt \tag{6.24}$$

for $\tau > \tau_0$.

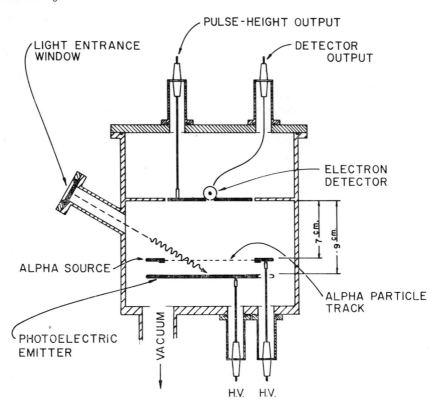

**Figure 6.10** Swarm apparatus for the measurement of electron attachment coefficient and drift velocity (Christophorou and coworkers, 1965)

The quantity $V(\tau)$ has a maximum (pulse height) $V(\tau')$ at $\tau = \tau' \geqslant \tau_0$ which is obtained by integration of Equation (6.24). The pulse height is a function of $\tau_0/t_1$ and $f$. From curves or tables giving the pulse height as a function of $\tau_0/t_1$ at various $f$, the attachment coefficient is calculated from the experimentally measured quantities $V(\tau')$ and $\tau_0$. The effects due to negative ions have been assumed negligible in this treatment.

The electron drift velocity $w$ is obtained simultaneously by measuring the time interval between the release (at the beginning of the drift space) of a swarm of electrons from a photoelectric emitter (see Figure 6.10) and the arrival of the same electron swarm at a Geiger–Müller electron detector at the end of the drift space (Chapter 4).

(vi) *Avalanche methods.* Electron avalanche methods have been employed to study electron attachment at relatively high $E/P$. Under a strong uniform electric field and in the absence of electron attachment, an electron travelling a distance $d$ through a gas in the field direction will, on the average, produce $e^{\alpha_T d}$ additional electrons. The quantity $\alpha_T$ is the Townsend primary (first) ionization coefficient defined as the mean number of ion pairs produced per electron per centimeter drift in the field direction. If in addition to ionization by electron impact, secondary electrons are produced as well, the steady-state growth of prebreakdown currents between electrodes separated by a distance $d$ at any particular $E/P$ is (Townsend, 1915, 1947; Loeb, 1955; Prasad and Craggs, 1962)

$$I = I_0 \frac{e^{\alpha_T d}}{1 - \gamma(e^{\alpha_T d} - 1)} \tag{6.25}$$

where $I_0$ is an externally generated current and $\gamma$ is Townsend's secondary coefficient. Equation (6.25) holds for low current densities and when diffusion losses are unimportant. When electron attachment by either a resonance or a pair-production process takes place, Equation (6.25) has to be modified. If $\alpha'$ and $\delta$ are the probabilities per electron per centimeter of drift in the field direction for resonance attachment and ion-pair production, respectively, then the Townsend current growth equation is modified to (Harrison and Geballe, 1953; Prasad and Craggs, 1962).

$$I = I_0 \frac{\left( \dfrac{(\alpha_T + \delta)}{(\alpha_T - \alpha')} e^{(\alpha_T - \alpha')d} - \dfrac{(\alpha' + \delta)}{(\alpha_T - \alpha')} \right)}{\left( 1 - \gamma \left( \dfrac{\alpha_T + \delta}{\alpha_T - \alpha'} \right) \{ e^{(\alpha_T - \alpha')d} - 1 \} \right)} \tag{6.26}$$

When $\delta$ is small, which is true at not too high $E/P$ (the ion-pair process occurs at relatively high energies compared to the resonance attachment process) Equation (6.26) reduces to

$$I = I_0 \frac{\left( \left( \dfrac{\alpha_T}{\alpha_T - \alpha'} \right) e^{(\alpha_T - \alpha')d} - \dfrac{\alpha'}{(\alpha_T - \alpha')} \right)}{\left( 1 - \dfrac{\gamma \alpha_T}{(\alpha_T - \alpha')} \{ e^{(\alpha_T - \alpha')d} - 1 \} \right)} \tag{6.27}$$

The coefficients $\alpha_T$, $\alpha'$ and $\gamma$ are, then, evaluated simultaneously by a process of curve fitting. The method by its very nature lacks accuracy (see discussions by Prasad (1960); Bhalla and Craggs (1960)). The Townsend avalanche method as well as other pulsed avalanche methods are treated in detail by Prasad and Craggs (1962). Avalanche methods have been employed to the study of a number of gases including oxygen (Harrison and Geballe, 1953; Burch and Geballe, 1957; Prasad and Craggs, 1961; Freely and Fisher, 1964; Frommhold, 1964; Sukhum, Prasad and Craggs, 1967), carbon monoxide (Bhalla and Craggs, 1961), carbon dioxide (Bhalla and Craggs, 1960), sulphur hexafluoride (Bhalla and Craggs, 1962) and chlorine (Božin and Goodyear, 1967).

## 6.3.2 *Electron Beam Methods*

### 6.3.2.1 Introduction

Electron beam experiments, in contrast with swarm experiments, deal with nearly monoenergetic electrons and single collision processes (pressure range $10^{-6}$ to $10^{-3}$ torr). Until recently beam experiments were restricted to electron energies above $\sim 2$ eV which excluded studies in the interesting range below this energy. Most electron–molecule collision processes have a sharp dependence on the kinetic energy of the incident electron. The energy widths of the dissociative electron attachment resonances are of the order of 0.2 to 2 eV (Section 6.4). Thus, to obtain the true dependence of the cross section for the various resonance electron attachment processes a narrow electron energy distribution is required. The retarding potential difference method (RPD) (Fox and coworkers, 1955) has considerably improved the electron beam method allowing an energy resolution of $\sim 0.1$ eV to be generally obtained. Since in such quasi-monoenergetic electron beams, the energy spread is of the order of 0.1 eV, the true shape of resonances is not revealed when the latter varies rapidly with energy. One should note here the advantage of the swarm method for studies at epithermal energies. In a thermal electron gas at 300°K, $\sim 90\%$ of the electrons have energies below 0.08 eV, the thermal spread of energy being $\sim 0.04$ eV.

In addition to having a narrow electron pulse, a correct energy scale is a basic quality of beam experiments. Energy scale calibrations have been, in most cases, relative, i.e. the energy scale has been calibrated by comparison with the energy of well-established resonances, or by use of the energies of certain well-known electronic transitions of the gas under study or of an auxiliary gas introduced in the collision chamber for this purpose. Because of the difficulty in establishing an accurate energy scale for negative-ion studies, a number of investigators have used the property of $SF_6$ to form $SF_6^-$ at essentially zero energy (Hickam and Fox, 1956) to establish the electron energy scale. However, it has been pointed out by Schulz (1960b) and Christophorou and coworkers (1965) that the energy scale established using $SF_6^-$ is in most cases underestimated. The potential along the path of the electron beam in the collision chamber must be considered in detail. If, for example, the potential along the path of the electron beam is positive with respect to the potential of the entrance hole of the collision chamber, serious errors in the energy scale can arise (Christophorou and coworkers, 1965). An independent method of establishing the energy scale has been developed by Christophorou and coworkers (1965) and will be discussed in Section 6.3.3. Although electron beam experiments are relatively easy to interpret compared with swarm experiments, they are more difficult to perform. To the experimental problems of energy resolution and energy scale calibration, a number of other difficulties can be added such as: discriminating effects in detection efficiency when negative ions possess large or varied amounts of kinetic energy or have anisotropic angular distributions especially in experiments with small acceptance angles. Further, space-charge effects, errors in current and pressure measurements, scattered electrons*, and poisoning of the electron source by electronegative gases are common difficulties in electron beam studies. Thus, despite studies (e.g. Hagstrum, 1951) of the appearance potential of negative ions formed by dissociative electron attachment or by ion-pair production, no serious attempt to measure negative ion currents had been made until quite recently. Until the middle 1950's virtually no absolute measurements of low-pressure, single-collision condition electron attachment cross sections had been made.

Electron beam experiments have been adapted to identify the products of electron–molecule interactions and to measure their kinetic energies. The various ion yields are commonly measured in relative units, and the determination of electron attachment cross sections is essentially limited to total ionization experiments. This shortcoming can be overcome by incorporating swarm data with beam negative ion data, as will be discussed in Section 6.3.3.

### 6.3.2.2 Measurement of negative ion currents and electron attachment cross sections as a function of electron energy

Several techniques are in use for negative ion studies under single-collision conditions. The basis of these techniques is essentially the same: a well-collimated monoenergetic electron beam passes through an ionization chamber containing the gas to be investigated at a low pressure†, and upon

---

*In most total ionization experiments, elastically and inelastically scattered electrons can contribute to the negative ion current which results in an overestimate of the electron attachment cross section.
†High-pressure mass spectrometric investigations have also been initiated and allowed studies of secondary processes leading to ions which are not produced in the single collision low-pressure beam experiments.

emerging from the collision chamber it is collected by an electron trap. In passing through the gas a minor portion of the electron beam undergoes interactions with the gas molecules, and the ions formed are either totally collected or extracted from the collision region to be mass analysed. Three types of beam experiments will now be discussed briefly:

(i) *Lozier tube.* The apparatus was originally designed by Lozier (1933, 1934) and formed the basis of many beam experiments not incorporating mass analysis. Schematically the Lozier tube is represented in Figure 6.11 (Tozer, 1958). Electrons from the filament F are accelerated through the slits A, B and C to the desired energy and they pass through the electric field-free collision chamber D. Axial magnetic fields ($\sim 200$ Gauss) are used to confine the electrons to a narrow beam. The emerging beam from the collision chamber is collected at E. The pressures used are low so that multiple collisions do not occur and the attenuation of the incident electron beam is not appreciable. *I* represents a set of vanes concentric with and perpendicular to the chamber axis. The vanes I, as well as C and L, are at ground potential. An ion draw-out potential is applied to the outer set of vanes O and aids the collection of negative ions by the cylindrical electrode H surrounded by the guard electrode G. Only those negative ions are collected whose initial direction is within $\sim 12°$ of the perpendicular to the electron beam. For the measurement of the kinetic energy of the ions a retarding potential is applied to the collector.

Absolute total cross sections for the formation of negative ions are obtained by comparison of the yields of negative ions with known positive ion yields. Reversing the draw-out field, the positive ion current at some moderate electron energy is compared with the negative ion current, and the absolute electron attachment cross section at an energy $\varepsilon_1$, $\sigma_a(\varepsilon_1)$, is obtained from:

$$\frac{\sigma_a(\varepsilon_1)}{\sigma_T(\varepsilon_2)} = \frac{i_-(\varepsilon_1)}{i_+(\varepsilon_2)} \tag{6.28}$$

where $\sigma_T(\varepsilon_2)$ is the total ionization cross section at $\varepsilon_2$ (known from independent measurements), $i_+(\varepsilon_2)$ is the saturated positive ion current at $\varepsilon_2$, and $i_-(\varepsilon_1)$ is the saturated negative ion current at $\varepsilon_1$, corresponding to the same electron current. The magnitude and the shape of a dissociative attachment cross section $\sigma_{da}(\varepsilon)$ is dependent upon kinetic and angular distribution discriminations as well as on elastically and inelastically scattered electrons. Further, it depends on the accuracy of the determination of the cross section for the positive ions.

(ii) *Total ionization methods.* Many experiments have employed a total ionization apparatus to determine absolute electron attachment cross sections. An apparatus similar to the Lozier tube but without the vanes

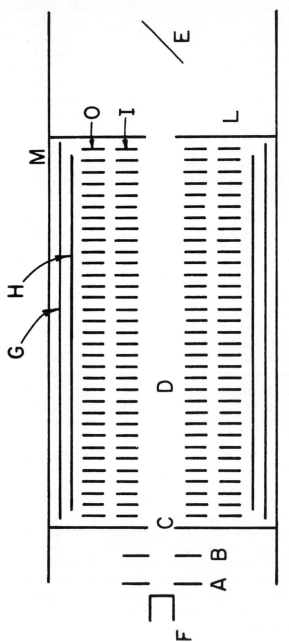

**Figure 6.11** Schematic diagram of the Lozier tube (Tozer, 1958)

perpendicular to the electron beam (which might seriously affect negative ion collection) was used by Buchel'nikova (1959) in connection with an RPD gun. Her results on electron attachment cross sections for $O_2$, $H_2O$ and a number of halogenated compounds are discussed in Section 6.4. Another apparatus of this type was employed by Schulz (1959a, b, 1960a, 1961, 1962). Figure 6.12 shows a schematic diagram of Schulz's total ionization apparatus which employs an RPD source and is free of kinetic energy and angular discrimination effects (see also Figure 5.5 in Chapter 5). The ion collector C is held at a slightly positive potential with respect to ground and a slight field penetration through the grid G into the ionization chamber ensures complete ion collection. The penetrating field is reversed when the kinetic energy of the ions is to be measured. The collector plate is then used as a retarding electrode. Elastically and inelastically scattered electrons may affect the measured ion currents in this method.

(iii) *Mass spectrometric methods.*   Mass spectrometric investigations of negative ions, pioneered by Hagstrum (1947, 1951), were performed mostly on simple rather than on complex molecules. It is virtually impossible to determine electron attachment cross sections or ion kinetic energies by mass spectrometers. However, appearance potentials and negative ion yields as a function of electron energy can be determined with precision. Many mass spectroscopic studies employed the RPD method for improved energy resolution. The resolution of the quasi-monoenergetic electron beam can be greatly improved by avoiding the large contribution to the spreading of the electron pulse originating from electrons reflected in the source. This latter contribution can be greatly diminished by electrodepositing platinum black on all of the electrodes in the collision region (Christophorou and coworkers, 1965). The spreading in the electron energy distribution plays a major role in obtaining the true shape of the attachment cross section for resonance processes, as opposed to positive ions where such energy spread is generally unimportant. Dissociative and non-dissociative electron attachment resonances often have as narrow a resonance width as a few tenths of an electron volt (Section 6.4). An energy spread of several tenths of an electron volt in the electron pulse can, therefore, seriously alter the observed cross sections for narrow resonances and can extend their 'tails'. In cases where the width of the ion current, $I(\varepsilon)$, at half-height is comparable to that of the electron pulse, $I_e(\varepsilon)$, it is necessary to unfold $I(\varepsilon)$ in order to approximate better the true shape of the cross section. Christophorou, Compton and Dickson (1968) in their studies of dissociative electron attachment in hydrogen halides and their deuterated analogues employed an unfolding procedure similar to that of Owen and Primakoff (1948) and Freedman and coworkers (1956).

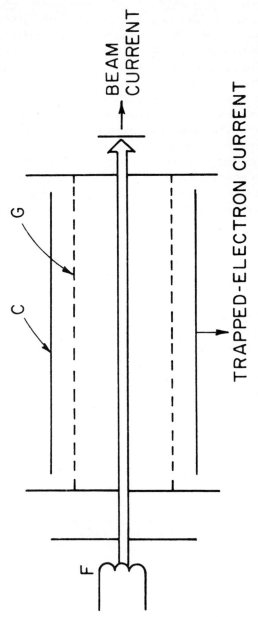

**Figure 6.12** Schematic diagram of Schulz's total ionization apparatus (see also Figure 5.5 in Chapter 5)

### 6.3.3  The Swarm-Beam Technique

There are certain unique advantages in the combination of electron swarm and electron beam techniques. Such a swarm–beam combination has been performed (Christophorou and coworkers, 1965) and has provided a new tool for the determination of dissociative electron attachment cross sections (see results in Section 6.4). Briefly, from electron swarm experiments the absolute rate of electron attachment, $\alpha(E/P) \times w(E/P)$, is obtained; its relation to the electron swarm energy distribution function, $f(\varepsilon, E/P)$, and the attachment cross section, $\sigma_a(\varepsilon)$, is given by Equation (6.10). From electron beam experiments the negative ion yields, $I(\varepsilon)$, as a function of electron energy are obtained in relative units, and identification of the products of electron-molecule collisions is made. Provided that the width of the electron beam is small compared with the width of the negative ion current* and that the ions involved in the two experiments are the same, a trial function can be constructed as

$$\sigma_{aj}(\varepsilon) = K_j T_j I(\varepsilon) \tag{6.29}$$

where $T_j$ is an operator which simply shifts the $I(\varepsilon)$ curve along the $\varepsilon$ axis to correct for any uncertainties in the original energy calibration, and $K_j$ is a constant for each $T_j$ which transforms the $I(\varepsilon)$ curves from relative to absolute cross section units. To obtain both $T_j$ and $K_j$, Equation (6.29) is introduced into Equation (6.10) and a double least-squares procedure is performed, where one obtains for each $i$th value of $E/P$ the best fit between the calculated and the experimental rates through Equation (6.30):

$$\frac{\mathrm{d}}{\mathrm{d}T_j} \left\{ \sum_i [\, N (2/m)^{1/2} K_j \int_0^\infty \varepsilon^{1/2} T_j I(\varepsilon) f_i(\varepsilon) \, \mathrm{d}\varepsilon - \alpha_i w_i \,]^2 \right\} = 0 \tag{6.30}$$

Equation (6.30) forms the basis of the swarm–beam technique. It combines the information from the beam experiments ($I(\varepsilon)$) with that from the swarm experiments ($\alpha \times w$) and provides $K_j$ (thus $\sigma_a(\varepsilon)$) and $T_j$ (thus energy scale calibration). The method is very sensitive in determining the absolute magnitude of the cross section and in establishing the correct energy scale (Christophorou and coworkers, 1965, 1966, 1968).

---

*If the width of the electron pulse cannot be neglected, an unfolding procedure prior to the application of the swarm-beam combination is performed so that the unfolded $I(\varepsilon)$ curves are used in the swarm-beam equations.

## 6.4 Resonance Dissociative Electron Attachment

### 6.4.1 *Theoretical Considerations*

The physical grounds of electron attachment to molecules were first discussed by Bloch and Bradbury (1935) and later by Massey (1950). Electron attachment was considered to take place as a result of a breakdown in the Born–Oppenheimer approximation, the influence of the kinetic energy of the nuclei on the electron wave functions giving rise to transitions which result in electron capture. More recently the effect of nuclear motion in dissociative electron attachment was investigated by Stanton (1960) and Chen (1963, 1966).

An alternative, very successful description of dissociative electron attachment to molecules which treats the process as a resonance phenomenon has been developed recently (Bardsley, Herzenberg and Mandl, 1964, 1966; O'Malley, 1966; Chen, 1967, 1969; Chen and Peacher, 1967). In this treatment the mechanism of dissociative electron attachment is assumed to depend exclusively on the resonance ignoring, in general, non-adiabatic terms in the Hamiltonian. However, both direct and non-adiabatic transitions appear to play an important role in dissociative attachment (Chen, 1966, 1969; O'Malley, 1966). It should be noted that the role of an unstable intermediate negative ion in dissociative attachment was first investigated by Holstein (1951) who introduced the concept of the *survival probability*. That is, due to the transient nature of the negative ion, the captured electron may be emitted prior to the dissociation of the negative ion intermediate, the survival probability accounting for the possibility that the electron will remain captured while the molecular ion is dissociating along the repulsive curve in the autoionization region (Figure 6.1). The theoretical aspects of dissociative electron attachment have been reviewed by many workers in connection with other resonance phenomena in electron–atom and electron–molecule collisions (see, for example, Bardsley and Mandl, 1968; Chen, 1969; also Chapters 4, 5 and 6).

The resonance scattering theory has provided a very convenient expression for the cross section for dissociative electron attachment $\sigma_{da}$, which for a diatomic molecule AX, initially in the $v = 0$ vibrational level of the ground electronic state takes the form (O'Malley, 1966)

$$\sigma_{da}(\varepsilon)_{v=0} = \frac{4\pi^{3/2}}{k_i^2} \bar{g} \frac{\Gamma_{\bar{a}}}{\Gamma_d} \left[ \exp\left( \frac{(\frac{1}{2}\Gamma_a)^2 - (\bar{E}_0 - \varepsilon)^2}{(\frac{1}{2}\Gamma_d)^2} \right) \right] e^{-\rho(\varepsilon)} \tag{6.31}$$

In Equation (6.31) $k_i^2 = (2m/\hbar^2)\varepsilon$, $m$ is the electron mass, $\varepsilon$ is the electron energy, $2\pi\hbar$ is Planck's constant, $\bar{g}$ is a statistical factor, $\Gamma_a$ is the total autoionization width, $\Gamma_{\bar{a}}$ is the partial autoionization width, $\Gamma_d$ is the experimentally determined dissociative attachment cross-section width,

$\bar{E}_0 = \varepsilon_0 + \frac{1}{2}\hbar\omega$, $\varepsilon_0$ is the electron energy at the peak of $\sigma_{da}(\varepsilon)$, $\frac{1}{2}\hbar\omega$ is the zero point energy, and $e^{-\rho(\varepsilon)}$ is the *survival probability*. The exponent $\rho(\varepsilon)$ is expressed as

$$\rho(\varepsilon) = \int_{R_\varepsilon}^{R_C} \frac{\Gamma_a(R)}{\hbar} \frac{dR}{v(R)} \qquad (6.32)$$

where $R_\varepsilon$ is the interatomic separation at which electrons of energy $\varepsilon$ reach the negative ion state (Figure 6.1), $R_c$ is the value of $R$ at the crossing point between the negative ion and neutral molecule potential energy curves, and $v(R)$ is the relative velocity of separation of A and $X^-$.

A simplified form of expression (6.31) is

$$\sigma_{da}(\varepsilon) = \sigma_0 e^{-\tau_s/\tau_a} \qquad (6.33)$$

In Equation (6.33) $\sigma_0$ is the cross section for the formation of $AX^{-*}$,

$$\tau_s \left( = \int_{R_\varepsilon}^{R_C} \frac{dR}{v(R)} \right)$$

is the time required for A and $X^-$ to separate from the point of formation at $R_\varepsilon$ to the crossing point at $R_c$, $\tau_a (= \hbar/\bar{\Gamma}_a)$ is the mean autoionization lifetime, and $\bar{\Gamma}_a$ is the average autoionization width. Although expression (6.33) is a very convenient relation to apply to analyses of experimental data, it may not, as pointed out by Chen and Peacher (1967), always be valid. It should also be noted that due to the energy-dependent factors in Equation (6.31) the energy $\varepsilon_{max}$ at which the measured cross section peaks will not occur at the actual peak energy $\varepsilon_0$. If $\rho(\varepsilon)$ (Equation (6.31)) is not varying too rapidly, $\varepsilon_{max}$ will be given approximately by (O'Malley, 1966)

$$\varepsilon_{max} = \varepsilon_0 - \tfrac{1}{8}\Gamma_d^2(\rho' + 1/\varepsilon) \qquad (6.34)$$

where $\rho' = d\rho/d\varepsilon$ is always positive. It is only when $\rho'$ and $\Gamma_d$ are very small that the observed peak energies $\varepsilon_{max}$ coincide with their true values $\varepsilon_0$ (O'Malley, 1966; Christophorou and Stockdale, 1968).

Let us now turn our attention to the dependence of the dissociative attachment cross section on the reduced mass, $M_r$, of the separating fragments, A, $X^-$. It can be seen from Equations (6.31) and (6.33) that the peak value of the dissociative attachment cross section should depend on $M_r$ in two ways. Firstly, it depends on $M_r$ through the survival probability. Since the different isotopic species experience the same forces, the relative velocities of separation are proportional to $M_r^{-\frac{1}{2}}$ and hence the separation times $\tau_s$—thus the exponent in the survival probability (Equation (6.32))—will be proportional to $M_r^{\frac{1}{2}}$. This can lead to very large isotope effects, the dissociative attachment cross section being smaller for the heavier isotope. Secondly, it depends on $M_r$ through the mass-dependent quantity $\sigma_0$ (Equation (6.33)). Here the reduced mass enters in two places through the vibrational amplitude. The appearance of the vibrational amplitude in the Gaussian (Equation (6.31)) not only leads to an isotope effect in the width of $\sigma_{da}$ but also in its magnitude which is of the same form, $\exp(\text{const } M_r^{\frac{1}{2}})$, as for the survival factor. At the peak energy the constant is positive leading to a higher cross section for the heavier isotope. For the survival probability, the exponent is negative but it has an identical mass dependence. The isotope effects resulting from the Gaussian factor are in general, small. They, however, may become important in cases such as the vertical onset case (Section 6.2.2). Additionally, the term preceding the Gaussian factor in Equation (6.31) introduces an $M_r^{\frac{1}{4}}$ dependence of $\sigma_{da}$. If we neglect the effect of the Gaussian factor, we may write

$$\sigma_{da} \propto M_r^{\frac{1}{4}} \exp\left(-\text{const } M_r^{\frac{1}{2}}\right) \tag{6.35}$$

When the survival probability is small (large $\bar{\Gamma}_a$), autoionization is important and the second term in Equation (6.35) dominates. Large isotope effects are observed in such a case. When the survival probability is large (small $\bar{\Gamma}_a$) autoionization is less important and the $M_r^{\frac{1}{4}}$ term may well account for the isotope effect in $\sigma_{da}$. It is to be noted that these considerations are based on a diatomic-molecule-treatment and further that no *a priori* prediction can be made as to which of the above mass-dependent factors is the most important one. Experimentally (Section 6.4.2) three classes of isotope effects in $\sigma_{da}$ have been observed:

(i) *Large direct.* $H_2$, HD, $D_2$ (Schulz and Asundi, 1965, 1967; Rapp, Sharp and Briglia, 1965). In these systems, especially for the 3.5 eV process (Section 6.4.2.1 A) autoionization is very important ($\bar{\Gamma}_a$ (thus, $\rho(\varepsilon)$) is very large). The observed large isotope effect has been explained by the effect of the reduced mass on the survival probability (Demkov, 1965; Schulz and Asundi, 1967).

(ii) *Small direct.* $H_2O$, $D_2O$ (Compton and Christophorou, 1967), HCl, DCl, HBr, DBr, HI, DI (Christophorou, Compton and Dickson, 1968). Here $\sigma_{da}$ has been found to vary approximately as the inverse of the square root of the reduced mass. The competition between autoionization and dissociation could account for these observations, the times of separation $\tau_s$ being proportional to $M_r^{\frac{1}{2}}$. However, it is not clear as to why, in this case, the second term in Equation (6.35) will determine alone the observed isotope effect. Compton and Christophorou (1967) have noted that, alternatively, this small direct isotope effect can be explained without considering auto-ionization to be in competition with dissociation by applying first order perturbation theory and taking the kinetic energy operator as the perturbation. This treatment is valid only in cases where autoionization is relatively unimportant (Christophorou and Stockdale, 1968). Similar small direct isotope effects have been observed for $NH_3$ and $ND_3$ (Sharp and Dowell, 1969; Compton, Stockdale and Reinhardt, 1969).

(iii) *Small inverse.* $CH_4$, $CD_4$ (Sharp and Dowell, 1967). This small inverse isotope effect ($\sigma_{H^-}(CH_4) \simeq 0.8\sigma_{D^-}(CD_4)$) has been explained (O'Malley, 1967b; Christophorou, Compton and Dickson, 1968) as being due to the preexponential term in Equation (6.35). The $M_r^{\frac{1}{4}}$ dependence, as discussed earlier in this section, is explicitly predicted by Equation (6.31) when $\bar{\Gamma}_a$ and $\rho(\varepsilon)$ are small. It is noted that although the ratio of the peak cross section values is proportional to $[(M_r)_D/(M_r)_H]^{\frac{1}{4}}$ the energy integrated cross sections for the two isotopic species may be the same.

## 6.4.2 *Dissociative Electron Attachment Cross Sections*

### 6.4.2.1 Diatomic molecules

A. $H_2$, HD and $D_2$. Dissociative electron attachment to $H_2$ has been studied by a number of workers using total ionization methods. The cross section for $H^-/H_2$ shows main resonance peaks at 3.75 eV (Schulz and Asundi, 1965, 1967) at $\sim 10.1$ eV and at 14.0 eV (Schulz, 1959a; Rapp, Sharp and Briglia, 1965). The two lower peaks correspond to states asymptotic to $H(1s) + H^-$, while the peak at 14 eV corresponds to an excited state asymptotic to $H(2s) + H^-(1s)^2$ since the ions from this state have very low kinetic energy (Khvostenko and Dukel'skii, 1958). The minimum energy required to produce $H(2s) + H^-(1s)^2$ is 13.9 eV and it appears that a vertical onset occurs at that energy for the sharp peak in $H_2$, HD and $D_2$ with the actual spread of the electron beam rounding off the threshold. The 3.75 eV process also has a very sharp onset and peaks sharply close to it. It is clearly a vertical onset case where dissociation of $H_2^-{}^*$ becomes energetically possible at energies $\geqslant 3.75$ eV.

The peak energy, $\varepsilon_{max}$, peak cross section, $\sigma_{da}(\varepsilon_{max})$, cross section energy onset, and energy width of the resonance (full width at half height) for $H_2$, HD and $D_2$ are summarized in Table 6.3. The data of Schulz (1959a) and Rapp, Sharp and Briglia (1965) for $H^-$ from $H_2$, $H^-$ (or $D^-$) from HD, and $D^-$ from $D_2$ are plotted as a function of the electron energy for the 10 and 14 eV processes in Figure 6.13. In the same figure the data of Schulz and Asundi (1967) for the 3.75 eV process are also plotted. A peak at 6.8 eV in the study of $H_2$ by Schulz (1959a) has been attributed to $H^-$ from $H_2O$ (present as an impurity in his experiment).

The $H^-$ ions produced by electrons with energies between 8 to 12.5 eV have about 3.5 eV of initial energy (Schulz, 1959a). It is seen from Figure 6.13 that in this energy region there is evidence that the cross section exhibits some structure. The rising portion of the cross section above 14 eV has been originally attributed (Khvostenko and Dukel'skii, 1958) to ion-pair formation ($H_2 + e \rightarrow H^+ + H^- + e$). However, the work of Rapp, Sharp and Briglia (1965) indicates that the rising portion of the cross section starts at least 1.5 eV below the minimum energy of 17.2 eV for ion-pair formation. The energy scales may probably be off, although the possibility exists that

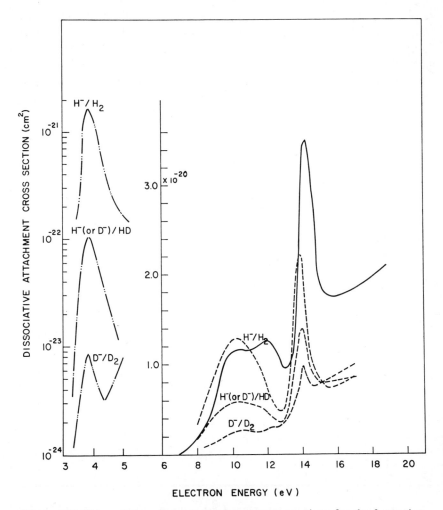

**Figure 6.13** Dissociative electron attachment cross sections for the formation of $H^-$ from $H_2$, $H^-$ (or $D^-$) from HD and $D^-$ from $D_2$. Curves – – – (Rapp, Sharp and Briglia, 1965); curve ——— (Schulz, 1959a); curves —··—··— (Schulz and Asundi, 1967)

scattered electrons may be contributing to the negative ion current since in both the work of Schulz (1959a) and Rapp, Sharp and Briglia (1965) mass analysis was not provided. This may partially explain the large discrepancy between the absolute magnitudes of the cross sections near 14 eV as reported by Schulz (1959a) and by Rapp, Sharp and Briglia (1965).

The data on the dissociative electron attachment to $H_2$, HD and $D_2$ have been the subject of many recent theoretical investigations and have served as a prototype in dissociative electron attachment studies. Potential energy diagrams are shown in Figure 6.14 (Schulz and Asundi, 1967). The ground state $^1\Sigma_g^+$ of $H_2$, the ground compound state $^2\Sigma_u^+$ of $H_2^-$ and the repulsive compound state $^2\Sigma_g^+$ and the bound compound state $^2\Sigma_g^+$ of $H_2^-$ (Taylor and Williams, 1965) are shown in the figure. At electron energies in the range from 1 to 4 eV the compound state $^2\Sigma_u^+$ can be reached (Taylor and

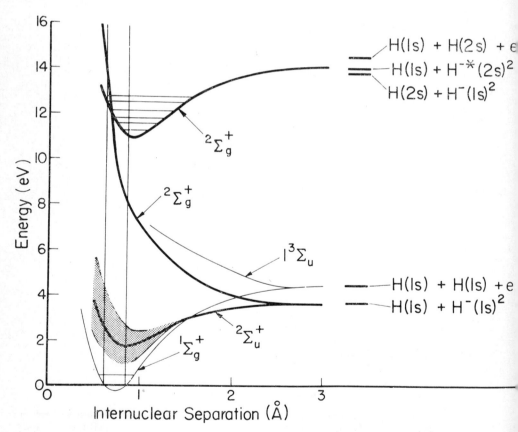

**Figure 6.14** Schematic potential energy curves for $H_2$ and $H_2^-$ (Schulz and Asundi, 1967) (see text for discussion)

Harris, 1963; Taylor and Gerhauser, 1964). Since the formation of $H^-$ becomes energetically possible at incident electron energies greater than 3.75 eV, at energies below this value the compound state $^2\Sigma_u^+$ will autoionize decaying down to the lowest or higher vibrational levels of the $^1\Sigma_g^+$ state of $H_2$. The stippled area surrounding the solid curve represents the finite width of the state $^2\Sigma_u^+$, which is a function of the internuclear separation $R$. See Chapters 4 and 5 as to the role of the $^2\Sigma_u^+$ compound state in the elastic and inelastic scattering channel.

The dissociative electron attachment process at $\sim 10$ eV has been attributed (Schulz and Asundi, 1967) to the repulsive $^2\Sigma_g^+$ compound state. The bound $^2\Sigma_g^+$ state leads to resonances in the elastic and inelastic channels and is probably responsible for the structure observed in the formation of $H^-$ around 12 eV. As discussed in Chapters 4 and 5, evidence for two series of overlapping resonances in the energy range 11 to 13 eV has been provided from electron scattering experiments. Additional evidence for overlapping among the resonances of $H_2^-$ in the energy region 11 to 13 eV has been provided by the work of Dowell and Sharp (1968). They observed structure in the $H^-/H_2$ cross section in the energy region 11.2 to 12.5 eV, which they attributed to the vibrational spacing of the compound $H_2^-$ states. Here, then, dissociative attachment proceeds as an indirect process (Figure 6.14). The incident electron is first captured into the vibrational levels of the attractive compound negative-ion state, and subsequently as a result of the overlapping of the attractive compound state with a continuum of compound negative ion states, a radiationless intramolecular decomposition takes place. Dowell and Sharp (1968) pointed out that to explain their results it is essential to consider the overlapping of the stable $^2\Sigma_g^+$ compound state of $H_2^-$ with the unstable $^2\Sigma_g^+$ compound state of $H_2^-$ in the energy range 11 to 13 eV. Calculations of the ground $^2\Sigma_u^+$ and lowest $^2\Sigma_g^+$ compound $H_2^-$ states have been performed by Bardsley, Herzenberg and Mandl (1966) and Chen and Peacher (1968).

*Isotope effect.* It is seen from Figure 6.13 and Table 6.3 that the dissociative attachment cross section which results from the compound state $^2\Sigma_u^+$ exhibits an exceptionally large isotope effect. This is in contrast to the relatively small isotope effect (a factor of two) for the dissociative electron attachment at $\sim 10$ and 14 eV. The fact that a large isotope effect exists for the 3.75 eV process implies that the mean autoionization width of $^2\Sigma_u^+$ is considerably larger (its mean lifetime considerably smaller) than either the repulsive or the bound $^2\Sigma_g^+$ states of $H_2^-$. The heavier atom requires more time to reach the stable region with respect to autodetachment, and thus the cross section is smaller. Schulz and Asundi (1967) assumed the simple expression (6.33) for the dissociative attachment cross section, and from the measured cross sections for $H_2$, HD and $D_2$ they obtained for the 3.75 eV process a cross section for compound negative ion formation, $\sigma_0$, equal to $7 \times 10^{-16}$ cm$^2$ and a mean autoionization lifetime for the $^2\Sigma_u^+$ state of $H_2^-$ equal to $1 \times 10^{-15}$ sec on the assumption that the mean autoionization width, $\bar{\Gamma}_a$, is the same for the three molecules. Because of the small autoionization lifetime and the relatively long time it takes for the dissociation fragments to separate, the system remains in the unstable region longer and the survival probability (and thus the cross section) becomes smaller. Demkov (1965) explained these isotope effects similarly, i.e. as coming directly from the survival probability factor in the expression for the cross section, assuming the cross section for formation of the compound state insensitive to the mass. However, as stated in Section 6.4.1, this may not be entirely correct.

    B. HCl, DCl, HBr, DBr, HI, DI.    There exists considerable early swarm work on some of the hydrogen halides (Bailey and Higgs, 1929; Bailey and Duncanson, 1930; Bradbury, 1934a; Healey and Reed, 1941) which appears to be inconsistent due mainly to unknown energy distributions in the electron swarms. There exists, also, some beam work on HF (Frost and McDowell, 1958a), HCl (Barton, 1927; Nier and Hanson, 1936; Gutbier and Neuert, 1954; Reese, Dibeler and Mohler, 1956; Fox, 1957; Frost and McDowell, 1958a; Buchel'nikova, 1959; Thorburn, 1959), HBr (Gutbier and Neuert, 1954; Reese, Dibeler and Mohler, 1956; Frost and McDowell, 1958a; Buchel'nikova, 1959), and HI (Frost and McDowell, 1958a). Absolute cross sections for the formation of Cl$^-$ from HCl and Br$^-$ from HBr have been reported by Buchel'nikova (1959), while more recently Christophorou, Compton and Dickson (1968) have reported dissociative electron attachment cross sections for HCl, DCl, HBr, DBr, HI and DI. Christophorou and coworkers applied the swarm–beam method for the determination of the dissociative attachment cross sections and the energy scale calibration, and have corrected their results for the finite width of the electron pulse. The results of Christophorou and coworkers are plotted in Figure 6.15. It is possible to use the shapes of the cross sections in Figure 6.15 to draw potential energy curves for the AX$^-$* (where A = H, D and X = halogen) compound state in the Franck–Condon region.

    The results of Christophorou and coworkers are listed with those of Buchel'nikova in Table 6.3. The cross sections for HCl and HBr as obtained by Christophorou and coworkers and by Buchel'nikova are not in agreement. The ratio of $\sigma_{da}(\varepsilon_{max})$ for HCl as obtained by Christophorou and coworkers to $\sigma_{da}(\varepsilon_{max})$ as obtained by Buchel'nikova is 5.0, while the similar ratio for HBr is 4.7. Since the ratio of the energy integrated cross section for HCl to HBr as obtained by Buchel'nikova is 9.6, and as obtained by Christophorou and coworkers it is 10, it seems possible that the cross sections of Buchel'nikova, obtained with a total ionization apparatus, are smaller because of incomplete ion collection.

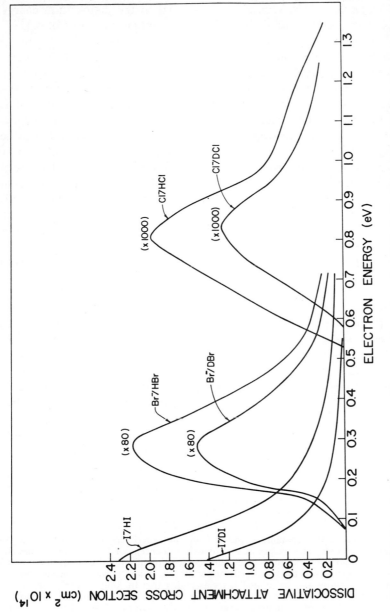

**Figure 6.15** Dissociative electron attachment cross sections for HCl, DCl; HBr, DBr; HI, DI (Christophorou, Compton and Dickson, 1968)

The appearance potential for $Cl^-$ from HCl deserves some comment. The minimum energy at which $Cl^-$ ions can appear from dissociative attachment to HCl molecules in the ground state is equal to the dissociation energy of HCl less the electron affinity of chlorine. Accepting the value of 4.43 eV for the dissociation energy of HCl, an appearance potential of 0.64 eV (0.63 ± 0.05, Christophorou and coworkers (1968); 0.62 ± 0.05 eV, Frost and McDowell (1958a); 0.66 ± 0.02, Fox (1957)) is not compatible with an electron affinity of 3.61 eV for Cl obtained from photodetachment experiments (Berry and Reiman, 1963). The measured appearance potential is 0.18 eV lower than the calculated minimum appearance potential and the difference is believed (Christophorou, Compton and Dickson, 1968) to be well outside the experimental error. One possible explanation for this discrepancy might be the effect of initial population of rotational states at room temperature, the effect being due to initial population of vibrational levels is considered to be unlikely. According to Chen and Peacher (1967) for attachment very close to the threshold, where the kinetic energy of the separating fragments is small, the survival probability can increase considerably with increasing rotational quantum number.

A small direct isotope effect in the dissociative attachment of electrons to HX and DX (X = halogen) molecules is clearly evident from the data of Christophorou, Compton and Dickson. As seen from Table 6.3, the cross section is always smaller for the heavier isotope, while the ratio of the cross sections (both the peak cross section values and the energy integrated values) for isotopic species is well contained within the square root of the inverse ratio of the reduced masses of the dissociation products. This finding is in accord with that of Compton and Christophorou (1967) on $H_2O$ and $D_2O$ (see later this section), but the isotope effect for this group of molecules is small compared to that observed in the case of $H_2$, HD and $D_2$ (see discussion in Section 6.4.1)*.

An attachment process other than that leading to direct dissociation has been observed by Christophorou and coworkers in HCl, DCl, HBr and DBr at near thermal electron energies. No parent negative ion was found in a

---

*Christophorou, Compton and Dickson (1968) assumed that the potential energy curve for HX and DX (X = halogen) molecules is represented by a Morse function and that for a purely repulsive $AX^{-*}$ state by

$$V(R) = A/R^2 + B$$

They, then, evaluated the constants A and B from the experimental data in the Franck-Condon region and from the values of the dissociation energy of HX and DX molecules and the electron affinity of X atoms. This allowed determination of the separation time $\tau_s$. On the assumption that $\tau_a$ is the same for isotopic species, $\sigma_0$ and $\tau_a$ were deduced from the measured cross sections and calculated $\tau_s$ by assuming that the only isotopic effect on $\sigma_0$ is that due to the zero point wave function and by making use of Equation (6.33). Both $\tau_s$ and $\tau_a$ were found to be of comparable magnitudes with values of $\sim 4 \times 10^{-15}$ sec. Since the ratio $\tau_s/\tau_a$ for these systems does not differ appreciably, the increase in $\sigma_{da}$ in going from HCl to HBr to HI must lie in $\sigma_0$. A value equal to $5 \times 10^{-17}$ cm$^2$ and $1 \times 10^{-15}$ cm$^2$ was estimated for $\sigma_0$ for HCl and HBr, respectively.

mass spectrometer and direct primary dissociative electron attachment was excluded on energetic grounds. The process most probably involves more than one HX or DX molecule and it leads to the formation of a complex negative ion which subsequently may decay into various fragments.

C. CO. Dissociative electron attachment cross sections for $O^-$ from CO are given in Figure 6.16 and the available data are summarized in Table 6.3. Among the early studies of dissociative electron attachment to CO are those of Lozier (1934) and Hagstrum and Tate (1941). Hagstrum (1955) characterized the process as

$$CO(X^1\Sigma^+) + e \rightarrow C(^3P) + O^-(^2P) \tag{6.36}$$

Other processes yielding $O^-$ occur at $\sim 21$ and 23 eV and are of a continuous nature involving ion pairs. Chantry (1968) measured the kinetic energy distribution of the $O^-$ ions produced by dissociative attachment to CO in a beam experiment. His results are in agreement with Equation (6.36), namely that the single peak in the dissociative attachment cross section arises predominantly from the production of $O^-$ and ground-state carbon atoms $(^3P)$. At electron energies greater than 10.9 eV, Chantry (1968) observed a second peak in the kinetic energy distribution arising from the production of $O^-$ and a carbon atom in its first excited state $(^1D)$. Taking for the maximum of the cross section for the process (6.36) the value of $2 \times 10^{-19}$ cm$^2$, Chantry estimated a peak cross section value for the dissociation of CO giving $O^-$ and $C(^1D)$ equal to $9.5 \times 10^{-21}$ cm$^2$.

If we accept for the dissociation energy of CO the value of 11.11 eV and for the electron affinity of atomic oxygen the value of 1.46 eV, the minimum energy required to produce $C + O^-$ from CO in the ground state is 9.65 eV. Experimentally the cross section for $O^-$ maximizes at $\sim 10.0$ eV and has a sharp drop-off at lower energies, with an onset around 9.3 eV (see Table 6.3 and inset in Figure 6.16). Since $O^-$ from CO is not energetically possible below 9.65 eV, the tailing of the $O^-$ from CO resonance at lower energies must be due to the electron energy spread, assuming that the threshold is not affected by vibrational and rotational excitation. It thus appears that a vertical onset occurs in CO and a suggested possible shape (Rapp and Briglia, 1965) is shown by the broken line in the inset of Figure 6.16. According to Rapp and Briglia, this hypothetical cross section is commensurate with the observed cross section if the energy spread is taken into account. Evidence for a vertical onset in CO has also been provided by the work of Christophorou and Stockdale (1968).

It might be noted that the cross section for $O^-$ from CO obtained by Schulz (1962) (curve B, Figure 6.16) begins at 6 eV. Schulz (1962) pointed out that this may be due to metastable CO molecules hitting the grid and releasing secondary electrons which are collected on the ion collector. Schulz (1962) has also pointed out that the cross section data of Craggs and

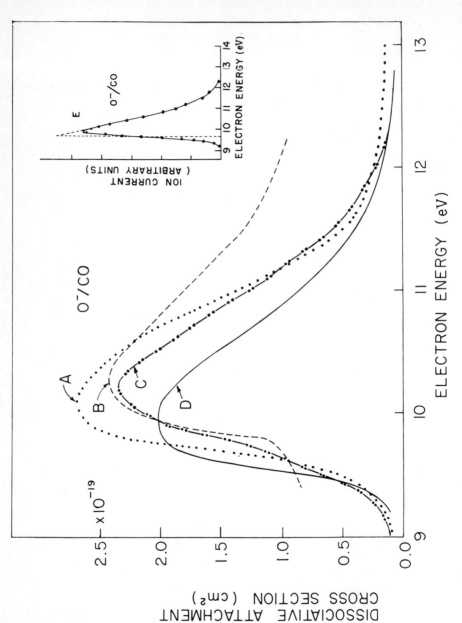

**Figure 6.16** Dissociative electron attachment cross section for O⁻ from CO as a function of electron energy.
A – Craggs and Tozer (1958). B – Schulz (1962). C – Asundi, Craggs and Kurepa (1963). D – Rapp and Briglia (1965).
E – Figure E gives O⁻/CO obtained by the RPD method with a retarding potential difference of 0.1 eV (Rapp and
Briglia, 1965). The broken curve indicates the possible actual shape of the cross section which could be obtained

Tozer (1958) are probably high because of the higher values for the positive ion cross sections used by Craggs and Tozer to normalize their $O^-$ from CO currents (see discussion in Schulz (1962)).

D. NO. Formation of $O^-$ in NO by resonance dissociative electron attachment has been investigated by Tate and Smith (1932), Hagstrum (1951), and Rapp and Briglia (1965). The dissociative electron attachment cross section of Rapp and Briglia (1965) is plotted in Figure 6.17 as a function of electron energy. Some other data are listed in Table 6.3. From Figure 6.17 it is seen that the cross section exhibits two overlapping peaks which may suggest two potential energy curves in the Franck–Condon region. To account for the second hump, Dorman (1966) postulated that it corresponds to the production of $O^-$ and a nitrogen atom N in its second ($^2P$) excited state. However, Chantry (1968) from kinetic energy measurements argued that the dissociative attachment process in NO produces exclusively N atoms

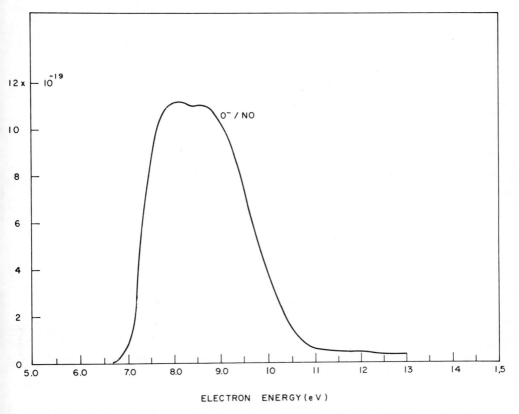

**Figure 6.17** Dissociative electron attachment cross section for $O^-$ from NO as a function of electron energy (Rapp and Briglia, 1965)

**Table 6.3** Data on dissociative electron attachment to diatomic molecules

| Molecule | Products | Onset (eV) | Peak energy (eV) | Peak cross section (cm²) | Full width at ½ height (eV) | Method | Ref.[h] | D(AX)[a] | Comments |
|---|---|---|---|---|---|---|---|---|---|
| $H_2$ | $H^- + H$ | $3.73 \pm 0.07$ | $3.75 \pm 0.07$ | $\sim 2 \times 10^{-21}$ | | TI[b] | 1 | 4.476 | Very sharp onset at 3.75 eV and very steeply rising cross section peaking close to its onset. |
| | | — | $3.75 \pm 0.07$ | $1.6 \times 10^{-21}$ | | TI | 2 | | |
| | | | $\sim 10.5$ | $1.2 \times 10^{-20}$ | | TI | 3 | | The cross section exhibits a plateau around 10 eV with some structure. $\Gamma_d$ is a rough estimate. |
| | | | 10.1 | $1.25 \times 10^{-20}$ | $\sim 3.6$ | TI | 4 | | |
| | $H^- + H^*$ | $13.8 \pm 0.2$ | $14.2 \pm 0.1$ | $3.5 \times 10^{-20}$ | $\sim 1.5$ | TI | 3 | | Sharp resonance; H* is in its $n = 2$ excited state. Background current high. |
| | | — | 13.95 | $2.10 \times 10^{-20}$ | 1.4 | TI | 4 | | |
| HD | $H^- + D$ or $D^- + H$ | — | 3.75 | $1.0 \times 10^{-22}$ | — | TI | 2 | 4.511 | See comments for corresponding processes in $H_2$. |
| | | — | $\sim 10.3$ | $0.62 \times 10^{-20}$ | $\sim 3.2\text{-}4.1$ | TI | 4 | | |
| | $H^- + D^*$ or $D^- + H^*$ | — | 13.95 | $1.50 \times 10^{-20}$ | 1.4 | TI | 4 | | |
| $D_2$ | $D^- + D$ | — | 3.75 | $<1.0 \times 10^{-22}$ | — | TI | 1 | 4.554 | See comments for corresponding processes in $H_2$. |
| | | — | 3.75 | $8 \times 10^{-24}$ | — | TI | 2 | | |
| | | — | 10.6 | $0.26 \times 10^{-20}$ | 3.0 | TI | 4 | | |
| | $D^- + D^*$ | | 14.0 | $\sim 1.0 \times 10^{-20}$ | 1.6 | TI | 4 | | |
| HCl | $Cl^- + H$ | $0.46 \pm 0.02$ | 0.6 | $3.9 \times 10^{-18}$ | — | TI | 5 | 4.430 | Corrected for finite width of electron pulse. (Ref. 6). |
| | | $0.63 \pm 0.05$ | 0.81 | $1.95 \times 10^{-17}$ | 0.31 | SB[c] | 6 | | |
| HBr | $Br^- + H$ | $0.43 \pm 0.01$ | 0.5 | $5.8 \times 10^{-17}$ | — | TI | 5 | 3.754 | Corrected for finite |

*(The table on this page is printed rotated 90°. Its column headers are cut off at the top of the page; the readable fragments are "Width of electron pulse. (Ref. 6)." and "Relative absolute.")*

| Molecule | Products | Onset (eV) | (eV) | Cross section (cm²) | Width (eV) | Method | Ref | EA / value | Remarks |
|---|---|---|---|---|---|---|---|---|---|
| (DC) | (C+D) | 0.03±0.03 | 0.61 | | 0.23 | SB | 6 | | |
| DBr | Br+D | 0.11 | 0.28 | $1.87 \times 10^{-16}$ | 0.23 | SB | 6 | $11.1^{e}$ | Relative absolute. |
| DI | I+D | ~0 | ~0 | $1.4 \times 10^{-14}$ | — | SB | 6 | $9.605^{e}$ | |
| CO | O+C | 9.35 | 10.1 | $2.7 \pm 0.3 \times 10^{-19}$ | 1.3 | $LT^{d}$ | 7 | | |
| | | 9.4 | 10.1 | $1.6 \pm 0.3 \times 10^{-19}$ | ~1.4 | TI | 8 | | |
| | | — | 10.1 | $2.18 \pm 0.3 \times 10^{-19}$ | 1.3 | LT | 9 | | |
| | | — | 9.9 | $2.02 \times 10^{-19}$ | 1.3 | LT | 10 | | |
| | | — | 9.9 | $2.3 \times 10^{-19}$ | 1.3 | TI | 9 | | |
| NO | O+N | ~6.6 | 8.1 | $1.12 \times 10^{-18}$ | 2.5 | TI | 10 | 6.5 | There exists a plateau in the region from 8 to 9 eV with two peaks at 8.1 and 8.6 eV. |
| $O_2$ | O+O | 4.4 | 6.7 | $1.3 \pm 0.2 \times 10^{-18}$ | 2.2 | TI | 8 | | |
| | | 4.6±0.15 | 6.5 | $1.3 \pm 0.2 \times 10^{-18}$ | 2.0 | — | 9 | | |
| | | 4.72±0.15 | 6.9 | $1.34 \pm 0.1 \times 10^{-18}$ | 2.0 | — | 9 | | |
| | | 4.3 | 6.5 | $1.41 \times 10^{-18}$ | 2.1 | TI | 10 | | |
| | | 4.7 | 6.7 | $2.25 \pm 0.3 \times 10^{-18}$ | | LT | 11 | | |
| | | 4.9 | 6.7 | $1.5 \times 10^{-18}$ | 2.1 | SB | 12 | 5.08 | Relative absolute. |
| $I_2$ | I+I | 0.03±0.03 | 0.03±0.03 | $3.9 \times 10^{-16}$ | | $M^{f}$ | 13 | | |
| | | ~0 | | | | TI | 14 | 1.5417 | |
| | | 0.03±0.03 | 0.03±0.03 | $3.0 \times 10^{-15}$ | ~0.01 | $MS^{g}$ | 15 | | |
| | | — | — | $1.7 \times 10^{-17}$ | | M | 16 | | |

---

[a] For electron affinity values of X, see Chapter 7. Dissociation energies were obtained from Herzberg (1950).
[b] Total ionization method.
[c] Swarm-beam method.
[d] Lozier tube.
[e] See Herzberg (1950) for discussion of this value.
[f] Microwave.
[g] Mass spectrometer.
[h] References

1. Schulz and Asundi (1965).
2. Schulz and Asundi (1967).
3. Schulz (1959a).
4. Rapp, Sharp and Briglia (1965).
5. Buchel'nikova (1959).
6. Christophorou, Compton and Dickson (1968).
7. Craggs and Tozer (1958).
8. Schulz (1962).
9. Asundi, Craggs and Kurepa (1963).
10. Rapp and Briglia (1965).
11. Craggs, Thorburn and Tozer (1957).
12. Christophorou and coworkers (1965).
13. Biondi (1958); at 300°K.
14. Fox (1958).
15. Biondi and Fox (1958).
16. Truby (1968); thermal value at 295°K.

in their first excited ($^2D$) state. Chantry obtained an onset equal to $7.5 \pm 0.1$ eV which agrees well with that (7.41 eV) calculated for the process

$$e + NO \rightarrow O^- + N^*(^2D) \tag{6.37}$$

from the accepted values of $D(NO) = 6.5$ eV (Herzberg, 1950), $EA(O) = 1.47$ eV (Branscomb and coworkers, 1958), and of the excitation energy of $N^*(^2D) = 2.383$ eV (Moore, 1949). Chantry did not observe any $O^-$ ions corresponding to the production of ground state or second excited state N atoms. In view of his results, it would appear that if two separate $NO^-$ potential energy curves are responsible for the structure in the cross section, they both lead to the same asymptote, namely $O^- + N^*(^2D)$. A transient $NO^-{}^*$ ion can also be formed in the ground state through vibrational excitation as has been discussed in Chapter 5.

Non-dissociative electron attachment to NO at low electron energies has been studied by Bradbury (1934a) by the electron swarm method. These results have been summarized by Loeb (1955). The low-energy electron attachment process in NO involves another NO molecule through the formation of $(NO)_2$ (Loeb, 1955), viz.

$$NO + NO \rightarrow (NO)_2$$
$$(NO)_2 + e \rightarrow NO^- + NO \tag{6.38}$$

The process, then, may be thought of as dissociative electron attachment to $(NO)_2$. The probability of attachment is, obviously, pressure dependent and increases with decreasing electron energy.

E. $O_2$.  Cross sections for resonance dissociative electron attachment to $O_2$ are shown in Figure 6.18. Some experimentally determined parameters are listed in Table 6.3. Early beam studies on $O^-$ from $O_2$ by Lozier (1934) and Hagstrum (1951) identified the process as

$$O_2 + e \rightarrow O(^3P) + O^-(^2P) \tag{6.39}$$

At higher energies ($\gtrsim 17$ eV) a continuous production of $O^-$ has been observed due to ion-pair processes. Mean attachment cross sections derived from swarm experiments have been reported by many workers (see, for example, Craggs, Thorburn and Tozer, 1957; Thompson, 1959; Prasad and Craggs, 1962). An excellent review of earlier work can be found in Loeb (1955). Non-dissociative electron attachment leading to $O_2^-$ has been studied extensively by the swarm method. These findings will be discussed in Section 6.5. Formation of a temporary compound $O_2^-{}^*$ by vibrational excitation has been discussed in Chapters 4 and 5.

Of special interest is the large temperature dependence of the cross section for $O^-$ from $O_2$ observed by Fite and coworkers (1963, 1965)[†]. A remarkable

---

[†]See also a recent study of dissociative attachment of electrons to hot oxygen by Henderson, Fite and Brackmann (1969).

shift in both the threshold and peak energy of the $O^-$ from $O_2$ resonance was observed (compared to 300°K) when the process was investigated at 2100°K. Further, both the width and the magnitude of the cross section were considerably larger at the higher temperature. It is clear that the dissociative attachment cross section should increase with initial vibrational state of $O_2$, since the electron can be captured while the $O_2$ molecule is at large internuclear separations. Vertical transitions from higher vibrational levels of $O_2$ are expected to affect both the threshold and width of the $O^-$ from $O_2$ resonance.

O'Malley (1967a) reproduced the experimental data of Fite and Brackmann (1963), assuming that dissociation proceeds via a single compound state of $O_2^-$ and considering only the rapid variation of the survival

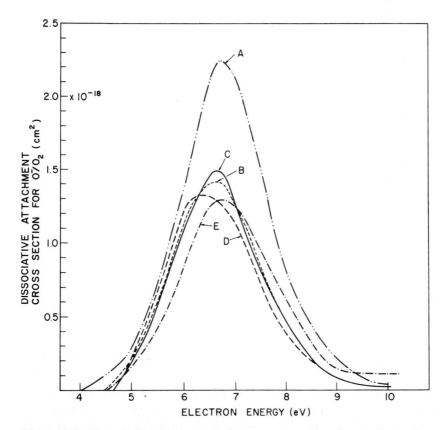

**Figure 6.18** Dissociative electron attachment cross section for the formation of $O^-$ from $O_2$ as a function of electron energy. (A – Craggs, Thorburn and Tozer (1957); B – Rapp and Briglia (1965); C – Christophorou and coworkers (1965); D – Buchel'nikova (1959); E – Schulz (1962))

probability with the initial distribution of vibrational states. He considered the initial distribution of rotational states to have a negligible effect. A comparison of O'Malley's calculation with the experimental results of Fite and Brackmann (1963) is given in Figure 6.19. The agreement is remarkable, although Chen (1969) argued that the expression for the cross section used by O'Malley is an oversimplified one with limited validity. Chen (1969) also pointed out that the use of a single state of $O_2^-$ is restrictive as there are many electronic states of $O_2^-$ which at large separations yield $O(^3P) + O^-(^2P)$. Similar pronounced temperature effects on $\sigma_{da}$ have been observed for $N_2O$ (Chaney and Christophorou, 1969; Chantry, 1969a).

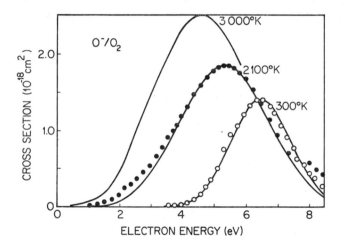

**Figure 6.19** Temperature dependence of the dissociative electron attachment cross section for $O^-$ from $O_2$. The solid curves are the calculated values of O'Malley (1967a); the points are the experimental results of Fite and Brackmann (1963)

Finally, mean attachment cross sections versus mean electron energy in dry air for an assumed velocity distribution (Maxwell or Druyvesteyn) have been summarized by Rees and Jory (1964). Earlier work includes that of Prasad (1959) and Prasad and Craggs (1960b).

F. $I_2$. Cross section data on the production of $I^-$ from $I_2$ are listed in Table 6.3. The earlier cross section values do not agree well. A comparison of Biondi and Fox's cross section for $I^-/I_2$ with the early data of Healey (1938) and Buchdahl (1941) has been given by Biondi and Fox (1958). The cross section of Biondi and Fox (1958) showed a sharp energy dependence at near thermal energies requiring the repulsive $I_2^-{}^*$ potential energy curve to cross that of $I_2$ at its minimum. However, Frost and McDowell (1958b, 1960) found $I^-$ from $I_2$ to have an onset at $0.03 \pm 0.03$ eV and to peak at $0.34 \pm 0.07$ eV. The $I^-$ ion current was quite asymmetrical. At electron energies $\geq 8.84 \pm 0.07$ eV they found $I^-$ from $I_2$ produced through an ion-pair process.

The fluorine negative-ion current from $F_2$ has an onset at energies $< 2$ eV (Thorburn, 1959), while $Cl^-$ from $Cl_2$ has an onset at $1.6 \pm 0.05$ and peaks at $2.4 \pm 0.1$ eV (Frost and McDowell, 1960). Frost and McDowell (1960) also reported $Br^-$ from $Br_2$ peaking at $0.03 \pm 0.03$ eV.

### 6.4.2.2 Triatomic molecules

A. $H_2O$, $D_2O$. The cross sections for the production of $H^-$ from $H_2O$ and $D^-$ from $D_2O$ are presented in Figure 6.20. The yields of $O^-$ from $H_2O$

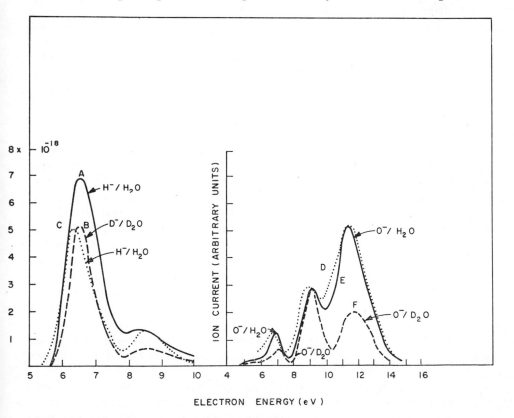

Figure 6.20 Dissociative electron attachment cross section for the production of $H^-$ from $H_2O$ and $D^-$ from $D_2O$ as a function of electron energy. Yield of $O^-$ from $H_2O$ and $D_2O$ as a function of electron energy. (A, B, E and F – Compton and Christophorou (1967); C – Buchel'nikova (1959); D – Dorman (1966))

and $D_2O$ are also shown in Figure 6.20; these were normalized at the second maximum. Mean attachment cross sections calculated from swarm and beam data have been reported by Crompton, Rees and Jory (1965). A summary of the results on the cross sections is given in Table 6.4. The cross section for the formation of $H^-$ from $H_2O$ and $D^-$ from $D_2O$ shows two major peaks whereas three major peaks are clearly resolved in the yield of $O^-$ from $H_2O$ and $D_2O$. The fact that $O^-$ but not $H^-$ from $H_2O$ is observed at 11.4 eV could be due to a competition between the decomposition paths of $H_2O^{-*}$ to give $O^-$ and decreasing $H^-$ with increasing energy (Figure 6.20). This would imply crossing of upper potential energy surfaces of $H_2O^{-*}$. From the shapes of the cross sections in Figure 6.20, it would seem possible to postulate the existence of three energy states at 6.5, 8.6 and $\sim 11.4$ eV for $H_2O^-$ and $D_2O^-$, with subsequent decomposition within the lifetime of the compound negative ion.

The observed isotope effects (Table 6.4 and Figure 6.20) suggest that the three compound negative ion states of $H_2O$ and $D_2O$ which are responsible for the production of $O^-$ have different lifetimes. Since $H^-$ from $H_2O$ is not observed at $\sim 11.4$ eV, the $H_2O^-$ state at this energy must be less stable. A competition in the formation of $O^-$ from $H_2O$ and $D_2O$, and $H^-$ from $H_2O$ and $D^-$ from $D_2O$ at $\sim 6.5$ and $\sim 9$ eV could alter the relative yields of the three $O^-$ peaks, but not the relative cross sections for the production of $H^-$ and $D^-$ since the $O^-$ yields are much smaller.

Isotope effects have been observed also in the magnitude and the width of the dissociative attachment cross section for the formation of $H^-$ from $H_2O$ and $D^-$ from $D_2O$ (Compton and Christophorou, 1967). The width at half-height of the $D^-$ from $D_2O$ resonance as observed by Compton and Christophorou was approximately 0.3 eV smaller than that of the $H^-$ resonance. This result is to be expected since the square of the ground-state vibrational wave function for the D–OD motion is narrower than that of the H–OH motion. Hurst and Stockdale (1964) analysed their swarm data on $H_2O$ and $D_2O$ representing $\sigma_a(\varepsilon)$ by a $\delta$ function at 6.5 eV and detected an isotope effect which was later found to be in reasonable agreement with that observed by Compton and Christophorou (1967) for $H^-$ from $H_2O$ and $D^-$ from $D_2O$. The ratio of the dissociative attachment cross section for $H^-$ from $H_2O$ and $D^-$ from $D_2O$ is well-contained within the square root of the inverse ratio of the reduced masses, as in the case of the hydrogen halides (see Section 6.4.1).

Negative hydroxyl ions ($OH^-$) are not observed in the electron bombardment of $H_2O$ at low pressures ($\sim 10^{-6}$ torr) for electron energies below $\sim 15$ eV.† At higher pressures, however, $OH^-$ becomes the most

---

†However, Doumont, Henglein and Jäger (1969) have reported seeing $OH^-$ as a primary product of electron attachment to $H_2O$ with structure in the $OH^-$ current plotted versus the electron energy.

abundant ion. The work of Muschlitz and Bailey (1956) and Muschlitz (1957) has shown that the $OH^-$ ions result from ion–molecule reactions involving $H^-$ (and possibly $O^-$) with $H_2O$. In agreement with this finding is the observation of Compton and Christophorou (1967) that the peaks in the $OH^-$ yield correlate with those of $H^-$ and $O^-$. A great deal may be learned about the transient $H_2O^{-*}$ and $D_2O^{-*}$ ions from the study of appropriate ion–molecule reactions.

The existence of long-lived (or stable) $H_2O^-$ ions is not, as yet, fully resolved. Bradbury (1934b) and Bradbury and Tatel (1934) reported electron attachment to $H_2O$ at very low electron energies, which increased with increasing $H_2O$ pressure and decreasing $E/P$. Although a somewhat similar result has been reported by Kuffel (1959), a number of other studies (e.g. Hurst, Stockdale and O'Kelly, 1963; Christophorou, 1965; Moruzzi and Phelps, 1966) failed to reproduce Bradbury's result. In this respect, it might be pointed out that increased attachment was observed in a mixture of water with an impurity (such as $O_2$) compared to the attachment in either component alone (Christophorou, 1965). This would suggest that in the mixture an additional attachment process occurs, which involves water molecule(s). Wobschall, Graham and Malone (1965) and Compton and Christophorou (1967) have observed $O_2^-$ formed by ion–molecule reactions in $H_2O$–$O_2$ mixtures. Compton and Christophorou (1967) reported that the $O_2^-$ peaks corresponded exactly to the two $OH^-$ (and consequently $H^-$) ion current peaks, but no $H_2O^-$ ions were detected in the energy region reported by Wobschall, Graham and Malone (1965). Clustering of positive ions in $H_2O$ vapour has been observed both in the ionosphere (Narcisi and Bailey, 1965) and in the laboratory (Munson and Tyndall, 1939; Kebarle and Hogg, 1964). Moruzzi and Phelps (1966) reported that in $H_2O$–$O_2$ mixtures at low $E/P$, $O_2^-$ is initially formed by a three-body attachment process (Section 6.5) and subsequently $O_2^-$ forms clusters with various numbers of water molecules. They observed negative ions of the form $(H_2O)_nO_2^-$ from $n = 1$ to $5$.

**B. $CO_2$.** Dissociative electron attachment cross sections for $O^-$ from $CO_2$ are presented in Figure 6.21. Two well-separated resonances exist, one at $\sim 4.1$ and the other at $8\,eV$. Cross section data are summarized in Table 6.4. Other than the absence of the low-energy resonance in the cross section data of Craggs and Tozer (1960) the cross sections for $O^-$ from $CO_2$ are in good agreement. Craggs and Tozer (1960) attributed the 8 eV process to

$$e + CO_2 \rightarrow CO(X^1\Sigma^+) + O^-(^2P) \tag{6.40}$$

Asundi and Craggs (1964a) described the $\sim 4.1$ eV process in a similar fashion with the $O^-$ ions having practically zero kinetic energy, whereas in the 8 eV process $O^-$ ions are formed with considerable kinetic energy.

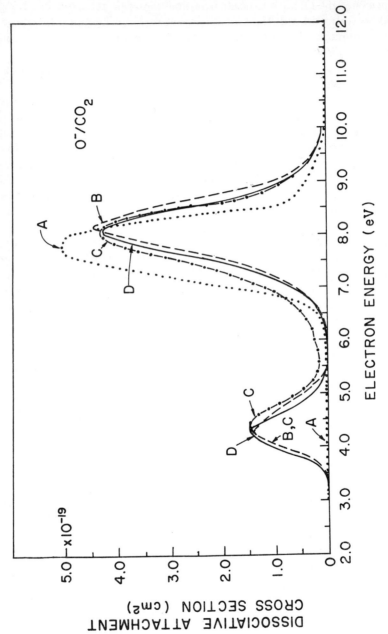

**Figure 6.21** Dissociative electron attachment cross section for the production of O⁻ from $CO_2$ as a function of electron energy. (A – Craggs and Tozer (1960); B – Schulz (1962); C – Asundi, Craggs and Kurepa (1963); D – Rapp and Briglia (1965))

The minimum energy to produce $O^- + CO$ from $CO_2$ is 4.1 eV. From Figure 6.21 and Table 6.4, however, it is seen that $O^-$ is observed at lower energies than the threshold. This may, in part, be due to the energy spread of the electron beam. The $\sim 4.1$ eV process is another example of a molecule for which the threshold may well be close to vertical at 4.1 eV in an experiment with high energy resolution. That the 4.1 eV process is a vertical onset case is further supported by the work of Christophorou and Stockdale (1968). A compound negative ion state around 3.8 eV with a lifetime of $\sim 10^{-15}$ sec has been evidenced from studies of vibrational excitation of $CO_2$ by electron impact (Boness and Schulz, 1968). Also, in a recent study of the formation of $O^-$ ions with zero kinetic energy from $CO_2$ by Schulz and Spence (1969), it was found that the onset energy shows a marked dependence on the gas temperature. The onset for $O^-/CO_2$ was shifted from 3.83 eV at 300°K to lower values as the gas temperature was increased, and extrapolated to 4.1 eV at 0°K, thus removing the apparent discrepancy in the onset value. These observations show that onset measurements must be extrapolated to 0°K to give meaningful results, especially when they are used to determine electron affinities. Lack of such a correction may explain the often higher electron affinity values obtained by electron impact (Chapter 7). The cause of the mentioned discrepancy on the threshold value is the effect of the internal excitation of molecular vibrational and rotational motion at the temperature of the experiment. When the lifetime of the compound negative ion state which is involved in the formation of the stable fragment-negative ions is very short, the dissociative attachment cross section is strongly enhanced when the neutral molecule is initially vibrationally (O'Malley, 1967a) or rotationally (Chen and Peacher, 1967) excited, thus decreasing the resonance energy onset with increasing gas temperature. This situation is similar to that we have discussed earlier in this section for $O_2$ and to that presented below for $N_2O$.

C. $N_2O$.  Much work has been done recently on the electron attachment processes in nitrous oxide ($N_2O$). The various processes occurring, however, are not yet fully understood. Oxygen negative ions ($O^-$) produced by dissociative electron attachment to $N_2O$ (i.e. $e + N_2O \rightarrow O^- + N_2$) have been identified by electron beam methods (Schulz, 1961; Curran and Fox, 1961). The yield of $O^-$ from $N_2O$ begins at $\sim 0.0$ eV (Schulz, 1961; Curran and Fox, 1961) and shows (Schulz, 1961; Curran and Fox, 1961; Rapp and Briglia, 1965) a broad peak—with some structure (Schulz, 1961)—extending from $\sim 0.0$ eV to $\sim 1.2$ eV, and another and more intense peak at 2.3 eV. The relative heights of the two peaks as reported by Schulz and by Rapp and Briglia differ appreciably and this may be due to different source conditions, since recent work by Chaney and Christophorou (1969) and by Chantry (1969a) has shown that the first peak, but not the second, is sensitive to temperature. The low-energy side of the $O^-/N_2O$ cross section shown in Figure 6.22 has been attributed (Schulz, 1961; Chaney and Christophorou 1969) to dissociative electron attachment from various vibrational levels of $N_2O$. Dissociative attachment from higher vibrational levels of $N_2O$ enhances the cross section and thus causes a shift in the apparent onset to lower energies. This effect can explain (Kaufman, 1967; Chaney and Christophorou, 1969) the result that $O^-$ appeared at lower energies than the energetically allowed threshold of 0.21 eV in the beam work of Schulz, Curran and Fox and Rapp and Briglia.

The main intense $O^-/N_2O$ peak at 2.3 eV (Figure 6.22) has been interpreted (Schulz, 1961) as dissociative attachment via a compound negative

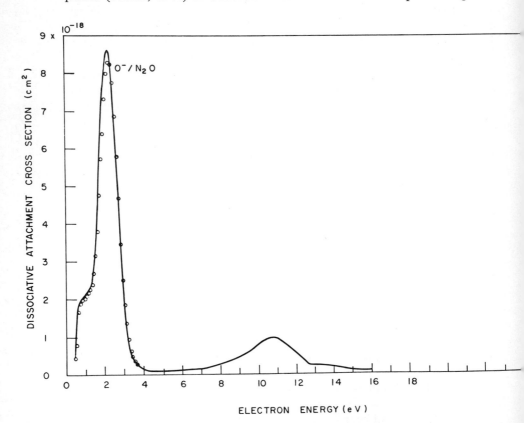

**Figure 6.22** Dissociative electron attachment cross section for the formation of $O^-$ from $N_2O$ a function of electron energy. (Solid line – Rapp and Briglia (1965); open circles – Chaney Christophorou (1969)). Chaney and Christophorou obtained the plotted cross sections by applica of the swarm–beam technique using their attachment rates at 373 °K and the shape of the $O^-/N$ current as given by Rapp and Briglia (1965). The cross section for the production of $O^-$ from N at 373 °K has a peak value of $8.3 \times 10^{-18}$ cm$^2$ at 2.3 eV (Chaney and Christophorou, 1969)

ion state at this energy which can either autoionize (elastically or inelastically) or dissociate into $O^- + N_2^{(*)}$. Both the inelastic scattering cross section and the negative ion cross section have the same shape as they both proceed through the same temporary negative ion intermediate (see results on elastic and inelastic scattering in Chapters 4 and 5). Chantry (1969c) reported that the intensity of the high-energy weak peak around 10 eV (Figure 6.22) depended quadratically on pressure, and attributed the formation of $O^-$ from $N_2O$ in this energy region to an indirect process namely, to electrons which have lost a portion of their energy to electronic excitation of $N_2O$ and were left subsequently with appropriate energy to be captured dissociatively by $N_2O$ through the resonance process at 2.3 eV. A similar process may take place for other systems.

The formation of $N_2O^-$ has been a subject of recent scientific scrutiny. Some of the recent work has been modivated by a suggestion made by Ferguson, Fehsenfeld and Schmeltekopf (1967) (see also Ferguson, 1968) that the formation of $N_2O^-$ requires a substantial geometrical deformation of the neutral $N_2O$. The geometrical configuration of many polyatomic molecules depends to a large extent on the number of valence electrons they possess. Triatomic molecules not containing hydrogen, for example, with 16 or fewer valence electrons are linear in their ground electronic states, whereas such molecules with 17 or more valence electrons are bent in the ground state. On this basis, then, electron attachment to a linear molecule such as $N_2O$ will require a geometrical change to form a ground state molecular negative ion and hence parent negative ion formation (e.g. $N_2O^-$) would require some activation energy.

No evidence has been presented from beam studies as to the existence of $N_2O^-$ at low energies ($< 1$ eV). Paulson (1966) has reported the formation of $N_2O^-$ in an ion–molecule reaction process ($O^- + N_2O \rightarrow N_2O^- + O$). Also, Chantry (1969b) has reported that $N_2O^-$ is formed as a tertiary ion via ion–molecule reactions in $N_2O$. Negative ion-molecule reactions in $N_2O$ have been studied by Moruzzi and Dakin (1968), but these authors reported no $N_2O^-$ at low $E/N$. Phelps and Voshall (1968) found in high pressure swarm experiments with $N_2O$ that at low $E/N$ (volts cm$^2$) the values of $\alpha'/N$ (cm$^2$) increase with increasing $N_2O$ density and decrease with increasing $E/N$. This was interpreted by Phelps and Voshall as suggestive of a low-energy (at or near thermal energies), three-body electron attachment process leading to $N_2O^-$ (i.e. $e + N_2O + N_2O \rightarrow N_2O^- + N_2O + energy$). In agreement with this finding is the work of Warman and Fessenden (1968) who have shown that at energies below $\sim 0.2$ eV the attachment is a three-body process. They obtained a rate for the reaction

$$e + N_2O + M \rightarrow N_2O^- + M + energy$$

equal to $(5.6 \pm 0.2) \times 10^{-33}$ cm$^6$/sec which compares well with the value of

$(6 \pm 1) \times 10^{-33}$ cm$^6$/sec obtained by Phelps and Voshall (1968). These results were taken as evidencing the formation of $N_2O^-$ as an initial product of thermal electron capture in $N_2O$, and that the barrier towards molecular ion formation does not appear to be as great as suggested by Ferguson, Fchscnfeld and Schmeltekopf (1967). However, in these experiments the question arises as to the possibility that $O_2$, present as an impurity in $N_2O$, might have been responsible for the observed attachment rates.

A more direct evidence for the formation of $N_2O^-$ and the need for an activation energy for its formation has been presented by Chaney and Christophorou (1969). These workers have studied $N_2O$ in mixtures with $N_2$ and they have found that the attachment rates increased linearly with total pressure. Since the attachment rates extrapolated to zero total pressure were found to be characteristic of the formation of $O^-/N_2O$, Chaney and Christophorou have attributed the observed increase in attachment to an apparent three-body attachment process, namely,

$$e + N_2O + N_2 \overset{k_3}{\rightarrow} N_2O^- + N_2 + \text{energy} \tag{6.41}$$

The apparent three-body attachment rate coefficient $k_3$ is shown in Figure 6.23. The observed increase in $k_3$ with increasing electron energy suggests the need for an activation energy for the formation of $N_2O^-$ which on the basis of Figure 6.23 is around 1 eV.

D. $NO_2$, $O_3$. The mass-spectroscopic data of Curran (1961a) on $NO_2$ are given in Table 6.4. The cross sections quoted are crude estimates. More recently Mahan and Walker (1967) have measured the attachment rate for thermal electrons at 300°K in $NO_2$ in the presence of several inert gases. They reported a value for the thermal attachment rate of the order of $3 \times 10^{-11}$ cm$^3$/molecule sec. They have found also that although the attachment rates were independent of pressure they depended on the nature of the inert gas. This finding led them to postulate that an $NO_2^-{}^*$ intermediate is formed which autoionizes, is stabilized in a collision, or is destructed in a collision with the inert gas. The dependence of the attachment rate on the nature of the inert gas shows that the cross section for stabilization and destruction of the unstable negative ion is a strong function of the stabilizing gas.

Curran (1961b) reported $O_3^+$ appearing at $12.89 \pm 0.1$ eV. From his data on the yield of $O^-$ and $O_2^-$ from $O_3$ he obtained a dissociation energy $\leqslant 1.02 \pm 0.15$ eV for $O_2$–O and an electron affinity for $O_2 \geqslant 0.58$ eV.

### 6.4.2.3 Polyatomic molecules

A. $CH_4$, $CD_4$. The primary ions formed in the dissociative attachment of electrons to $CH_4$ have been identified (Smith, 1937; Trepka and Neuert, 1963) as $H^-$ and $CH_2^-$ with small amounts of $C^-$ and $CH^-$.

**Table 6.4** Data on dissociative electron attachment to triatomic molecules

| Molecule | Products | Onset (eV) | Peak energy (eV) | Peak cross section (cm²) | Full width at ½ height (eV) | Method | Ref.[f] | $D(AX)$[a] (eV) | Comments |
|---|---|---|---|---|---|---|---|---|---|
| $H_2O$ | $H+OH$ | 5.45±0.09 | 6.4±0.1 | (4.8 ±1.5)×10⁻¹⁸ | | TI[b] | 1 | 5.113(H—OH) | A width of 0.8 eV can be obtained from the ion-current data for H⁻/H₂O of Schulz (1960a) for the first peak. |
| | | | 8.6±0.1 | (1.3 ±0.1)×10⁻¹⁸ | | TI | 1 | | |
| | | 5.7 ±0.2 | 6.5±0.1 | 6.9×10⁻¹⁸ | 1.05 | SB[c] | 2 | | |
| | | | 8.6±0.2 | 1.3×10⁻¹⁸ | | SB | 2 | | |
| $D_2O$ | $D+OD$ | 5.7 ±0.2 | 6.5±0.1 | 5.2×10⁻¹⁸ | 0.8 | SB | 2 | | |
| | | | 8.6±0.2 | 0.6×10⁻¹⁸ | | SB | 2 | | |
| $CO_2$ | $O+CO$ | 3.6 ±0.15 | 4.3 | (1.29±0.2)×10⁻¹⁹ | 0.9 | LT[d] | 3 | 5.453(O—CO) | |
| | | 3.6 ±0.15 | 4.55 | (1.48±0.1)×10⁻¹⁹ | 1.0 | — | 3 | | |
| | | 3.85±0.1 | 4.4 | (1.5 ±0.2)×10⁻¹⁹ | 0.9 | TI | 4 | | |
| | | — | 4.3 | 1.5 ×10⁻¹⁹ | 0.9 | TI | 5 | | |
| | | — | 8.1 | 4.3 ×10⁻¹⁹ | 1.1 | TI | 5 | | |
| | | 6.6 ±0.10 | 8.0 | (4.3 ±0.3)×10⁻¹⁹ | 1.3 | — | 3 | | |
| | | 6.75±0.15 | 8.3 | (4.55±0.6)×10⁻¹⁹ | 1.3 | LT | 3 | | |
| | | 6.6 ±0.10 | 8.2 | (4.5 ±0.7)×10⁻¹⁹ | 1.1 | TI | 4 | | |
| | | 6.7 | 7.8 | (5.07±0.5)×10⁻¹⁹ | | LT | 6,7 | | |
| $N_2O$ | $O+N_2$ | <0.4 | 2.25 | 8.6×10⁻¹⁸ | 1.1 | TI | 5 | 1.677(O—N₂) | |
| | | — | 11 | 9 ×10⁻¹⁹ | | TI | 5 | 4.930(N—NO) | |
| $NO_2$ | $O+NO$ | 1.35±0.05 | 1.9 | ~10⁻¹⁸ | | MS[e] | 8 | 3.115(O—NO); 4.505(N—NO₂) | |
| $O_3$ | $O+O_2$ | 0-0.05 | — | 10⁻¹⁷ – 10⁻¹⁸ | | MS | 9 | 1.04(O₂—O) | Cross sections are crude estimates. |
| | $O_2+O$ | 0.42±0.03 | — | 10⁻¹⁷ – 10⁻¹⁸ | | MS | 9 | | |

[a] From G. Herzberg (1966). For electron affinities of atoms, see Chapter 7.
[b] Total ionization method.
[c] Swarm–beam method.
[d] Lozier tube.
[e] Mass spectrometer.
[f] References:

1. Buchel'nikova (1959).
2. Compton and Christophorou (1967).
3. Asundi, Craggs and Kurepa (1963).
4. Schulz (1962).
5. Rapp and Briglia (1965).
6. Craggs and Tozer (1958).
7. Craggs and Tozer (1960).
8. Fox (1960).
9. Curran (1961b).

The $CH_3^-$ ion was observed as a product in ion-pair formation above $\sim 20$ eV, but not in dissociative electron attachment.

Dissociative electron attachment to $CH_4$ and $CD_4$ has been studied recently by Sharp and Dowell (1966, 1967). Their results are shown in Figure 6.24. From a mass analysis of the negative ions, Sharp and Dowell concluded that the cross section for $CH_4$ in the region from 7.5 to 13.5 eV

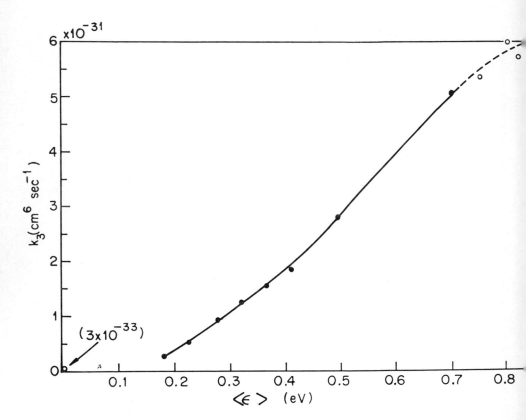

**Figure 6.23** Three-body electron attachment rate coefficient for $N_2O-N_2$ mixtures as a function mean electron energy (Chaney and Christophorou, 1969). The mean electron energies were obtair from the distribution functions of Phelps and coworkers (1968)

is composed of a main peak (at $\sim 10.5$ eV) due to $CH_2^-$ and a shoulder on its low-energy side (at $\sim 9.5$ eV) due to $H^-$. The intensity of the high-energy peak (due to $CH_2^-$) decreases upon increasing deuteration of methane and is absent for $CD_4$, the cross section of which consists of only $D^-$. An upper limit equal to $3.6 \times 10^{-22}$ cm$^2$ was estimated by Sharp and Dowell (1966, 1967) for the formation of $CD_2^-$ from $CD_4$. Approximate values of the peak cross

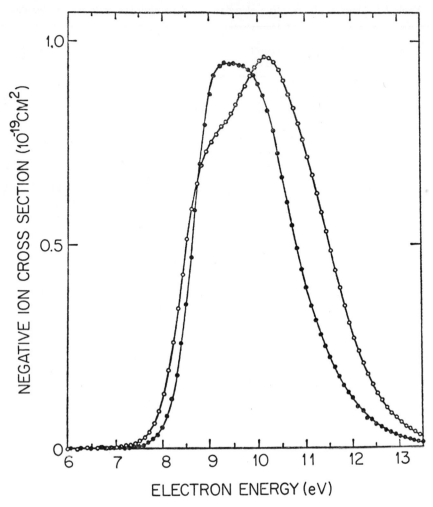

**Figure 6.24** Dissociative electron attachment cross sections in $CH_4$ (open circles) and $CD_4$ (full circles) (Sharp and Dowell, 1967). Two overlapping, unresolved peaks are seen to be present in $CH_4$. The lower energy peak is due to $H^-$ and the higher energy one to $CH_2^-$. A single peak due to $D^-$ is present in $CD_4$

sections and peak energies, taken from the graphs given by Sharp and Dowell, are listed in Table 6.5 for $CH_4$, $CH_3D$, $CH_2D_2$, $CHD_3$ and $CD_4$.

In addition to the large isotope effect in the production of $CH_2^-$, an inverse isotope effect has been observed for the production of $H^-$ from $CH_4$ and $D^-$ from $CD_4$. The cross section for $H^-$ from $CH_4$ was found

**Table 6.5** Data on dissociative electron attachment to polyatomic molecules

| Molecule | Ion | Onset (eV) | Peak energy (eV) | Peak cross section (cm$^2$) | Method[h] | Ref.[i] |
|---|---|---|---|---|---|---|
| $BCl_3$ | $Cl^-$ (?) | — | 0.4 | $2.8 \times 10^{-17}$ | TI[a] | 1 |
| $PH_3$ | $P^-$ | $5.7 \pm 0.2$ | — | $3 \times 10^{-19}$ | MS | 2 |
| | $PH^-$ | $2.1 \pm 0.2$ | — | $2 \times 10^{-18}$ | MS | 2 |
| | $PH_2^-$ | $2.3 \pm 0.1$ | 2.8 | $4 \times 10^{-18}$ | MS | 2 |
| $NH_3$ | $H^-$ | $5.3 \pm 0.1^b$ | 5.65 | $1.4 \times 10^{-18}$ | TI; MS | 3 |
| | $NH_2^-$ | $5.45 \pm 0.05^b$ | 5.65 | $1.5 \times 10^{-18}$ | TI; MS | 3 |
| | $H^-$ | — | 10.5 | $3.8 \times 10^{-19}$ | TI; MS | 3 |
| | $NH_2^-$ | — | 10.5 | $4.6 \times 10^{-20}$ | TI; MS | 3 |
| $ND_3$ | $D^-$ | — | 5.65 | $1.4 \times 10^{-18}$ | TI; MS | 3 |
| | $ND_2^-$ | — | 5.65 | $1.2 \times 10^{-18}$ | TI; MS | 3 |
| | $D^-$ | — | 10.5 | $3.4 \times 10^{-19}$ | TI; MS | 3 |
| | $ND_2^-$ | — | 10.5 | $3.2 \times 10^{-20}$ | TI; MS | 3 |
| $AsH_3$ | $As^-$ | $5.3 \pm 0.2$ | — | $9 \times 10^{-19}$ | MS | 2 |
| | $AsH^-$ | $2.0 \pm 0.1$ | — | $3 \times 10^{-18}$ | MS | 2 |
| | $AsH_2^-$ | $2.0 \pm 0.1$ | $\sim 2.6$ | $6 \times 10^{-18}$ | MS | 2 |
| $SbCl_3$ | $Cl^-$ | $\sim 0.01$ | — | $1.8 \times 10^{-16}$ | TI | 4 |
| $CH_4$ | $H^-, CH_2^-$ | — | $\sim 10$ | $\sim 0.95 \times 10^{-19}$ | TI | 5, 6 |
| $CH_3D^c$ | $H^-(D^-), CH_2^-$ | — | $\sim 10$ | $\sim 10^{-19}$ | TI | 6 |
| $CH_2D_2^c$ | $H^-(D^-), CH_2^-$ | — | $\sim 10$ | $\sim 10^{-19}$ | TI | 6 |
| $CHD_3^c$ | $H^-(D^-)$ | — | $\sim 10$ | $\sim 10^{-19}$ | TI | 6 |
| $CD_4$ | $D^-$ | — | $\sim 9.5$ | $\sim 0.95 \times 10^{-19}$ | TI | 5, 6 |
| $CH_2Cl_2$ | $Cl^-$ (?) | — | 0.038 | $1.48 \times 10^{-18}$ | Swarm[d] | 7 |
| | | — | 0.45 | $3.18 \times 10^{-18}$ | Swarm[d] | 7 |
| $CHCl_3$ | $Cl^-$ (?) | — | 0.038 | $3.66 \times 10^{-16}$ | Swarm[d] | 7 |
| | | — | 0.215 | $7.32 \times 10^{-16}$ | Swarm[d] | 7 |
| $CF_2Cl_2$ | $Cl^-$ | — | 0.15 | $5.4 \times 10^{-17}$ | TI[a] | 1 |
| | | — | 0.15 | $11.0 \times 10^{-17}$ | SB[a] | 8 |
| $CFCl_3$ | $Cl^-$ | — | $\sim 0.0$ | $9.5 \times 10^{-15}$ | SB[a] | 8 |
| $CF_3I$ | $I^-$ (?) | — | 0.05 | $7.8 \times 10^{-17}$ | TI[a] | 1 |
| | | — | 0.9 | $3.2 \times 10^{-17}$ | TI[a] | 1 |
| $CCl_4$ | $Cl^-$ (?) | — | 0.02 | $1.3 \times 10^{-16}$ | TI[a] | 1 |
| | | — | 0.60 | $1.0 \times 10^{-16}$ | TI[a] | 1 |
| | | — | $\sim 0.0$ | $1.6 \times 10^{-14}$ | SB[a] | 8 |
| | | — | 0.78 | $5.2 \times 10^{-16}$ | SB[a] | 8 |
| | | — | 0.038 | $2.62 \times 10^{-14}$ | Swarm[d] | 7 |
| $SiH_4$ | $Si^-$ | $8.2 \pm 0.2$ | $9.15^e$ | $2 \times 10^{-19}$ | MS | 2 |
| | $SiH^-$ | $7.9 \pm 0.2$ | $8.9^e$ | $7 \times 10^{-19}$ | MS | 2 |
| | $SiH_2^-$ | $8.0 \pm 0.2$ | $8.9^e$ | $1 \times 10^{-18}$ | MS | 2 |
| | $SiH_3^-$ | $7.0 \pm 0.2$ | $8.4^e$ | $1.6 \times 10^{-18}$ | MS | 2 |
| $C_2HCl_3$ | $Cl^-$ (?) | — | 0.39 | $2.84 \times 10^{-16}$ | Swarm[d] | 7 |
| $SF_6$ | $SF_6^-$ | $\sim 0.0$ | $< 0.05$ | $\sim 10^{-15}$ | MS | 9 |
| | $SF_5^-$ | $\sim 0.0$ | $< 0.1$ | $\sim 10^{-17}$ | MS | 9 |
| | $SF_6^-$ | $\sim 0.0$ | $0.03 \pm 0.03$ | $\sim 1.3 \times 10^{-15}$ | TI | 10 |
| | $SF_6^-$ | $\sim 0.0$ | $\sim 0.0$ | $5.7 \times 10^{-16}$ | TI | 1 |
| | $SF_5^-$ | $\sim 0.0$ | $\sim 0.0$ | | TI | 1 |
| | $F^-$ | $\sim 0.0$ | $\sim 0.0$ | | TI | 1 |
| | $SF_6^-$ | $\sim 0.0$ | 0.1 | $\sim 2.2 \times 10^{-16}$ | TI | 11 |

**Table 6.5** (*continued*)

| Molecule | Ion | Onset (eV) | Peak energy (eV) | Peak cross section (cm²) | Method[h] | Ref.[i] |
|---|---|---|---|---|---|---|
| | $SF_6^-$ | 0.0 | Thermal | $2.07 \times 10^{-14}$ (for $\langle \varepsilon \rangle = 0.03$ eV) | SB[h] | 12 |
| | | — | Thermal | $2.6 \times 10^{-14}$ (average) | Microwave | 13 |
| | $SF_5^-$ | 0.0 | $\sim 0.37$ | $4.9 \times 10^{-17}$ | SB | 12 |
| $1,1,1C_2H_3Cl_3$ | $Cl^-$ (?) | | 0.038 | $1.51 \times 10^{-15}$ | Swarm[d] | 7 |
| $1,1,2C_2H_3Cl_3$ | $Cl^-$ (?) | | 0.42 | $1.06 \times 10^{-16}$ | Swarm[d] | 7 |
| $C_2F_6$ | $F^-$ | 2.2 | 3.75 | $1.76 \times 10^{-17}$ | MS | 14 |
| | $CF_3^-$ | 2.8 | 3.75 | $0.46 \times 10^{-17}$ | MS | 14 |
| | | 2.93 | 4.51 | $2.06 \times 10^{-17}$ | TI[g] | 15 |
| $C_3F_8$ | $F^-$ | 1.8 | — | $3.65 \times 10^{-16}$ | MS | 14 |
| | $CF_3^-$ | 2.2 | — | $2.3 \times 10^{-17}$ | MS | 14 |
| | | 2.2 | — | $2.38 \times 10^{-16}$ | TI[g] | 15 |
| | $C_2F_5^-$ | 2.1 | — | $2.7 \times 10^{-17}$ | MS | 14 |
| | ? | 5.4 | — | $1.1 \times 10^{-17}$ | TI | 15 |
| $C_7F_{14}$ | $C_7F_{14}^-$ ($C_7F_{13}^-$, $C_6F_{11}^-$) | $\sim 0$ | 0.15 | $7.5 \times 10^{-15}$ | TI | 10 |

[a] See discussion in Ref. 8.
[b] See Reference 3 for a comparison of this value with other beam data.
[c] The cross section curves exhibit two partially unresolved peaks due to $H^-(D^-)$ and methylene negative ions. The methylene negative ion decreases upon deuteration.
[d] Mean velocity weighted total cross sections; see Reference 7.
[e] Private communication to the author by Prof. H. Neuert (1968).
[f] See Reference 12 for the energy dependence of the cross section at electron energies $\leq 0.11$ eV.
[g] No distinction could be made between $F^-$ and $CF_3^-$.
[h] Symbols as in Table 6.4.
[i] References:
  1. Buchel'nikova (1959)
  2. Ebinghaus and coworkers (1964).
  3. Sharp and Dowell (1969).
  4. Grob (1963).
  5. Sharp and Dowell (1967).
  6. Sharp and Dowell (1966).
  7. Blaunstein and Christophorou (1968).
  8. Christophorou and Stockdale (1968).
  9. Hickam and Fox (1956).
  10. Asundi and Craggs (1964b).
  11. Rapp and Briglia (1965).
  12. Christophorou, McCorkle and Carter (1970).
  13. Mahan and Young (1966).
  14. Bibby and Carter, quoted in Reference 15.
  15. Kurepa (1965).

to be 0.8 times that for $D^-$ from $CD_4$. This observation of Sharp and Dowell can be reconciled from the resonance scattering theory where $\sigma_{da}$ is proportional to $M_r^{\frac{1}{4}}$ when $\bar{\Gamma}_a$ and $\rho(\varepsilon)$ are small (see discussion in Section 6.4.1).

An investigation of negative ion formation in $C_2H_6$, $C_3H_8$, $n$–$C_4H_{10}$, $i$–$C_4H_{10}$, $C_2H_4$, $C_3H_6$, 1–$C_4H_8$, $i$–$C_4H_8$, $C_2H_2$, $C_6H_6$, $CH_3OH$ and $C_2H_5OH$ has been made by Trepka and Neuert (1963). For $H^-$ from the hydrocarbons $CH_4$, $C_2H_6$, $C_3H_6$, $n$–$C_4H_{10}$, see also Dorman (1966).

B. $NH_3$, $ND_3$, $PH_3$, $AsH_3$ and $SiH_4$. Dissociative electron attachment in ammonia has been studied using the swarm method by Bailey and Duncanson (1930) and Bradbury (1934a) and using the beam method by Mann, Hustrulid and Tate (1940), Kraus (1961), Melton (1966), Dorman (1966) and Collin, Hubin-Franskin and D'or (1968). Attachment cross sections for $NH_3$ and $ND_3$ have also been reported recently (Sharp and Dowell, 1969; Compton, Stockdale and Reinhardt, 1969). The cross section maximum values and peak energies are listed in Table 6.5. Two peaks were observed at 5.65 and 10.5 eV, both containing $H^-(D^-)$ and $NH_2^-(ND_2^-)$. A small direct isotope effect analogous to that for the hydrogen halides and water has been found in the negative ion production, except for the ratio $H^-/D^-$ at 5.65 eV where the cross sections for the two isotopes were found equal (Sharp and Dowell, 1969; see Table 6.5).

Cross sections for some fragment ions produced by resonance dissociative electron attachment to $PH_3$, $AsH_3$ and $SiH_4$ were obtained by Ebinghaus and coworkers (1964) and are summarized in Table 6.5. For information on the energy dependence of the various ions, see Ebinghaus and coworkers (1964). Kraus (1961) also reported on negative ion formation in $CO_2$, $H_2S$, $CS_2$ and $SO_2$.

C. *Halogenated polyatomic molecules.* Electron attachment cross sections for a number of halogenated compounds are listed in Table 6.5. In Figure 6.25 electron attachment rates as a function of mean electron energy are shown for a number of chlorinated hydrocarbons. These data can be used in connection with Equation (6.12) to obtain mean velocity-weighted total attachment cross sections. Although in these experiments no mass analysis was incorporated, halogen negative ions are expected to be the predominant ions at these energies (Christophorou and Carter, 1968; MacNeil and Thynne, 1968; also references at the end of this section). It is seen (Figure 6.25; see also Blaunstein and Christophorou, 1968) that for some of these compounds the cross sections are strongly peaking functions at or close to zero (generally $\lesssim 1$ eV) and are very large. From their shapes it is suggested that the compound negative ion state crosses the neutral molecule potential energy surface of the respective system at or close to its minimum and is purely repulsive in the Franck–Condon region. It is further noted (see Tables 6.5 and 6.6 and also Blaunstein and Christophorou, 1968) that for a given compound the cross section increases greatly upon changing the substituent from Cl to Br to I. Also, multiple halogen substitution greatly enhances the dissociative attachment cross section.

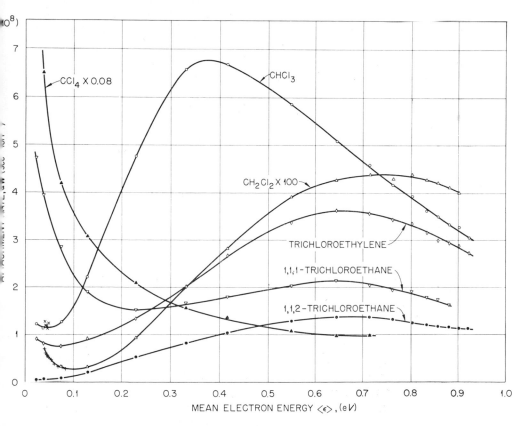

**Figure 6.25** Electron attachment rates as a function of mean electron energy for a number of chlorinated aliphatic hydrocarbons (Christodoulides and Christophorou, 1970). × × × are for $C_2H_4$ as a carrier gas and the rest of the data are for the carrier gas $N_2$. The mean electron energies were determined as shown in Table 6.1

It has to be pointed out here that the magnitude of the dissociative attachment cross sections for some of the compounds listed in Table 6.5 at energies $\lesssim 1$ eV is such as to make dissociative electron attachment a very efficient process for the removal of subexcitation electrons. Further, the magnitude and sharpness of the cross sections for certain of these compounds make them uniquely suited for detecting (scavenging) low-energy electrons.

Electron scavengers, so far, have been restricted to very few substances. In particular, one should notice the very large cross section at thermal energies for $CCl_4$. Christodoulides and Christophorou (1970) have applied the procedure described in Section 6.3.1.4 to the attachment rates as a function of $E/P$, measured for this molecule in mixtures with $N_2$ by Blaunstein and Christophorou (1968), and have found that for electron energies below 0.55 eV the attachment cross section can be approximated very well by $A_\gamma/\varepsilon^\gamma$ with values for the constants $A_\gamma$ and $\gamma$ equal to $3.6 \times 10^{-16}$ ((eV)$^\gamma$ cm$^2$) and $\gamma = 1.226$. This gives a value for $\sigma_a$ for $CCl_4$ at 0.03 eV equal to $2.2 \times 10^{-14}$ cm$^2$.

**Table 6.6** Data on dissociative electron attachment to halogenated benzene derivatives[a]

| Molecule | Ion | Peak energy (eV) | Peak cross section (cm$^2$) | Full width at $\frac{1}{2}$ height (eV) |
|---|---|---|---|---|
| $C_6H_5Cl$ | $Cl^-$ | 0.86 | $1.4 \times 10^{-17}$ | 0.6 |
| $o\text{-}C_6H_4Cl_2$ | $Cl^-$ | 0.36 (?) | $4.3 \times 10^{-16}$ | 0.52 |
| $o\text{-}C_6H_4CH_3Cl$ | $Cl^-$ | 1.10 | $2.2 \times 10^{-17}$ | 0.63 |
| $C_6H_5Br$ | $Br^-$ | 0.84 | $9.6 \times 10^{-17}$ | 0.68 |
| $C_6D_5Br$ | $Br^-$ | 0.80 | $1.04 \times 10^{-16}$ | 0.64 |
| $o\text{-}C_6H_4CH_3Br$ | $Br^-$ | 0.95 | $6.0 \times 10^{-17}$ | 0.75 |

[a] Christophorou and coworkers (1966)—swarm–beam data.

In Figure 6.26 electron attachment rates are given for the aliphatic hydrocarbons $C_2H_5Br$, $n\text{-}C_4H_9Br$, and $n\text{-}C_6H_{13}Br$. These data provide additional support to that presented in Section 6.3.1.2 for the correctness of the electron energy distribution functions used by Christophorou and coworkers for the carrier gases $C_2H_4$, $N_2$, and Ar. A swarm-beam study of the compounds in Figure 6.26 and of $n\text{-}C_3H_7Br$ and $n\text{-}C_5H_{11}Br$ has been made (Christophorou and coworkers, 1970b) and the dissociative attachment cross sections for the formation of $Br^-$ from these molecules have been determined and are presented in Figure 6.27. The magnitudes and the widths of these cross-section functions have provided evidence that dissociative electron attachment to such compounds can be understood by considering the $C_nH_{2n+1}Br$ system as a diatomic-like A–X ($A = C_nH_{2n+1}$, $X = Br$) molecule.

For further information on negative ion formation in non-aromatic halogenated polyatomics, see Craggs, McDowell and Warren (1952); Marriott, Thorburn and Craggs (1954); Reese, Dibeler and Mohler (1956); Hickam and Berg (1958, 1959); Buchel'nikova (1959); Curran (1961a); Fox and Curran (1961); Dorman (1966); Khvostenko, Sultanov and Furley

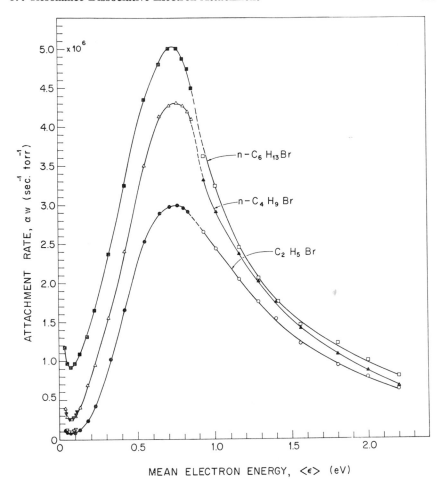

**Figure 6.26** Electron attachment rates as a function of mean electron energy for $C_2H_5Br$, $n$-$C_4H_9Br$, $n$-$C_6H_{13}Br$. (Christodoulides and Christophorou, 1970). Note the excellent agreement between the data in the three carrier gases $C_2H_4$, $N_2$ and Ar. $C_2H_5Br$: $\triangledown$, $\bullet$, $\circ$ are data obtained independently in the carrier gases $C_2H_4$, $N_2$ and Ar, respectively. $n$-$C_4H_9Br$: $\blacktriangledown$, $\triangle$, $\blacktriangle$ are data obtained independently in the carrier gases $C_2H_4$, $N_2$ and Ar, respectively. $n$-$C_6H_{13}Br$: $\blacksquare$, $\square$ are data obtained independently in the carrier gases $N_2$ and Ar, respectively

(1967); Wentworth, Becker and Tung (1967); Christophorou and Stockdale (1968); Blaunstein and Christophorou (1968); and Christodoulides and Christophorou (1970). Also, Lee (1963) has studied electron attachment in a large number of halogenated compounds using the swarm method.

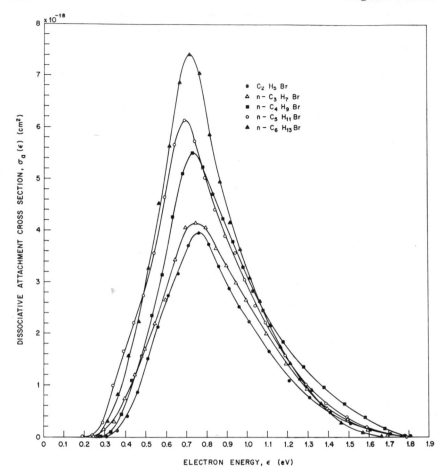

**Figure 6.27** Dissociative electron attachment cross sections for brominated hydrocarbons of the form n-$C_nH_{2n+1}Br$ ($2 \leqslant n \leqslant 6$) (Christophorou and coworkers, 1970b)

Sulphur hexafluoride ($SF_6$) is one of the most intensively investigated halogenated compounds. It has been studied by both the electron swarm method (Hasted and Berg, 1965; Mahan and Young, 1966; Compton and coworkers, 1966; and Christophorou, McCorkle and Carter, 1970) and the electron beam method (Ahearn and Hannay, 1953; Hickam and Fox, 1956; Buchel'nikova, 1959; Curran, 1961a, 1963; Fox and Curran, 1961; Edelson, Griffiths and McAfee, 1962; Asundi and Craggs, 1964b; Rapp and Briglia, 1965; and Compton and coworkers, 1966). Multiple fragmentation of $SF_6^-*$ has been observed at low-electron energies and ions such as $SF_6^-$,

$SF_5^-$, $F_2^-$ and $F^-$ have been identified. At higher energies $SF_4^-$ (at $\sim$5 eV) and $SF_3^-$ (at $\sim$10 eV) were detected (Edelson, Griffiths and McAfee, 1962). The attachment cross sections at thermal energies, obtained by the various techniques, are summarized in Table 6.5 (see also Table 6.8). The yield, in relative units, of the $SF_6^-$ and $SF_5^-$ ions as a function of electron energy is shown in Figure 6.28. The width of the $SF_6^-$ resonance shown in Figure 6.28 is instrumental. Schulz (1969) has found the width of the observed peak due to this resonance to be instrumental even for an energy resolution of 0.015 eV. Further, Chen and Chantry (1969) reported recently that the $SF_5^-$ current at room temperature exhibits another peak coincident

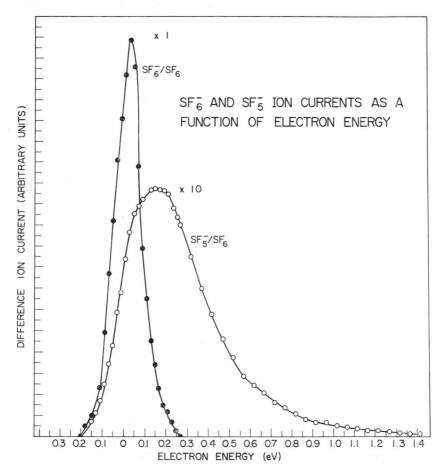

**Figure 6.28** Negative ion yield for the production of $SF_5^-$ and $SF_6^-$ from $SF_6$ (Compton and coworkers, 1968). Note that the width of $SF_6^-$ is instrumental (see text)

in energy with that of $SF_6^-$. The intensity of this peak increased with temperature; at 500°K the $SF_5^-$ current at ~0.0 eV was reported by Chen and Chantry to exceed the $SF_6^-$ current. However, at room temperature $SF_6^-$ is the predominant ion (Chen and Chantry, 1969; Fehsenfeld, 1969). Christophorou, McCorkle and Carter (1970) have found that for electron energies below 0.11 eV the attachment cross section for $SF_6^-$ at room temperature varies as $\varepsilon^{-\gamma}$ where $\gamma = 1.12$ (see Table 6.2). Also from their analysis Christophorou, McCorkle and Carter found that at 298°K $SF_5^-$ showed a pronounced peak at 0.37 eV with a cross-section value at the peak energy equal to $4.9 \times 10^{-17}$ cm$^2$.

The $SF_6^-$ ion has been found (Edelson, Griffiths and McAfee, 1962; Compton and coworkers, 1966) to be metastable with respect to electron ejection and its lifetime against autoionization has been measured. The lifetime measurement in connection with swarm data will be discussed in Section 6.5. As discussed in Chapter 5 due to the sharp and large cross section of $SF_6^-$ at thermal energies, $SF_6$ has been used by a number of workers as a detector of near-zero energy electrons ($\lesssim 0.03$ eV) in studies of electronic excitation of atoms and molecules at energies close to their electronic excitation thresholds.

D. *Halogenated benzene derivatives.* Dissociative electron attachment to halogenated benzene derivatives has been investigated by Christophorou and coworkers (1966) using the swarm–beam method. Their results are summarized in Table 6.6 and are plotted in Figure 6.29. Earlier swarm studies on chlorobenzene and bromobenzene have been performed by Stockdale and Hurst (1964). The dissociative attachment resonances are quite broad (Table 6.6); their peak energies range from ~1 eV down to thermal energies. The magnitude of the cross section increases and the peak energy decreases rapidly in changing the substituent from Cl to Br (to I). This change correlates with the increase in resonance interaction between the halogen and the benzene ring. No parent negative ions have been observed (Christophorou and coworkers, 1966) in the mass spectrometer for the compounds listed in Table 6.6.

Christophorou and coworkers (1966) have also reported on negative ion formation in *m*- and *p*-$C_6H_4Cl_2$, $C_6H_5I$, $C_6H_5NO_2$ and *o*- and *m*-$C_6H_4CH_3NO_2$. The latter three compounds form parent negative ions at thermal energies which were found to be metastable (see Section 6.5 for the thermal rate of attachment to $C_6H_5NO_2$ and the lifetime of $C_6H_5NO_2^-$* and *o*–, *m*– and *p*–$C_6H_4CH_3NO_2^-$*).† Two major $NO_2^-$ peaks were observed by Christophorou and coworkers (1966) in the study of $C_6H_5NO_2$, one

---

†A study of negative ion formation by electron impact of a number of $NO_2$ containing compounds has been made by Jäger and Henglein (1967). These authors observed parent negative ions in nitrobenzene and tetranitromethane.

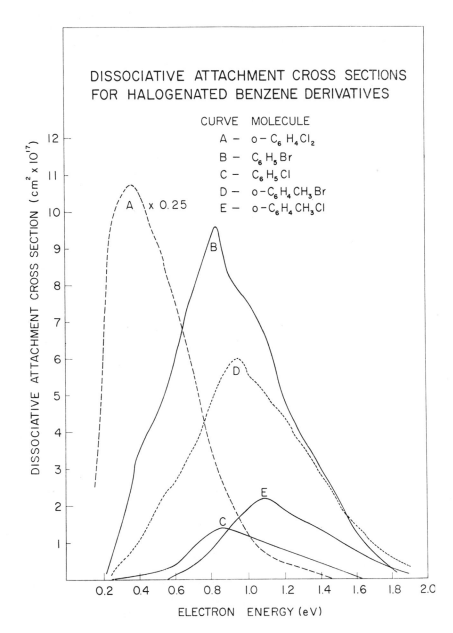

**Figure 6.29** Dissociative attachment cross sections for halogenated benzene derivatives (Christophorou and coworkers, 1966)

peaking at 1.06 eV and the other at 3.53 eV. Similar results have been obtained for $o-$ and $m$–nitrotoluene. The position and relative intensities of the two $NO_2^-$ peaks, however, differ for the three molecules ($C_6H_5NO_2$, $o-$ and $m-C_6H_4CH_3NO_2$), and this difference has been attributed by Christophorou and coworkers (1966) to interactions between the $CH_3$ and $NO_2$ groups.

As discussed in Chapter 5, intense electron energy-loss processes below the first electronic excitation potential take place in benzene and a number of its derivatives. The energy-loss resonances for $C_6H_5Cl$, $C_6H_5Br$, $o-C_6H_4CH_3Cl$ and $o-C_6H_4Cl_2$ were found by Compton, Christophorou and Huebner (1966) to agree both in shape and peak energy with the dissociative attachment resonances, a result strongly indicating that both resonances proceed via a common temporary negative ion state. The relevance of the temporary negative ion states, observed in these molecules, to the molecular electron affinity and vertical attachment energy are discussed in Chapter 7. It should be noted that the data in Figure 6.29 suggest some structure. Certainly such structure would be consistent with that observed by Boness and coworkers (1967) for the benzene compound negative ion in a transmission experiment.

### 6.4.2.4 Dependence of the magnitude of the dissociative attachment cross section on the resonance energy

The dissociative attachment peak cross section, $\sigma_{da}(\varepsilon_{max})$, for a number of molecules listed in Tables 6.3 to 6.6 is plotted in Figure 6.30 as a function of the peak (resonance) energy $\varepsilon_{max}$. This plot helps demonstrate the large variation of $\sigma_{da}(\varepsilon)$ from molecule to molecule* and the strong dependence of $\sigma_{da}(\varepsilon)$ on $\varepsilon_{max}$. The figure also shows the large decrease in $\sigma_{da}$ when $\varepsilon_{max}$ lies at or exceeds the energy of electronic levels of the neutral molecule, suggesting increased autoionization due to the decay of the compound negative ion to an electronically excited neutral molecule and a slow free electron. In actual fact, based on the experimental data plotted in Figure 6.30, one may distinguish (Christophorou and Stockdale, 1968) three groups of molecules: (i) those where $\varepsilon_{max}$ is less than the energy, $E_N$, of known electronic states of the neutral molecule, and the negative ion state is purely repulsive or contains very small minima in the Franck–Condon region[†]; (ii) those where $\varepsilon_{max} \geqslant E_N$; and (iii) those with, often, exceptionally small $\sigma_{da}(\varepsilon)$ for which a vertical onset for dissociative attachment occurs. Christophorou and Stockdale (1968) pointed out that for molecules in group (i) (see Figure 6.30), $\sigma_{da}(\varepsilon_{max})$ varies almost as $\varepsilon_{max}^{-1}$, while for group (ii) $\sigma_{da}(\varepsilon_{max})$ is a much stronger decreasing function of the resonance energy $\varepsilon_{max}$.

---

*Hence one should be careful of impurities in such studies.
†Such a situation seems very likely from the experimentally known asymptotic limits of the negative ion and ground-state molecule potential energy curves and the shape and position of the product negative ion current curves.

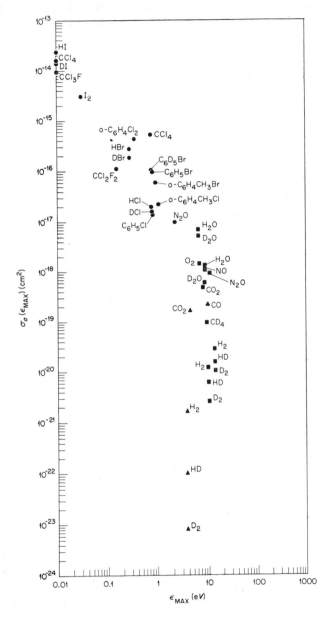

**Figure 6.30** Maximum value of the dissociative attachment cross section, $\sigma_{da}$ $(\varepsilon_{max})$, as a function of the energy, $\varepsilon_{max}$, at the peak of the cross section $\varepsilon_{max}$. (Christophorou and Stockdale, 1968). Additional recent work (e.g. on $NH_3$ and $ND_3$ – Sharp and Dowell (1969); halogenated aliphatic hydrocarbons – Blaunstein and Christophorou, 1968) is in general agreement with the trend shown in this figure

For molecules in group (i) the effect of autoionization on $\sigma_{da}(\varepsilon_{max})$ is generally small (this results to, generally, small isotope effects), and $\sigma_{da}(\varepsilon)$ is mainly determined by the magnitude of $\sigma_0$. One may, then, argue that for this group of molecules the variation of $\sigma_{da}(\varepsilon_{max})$ with $\varepsilon_{max}$ reflects the variation of $\sigma_0$ with $\varepsilon$. For group (ii) and especially group (iii) the effect of auto-ionization on $\sigma_{da}(\varepsilon_{max})$ is expected to be large and so are the isotope effects.

## 6.5 Resonance Non-Dissociative Electron Attachment

### 6.5.1 *Introduction*

In Section 6.4 we discussed the resonance dissociative electron attachment process in which stable negative ions and neutral atomic or molecular fragments are formed. We visualized dissociative electron attachment as proceeding via a temporary negative ion, often with as short a lifetime as $10^{-15}$ sec. The unstable (excited) ion decays through decomposition (sometimes leading to multiple fragmentation) or autodetachment, resulting in elastic or inelastic electron scattering. When dissociative electron attachment is not energetically possible, autodetachment is the dominant decay channel if the attachment process remains an isolated event. Diatomic negative ions formed by unimolecular electron attachment have never been detected directly because of their short lifetimes. For molecules with a positive electron affinity, the temporary negative ion can lose its excess energy and form a stable parent negative ion, the process of stabilization being in competition with those of autodetachment and dissociation when the latter is energetically possible.

The stabilization of an excited negative ion is a function of its lifetime and the cross section for stabilization by collision or radiation. The most effective mode of stabilization (see also discussion in Section 6.2) is that through collision with molecules of the same or of different kind. The collisional stabilization process depends on the nature and volume density of the stabilizing agent. When the lifetime of the negative ion is long (say, of the order of $10^{-5}$ sec) compared to the time between collisions (for pressures of a few hundred torr this time is of the order of $10^{-12}$ sec), complete stabilization of the excited negative ion is possible and the measured attachment rates will be representative of the forward reaction $(AX + e \rightarrow AX^{-*} \rightarrow AX^{-})$ and will not be complicated by the reverse transition $(AX^{-*} \rightarrow AX^{(*)} + e)$. It has to be pointed out here, however, that in high pressure (swarm) experiments the excited ion in addition to being stabilized, autodissociated or autoionized may undergo collisional detachment or collision-induced dissociation. The latter two processes complicate high-pressure electron attachment experiments. Electron swarm and electron beam techniques have been employed to the study of detachment of electrons from negative ions in collisions with molecules. This and other electron detachment processes will be discussed in Chapter 7. For a discussion of the rates of reaction in two- and three-body encounters, see Section 4.1.2.6. We shall now discuss briefly the process of non-dissociative electron attachment to diatomic and polyatomic molecules.

### 6.5.2 Non-Dissociative Electron Attachment to Diatomic and Triatomic Molecules

The information on non-dissociative electron attachment to diatomic and triatomic molecules is very limited. In Section 6.4 we elaborated on non-dissociative electron attachment to NO, $H_2O$ and $N_2O$. In this section our discussion will be restricted to $O_2$ which has been the subject of many investigations since the 1930's. For a summary of the earlier work on electron attachment to oxygen molecules to form $O_2^-$ ions, see Loeb (1955, 1956) and Chanin, Phelps and Biondi (1962). The early work of Bradbury (1933), Doehring (1952) and Herreng (1952) was done with relatively low pressures (of the order of a few torr) and led to the conclusion that the attachment process exhibited a two-body pressure dependence. More recent work by Hurst and Bortner (1959a, b), Chanin, Phelps and Biondi (1959, 1962), Pack and Phelps (1966a, b), Bouby and Abgrall (1967) and Stockdale, Christophorou and Hurst (1967) has shown that the attachment at low energies ($\lesssim 1$ eV) involves a three-body process.

Bloch and Bradbury (1935) suggested that the attachment of near-thermal energy electrons to $O_2$ might be explained in terms of a two-step process involving one electron and two $O_2$ molecules. The first step in the Bloch–Bradbury mechanism is the attachment of a low-energy electron to $O_2$ in the ground vibrational state ($v = 0$) of its ground electronic state ($^3\Sigma_g^-$), leading to the formation of an unstable vibrationally excited negative ion in its $v = 1$ vibrational state, i.e.

$$(O_2)_{v=0} + e \rightleftarrows (O_2^-)^*_{v=1} \tag{6.42}$$

The second step in the Bloch–Bradbury mechanism concerns the vibrational de-excitation of $O_2^-$* in collision with a third body S, i.e.

$$(O_2^-)^*_{v=1} + S \rightleftarrows (O_2^-)_{v=0} + S^{(*)} + \text{energy} \tag{6.43}$$

In the absence of collision, the system will autoionize since stabilization by radiation is very small. In the light of recent information (Chapter 7) concerning the electron affinity of $O_2$, the Bloch–Bradbury picture of the attachment process must be modified to place the negative ion in a higher vibrational state than $v = 1$ as is indicated by the potential curves in Figure 6.31. Process (6.42) is then expressed as

$$(O_2)_{v=0} + e \rightleftarrows (O_2^-)_{v=m}^* \tag{6.44}$$

where $m > 1$.

Recent studies on $O_2$ at energies $< 1$ eV, referred to in the beginning of this section, have shown that electron attachment to $O_2$ molecules is consistent with a three-body attachment reaction, viz.

$$(O_2)_{v=0} + e + S \rightleftarrows (O_2^-)_{v=0} + S^{(*)} + \text{energy} \tag{6.45}$$

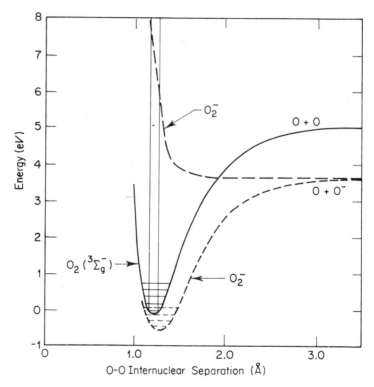

**Figure 6.31** Hypothetical potential energy curves for $O_2$ and $O_2^-$ based on available spectroscopic and negative ion data (Biondi, 1963; Chanin, Phelps and Biondi, 1962)

The third body S can either be another $O_2$ molecule or a different atom or molecule. In the case where S is an atom the excess energy appears as translational; when S is a molecule the excess energy may also appear as internal energy of S. In Equation (6.45) we have indicated that the $O_2^-$ ion is formed in its ground vibrational state since Pack and Phelps (1966a) in a swarm experiment with pure $O_2$ have been unable to convert the molecular oxygen negative ion into a more stable one by increasing the $O_2$ density ($O_2^-$ sustained $\sim 10^8$ collisions). The results of Chanin, Phelps and Biondi (1962) on the dependence of the attachment coefficient $\alpha (=\alpha'/P)$ on $E/P$ and $O_2$ pressure have shown that at $E/P \gtrsim 3$ volts/cm torr, $\alpha$ is independent of $O_2$ pressure, while below this $E/P$ value the attachment coefficient increases linearly with $O_2$ pressure. The high-energy attachment process is the two-body dissociative electron attachment discussed in Section 6.4, which does not require collisional stabilization, while the low-energy one is the three-body attachment process which leads to parent $O_2^-$ ions.

If $n_e$, $n_{O_2}$, $n_S$ are, respectively, the number density of electrons, oxygen molecules, and third-body atoms or molecules, $v_a$ is the overall attachment frequency and $k$ is the three-body attachment coefficient for the reaction (6.45), then the rate of change of electron density due to attachment in an $O_2$–S mixture is

$$\left. \frac{dn_e}{dt} \right|_a = -v_a n_e = -k_S n_S n_{O_2} n_e - k_{O_2} n_{O_2}^2 n_e \qquad (6.46)$$

and the attachment rate (in $\sec^{-1}$ torr$^{-1}$) is

$$\alpha w = N^2 P [f_2 k_s + f_1 k_{O_2}] \qquad (6.47)$$

In Equation (6.47) $N$ is the number of molecules per $cm^3$ at 1 torr, $P$ is the total pressure in torr, and $f_1$ and $f_2$ are the molar fractions of $O_2$ and S, respectively. The recent measurements on the thermal electron attachment to $O_2$ in $O_2$–$C_2H_4$ mixtures by Stockdale, Christophorou and Hurst (1967) are in agreement with Equation (6.47). Similarly, Bouby and Abgrall (1967) found agreement with Equation (6.47) and the three-body attachment coefficients obtained in these experiments are listed in Table 6.7.

Pack and Phelps (1966a) extended the work of Chanin, Phelps and Biondi (1962) on $O_2$ to higher temperatures and oxygen densities. Their results on the three-body attachment coefficient for pure $O_2$ are shown in Figure 6.32 as a function of the characteristic energy for various temperatures. It is seen that for all temperatures the three-body attachment coefficient reaches a maximum at an average electron energy of $\sim 0.1$ eV, which suggests that $\sim 0.1$ eV energy is required to reach the lowest state of $O_2^-$*.

Recent studies on $O_2$ at low energies elaborated also on the effectiveness of various gases in acting as a third body, S, to stabilize $O_2^-$* in $O_2$–S mixtures. Figure 6.33 shows the results of Chanin, Phelps and Biondi (1962) for $O_2$, $N_2$ and He. It is clearly seen that the monatomic gas He is much less effective in stabilizing $O_2^-$* than either $N_2$ or $O_2$. The effectiveness of a number of other molecules can be seen from the data collected in Table 6.7. The three-body attachment coefficients listed are for thermal (300°K) energies and they do not seem to correlate with any particular molecular parameter, although it is worth noting that the values of the three-body attachment coefficient for the polar molecules $H_2O$, $H_2S$, $NH_3$, $CH_3OH$, and especially for the highly polar molecule $(CH_3)_2CO$, are large.

Table 6.7 Three-body attachment coefficients at 300°K
for the reaction $e + O_2 + S \rightarrow O_2^- + S^{(*)} + $ energy

| Third body, S | Three-body attachment coefficient $(10^{-30}$ cm$^6$/sec) | Ref.[b] |
|:---:|:---:|:---:|
| He | $\sim 0.03$ | 1 |
| $N_2$ | $\sim 0.1$ | 1 |
| $O_2$ | 2 | 2 |
| | 2.6 | 4 |
| $H_2O$ | 14 | 3 |
| | 14 | 4 |
| | 15.2 | 5 |
| $H_2S$ | 10 | 5 |
| $CO_2$ | 3.1 | 3[a] |
| | 3.2 | 5 |
| $NH_3$ | 7.5 | 5 |
| $C_2H_4$ | 3.2 | 4 |
| | 1.7 | 5 |
| $CH_3OH$ | 9.6 | 5 |
| $(CH_3)_2CO$ | $> 35$ | 5 |
| $C_6H_6$ | 18 | 5 |

[a] The three-body attachment coefficients at 477 and 525°K are, respectively, 4.0 and $2.8 \times 10^{-30}$ cm$^6$/sec.
[b] References:
1. Chanin, Phelps and Biondi (1962).
2. Pack and Phelps (1966a).
3. Pack and Phelps (1966b).
4. Stockdale, Christophorou and Hurst (1967).
5. Bouby and Abgrall (1967).

The inverse of the three-body attachment process is the collisional detachment of the attached electron. Such a process, as stated earlier, may take place especially at high pressures. Pack and Phelps (1966a) used their drift tube apparatus to investigate electron detachment from molecular oxygen negative ions. Thermal attachment and detachment coefficients in molecular oxygen as a function of gas temperature are shown in Figure 6.34.

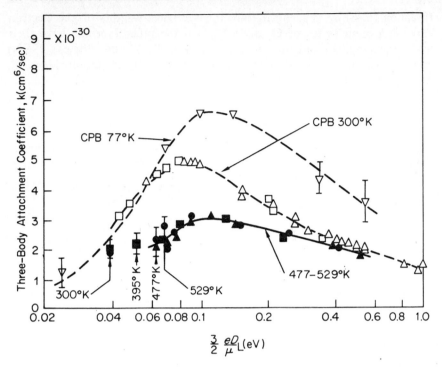

■, ●, and ▲ are 477°, 523° and 529°K, respectively. The open triangles
and squares are from Chanin, Phelps and Biondi (1962) CPB (∇, △ and □
are $O_2$ + He (77°K), $O_2$ (300°K), and $O_2$ + He (300°K), respectively).

**Figure 6.32** Three-body attachment coefficient for $O_2$ versus
the energy $\dfrac{3}{2}\dfrac{eD_L}{\mu}$ for various gas temperatures (Pack and
Phelps, 1966a). The solid circles with error bars were obtained
from $O_2$–$CO_2$ mixtures for thermal electrons at 300°, 477°, and
529°K. The solid squares with error bars were obtained with
$O_2$–$H_2O$ mixtures for thermal electrons at 300 and 395°K. The
solid points without error bars were derived from an analysis of
the high-pressure data at each gas temperature

Expressing the equilibrium constant, $K$, for the reaction $e + 2O_2 \rightleftarrows O_2 + O_2^-$
by

$$K = \frac{v_d[O_2]}{v_a} = \frac{[e][O_2]}{[O_2^-]} = \frac{g_e g_{O_2}}{g_{O_2^-}}\left(\frac{2\pi m kT}{h^2}\right)^{3/2} e^{-(EA)/kT} \qquad (6.48)$$

Pack and Phelps (1966a) deduced from the measured attachment and detachment coefficients an electron affinity, $EA$, for $O_2$ equal to $0.43 \pm 0.02$ eV. In

Equation (6.48) $g_e$, $g_{O_2}$ and $g_{O_2^-}$ are the statistical weights of the electron and the ground states of $O_2$ and $O_2^-$, and the other symbols have their usual meaning. In solving Equation (6.48), Pack and Phelps (1966a) assumed that the rotational and vibrational states of $O_2$ and $O_2^-$ are identical, that the $^2\Pi_g$ state is the only $O_2^-$ state occupied, and that $g_{O_2}/g_{O_2^-} = 3/4$.

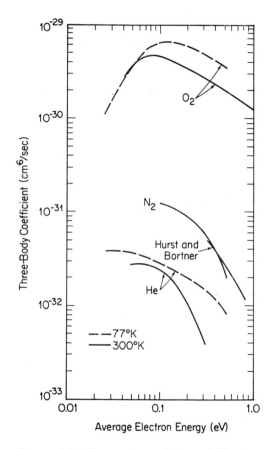

**Figure 6.33** Comparison of the stabilization efficiencies of $O_2$, $N_2$ and He in three-body electron attachment to $O_2$ (Chanin, Phelps and Biondi, 1962)

Moruzzi and Phelps (1966) surveyed negative ion–molecule reactions in a number of gases and gas mixtures using a small rf mass spectrometer coupled to a high pressure drift tube. In pure $O_2$ for low electron energies ($E/P \leqslant 2$ volts/cm torr) $O_2^-$ was the only negative ion found. The variation

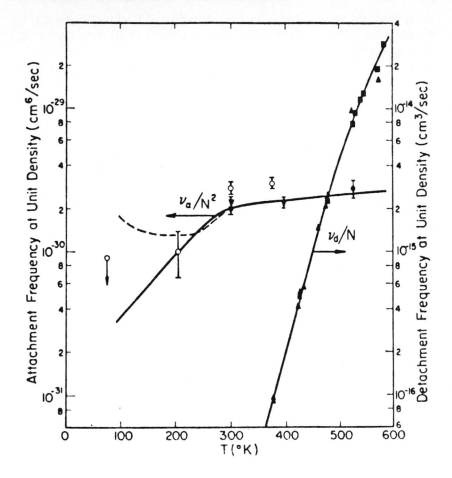

▼ $H_2O+O_2$ , ▲ low pressure, ■ high pressure; ● $\left[CO_2+O_2\right.$ [Pack and Phelps (1966a)$\left.\right]$. ○ Chanin, Phelps, and Biondi (1962). −−− Lint, Wikner, and Trueblood (1960).

**Figure 6.34** Thermal attachment and detachment coefficients in $O_2$ versus gas temperature

of the intensity of the $O_2^-$ signal with $O_2$ pressure again was consistent with a three-body attachment process. At energies corresponding to $E/P$ values $>2$ volts/cm torr, $O^-$, $O_2^-$ and $O_3^-$ ions were detected, the first ion produced by resonance dissociative attachment and the latter two ions by ion–molecule reactions. The results of Moruzzi and Phelps (1966) are given in Figure 6.35. Finally, Prasad (1965) gave a value of $10^{-9}$ sec as a rough estimate for the lifetime of $O_2^- *$.

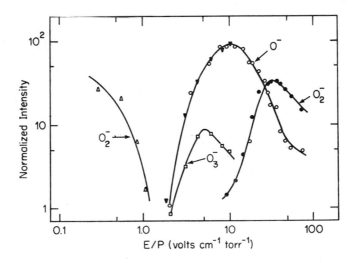

**Figure 6.35** Relative ion intensity versus $E/P$ in $O_2$ (Moruzzi and Phelps, 1966). The $O_2^-$ ions observed at low $E/P$ are formed in a three-body attachment process and do not undergo any ion–molecule reactions. At higher energies $O^-$ is initially formed in a dissociative attachment process and subsequently undergoes conversion to $O_3^-$ and $O_2^-$. $\circ$, $\bullet - P = 1.76$ torr; $\square$, $\triangle$, $\blacktriangle - P = 3.37$ torr

### 6.5.3 Non-Dissociative Electron Attachment to Polyatomic Molecules

#### 6.5.3.1 Long-lived parent negative ions in fluorinated polyatomic molecules

A number of molecules have been reported to attach electrons by unimolecular electron attachment for times greater than $10^{-6}$ sec (Table 6.8). Edelson, Griffiths, and McAfee (1962) first examined long-lived negative ions and showed that the $SF_6^- *$ ion was metastable and subject to auto-ionization with a mean lifetime of $\sim 10^{-5}$ sec. The suggestion was also made (Hickam and Fox, 1956; Asundi and Craggs, 1964b) that for complex molecules the energy of the attached electron is shared with the many molecular degrees of freedom resulting in autoionization lifetimes which are long enough to allow observation of the metastable ion with a mass spectrometer.

**Table 6.8** Thermal electron attachment rates and parent negative ion lifetimes for polyatomic molecules

| Molecule | Attachment rate $(sec^{-1} torr^{-1})$ | Temperature $(°K)$ | Ref.[e] | Lifetime $(\mu sec)$ | Ref.[e] |
|---|---|---|---|---|---|
| *Fluorinated Compounds* | | | | | |
| $SF_6$ | $1.0 \times 10^{10}$ | ~300 | 1 | 25 | 5 |
| Sulphurhexafluoride | $8.9 \times 10^{9a}$ | ~300 | 2 | | |
| | $8.7 \times 10^{9}$ | ~300 | 3 | 32 | 6 |
| | $7.8 \times 10^{9}$ | ~300 | 4 | | |
| $C_4F_6$ | | | | 6.9 | 7 |
| Perfluorocyclobutene | | | | | |
| $C_4F_8$ | | | | 12.0 | 7 |
| Octafluorocyclobutane | | | | | |
| $C_6F_6$ | | | | 12.0 | 7 |
| Hexafluorobenzene | | | | | |
| $C_6F_5CF_3$ | | | | 12.2 | 7 |
| Octafluorotoluene | | | | | |
| $C_5F_8$ | | | | 50 | 7 |
| Perfluorocyclopentene | | | | | |
| $C_6F_{10}$ | | | | 113 | 7 |
| Perfluorocyclohexene | | | | | |
| $C_6F_{12}$ | | | | 450 | 7 |
| Perfluorocyclohexane | | | | | |
| $C_7F_{14}$ | $3.46 \times 10^{9}$ | ~300 | 1 | 793 | 7 |
| Perfluoromethylcyclo-hexane | $2.81 \times 10^{9}$ | ~300 | 4 | | |
| $CF_3COCF_3$ | | | | ~ 60 | 17 |
| Hexafluoroacetone | | | | | |
| *Aromatic Hydrocarbons* | | | | | |
| $C_6H_6$ | $< 3 \times 10^{3}$ | 298 | 8 | | |
| Benzene | | | | | |
| $C_{10}H_8$ | $\leqslant 10^{4}$ | 343–398 | 8[b] | | |
| Naphthalene | | | | | |
| $C_{10}H_8$ | $1 \times 10^{9}$ | 423–488 | 9[b] | $7 \pm 1.5$ | 9 |
| Azulene | | | | | |
| $C_{14}H_{10}$ | $1.52 \times 10^{8}$ | 383–479 | 8[b] | $>1$ | 10 |
| Anthracene | $1.45 \times 10^{8}$ | | 11 | | |
| $C_{14}H_{10}$ | $7 \times 10^{5}$ | 453–473 | 8[b] | $>1$ | 10 |
| Phenanthrene | | | | | |
| $C_{18}H_{12}$ | $2.74 \times 10^{7}$ | 443–483 | 8[b] | | |
| Triphenylene | | | | | |
| $C_{18}H_{12}$ | $3.25 \times 10^{7}$ | 473–483 | 8[b] | | |
| Chrysene | | | | | |
| $C_{18}H_{12}$ | $1.47 \times 10^{8}$ | | 11 | | |
| 3,4-Benzophenanthrene | | | | | |
| $C_{18}H_{12}$ | $6.64 \times 10^{8}$ | 393–473 | 8[b] | | |
| 1,2-Benzanthracene | $2.11 \times 10^{9}$ | | 11 | | |
| $C_{16}H_{10}$ | $1.4 \times 10^{7}$ | 443–483 | 8[b] | $>1$ | 10 |
| Pyrene | $4.4 \times 10^{8}$ | | 11 | | |
| $C_{20}H_{12}$ | $3.4 \times 10^{8}?$ | 473–483 | 8[b] | | |
| Perylene | | | | | |
| *Oxygen-Containing Organic Molecules* | | | | | |
| $C_6H_5NO_2$ | $2.1 \times 10^{7}?$ | ~300 | 5 | 40 | 5 |
| Nitrobenzene | | | | | |
| $(CHO)_2$ | $1 \times 10^{8}$ | | 12 | 2.5 | 12 |

Table 6.8  (*continued*)

| Molecule | Attachment rate ($sec^{-1}$ $torr^{-1}$) | Temperature (°K) | Ref.[e] | Lifetime ($\mu sec$) | Ref.[e] |
|---|---|---|---|---|---|
| Glyoxal | | | | | |
| $(CH_3CO)_2$ | $1.1 \times 10^9$ | | $13^c$ | 12 | 12 |
| Biacetyl | $1.4 \times 10^9$ | | | | |
| | $1.1 \times 10^9$ | | | | |
| $CH_3-(CO)_2-C_2H_5$ | $3 \times 10^9$ | | $13^d$ | | |
| 2,3-Pentanedione | $1.7 \times 10^9$ | | | | |
| $C_6H_4O_2$ | $1.2 \times 10^5$ | $\sim 300$ | 14 | | |
| *p*-Benzoquinone | | | | | |
| $C_{10}H_6O_2$ | $1.85 \times 10^9$ | 348 | 15 | See Ref. 15. Lifetime varies with electron energy (Section 7.1.1). | |
| 1,4-Naphthoquinone | | | | | |
| $C_{10}H_7CHO$ | $3.9 \times 10^8$ | | 16 | 7.5 | 6 |
| 2-Naphthaldehyde | | | | | |
| $C_{10}H_7CHO$ | $3.35 \times 10^8$ | | 16 | 16 | 6 |
| 1-Naphthaldehyde | | | | | |
| $C_6H_5CH=CHCHO$ | $2.0 \times 10^8$ | | 16 | 12 | 6 |
| Cinnamaldehyde | | | | | |
| $C_6H_4CH_3NO_2$ | | | | 13 | 6 |
| *o*-Nitrotoluene | | | | | |
| $C_6H_4CH_3NO_2$ | | | | 18.8 | 6 |
| *m*-Nitrotoluene | | | | | |
| $C_6H_4CH_3NO_2$ | | | | 14. | 6 |
| *p*-Nitrotoluene | | | | | |

[a]  See Reference 2 for a discussion of other data.
[b]  Attachment rates did not change with temperature in the range quoted.
[c]  The three values are for the carrier gases $C_2H_4$, $CO_2$ and $CH_3OH$, respectively.
[d]  The first value is for the carrier gas $C_2H_4$ and the second for the carrier gas $CH_3OH$.
[e]  References:
 1. Mahan and Young (1966).
 2. Christophorou, McCorkle and Carter (1970).
 3. Fehsenfeld (1969).
 4. Chen, George and Wentworth (1968).
 5. Compton and coworkers (1966).
 6. Christophorou and coworkers (1970a).
 7. Naff, Cooper and Compton (1968).
 8. Christophorou and Blaunstein (1969).
 9. Chaney and coworkers (1969, 1970).
 10. Von Ardenne, Steinfelder and Tümmler (1961).
 11. Wentworth, Chen and Lovelock (1966).
 12. Compton and Bouby (1967).
 13. Bouby, Fiquet-Fayard and Abgrall (1965).
 14. Christophorou, Carter and Christodoulides (1969).
 15. Collins and coworkers (1969, 1970).
 16. Wentworth and Chen (1967).
 17. Collins, Christophorou and Carter (1970). Harland and Thynne (1969) reported that $CF_3COCF_3^-$ ions form at thermal energies with a cross section of $\sim 59$ times less than that for $SF_6^-$.

Sulphur hexafluoride ($SF_6$) is one of the few polyatomic molecules which have been extensively investigated in negative ion studies; it is of considerable importance as a gaseous dielectric. The thermal and near thermal electron attachment process in $SF_6$ has been described as

$$SF_6 + e \underset{\tau_a^{-1}}{\overset{\sigma_o}{\rightleftharpoons}} SF_6^{-*} \underset{p_{da}}{\overset{p_{st}}{<}} \begin{array}{c} SF_6^- \\ \\ SF_5^- + F \end{array} \qquad (6.49)$$

The yields of $SF_6^-$ and $SF_5^-$ at thermal and near thermal energies have been presented in Section 6.4 (Figure 6.28). Both ions have been identified in the mass spectrometer. The $SF_5^-$ ion is less abundant than $SF_6^-$ at room temperature. It peaks at a higher energy ($\sim 0.37$ eV) and spreads over a wider energy range (Figure 6.28) (see also discussion in Section 6.4.2.3C).

Measurements of the mean autoionization lifetime ($> 10^{-6}$ sec) have been made with time-of-flight mass spectrometers (e.g. Edelson, Griffiths and McAfee, 1962; Compton and coworkers, 1966; Christophorou, Carter and Christodoulides, 1969; Chaney and coworkers, 1970; Christophorou and coworkers, 1970). Figure 6.36 shows the experimental arrangement used by Compton and coworkers (1966). The principle of the method is as follows: Assume that $N_0$ metastable negative ions leave the ion source with velocity $v$ at time $t = 0$. These travel along the flight tube, and some autoionize, but they are all counted at the end of the flight tube. At a distance $L$ down the flight path a 'flat-top' potential barrier is applied which decelerates (or deflects) the negative ions and leaves the neutral molecules unaffected. The potential barrier separates the negative ions from the neutral molecules after a time $t = L/v$ and, therefore, permits a measurement of the number of ions, $N^-(t)$, which have not decayed after a time $t$, and the number of ions, $N^0(t)$, which have decayed into neutral molecules. If an exponential decay law is assumed for the autodetachment of electrons, the mean lifetime can be determined from

$$\tau = \frac{t}{\ln\left[(N^- + N^0)/N^-\right]} \qquad (6.50)$$

Alternatively, the time $t$ between the ion source and the 'flat-top' separator can be varied, and the slope of the $\ln(N^-/N_0)$ versus $t$ straight line would yield the decay constant $\lambda(=1/\tau)$.

In Figure 6.37 we show an example of this procedure (Christophorou and coworkers, 1970a) for a number of molecules (see Table 6.8 for values of $\tau$). When lifetimes are determined with time-of-flight mass spectrometers, care must be taken that neutral molecules are counted with the same efficiency as negative ions of the same kinetic energy, collisional detachment or charge transfer does not take place, detachment by the grids in the path of the ion beam is negligible, and spontaneous dissociation or collision-induced

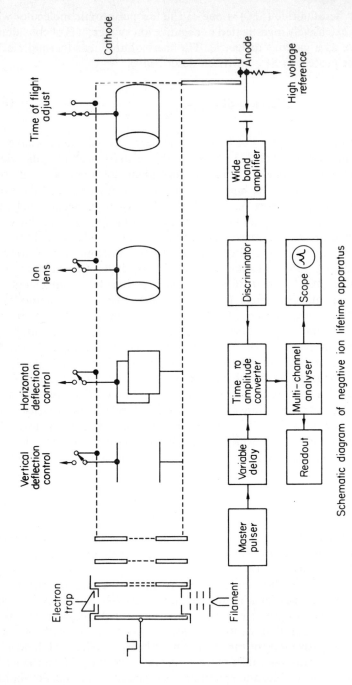

Schematic diagram of negative ion lifetime apparatus

**Figure 6.36** Schematic diagram of the arrangement of Compton and coworkers (1966) for the measurement of negative ion lifetimes

dissociation of the parent negative ion can be neglected. Lifetimes of long-lived negative ions of fluorinated molecules, measured with time-of-flight mass spectrometers, are listed in Table 6.8 and tend to indicate an increase in $\tau$ with increasing number of degrees of freedom. In Table 6.8 the available thermal electron attachment rates are also included.

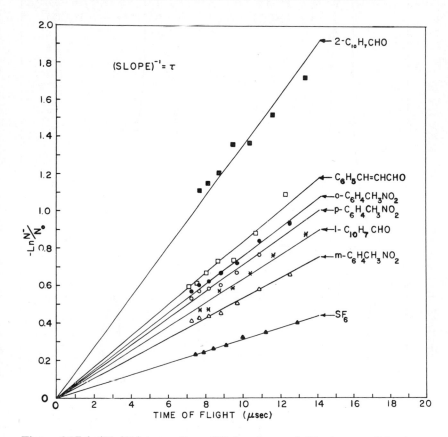

**Figure 6.37** $\ln(N^-/N_0)$ versus time of flight $t$ (see text). The inverse of the slopes yields the mean autodetachment lifetime for an assumed exponential decay (Christophorou and coworkers, 1970a)

### 6.5.3.2 Long-lived parent negative ions in aromatic and other organic molecules

Parent negative ions have been observed in a large number of organic vapours. The relative abundance of some of these metastable ions is shown as a function of electron energy in Figures 6.38 and 6.39. For comparison the $SF_6^-$ current is plotted with each resonance (the two currents in

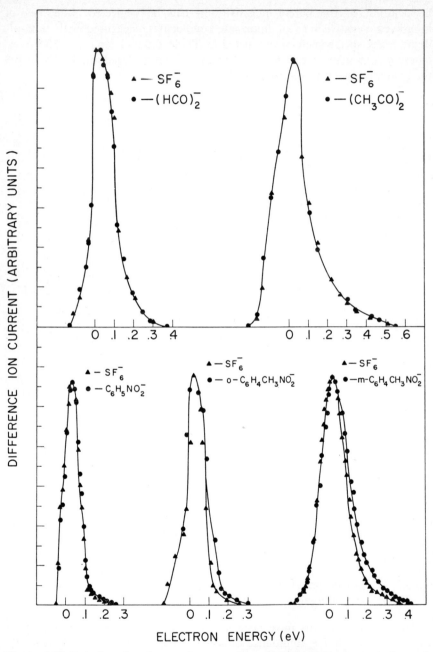

**Figure 6.38** Relative abundance of metastable parent negative ions at thermal and epithermal electron energies. The observed shapes are not real but instrumental. The negative ion currents have been normalized at the corresponding maxima (Christophorou and Compton, 1967)

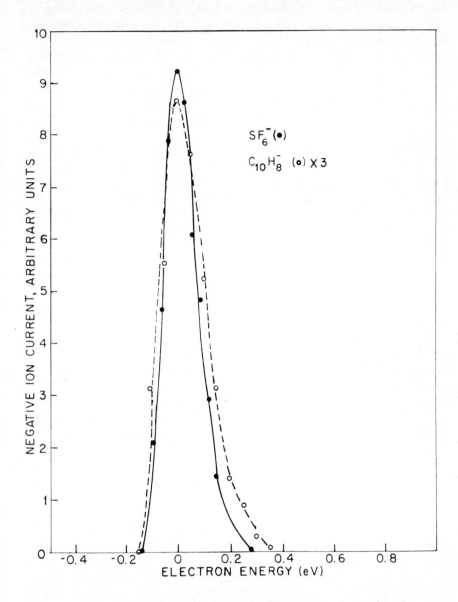

**Figure 6.39** Relative abundance of metastable negative ions of azulene at thermal and near thermal energies compared with that for $SF_6$ (Chaney and coworkers, 1970). The observed shapes are instrumental.

Figure 6.38 have been normalized at the peak). The widths of the resonances shown in Figures 6.38 and 6.39 are essentially identical with that of $SF_6^-$ and thus they are instrumental. Measurements of both the thermal attachment rates in high-pressure swarm experiments and the mean autoionization lifetimes in low pressure beam experiments have been made for a number of organic molecules (Table 6.8). Measurements of the temperature dependence of the attachment rate at thermal energies have also been made (Hendrick, Christophorou and Hurst, 1968; Christophorou and Blaunstein, 1969).

The variation of the attachment rate $\alpha w$ with the mean electron energy $\langle \varepsilon \rangle$ for 1,4-naphthoquinone, anthracene, and 1,2-benzanthracene is shown in Figure 6.40. In the same figure the variation of the attachment cross section with the electron energy $\varepsilon$ is also shown for these molecules. The cross sections were determined by Christophorou, McCorkle and Carter (1970) using the procedure outlined in Section 6.3.1.4. From the data shown in Figure 6.40, it is seen that the attachment cross sections increase sharply at thermal energies and—even for the aromatic hydrocarbons—are very large. This distinct feature—a large and sharply increasing attachment cross section at thermal electron energies—seems to be characteristic of polyatomic molecules which form long-lived parent negative ions by electron capture in the field of the ground electronic state.

Mass spectra of naphthalene, anthracene, tetracene, phenanthrene, pyrene and coronene have been studied by von Ardenne, Steinfelder and Tümmler (1961). In each of these molecules they found that the most intense signal lies exactly at the molecular weight corresponding to the parent mass, or because of splitting of an H atom around one mass unit below.

The cross sections for non-dissociative electron attachment to polyatomic molecules are seen to be, in most cases, very large (for $s$-wave capture the maximum thermal (300 °K) attachment rate is $1.38 \times 10^{10}$ sec$^{-1}$ torr$^{-1}$), and from the data listed in Table 6.8 it is clear that complex polyatomic molecules can possess large attachment cross sections and at the same time have long autoionization lifetimes.

In general, large attachment cross sections and long lifetimes can be reconciled in two manners: (i) weak transitions from the initial state of the neutral molecule directly to a large number of final states of the ion, and (ii) the transition from the neutral molecule to the negative ion is strongly allowed but the excess energy is distributed among the various internal degrees of freedom. Under assumption (i) large cross sections result because of the large number of final states, and long lifetimes are accounted for by the weak coupling between the initial and any of the final states. Under assumption (ii) large cross sections result from the large transition probability, and the long lifetimes are due to the time required for the system to return to a configuration which will lead to autodetachment.

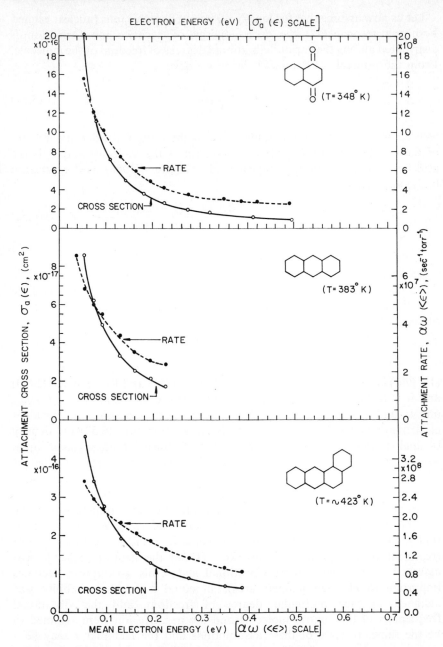

**Figure 6.40** Electron attachment cross section as a function of electron energy and electron attachment rate as a function of mean electron energy for 1,4-naphthoquinone, anthracene, and 1,2-benzanthracene (Christophorou, McCorkle and Carter, 1970)

Let us now assume that for long-lived parent negative ions (nuclear excited Feshbach resonances*) the excess energy of negative ion formation is distributed among the various vibrational degrees of freedom of the molecule. From the principle of detailed balance we have

$$\tau = \frac{\rho}{\rho^0} \frac{1}{v\sigma_a} \tag{6.51}$$

where $\tau$ is the negative ion lifetime, $\rho^-/\rho^0$ is the ratio of the density of states of the negative ion to that of the electron plus the molecule and $\sigma_a$ is the attachment cross section. Compton and coworkers (1966) used expression (6.51) and arrived at

$$\sigma_a(\varepsilon) = \frac{\pi^2\hbar^3}{\tau m\varepsilon_0^{\frac{1}{2}}\varepsilon_i^{\frac{1}{2}}} \frac{1}{\displaystyle\prod_{i=1}^{N} h\nu_i} \frac{[EA+\varepsilon_i+(1-\beta\omega_0)\varepsilon_z]^{N-1}}{(N-1)!} \tag{6.52}$$

which relates the measured electron attachment cross section, $\sigma_a(\varepsilon)$, and the mean negative ion lifetime, $\tau$, to the molecular electron affinity, $EA$, and the vibrational degrees of freedom, $N$. $\nu_i$ are the vibrational frequencies of the molecule,

$$\varepsilon_z = \tfrac{1}{2}\sum_{i=1}^{N} h\nu_i$$

$(1-\beta\omega_0)$ is a correction factor described by Whitten and Rabinovitch (1963) and it is to be evaluated for each molecule and each value of $(\varepsilon_i + EA)/\varepsilon_z$, and $\varepsilon_0$ and $\varepsilon_i$ are, respectively, the energies of the outgoing and incident electrons (equal by assumption). In deriving expression (6.52) the density of final states was taken to be the number of molecular states (taken equal to one) times the density of states of the free electron (taken equal to $(m^2v)/(\pi^2\hbar^3)$) while the density of states of the negative ion was calculated from prescriptions given by Rabinovitch and Diesen (1959).

Equation (6.52) is interesting as it relates four basic parameters; $\tau$, $\sigma_a(\varepsilon)$, $EA$ and $N$. The product of $\tau$ and $\sigma_a(\varepsilon)$ is very sensitive to $EA$ and indicates that $EA$ can be obtained from measurements of $\tau$ and $\sigma_a(\varepsilon)$. In the simple treatment of Compton and coworkers (1966) the neutral molecule was assumed to be in its lowest vibrational state (this assumption becomes important at elevated temperatures), an equal *a priori* probability was assigned to each molecular degree of freedom, and the vibrational frequencies of the neutral molecule and the negative ion were assumed to be the same. In spite of these assumptions and the fact that $\tau$ may be a

---

*See Bardsley and Mandl (1968) for a definition of the various types of resonances in electron–molecule collisions; see also Section 5.1.4.3.

strong function of the electron energy, application of Equation (6.52) in connection with the experimental data on $\tau$ and $\sigma_a(\varepsilon)$ gave reasonable values for the electron affinity of a number of polyatomic molecules (Chapter 7). These values must be considered as lower limits to *EA* (see further discussion in Compton and coworkers (1966) and Klots (1967); see also Chaney and coworkers (1970); Chaney and Christophorou (1970) and Christophorou and coworkers (1970a)).

### 6.5.3.3 Long-lived parent negative ions in the field of excited electronic states

The parent negative ions discussed in Sections 6.5.3.1 and 6.5.3.2 have the largest probability of occurrence at $\sim 0.0$ eV and are formed by electron capture in the field of the ground electronic state of the complex molecule. Recently, Christophorou and coworkers have presented experimental evidence showing that polyatomic parent negative ions can be formed by

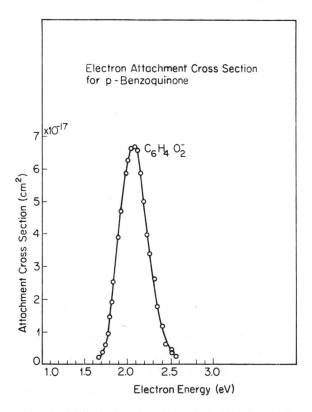

**Figure 6.41** Electron attachment cross section versus electron energy for *p*-benzoquinone at 2.1 eV (see text) (Christophorou, Carter and Christodoulides, 1969)

electron capture in the field of an excited electronic state with the parent ion so formed surviving autoionization for times longer than 1 μsec.

The results of Christophorou, Carter and Christodoulides (1969) on p–benzoquinone have been presented in Section 6.3.1.2 in connection with carrier gas distribution functions (Figure 6.5). The peak in the attachment rate at ~2 eV has been associated with electron capture in the field of the lowest triplet state $T_1$ resulting from an $n \rightarrow \pi^*$ transition at ~2.31 eV. This process is shown in Figure 6.41. The cross section scale and energy scale calibration have been obtained through the swarm–beam technique. At the peak energy (at 2.1 eV) the attachment cross section is equal to $6.7 \times 10^{-17}$ cm$^2$ and the mean autoionization lifetime equal to 38 μsec. The

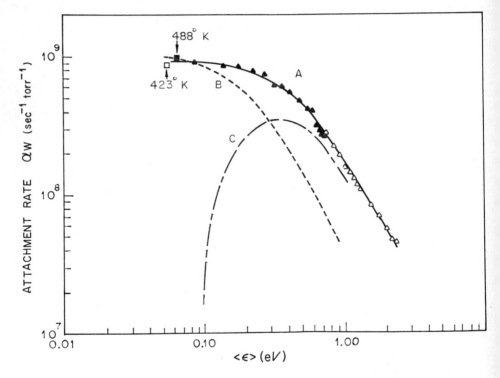

Curve A: Measured attachment rates for (i) $C_2H_4$, 500 torr, □ 423°K and ■ 488°K;

(ii) $N_2$, 750 torr, ▲ 423°K and 488°K;

(iii) Ar, 1000 torr, △ 423°K and 488°K.

Curve B: Calculated attachment rates using a thermal peaking cross section

such as shown in Fig. 6.39. Curve C: Curve B subtracted from curve A.

**Figure 6.42** Electron attachment rates versus mean electron energy for azulene in the carrier gases $C_2H_4$, $N_2$, and Ar (Chaney and coworkers, 1970)

parent negative ion lifetime decreases from 48 μsec at 1.7 eV to 8 μsec at 3.2 eV (see Collins and coworkers, 1969, 1970 and also Figure 7.1). The energy difference $2.31 - 2.1 = 0.21$ eV may be taken to suggest that the electron affinity of $p$–benzoquinone in the first excited triplet state is $\sim 0.2$ eV.

A similar process to that observed for $p$–benzoquinone has been found for azulene (Chaney and coworkers, 1969, 1970). The attachment rates as a function of the mean electron energy for azulene are shown in Figure 6.42. The variation of $\alpha w$ with $\langle \varepsilon \rangle$ shown in Figure 6.42 is distinctly different from that for other aromatic molecules known to capture electrons via a non-dissociative process at thermal energies (Sections 6.5.3.1 and 6.5.3.2). While the thermal peaking attachment rate for other aromatic molecules rapidly decreases when $\langle \varepsilon \rangle$ exceeds its thermal value, the same is seen not to be true for azulene. Chaney and coworkers (1969, 1970) have shown that the measured attachment rates (curve A, Figure 6.42) are composed of two parts, B and C, the former due to an electron attachment process peaking at thermal energies and the latter due to an electron attachment process peaking at $\sim 0.35$ eV. They have associated the 0.35 eV process with electron attachment in the field of the first triplet state of azulene. But this latter process was not observed in the time-of-flight mass spectrometer, and hence the lifetime of the compound negative ion at this energy must be shorter than 1 μsec.

## 6.6 References

A. J. Ahearn and N. B. Hannay (1953). *J. Chem. Phys.*, **21**, 119.

R. K. Asundi and J. D. Craggs (1964a). In M. R. C. McDowell (Ed.), *Atomic Collision Processes*, North Holland Publishing Co., Amsterdam, p. 549.

R. K. Asundi and J. D. Craggs (1964b). *Proc. Phys. Soc.* (*London*), **83**, 611.

R. K. Asundi, J. D. Craggs and M. V. Kurepa (1963). *Proc. Phys. Soc.* (*London*), **82**, 967.

M. von Ardenne, K. Steinfelder and R. Tümmler (1961). *Angew. Chem.*, **73**, 136.

V. A. Bailey (1925). *Phil. Mag.*, **50**, 825.

V. A. Bailey and W. E. Duncanson (1930). *Phil. Mag.*, **10**, 145.

V. A. Bailey and A. J. Higgs (1929). *Phil. Mag.*, **7**, 277.

J. B. Bannon and H. L. Brose (1928). *Phil. Mag.*, **6**, 817.

J. N. Bardsley, A. Herzenberg and F. Mandl (1964). In M. R. C. McDowell (Ed.), *Atomic Collision Processes*, North Holland Publishing Co., Amsterdam, p. 415.

J. N. Bardsley, A. Herzenberg and F. Mandl (1966). *Proc. Phys. Soc.* (*London*), **89**, 321.

J. N. Bardsley and F. Mandl (1968). *Reps. Progr. Phys.*, XXXI (part 2), 471.

H. A. Barton (1927). *Phys. Rev.*, **30**, 614.

D. R. Bates and H. S. W. Massey (1943). *Phil. Trans. Roy. Soc.* (*London*), **239A**, 269.

G. Bekefi and S. C. Brown (1958). *Phys. Rev.*, **112**, 159.

R. S. Berry and C. W. Reiman (1963). *J. Chem. Phys.*, **38**, 1540.

M. S. Bhalla and J. D. Craggs (1960). *Proc. Phys. Soc.* (*London*), **76**, 369.

M. S. Bhalla and J. D. Craggs (1961). *Proc. Phys. Soc. (London)*, **78**, 438.
M. S. Bhalla and J. D. Craggs (1962). *Proc. Phys. Soc. (London)*, **80**, 151.
M. M. Bibby and G. Carter (1965). Quoted in Kurepa.
M. A. Biondi (1958). *Phys. Rev.*, **109**, 2005.
M. A. Biondi (1963). In L. Marton (Ed.), *Advances in Electronics and Electron Physics*, Vol. 18, Academic Press, New York, pp. 67–165.
M. A. Biondi and S. C. Brown (1949). *Phys. Rev.*, **75**, 1700.
M. A. Biondi and R. E. Fox (1958). *Phys. Rev.*, **109**, 2012.
R. P. Blaunstein and L. G. Christophorou (1968). *J. Chem. Phys.*, **49**, 1526.
F. Bloch and N. E. Bradbury (1935). *Phys. Rev.*, **48**, 689.
G. Boldt (1959). *Z. Physik*, **154**, 319, 330.
M. J. W. Boness, I. W. Larkin, J. B. Hasted and L. Moore (1967). *Chem. Phys. Letters*, **1**, 292.
M. J. W. Boness and G. J. Schulz (1968). *Phys. Rev. Letters*, **21**, 1031.
T. E. Bortner and G. S. Hurst (1958). *Health Phys.*, **1**, 39.
L. Bouby and H. Abgrall (1967). In *Proceedings 5th International Conference on the Physics of Atomic and Electronic Collisions, Leningrad, U. S. S. R., July 1967*, Publishing House 'Nauka', p. 584.
L. Bouby, F. Fiquet-Fayard and H. Abgrall (1965). *C. R. Acad. Sci. Paris*, **261**, 4059.
S. E. Božin and C. C. Goodyear (1967). *Brit. J. Appl. Phys.*, **18**, 49.
N. E. Bradbury (1933). *Phys. Rev.*, **44**, 883.
N. E. Bradbury (1934a). *J. Chem. Phys.*, **2**, 827.
N. E. Bradbury (1934b). *J. Chem. Phys.*, **2**, 835.
N. E. Bradbury and H. E. Tatel (1934). *J. Chem. Phys.*, **2**, 835.
L. M. Branscomb (1957). In L. Marton (Ed.), *Advances in Electrics and Electron Physics*, Vol. IX, Academic Press, New York, pp. 43–94.
L. M. Branscomb (1962). In D. R. Bates (Ed.), *Atomic and Molecular Processes*, Academic Press, New York, pp. 100–140.
L. M. Branscomb, D. S. Burch, S. J. Smith and S. Geltman (1958). *Phys. Rev.*, **111**, 504.
R. Buchdahl (1941). *J. Chem. Phys.*, **9**, 146.
I. S. Buchel'nikova (1959). *Soviet Phys.—JETP*, **8**, 783.
N. S. Bučel'nikova (1960). *Fortschr. Physik*, **8**, 626.
D. S. Burch and R. Geballe (1957). *Phys. Rev.*, **106**, 183.
N. P. Carleton and L. R. Megill (1962). *Phys. Rev.*, **126**, 2089, L. R. Megill, private communication.
E. L. Chaney and L. G. Christophorou (1969). *J. Chem. Phys.*, **51**, 883.
E. L. Chaney and L. G. Christophorou (1970). *Oak Ridge National Laboratory. Report* ORNL-TM 2613.
E. L. Chaney, L. G. Christophorou, P. M. Collins and J. G. Carter (1969). *22nd Gaseous Electronics Conference*, October 1969, Gatlinburg, Tenn., Conference Abstracts, Paper J.6, p. 40.
E. L. Chaney, L. G. Christophorou, P. M. Collins and J. G. Carter (1970). *J. Chem. Phys.*, **52**, 4413.
L. M. Chanin, A. V. Phelps and M. A. Biondi (1959). *Phys. Rev. Letters*, **2**, 344.
L. M. Chanin, A. V. Phelps and M. A. Biondi (1962). *Phys. Rev.*, **128**, 219.
P. J. Chantry (1968). *Phys. Rev.*, **172**, 125.
P. J. Chantry (1969a). *J. Chem. Phys.*, **51**, 3369.
P. J. Chantry (1969b). *J. Chem. Phys.*, **51**, 3380.
P. J. Chantry (1969c). *22nd Gaseous Electronics Conference*, October 1969, Gatlinburg, Tenn., Conference Abstracts, Paper J.2, p. 38.

P. J. Chantry, J. S. Wharmby and J. B. Hasted (1962). In H. Maecker (Ed.), *Proceedings of the Fifth International Conference on Ionization Phenomena in Gases, Munich*, 1961, Vol. I, North-Holland Publishing Co., Amsterdam, p. 630.

C. L. Chen and P. J. Chantry (1969). *22nd Gaseous Electronics Conference*, October 1969, Gatlinburg, Tenn., Conference Abstracts, Paper J.3, p. 39.

E. Chen, R. D. George and W. E. Wentworth (1968). *J. Chem. Phys.*, **49**, 1973.

J. C. Y. Chen (1963). *Phys. Rev.*, **129**, 202.

J. C. Y. Chen (1966). *Phys. Rev.*, **148**, 66.

J. C. Y. Chen (1967). *Phys. Rev.*, **156**, 12.

J. C. Y. Chen (1969). In M. Burton and J. L. Magee (Eds.), *Advances in Radiation Chemistry*, Wiley-Interscience, New York, Vol. I, pp. 245-376.

J. C. Y. Chen and J. L. Peacher (1967). *Phys. Rev.*, **163**, 103.

J. C. Y. Chen and J. L. Peacher (1968). *Phys. Rev.*, **167**, 30.

A. A .Christodoulides and L. G. Christophorou (1970). *J. Chem. Phys.*, (submitted).

L. G. Christophorou (1965). Unpublished results.

L. G. Christophorou and R. P. Blaunstein (1969). *Rad. Res.*, **37**, 229.

L. G. Christophorou and J. G. Carter (1968). Unpublished results.

L. G. Christophorou, J. G. Carter, E. L. Chaney and P. M. Collins (1970a). Paper presented at the *IVth International Congress of Radiation Research*, June 1970, Evian, France. (In press).

L. G. Christophorou, J. G. Carter and A. A. Christodoulides (1969). *Chem. Phys. Letters*, **3**, 237.

L. G. Christophorou, J. G. Carter, P. M. Collins and A. A. Christodoulides.(1970b). *J. Chem. Phys.* (In press).

L. G. Christophorou, E. L. Chaney and A. A. Christodoulides (1969). *Chem. Phys. Letters*, **3**, 363.

L. G. Christophorou and R. N. Compton (1967). *Health Phys.*, **13**, 1277.

L. G. Christophorou, R. N. Compton and H. W. Dickson (1968). *J. Chem. Phys.*, **48**, 1949.

L. G. Christophorou, R. N. Compton, G. S. Hurst and P. W. Reinhardt (1965). *J. Chem. Phys.*, **43**, 4273.

L. G. Christophorou, R. N. Compton, G. S. Hurst and P. W. Reinhardt (1966). *J. Chem. Phys.*, **45**, 536.

L. G. Christophorou, G. S. Hurst and W. G. Hendrick (1966). *J. Chem. Phys.*, **45**, 1081.

L. G. Christophorou, D. L. McCorkle and J. G. Carter (1970). *J. Chem. Phys.* (In press).

L. G. Christophorou and J. A. Stockdale (1968). *J. Chem. Phys.*, **48**, 1956.

L. W. Cochran and D. W. Forester (1962). *Phys. Rev.*, **126**, 1785.

J. E. Collin, M. J. Hubin-Franskin and L. D'or (1968). In E. Kendrick (Ed.), *Advances in Mass Spectrometry*, Vol .4, Institute of Petroleum, London, p. 713.

P. M. Collins, L. G. Christophorou and J. G. Carter (1970). *Oak Ridge National Laboratory Report* ORNL-TM 2614.

P. M. Collins, L. G. Christophorou, E. L. Chaney and J. G. Carter (1969). *22nd Gaseous Electronics Conference*, October 1969, Gatlinburg, Tenn., Conference Abstracts, Paper J.7, p. 41.

P. M. Collins, L. G. Christophorou, E. L. Chaney and J. G. Carter (1970). *Chem. Phys. Letters*, **4**, 646.

R. N. Compton and L. Bouby (1967). *C. R. Acad. Sci., Paris*, **264**, 1153.

R. N. Compton and L. G. Christophorou (1967). *Phys. Rev.*, **154**, 110.

R. N. Compton, L. G. Christophorou and R. H. Huebner (1966). *Phys. Letters*, **23**, 656.

R. N. Compton, L. G. Christophorou, G. S. Hurst and P. W. Reinhardt (1966). *J. Chem. Phys.*, **45**, 4634.

R. N. Compton, J. A. Stockdale and P. W. Reinhardt (1969). *Phys. Rev.*, **180**, 111.

J. D. Craggs and H. S. W. Massey (1959). In S. Flügge (Ed.), *Handbuch der Physik*, Vol. XXXVII, Springer-Verlag, Berlin, pp. 314–415.

J. D. Craggs, C. A. McDowell and J. W. Warren (1952). *Trans. Faraday Soc.*, **48**, 1093.

J. D. Craggs, R. Thorburn and B. A. Tozer (1957). *Proc. Roy. Soc. (London)*, **240A**, 473.

J. D. Craggs and B. A. Tozer (1958). *Proc. Roy. Soc. (London)*, **247A**, 337.

J. D. Craggs and B. A. Tozer (1960). *Proc. Roy. Soc. (London)*, **254A**, 229.

R. W. Crompton and R. L. Jory (1962). *Australian J. Phys.*, **15**, 451.

R. W. Crompton, J. A. Rees and R. L. Jory (1965). *Australian J. Phys.*, **18**, 541.

R. K. Curran (1961a). *J. Chem. Phys.*, **34**, 2007.

R. K. Curran (1961b). *J. Chem. Phys.*, **35**, 1849.

R. K. Curran (1963). *J. Chem. Phys.*, **38**, 780.

R. K. Curran and R. E. Fox (1961). *J. Chem. Phys.*, **34**, 1590.

F. J. Davis (1969). Private communication.

Yu. N. Demkov (1965). *Phys. Letters*, **15**, 235.

A. Doehring (1952). *Z. Naturforsch.*, **7a**, 253.

F. H. Dorman (1966). *J. Chem. Phys.*, **44**, 3856.

R. C. Dougherty (1969). *J. Chem. Phys.*, **50**, 1896.

M. Doumont, A. Henglein and K. Jäger (1969). *Z. Naturforsch.*, **24a**, 683.

J. T. Dowell and T. E. Sharp (1968). *Phys. Rev.*, **167**, 124.

V. M. Dukel'skii, V. V. Afrosimov and N. V. Fedorenko (1956). *Soviet Physics—JETP*, **3**, 764.

H. Von Ebinghaus, K. Kraus, W. Müller-Duysing, and H. Neuert (1964). *Z. Naturforsch.*, **19a**, 732.

D. Edelson, J. E. Griffiths and K. B. McAfee. Jr. (1962). *J. Chem. Phys.*, **37**, 917.

H. Ehrhardt, L. Langhans, F. Linder and H. S. Taylor (1968). *Phys. Rev.*, **A173**, 222.

A. G. Engelhardt and A. V. Phelps (1963). *Phys. Rev.*, **131**, 2115.

A. G. Engelhardt, A. V. Phelps and C. G. Risk (1964). *Phys. Rev.*, **135**, A1566.

F. C. Fehsenfeld (1969). *22nd Gaseous Electronics Conference*, October 1969, Gatlinburg, Tenn., Conference Abstracts, Paper J.4, p. 39.

E. E. Ferguson (1968). In L. Marton (Ed.), *Electronics and Electron Physics*, Vol. 24, Academic Press, New York, p. 1.

E. E. Ferguson, F. C. Fehsenfeld and A. L. Schmeltekopf (1967). *J. Chem. Phys.*, **47**, 3085.

F. H. Field and J. L. Franklin (1957). *Electron Impact Phenomena*, Academic Press, New York.

F. Fiquet-Fayard (1965). In M. Haissinsky (Ed.), *Actions Chimiques et Biologiques des Radiations*, Mason, Paris, pp. 31–104.

W. L. Fite and R. T. Brackmann (1963). In *Proceedings of Sixth International Conference on Ionization Phenomena in Gases*, Vol. I, Paris, p. 21.

W. L. Fite, R. T. Brackmann and W. R. Henderson (1965). In *Proceedings of the Fourth International Conference on the Physics of Electronic and Atomic Collisions*, Science Bookcrafters, Inc., Hastings-on-Hudson, New York, p. 100.

R. J. Fleming and G. S. Higginson (1963). *Proc. Phys. Soc. (London)*, **81**, 974.

Ya. M. Fogel', V. F. Kozlov and A. A. Kalmykov (1959). *Soviet Phys.—JETP*, **9**, 963.

R. E. Fox (1957). *J. Chem. Phys.*, **26**, 1281.

R. E. Fox (1958). *Phys. Rev.*, **109**, 2008.

R. E. Fox (1960). *J. Chem. Phys.*, **32**, 285.

R. E. Fox and R. K. Curran (1961). *J. Chem. Phys.*, **34**, 1595.

R. E. Fox, W. M. Hickam, D. J. Grove and T. Kjeldaas, Jr. (1955). *Rev. Sci. Instr.*, **26**, 1101.

M. S. Freedman, T. B. Novey, F. T. Porter and F. Wagner, Jr. (1956). *Rev. Sci. Instr.*, **27**, 716.

J. B. Freely and L. H. Fisher (1964). *Phys. Rev.*, **133**, A305.

J. H. Fremlin (1966). *Nature*, **212**, 1453.

L. Frommhold (1964). *Fortschr. Phys.*, **12**, 597.

D. C. Frost and C. A. McDowell (1958a). *J. Chem. Phys.*, **29**, 503.

D. C. Frost and C. A. McDowell (1958b). *J. Chem. Phys.*, **29**, 964.

D. C. Frost and C. A. McDowell (1960). *Can. J. Chemistry*, **38**, 407.

D. E. Golden and H. W. Bandel (1965). *Phys. Rev. Letters*, **10**, 1010.

V. E. Grob (1963). *J. Chem. Phys.*, **39**, 972.

H. von Gutbier and H. Neuert (1954). *Z. Naturforsch.*, **9a**, 335.

R. Haas (1957). *Z. Physik*, **148**, 177.

H. D. Hagstrum (1947). *Phys. Rev.*, **71**, 376.

H. D. Hagstrum (1951). *Revs. Mod. Phys.*, **23**, 185.

J. D. Hagstrum (1955). *J. Chem. Phys.*, **23**, 1178.

H. D. Hagstrum and J. T. Tate (1941). *Phys. Rev.*, **59**, 354.

P. Harland and J. C. J. Thynne (1969). *J. Phys. Chem.*, **73**, 2791.

M. A. Harrison and R. Geballe (1953). *Phys. Rev.*, **91**, 1.

J. B. Hasted (1964). *Physics of Atomic Collisions*, Butterworths, Washington.

J. B. Hasted and S. Berg (1965). *Brit. J. Appl. Phys.*, **16**, 74.

R. H. Healey (1938). *Phil. Mag.*, **26**, 940.

R. H. Healy and J. W. Reed (1941). *The Behaviour of Slow Electrons in Gases*, Amalgamated Wireless Ltd., Sydney.

H. G. M. Heideman, C. E. Kuyatt and G. E. Chamberlain (1966). *J. Chem. Phys.*, **44**, 335.

W. G. Hendrick, L. G. Christophorou and G. S. Hurst (1968). *Oak Ridge National Laboratory Report* ORNL-TM-1444.

W.R. Henderson, W.L. Fite and R.T. Brackmann (1969). *Phys. Rev.*, **183** (No. 1), 157.

P. Herreng (1952). *Cahiers Phys.*, **38**, 1.

G. Herzberg (1950). *Molecular Structure and Molecular Spectra I. Spectra of Diatomic Molecules*. 2 ed., D. van Nostrand Co., New York.

G. Herzberg (1966). *Molecular Structure and Molecular Spectra III. Electronic Spectra and Electronic Structure of Polyatomic Molecules*. D. van Nostrand Co., Princeton, New Jersey.

J. W. Hiby (1939). *Ann. Phys. (New York)*, **34**, 473.

W. M. Hickam and D. Berg (1958). *J. Chem. Phys.*, **29**, 517.

W. M. Hickam and D. Berg (1959). In J. D. Waldron (Ed.), *Advances in Mass Spectroscopy*, Pergamon Press, New York, p. 458.

W. M. Hickam and R. E. Fox (1956). *J. Chem. Phys.*, **25**, 642.

E. Holøien (1951). *Arch. Math. Naturvidenskab*, **51**, 81.

E. Holøien and J. Midtdal (1955). *Proc. Phys. Soc. (London)*, **68A**, 815.

T. Holstein (1951). *Phys. Rev.*, **84**, 1073.

C. A. Hurst and L. G. H. Huxley (1960). *Australian J. Phys.*, **13**, 21.

G. S. Hurst and T. E. Bortner (1959a). *Phys. Rev.*, **114**, 116.

G. S. Hurst and T. E. Bortner (1959b). *Rad. Res. Suppl.*, **1**, 547.

G. S. Hurst, J. A. Stockdale and L. B. O'Kelly (1963). *J. Chem. Phys.*, **38**, 2572.

L. G. H. Huxley and R. W. Crompton (1962). In D. R. Bates (Ed.), *Atomic and Molecular Processes*, Academic Press, New York, pp. 335–373.

L. G. H. Huxley, R. W. Crompton and C. H. Bagot (1959). *Australian J. Phys.*, **12**, 303.

E. Hylleraas (1950). *Astrophys. J.*, **111**, 209.

E. Hylleraas (1951). *Astrophys. J.*, **113**, 704.

K. Jäger and A. Henglein (1967). *Z. Naturforsch.*, **22a**, 700.

F. Kaufman (1967). *J. Chem. Phys.*, **46**, 2449.

P. Kebarle and A. M. Hogg (1964). *J. Chem. Phys.*, **42**, 798.

V. I. Khvostenko and V. M. Dukel'skii (1958). *Soviet. Phys.—JETP*, **6**, 657.

V. I. Khvostenko, A. Sh. Sultanov and I. I. Furley (1957). In *Abstracts of Vth International Conference on the Physics of Electronic and Atomic Collisions, Leningrad, U. S. S. R., July* 17-23, 1967, Publishing House 'Nauka', Leningrad, p. 586.

H. Kleinpoppen and V. Raible (1965). *Phys. Letters*, **18**, 24.

C. E. Klots (1967). *J. Chem. Phys.*, **46**, 1197.

K. von Kraus (1961). *Z. Naturforsch.*, **16a**, 1378.

E. Kuffel (1959). *Proc. Phys. Soc. (London)*, **74**, 297.

M. V. Kurepa (1965). *3rd Czechoslovak Conference on Electronics and Vacuum Physics Transactions, September* 1965, Prague, pp. 107–115.

C. E. Kuyatt, S. R. Mielczarek and J. A. Simpson (1964). *Phys. Rev. Letters*, **12**, 293.

C. E. Kuyatt, J. A. Simpson and S. R. Mielczarek (1965). *Phys. Rev.*, **138**, A385.

C. E. Kuyatt, J. A. Simpson and S. R. Mielczarek (1966). *J. Chem. Phys.*, **44**, 437.

T. G. Lee (1963). *J. Phys. Chem.*, **67**, 360.

V. A. J. van Lint, E. G. Wikner and D. L. Trueblood (1960). *Bull. Am. Phys. Soc.*, **5**, 122. See also General Atomic Division of General Dynamics Corporation, San Diego, California, Rept GACD-2461 (1961).

W. Lochte-Holtgreven (1951). *Naturwissenschaften*, **38**, 258.

L. B. Loeb (1955). *Basic Processes of Gaseous Electronics*, University of California Press, Berkeley and Los Angeles.

L. B. Loeb (1956). In S. Flügge (Ed.), *Handbuch der Physik*, Vol. XXI, Springer-Verlag, Berlin, pp. 445–503.

W. W. Lozier (1933). *Phys. Rev.*, **44**, 575.

W. W. Lozier (1934). *Phys. Rev.*, **46**, 268.

K. A. G. MacNeil and J. C. J. Thynne (1968). *Trans. Faraday Soc.*, **64**, 2112.

B. H. Mahan and I. C. Walker (1967). *J. Chem. Phys.*, **47**, 3780.

B. H. Mahan and C. E. Young (1966). *J. Chem. Phys.*, **44**, 2192.

M. M. Mann, A. Hustrulid and J. T. Tate (1940). *Phys. Rev.*, **58**, 340.

J. Marriott, R. Thorburn and J. D. Craggs (1954). *Proc. Phys. Soc. (London)*, **67B**, 437.

H. S. W. Massey (1950). *Negative Ions*, 2nd ed., Cambridge University Press, Cambridge.

H. S. W. Massey (1956). In S. Flügge (Ed.), *Handbuch der Physik*, Vol. XXXVI, Springer-Verlag, Berlin, pp. 307–408.

H. S. W. Massey and E. H. S. Burhop (1952). *Electronic and Ionic Impact Phenomena*, Oxford University Press, London.

E. W. McDaniel (1964). *Collision Phenomena in Ionized Gases*, John Wiley and Sons, New York.

R. H. McFarland (1964). *Phys. Rev.*, **136**, A1240.

J. W. McGowan and M. A. Fineman (1965). In *Defense Atomic Support Agency Report* DASA-GA-6699.

C. E. Melton (1966). *J. Chem. Phys.*, **45**, 4414.

M. G. Menendez and H. K. Holt (1966). *J. Chem. Phys.*, **45**, 2743.

C. E. Moore (1949). *Natl. Bur. Std. (U. S.)*, No. 467, **1**, 32.

J. L. Moruzzi and J. T. Dakin (1968). *J. Chem. Phys.*, **49**, 5000.

J. L. Moruzzi and A. V. Phelps (1966). *J. Chem. Phys.*, **45**, 4617.

N. F. Mott and H. S. W. Massey (1965). *The Theory of Atomic Collisions*, 3rd ed., Oxford University Press, London.

M. J. Mulcahy, M. C. Sexton and J. J. Lennon (1962). In H. Maecker (Ed.), *Proceedings of the Fifth International Conference on Ionization Phenomena in Gases, Munich*, 1961, Vol. 1, North Holland Publishing Co., Amsterdam, p. 612.

B. J. Munson and A. M. Tyndall (1939). *Proc. Roy. Soc. (London)*,**A172**, 28.

E. E. Muschlitz, Jr., (1957). *J. Appl. Phys.*, **28**, 1414.

E. E. Muschlitz, Jr. and T. L. Bailey (1956). *J. Phys. Chem.*, **60**, 681.

W. T. Naff, C. D. Cooper and R. N. Compton (1968). *J. Chem. Phys.*, **49**, 2784.

R. S. Narcisi and A. D. Bailey (1965). *J. Geophys. Res.*, **70**, 3687.

H. Neuert (1968). Private communication.

D. J. Nicholas, C. W. Trowbridge and W. D. Allen (1968). *Phys. Rev.*, **167**, 38.

A. O. Nier and E. E. Hanson (1936). *Phys. Rev.*, **50**, 722.

T. F. O'Malley (1966). *Phys. Rev.*, **150**, 14.

T. F. O'Malley (1967a). *Phys. Rev.*, **155**, 59.

T. F. O'Malley (1967b). *J. Chem. Phys.*, **47**, 5457.

G. E. Owen and H. Primakoff (1948). *Phys. Rev.*, **74**, 1406.

J. L. Pack and A. V. Phelps (1966a). *J. Chem. Phys.*, **44**, 1870.

J. L. Pack and A. V. Phelps (1966b). *J. Chem. Phys.*, **45**, 4316.

J. F. Paulson (1966). *Adv. Chem. Ser.*, **58**, 28.

A. V. Phelps (1968). Private communication.

A. V. Phelps and R. E. Voshall (1968). *J. Chem. Phys.*, **49**, 3246.

A. N. Prasad (1959). *Proc. Phys. Soc. (London)*, **74**, 33.

A. N. Prasad (1960). Ph.D. thesis, University of Liverpool.

A. N. Prasad (1965). In *Proceedings of the Seventh International Conference on Ionization Phenomena in Gases, Beograd, August 22-27, 1965*, Vol. I, p. 79.

A. N. Prasad and J. D. Craggs (1960a). In *Proceedings of the Fourth International Conference on Ionization Phenomena in Gases, Uppsala*, Vol. I, North Holland Publishing Co., Amsterdam, p. 142.

A. N. Prasad and J. D. Craggs (1960b). *Proc. Phys. Soc. (London)*, **76**, 223.

A. N. Prasad and J. D. Craggs (1961). *Proc. Phys. Soc. (London)*, **77**, 385.

A. N. Prasad and J. D. Craggs (1962). In D. R. Bates (Ed.), *Atomic and Molecular Processes*, Academic Press, New York, pp. 206–244.

B. S. Rabinovitch and R. W. Diesen (1959). *J. Chem. Phys.*, **30**, 735.

D. Rapp and D. D. Briglia (1965). *J. Chem. Phys.*, **43**, 1480.

D. Rapp, T. E. Sharp and D. D. Briglia (1965). *Phys. Rev. Letters*, **14**, 533.

J. A. Rees (1965). *Australian J. Phys.*, **18**, 41.

J. A. Rees and R. L. Jory (1964). *Australian J. Phys.*, **17**, 307.

R. M. Reese, V. H. Dibeler and F. L. Mohler (1956). *J. Res. Natl. Bur. Std. (U. S.)*, **57**, 367.

R. H. Ritchie and G. E. Whitesides (1961). *Oak Ridge National Laboratory Report* ORNL-3081; revised distributions by G. E. Whitesides and D. R. Nelson (1968). Private communication.

A. C. Riviere and D. R. Sweetman (1960). *Phys. Rev. Letters*, **5**, 560.

D. E. Rothe (1969). *Phys. Rev.*, **177**, 93.

G. J. Schulz (1959a). *Phys. Rev.*, **113**, 816.

G. J. Schulz (1959b). *Phys. Rev.*, **116**, 1141.

G. J. Schulz (1960a). *J. Chem. Phys.*, **33**, 1661.

G. J. Schulz (1960b). *J. Appl. Phys.*, **31**, 1134.
G. J. Schulz (1961). *J. Chem. Phys.*, **34**, 1778.
G. J. Schulz (1962). *Phys. Rev.*, **128**, 178.
G. J. Schulz (1963). *Phys. Rev. Letters*, **10**, 104.
G. J. Schulz (1964a). *Phys. Rev.*, **135**, A988.
G. J. Schulz (1964b). *Phys. Rev.*, **136**, A650.
G. J. Schulz (1964c). *Phys. Rev. Letters*, **13**, 583.
G. J. Schulz (1969). *Atomic and Molecular Physics Congress, University of Manchester, England, March* 1969.
G. J. Schulz and R. K. Asundi (1965). *Phys. Rev. Letters*, **15**, 946 (1965).
G. J. Schulz and R. K. Asundi (1967). *Phys. Rev.*, **158**, 25.
G. J. Schulz and D. Spence (1969). *Phys. Rev. Letters*, **22**, 47.
T. E. Sharp and J. T. Dowell (1966). In *Proceedings of the Fourteenth Annual Conference on Mass Spectrometry and Allied Topics, Dallas, Texas*, p. 37.
T. E. Sharp and J. T. Dowell (1967). *J. Chem. Phys.*, **46**, 1530.
T. E. Sharp and J. T. Dowell (1969). *J. Chem. Phys.*, **50**, 3024.
J. A. Simpson and U. Fano (1963). *Phys. Rev. Letters*, **11**, 158.
B. M. Smirnov and M. I. Chibisov (1966). *Soviet Phys.—JETP*, **22**, 585.
L. G. Smith (1937). *Phys. Rev.*, **51**, 263.
R. E. Stanton (1960). *J. Chem. Phys.*, **32**, 1348.
J. A. Stockdale, L. G. Christophorou and G. S. Hurst (1967). *J. Chem. Phys.*, **47**, 3267.
J. A. Stockdale and G. S. Hurst (1964). *J. Chem. Phys.*, **41**, 255.
W. K. Stuckey and R. W. Kiser (1966). *Nature*, **211**, 963.
N. Sukhum, A. N. Prasad and J. D. Craggs (1967). *Brit. J. Appl. Phys.*, **18**, 785.
D. R. Sweetman (1960). *Proc. Phys. Soc. (London)*, **76**, 998.
J. T. Tate and P. T. Smith (1932). *Phys. Rev.*, **39**, 270.
H. S. Taylor and J. Gerhauser (1964). *J. Chem. Phys.*, **40**, 244.
H. S. Taylor and F. E. Harris (1963). *J. Chem. Phys.*, **39**, 1012.
H. S. Taylor and J. K. Williams (1965). *J. Chem. Phys.*, **42**, 4063.
J. B. Thompson (1959). *Proc. Phys. Soc. (London)*, **73**, 821.
R. Thorburn (1959). *Proc. Phys. Soc. (London)*, **73**, 122.
J. S. Townsend (1915). *Electricity in Gases*, Oxford University Press, London.
J. S. Townsend (1947). *Electrons in Gases*, Hutchinson Scientific and Technical Publications, London.
B. A. Tozer (1958). *J. Electronics and Control*, **4**, 149.
B. A. Tozer, R. Thorburn and J. D. Craggs (1958). *Proc. Phys. Soc. (London)*, **72**, 1081.
L. von Trepka and H. Neuert (1963). *Z. Naturforsch.*, **18a**, 1295.
F. K. Truby (1968). *Phys. Rev.*, **172**, 24.
E. B. Wagner, F. J. Davis and G. S. Hurst (1967). *J. Chem. Phys.*, **47**, 3138.
J. M. Warman and R. W. Fessenden (1968). *J. Chem. Phys.*, **49**, 4719.
O. Weber (1958). *Z. Physik*, **152**, 281.
W. E. Wentworth, R. S. Becker and R. Tung (1967). *J. Phys. Chem.*, **71**, 1652.
W. E. Wentworth and E. Chen (1967). *J. Phys. Chem.*, **71**, 1929.
W. E. Wentworth, E. Chen and J. E. Lovelock (1966). *J. Phys. Chem.*, **70**, 445.
G. Z. Whitten and B. S. Rabinovitch (1963). *J. Chem. Phys.*, **38**, 2466.
R. Wildt (1939). *Astrophys. J.*, **89**, 295.
P. M. Windham, P. J. Joseph and J. A. Weinman (1958). *Phys. Rev.*, **109**, 1193.
D. Wobschall, J. R. Graham, Jr. and D. P. Malone (1965). *J. Chem. Phys.*, **42**, 3955.

# 7 Electron detachment from negative ions and electron affinity of atoms and molecules

In this chapter the various ways via which an electron can be detached from a negative ion and the binding energy of free electrons to atoms and molecules are discussed. Attention is focused on the following electron detachment processes: autodetachment, photodetachment, detachment by electron impact, and associative and non-associative collisional detachment. Other modes of destruction of negative ions such as detachment by electric fields and charge exchange reactions will be discussed very briefly.

The available information on electron affinities of atoms and molecules is summarized and discussed in the second part of the chapter. Such information is in most cases fragmentary. Most of the electron affinity values are still derived from approximate theoretical calculations, while the experimental results leave much to be desired.

## 7.1 Electron Detachment from Negative Ions

### 7.1.1 *Autodetachment*

In Chapters 4, 5 and 6 we discussed a number of interesting molecular negative ions which are not stable, but transient. Unstable, molecular negative ions have extremely varying lifetimes. On the basis of the magnitude of the temporary negative ion lifetime, $\tau$, three groups of temporary negative ions, formed by electron attachment in the field of either the ground or excited electronic states, can be distinguished: (i) extremely short-lived

$(10^{-15} \lesssim \tau \lesssim 10^{-13}$ sec)—these show up as resonances in electron scattering experiments or in dissociative electron attachment studies; (ii) moderately short-lived $(10^{-12} < \tau < 10^{-6}$ sec)—these can be stabilized in high-pressure experiments; (iii) long-lived negative ions $(\tau > 10^{-6}$ sec)—these can be detected in conventional mass spectrometers. The unstable negative ions are formed with excess energy which is necessarily above their ionization potential. Although molecular negative ions are formed from bound electronic states below the ionization potential they may be thought of as being in a superexcited state from which electron ejection is possible. Autoionization of a negative ion is referred to also as *autodetachment*.

A large number of parent negative ions observed either directly in low-pressure beam experiments or identified indirectly in high pressure swarm experiments have cross sections peaking at or close ($< 1$ eV) to zero electron energy (Chapter 6). A large autoionization lifetime has been measured for many polyatomic molecular negative ions and this has been explained as being a result of their many degrees of freedom, their positive electron affinity, and the distribution of the excess energy among their many vibrational degrees of freedom (Chapter 6). Autoionization lifetimes are crucially dependent upon the molecular structure and the total internal energy of the negative ion. Thus, they depend upon the initial vibrational population of the neutral molecule, and their measurement should be sensitive to the incident beam energy resolution and source conditions (temperature, for example). Although in the majority of the cases investigated (see Chapter 6), the autoionization lifetime has not been found to change with incident electron energy—this probably is because the attachment cross section for these cases in a sharply peaking function at $\sim 0.0$eV—variation of the negative ion lifetime with electron energy has been observed for a number of polyatomic negative ions. For example, the lifetime of the $p$–benzoquinone negative ion for the 2.1 eV resonance (Section 6.5.3.3) has been found to decrease from $\sim 50$ μsec at the low-energy side of the resonance to $\sim 8$ μsec at the high-energy side. The results of Collins and coworkers (1970) for the $p$–benzoquinone negative ion are shown in Figure 7.1 Similarly, Collins and coworkers (1970) found the lifetime of the negative ion of 1,4-naphthoquinone to decrease from $\sim 380$ μsec at thermal energies to $\sim 10$ μsec at $\sim 1$ eV (Figure 7.2). In this case we are dealing with a nuclear excited Feshbach resonance (see Sections 5.1.4.3 and 6.5). The results of Collins and coworkers on the yield of the parent negative ion as a function of electron energy indicated structure which was associated with skeletal vibrations. It might be noted that for this molecule the lifetime variation is probably more drastic than that actually indicated by the experiment since, due to detection difficulties, the energy resolution of Collins and coworkers was poor (see discussion in Collins and coworkers).

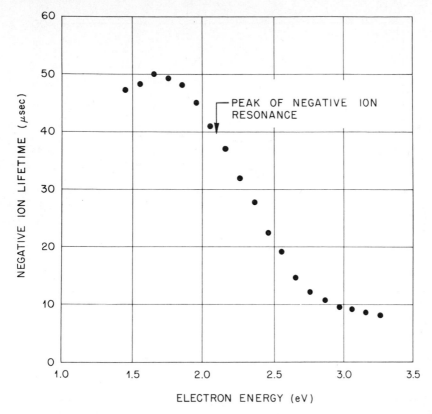

Variation of $p$-Benzoquinone Negative Ion Lifetime With Electron Energy.

**Figure 7.1** Variation of the lifetime of the $p$-benzoquinone metastable negative ion with electron energy (Collins and coworkers, 1970). The resonance capture process at 2.1 eV has been ascribed as a long-lived electron-excited-Feshbach resonance (see text)

### 7.1.2 *Photodetachment*

The photodetachment process

$$AX^- + h\nu \rightarrow AX + e \qquad (7.1)$$

can be thought of as the photoionization of the negative ion resulting in its destruction (Massey, 1950; Branscomb, 1957). It provides a uniquely precise way of obtaining electron affinities, *EA*, of atoms and vertical detachment energies, *VDE*, of molecules from the determination of the threshold of the

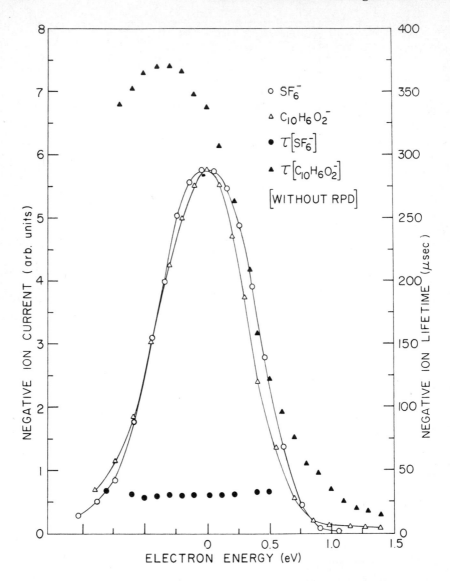

**Figure 7.2** Variation of the lifetime of the 1,4-naphthoquinone metastable negative ion with electron energy (Collins and coworkers, 1970). The resonance capture process peaks at ∼0.0 eV but it extends to electron energies well above thermal (∼1 eV); it has been associated with a nuclear-excited-Feshbach resonance (see text). Note that the lifetime variation is actually more drastic than indicated in the figure since as can be seen the energy resolution of these experiments, due to detection difficulties, was poor. Note also the constancy of the lifetime of the $SF_6^-{}^*$ within the energy range shown

photodetachment cross section, $\sigma_{pd}$. As stated in Section 6.2.2, the vertical detachment energy, although equal to the electron affinity for atoms, is, in general, different for molecules (see, however, Section 7.1.2.2). Measurement of the photodetachment cross section between specific initial and final states allows determination of the cross section for the reverse process between the same states, namely the radiative attachment of an electron to a neutral atom. Also, the measurement of photodetachment cross sections provides sensitive means to test theoretical approximations in atomic physics.

Now let us consider the interaction of an oscillating electric field of frequency $v$ with a negative ion possessing a dipole moment $\mu$. If $\rho(\varepsilon_e)$ is the density of states in the continuum per unit energy corresponding to an energy $\varepsilon_e(= hv - EA)$ of the ejected electron, the transition probability $p$ per unit time is (Branscomb, 1962)

$$p \propto v|\langle\psi_d|\mu|\psi_c\rangle|^2\rho(\varepsilon_e) \tag{7.2}$$

where $\psi_d$ and $\psi_c$ are the wave functions for the initial discrete state and the final continuum state, respectively. Since $\rho(\varepsilon_e) \propto \varepsilon_e^{\frac{1}{2}}$, the photodetachment cross section is

$$\sigma_{pd} \propto v\varepsilon_e^{\frac{1}{2}}|\langle\psi_d|\mu|\psi_c\rangle|^2 \tag{7.3}$$

Three formally equivalent expressions for the dipole moment matrix element, referred to as the dipole length, velocity, and acceleration formulae, have been used in theoretical calculations:

$$M = \int \psi_d^*(r_1 \ldots r_n) \sum_i z_i \psi_c(r_1 \ldots r_n) \, d\tau_1 \ldots d\tau_n \qquad \text{(dipole length)} \tag{7.4}$$

$$= \frac{1}{hv} \int \psi_d^*(r_1 \ldots r_n) \sum_i \frac{\partial}{\partial z_i} \psi_c(r_1 \ldots r_n) \, d\tau_1 \ldots d\tau_n \qquad \text{(dipole velocity)} \tag{7.5}$$

$$= \left(\frac{1}{hv}\right)^2 \int \psi_d^*(r_1 \ldots r_n) \sum_i \frac{z_i}{r_i^3} \psi_c(r_1 \ldots r_n) \, d\tau_1 \ldots d\tau_n \quad \text{(dipole acceleration)} \tag{7.6}$$

In the above expressions $z_i$ is the component of the radius vector $r_i$ of the $i$th electron parallel to the oscillating electric field. The three formulae test different regions of electron coordinate space. The dipole length formula emphasizes regions of large $r$, while the velocity and acceleration formulae emphasize, respectively, regions of intermediate and small values of $r$.

The threshold behaviour of the photodetachment cross section can be shown to depend on the angular momentum of the electron in the final free state, and because of the dipole selection rules this dependence can be correlated with the angular momentum vectors of the electron in the initial bound state (Massey, 1950; Branscomb, 1957). By representing the continuum function as a plane wave in a short-range static central field (Wigner, 1948; Massey, 1950; Branscomb and coworkers, 1958), the variation of $\sigma_{pd}$ can be expressed as a power series in $k$, raised to a power which is dependent upon the lowest angular momentum component of the continuum state, $l$, to which an electric dipole transition is allowed from the discrete initial state, namely

$$\sigma_{pd} \propto vk^{2l+1} \quad (a_0 + a_1 k^2 + a_2 k^4 + ...) \tag{7.7}$$

Here $k$ is the square root of the kinetic energy of the free electron and $a_{n(=0,1,2...)}$ are unspecified parameters. Hence, the cross section for photo-detaching an $s$ electron from an atomic negative ion—such as in the case of $H^-$ and alkali metal negative ions—can be expressed as (Branscomb, 1962)

$$\sigma_{pd}(s \rightarrow p) \propto v\varepsilon_e^{3/2}(a_0 + a_1 \varepsilon_e + a_2 \varepsilon_e^2 + ...) \tag{7.8}$$

while for $p$ electrons—such as in the case of $C^-$ and $O^-$ negative ions—the threshold behaviour of the photodetachment cross section is given by

$$\sigma_{pd}(p \rightarrow s) \propto v\varepsilon_e^{\frac{1}{2}}(a_0 + a_1 \varepsilon_e + a_2 \varepsilon_e^2 + ...) \tag{7.9}$$

From Equations (7.8) and (7.9) it is seen that the rate of change of $\sigma_{pd}$ with energy in the threshold region—from a zero value at the threshold—will be zero for the photodetachment of an $s$ electron and infinite for the photo-detachment of a $p$ electron. Similarly, Geltman (1958) has investigated the threshold energy dependence of $\sigma_{pd}$ for negative ions of diatomic molecules.

Although the energy range over which the above threshold laws hold is not known, it is clear that for $p$ electron shells, $\sigma_{pd}$ changes more rapidly from the threshold than for $s$ shells (compare $\sigma_{pd}$ for $H^-$ and $O^-$ near threshold in Figures 7.3 and 7.4). The predicted variation of $\sigma_{pd}$ with energy of the ejected electron in expressions (7.8) and (7.9) has been used by Branscomb and coworkers to interpret low-resolution experimental data near threshold for the determination of atomic electron affinities. Branscomb and Smith and their colleagues studied extensively the photodetachment of electrons from atomic and simple molecular negative ions using modulated crossed-beam techniques. In these experiments a beam of mass-analysed negative ions is intersected at right angles, in high vacuum, by an intense beam of filtered visible light (quasi-monoenergetic), and the current of free electrons produced by photon absorption is measured. A combination of weak electric and magnetic fields perpendicular to the ion and photon beams allowed collection of the photodetached electrons without significantly disturbing the ion beam. Use of single particle counting techniques for electron detection greatly increased the detection sensitivity. For experimental details, see Smith and Branscomb (1960) and Branscomb (1962).

### 7.1.2.1 Photodetachment of atomic negative ions

The photodetachment cross section for the negative ion of atomic hydrogen, $H^-$, is shown in Figure 7.3. The absolute integrated cross section was measured to a quoted accuracy of $\pm 10\%$ (Branscomb and Smith, 1955a; Smith and Burch, 1959a, b). The velocity formula using central field continuum functions with correction for exchange (John, 1960) or a variational procedure (Geltman and Krauss, 1960) gives best agreement with the experimental data normalized at $\lambda = 5280$Å. The normalized data also are consistent with the oscillator strength sum-rules (Branscomb, 1962; Chapter 3).

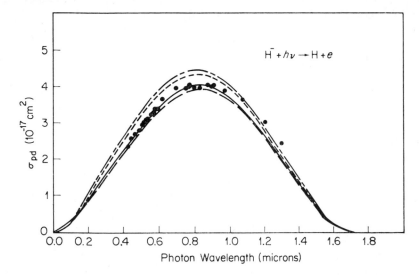

**Figure 7.3** Photodetachment cross section for $H^-$ (Branscomb, 1962); points: experimental values of Branscomb and Smith (1955a) and Smith and Burch (1959a, b). — — — —: calculations by Chandrasekhar and Elbert (1958). – – – – – –: calculations by Chandrasekhar (1958). ————: calculations by John (1960). — — calculations Geltman and Krauss (1960). The experimental points are normalized to the calculated values of John (1960) at 5280 Å (see text)

The photodetachment cross section for $O^-$, which is of special interest in problems of free-electron behaviour in the ionosphere, is shown in Figure 7.4. This figure combines the near threshold wavelength measurements (Smith and Branscomb, 1955a; Branscomb and Smith, 1955b; Branscomb and coworkers, 1958) with the measurement of the shape of the $O^-$ cross section above threshold (Smith, 1960). The relative measurements (solid line) above threshold were made to fit the predicted threshold law, increasing roughly as $\varepsilon_e^{\frac{1}{2}}$ for approximately 0.3 eV beyond threshold. The dashed line is chosen as a good fit to the experimental points. Of particular interest has been the precise determination of the wavelength of the absorption threshold for $O^-$ and the shape of the cross section near threshold. The first permits an accurate determination of $EA(O)$ corresponding to

$$O^-(^2P_{3/2}) + h\nu \rightarrow O(^3P_2) + e \quad (\varepsilon_e = 0) \tag{7.10}$$

and the latter offers a test for the possible existence of a very weakly bound excited state which might produce a resonance peak in the photodetachment continuum (Bates and Massey, 1943). The small additional bump in the curve just below the main threshold results from a small contribution from the $^2P_{1/2}$ configuration of $O^-$, lying slightly above the $^2P_{3/2}$ state.

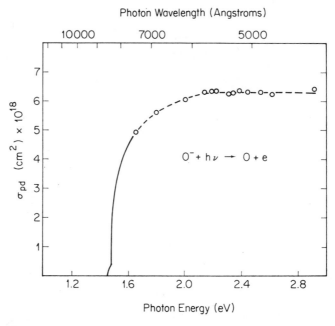

**Figure 7.4** Photodetachment cross section for $O^-$ (Smith, 1960) (see text)

The photodetachment spectrum of $C^-$ was measured by Seman and Branscomb (1962) in the visible region of the spectrum and relative to the spectrum of $O^-$. Using the known $O^-$ data of Smith (1960), they obtained the absolute values for the $C^-$ photodetachment cross section shown in Figure 7.5. The lower dashed line indicates absorption by an excited metastable state of the $C^-$ ion. These data yielded $EA(C) = 1.25 \pm 0.03$ eV.

Photodetachment spectra were obtained also for the atomic ions $S^-$ (Branscomb and Smith, 1956) and $I^-$ (Steiner, Seman and Branscomb, 1962, 1964). The $S^-$ continuum was similar in shape to that of $O^-$ with $EA(S) = 2.07 \pm 0.07$ eV. The recent experiments of Steiner (1968) on $I^-$ yielded a photodetachment cross section value for $I^-$ at 3470Å equal to $2.9 \pm 0.5 \times 10^{-17}$ cm². This value was obtained using a crossed-beam experiment and by comparison with photodetachment from $H^-$ at 9930Å. Steiner used his absolute cross section value at 3470Å to normalize the relative cross section data for the photodetachment for $I^-$ previously determined by Steiner, Seman and Branscomb (1964). The cross section so obtained was similar in shape with that calculated by Robinson and Geltman (1967), but the experimental values exceeded the calculated ones by about a factor of two (see also Rothe (1969)).

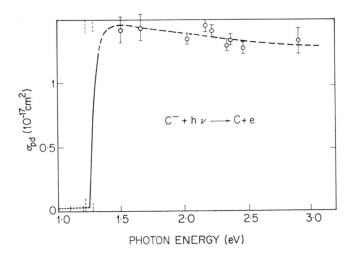

**Figure 7.5** Photodetachment cross section for $C^-$ (Seman and Branscomb, 1962)

Single-quantum photodetachment cross sections for $C^-$, $O^-$, $F^-$, $Si^-$, $S^-$, $Cl^-$, $Br^-$ and $I^-$ have been calculated by Robinson and Geltman (1967). Double-quantum photodetachment of $I^-$ at the ruby-laser wavelength has been measured by Hall, Robinson and Branscomb (1965).

As stated earlier, the most accurate method to measure the electron affinity of atoms is the determination of the absorption spectrum of the negative ion. The minimum photon energy for which photodecay of the negative ion is still possible corresponds to the binding energy of the electron to the negative ion, i.e. to the electron affinity of the atom involved. Electron affinities of atoms, obtained from the photodetachment work of Branscomb and colleagues, are given in Table 7.2. Some very precise measurements of the electron affinities of the halogen atoms have been made by another photodetachment method developed by Berry and coworkers (Berry, Reimann and Spokes, 1961, 1962; Berry and Reimann, 1963). In this method the ultraviolet absorption spectrum of gaseous atomic halogen negative ions is studied, the halogen ions being formed in a shock-wave-heated alkali halide vapour. From the absorption threshold the electron affinities for F, Cl, Br and I have been measured and are listed in Table 7.2. Additionally, Brehm, Gusinow and Hall (1967) measured the electron affinity of He $(2^3 S)$ from an energy analysis of the electrons which have been photodetached from a beam of negative helium ions upon the action of monochromatic (laser) light. Their value is listed in Table 7.2.

### 7.1.2.2 Photodetachment of simple molecular negative ions

Photodetachment spectra were studied for the molecular ions $OH^-$ and $OD^-$ (Smith and Branscomb, 1955b; Branscomb, 1966), $O_2^-$ (Burch, Smith and Branscomb, 1958), $SH^-$ (Steiner, 1968) and $NO_2^-$ (see Branscomb, 1962; Warneck, 1969). Attempts to study photodetachment of electrons from $CN^-$ and $C_2^-$ have been made by Branscomb and Pagel (1958) and from $[OH(H_2O)]^-$ by Golub and Steiner (1968). The photodetachment cross sections of Burch, Smith and Branscomb (1958) for $O_2^-$ are shown in Figure 7.6. The relative values for $O_2^-$ were made absolute by direct experimental comparison with the value for $O^-$ at 5400Å (2.30 eV). For reference, the $O^-$ photodetachment cross section is also shown in Figure 7.6

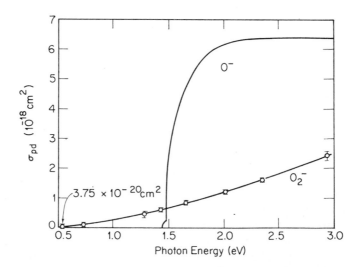

**Figure 7.6** Photodetachment cross section for $O_2^-$ (Burch, Smith and Branscomb, 1958)

(Branscomb, 1962). The experimental points were fitted to the threshold law with an onset of $0.15\pm0.05$ eV. Contrary to the sharp onsets of the photodetachment cross sections for $H^-$, $O^-$ and $C^-$, the $O_2^-$ photodetachment cross section function rises gradually from almost zero energy showing no sharp onset, making any determination of $EA(O_2)$ doubtful (the $EA(O_2)$ is considerably in doubt—see Table 7.3).

It might be instructive to point out again that since the photodetachment of diatomic molecular ions is governed by the Franck–Condon principle, the energy required to remove an electron by photodetachment may be different from the molecular electron affinity if the molecular ion potential function has a minimum at a larger internuclear separation than the minimum of the neutral molecule potential energy curve (see Chapter 6). Further, if the equilibrium nuclear separations of the neutral molecule and the negative ion are substantially different, a smearing-out of the spectrum can occur as a result of a relatively large number of final vibrational states taking part in the transition. This, in turn, conceals the true energy dependence of the electric dipole matrix element. The interpretation of photodetachment spectra of molecular negative ions may also be complicated by uncertainties as to the identity of the initial state occupied by the negative ion being studied. However, Branscomb (1966) argued on the basis of the cross section data for photodetachment of $OH^-$ and $OD^-$ ions shown in Figure 7.7 that the equilibrium distance $r_e$ for $OH^-$ and $OH$ are very similar,

$$r_e(OH^-) = r_e(OH)\pm0.002\text{Å}$$

and he proceeded to determine the electron affinity of the OH radical, which he found equal to $1.83\pm0.04$ eV. This value is slightly higher than the value of 1.75 eV reported earlier by Smith and Branscomb (1955b).

Additionally, Steiner (1968) found the photodetachment cross section of electrons from $SH^-$ to rise rapidly in the first 0.15 eV above threshold to a value of $1.9\pm0.4\times10^{-17}$ cm$^2$ and remain constant over the rest of the energy range investigated (0.75 eV above threshold). From the detailed shape of the photodetachment cross section versus wavelength, he derived the $SH^-$ structural parameters which were indistinguishable from those of SH. From the observed photodetachment threshold he obtained

$$EA(SH) = 2.319\pm0.010\text{ eV}$$

### 7.1.2.3 Radiative electron attachment to atomic systems

The inverse of the photodetachment process is the radiative attachment. The cross sections, $\sigma_{pd}$ and $\sigma_{ra}$, for the two processes are related through the principle of detailed balance (Massey, 1950),

$$\sigma_{pd} = \left(\frac{mcv}{h\nu}\right)^2 \left(\frac{g_0}{g_-}\right)\sigma_{ra} \tag{7.11}$$

where $m$ and $v$ are the electron mass and velocity, $hv$ is the photon energy, and $g_0$ and $g_-$ are the statistical weights of the neutral atom and negative ion, respectively. If it is assumed that the ground-state term of the negative ion

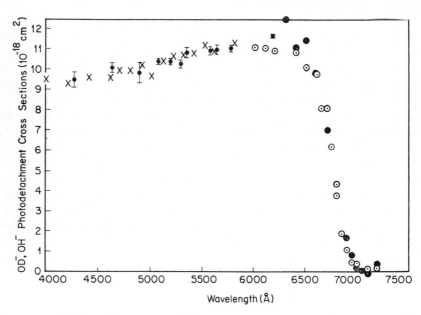

OH$^-$ ($\bullet$). OD$^-$ ($\odot$, X). The ($\dot{\text{I}}$) are data of Branscomb and Smith (1958).
The cross section values were obtained with respect to those for H$^-$
(Branscomb and Smith (1955a); Branscomb (1962).

**Figure 7.7** Photodetachment cross sections for OH$^-$ and OD$^-$ (Branscomb, 1966)

(on which $g_-$ depends) is known, then, from the measured photodetachment cross section for an atomic ion from the threshold to some energy $hv$, the radiative attachment cross section can be calculated for electrons whose energy ranges from zero to $\varepsilon_e (= hv - EA)$. Application of expression (7.11) to the case of H$^-$ and O$^-$ yielded the radiative attachment cross sections shown in Figures 7.8 and 7.9 (Branscomb, 1962). The cross section for H$^-$ was calculated by Branscomb (1962) from the photodetachment data of Smith and Burch (1959b), while that for O$^-$ was calculated from the photodetachment data of Smith (1960). It is seen from Figures 7.8 and 7.9 that the calculated radiative attachment cross sections are very small compared with those for other electron attachment processes (Chapter 6).

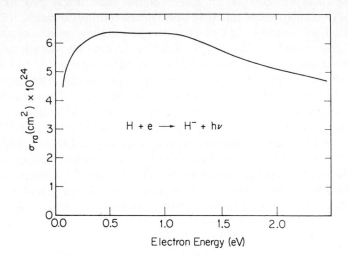

**Figure 7.8** Cross section for radiative attachment of electrons to atomic hydrogen calculated by detailed balancing from photodetachment data (see text)

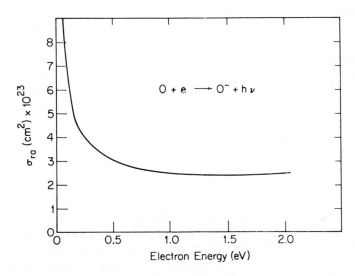

**Figure 7.9** Cross section for radiative attachment of electrons to atomic oxygen calculated by detailed balancing from photodetachment data (see text)

The small cross sections for the radiative attachment process made laboratory observations of radiative attachment continua difficult. Although observations of radiative capture continua in emission spectra of arcs and shocked gases have been reported in earlier years, it is only recently that such spectra have been quantitatively studied. Emission spectra due to radiative capture of low-energy electrons were reported for chlorine (Henning, 1962; Berry and David, 1964), bromine and iodine (Berry and David, 1964) and fluorine (Popp, 1965). Berry and David (1964) estimated radiative capture cross sections of electrons with energies 0.025 to 0.035 eV of $\sim 10^{-21}$ cm$^2$. Radiative capture intensity measurements for the chlorine, bromine, and iodine atoms have also been made by Rothe (1969) who used his intensity data to obtain photodetachment cross sections for Cl$^-$, Br$^-$ and I$^-$ in the wavelength region 3000 to 4000Å. For the case in which the halogen ion is left in the ground state ($^2P_{3/2}$), Rothe obtained photodetachment cross sections equal to $1.2 \times 10^{-17}, 2 \times 10^{-17}$ and $2.2 \times 10^{-17}$ cm$^2$ for Cl$^-$, Br$^-$ and I$^-$, respectively, near the photodetachment thresholds. The magnitudes and shapes of the photodetachment cross sections for Cl$^-$ and Br$^-$ as determined by Rothe (1969) agreed very well with the absorption data of Berry, Reimann and Spokes (1962), whilst Rothe's data on I$^-$ are in general agreement with the crossed-beam measurements of Steiner, Seman and Branscomb (1962, 1964) (see, also Steiner, 1968).

In spite of the fact that the cross sections for formation of negative ions by radiative attachment are small, the process is important for electron attachment to atoms at low pressures. An important region of absorption of the solar radiation continuum is known to be due to the hydrogen negative ion (Wildt, 1939; Pagel, 1956, 1959).

Equation (7.11) can be applied also to calculate the radiative attachment cross section for diatomic molecules for which the measured photo-detachment spectra are known to correspond to single initial and final vibrational states. To calculate $\sigma_{ra}$ from $\sigma_{pd}$ for a molecule, it is necessary to know the vibrational populations and the threshold vertical detachment energy. There is no certainty that this can be the case. In Figure 7.10 the monoenergetic radiative attachment coefficient is shown for O$_2$ for a number of assumed values of the vertical detachment energy (Branscomb, 1962). These curves are only correct if the photodetachment spectrum from which they have been calculated refers to single initial and final vibrational states. This is not certain for O$_2$ for which further difficulty arises in that no threshold in the photodetachment cross section is observed (see caption of Figure 7.10).

Finally, the threshold laws for photodetachment discussed earlier in this section can be converted through Equation (7.11) to threshold laws for radiative attachment. Thus the rate for electron attachment into an $s$ state (e.g. H) rises from zero as $k^2$ (i.e. as $\varepsilon_e$), while that into a $p$ state (e.g. O) is finite at the threshold. This is clearly seen in Figure 7.10 where the radiative attachment coefficients for H and O are shown for comparison with those for $O_2$ (see further discussion on threshold laws in Branscomb (1962)).

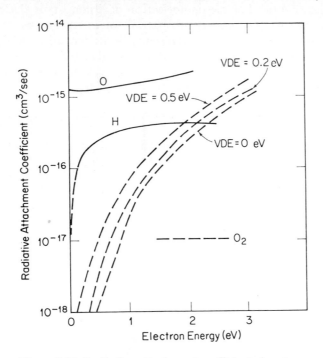

**Figure 7.10** Radiative attachment coefficients for atomic hydrogen and oxygen and illustrative curves for $O_2$ (Branscomb, 1962). The three curves for $O_2$ are for the three values of the vertical detachment energy indicated (see values for $EA(O_2)$ in Table 7.3). They are correct only if the experimental photodetachment spectrum from which they were calculated refers to single initial and final vibrational states

### 7.1.2.4 Angular distribution of electrons photodetached from negative ions

In Section 3.5.4.3 we elaborated on the angular distribution of electrons photoejected from neutral atoms. In the dipole approximation the angular distribution of photoelectrons for linearly polarized light has the general form (Cooper and Zare (1968); see also Cooper and Manson (1969))

$$d\sigma/d\Omega = (\sigma_T/4\pi)\,[1 + \beta P_2(\cos \Theta)] \tag{7.12}$$

where $P_2(\cos \Theta) = \frac{1}{2}(3 \cos^2 \Theta - 1)$, $\sigma_T$ is the total cross section, $\Theta$ is the angle between the direction of the ejected electron and the polarization of the incident light, and $\beta$ is an asymmetry parameter. For photoejection of an electron from an $s$-atomic orbital, as in $H^-$, the outgoing wave will be of only $p$ character and $\beta$ will be two, leading to a $\cos^2 \Theta$ behaviour of the angular distribution. For the photoejection of a non-$s$-type electron (from a $p$, $d$, or higher-order orbital) two partial waves represented by $l' = l \pm 1$ must be considered. In this case interference terms may arise and $\beta$ will differ from two giving rise to angular distributions other than $\cos^2 \Theta$. The value of $\beta$ will depend on the velocity of the outgoing electron and will approach two with increasing velocity (Cooper and Zare, 1968).

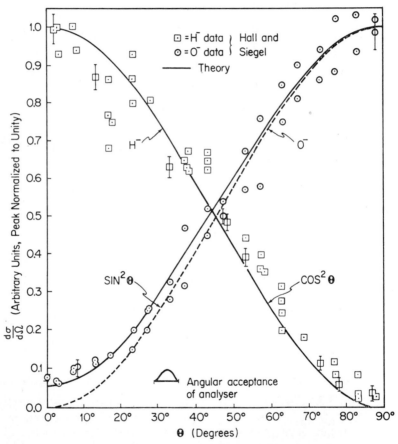

$H^-$: ▢ experimental data of Hall and Siegel (1968) at 4880 Å.

$O^-$: ⊙ experimental data of Hall and Siegel (1968) at 4880 Å.

——— : theoretical calculations by Cooper and Zare (1968).

**Figure 7.11** Angular distributions of photodetached electrons from $H^-$ and $O^-$ (Cooper and Zare, 1968)

Calculated and experimental photodetachment differential cross sections for H⁻ and O⁻ are given in Figure 7.11 (Cooper and Zare, 1968). The experimental measurements are those of Hall and Siegel (1968) obtained by passing a negative ion beam through the cavity of a high power argon-ion laser operated at $\lambda = 4880$Å. The theoretical results shown in the figure are those of Cooper and Zare (1968) and are seen to be in good agreement with experiment. The photodetachment differential cross section for $H^-(^1S)$ which has only a p-wave channel available shows a $\cos^2 \Theta$ behaviour. That for $O^-(^3P)$ (and C⁻, see Hall and Siegel, 1968) which has both s- and d-wave channels available is nearly pure $\sin^2 \Theta$ rather than $\cos^2 \Theta$. The dependence of the distribution on the velocity of the outgoing electron needs further investigation.

### 7.1.3 *Detachment by Electron Impact*

Recently the cross section for the reaction

$$A^- + e \rightarrow A + 2e \tag{7.13}$$

where A⁻ is an H⁻ or an O⁻ negative ion, has been measured using modulated crossed-beam techniques. The reaction (7.13) is most interesting since a repulsive Coulomb force exists between the incident and target particles which, of course, is absent in the photodetachment process. Tisone and Branscomb (1966) measured the absolute cross section for the process $H^- + e \rightarrow H + 2e$ at 100 eV incident electron energy and its energy dependence from 30 to 500 eV. They used a mass-analysed beam of 2.5 keV H⁻ ions ($\sim 4 \times 10^{-8}$A) which they bombarded by an electron beam ($\sim 30\,\mu$A) in a region of high vacuum ($\sim 5 \times 10^{-8}$ torr). The ions which survived the collision region were electrostatically swept into a Faraday cup and the neutral hydrogen atoms formed by process (7.13) were detected by a suitable detector. Also, measurements have been made of the cross section for electron detachment from H⁻ by electron impact, using the crossed-beam method, over a wider range of incident electron energies: 9 to 500 eV (Rundel, Harrison and Dance, 1967; Dance, Harrison and Rundel, 1967) and 8.4 to 488.4 eV (Tisone and Branscomb, 1968). The results of these workers are shown in Figure 7.12. The error bars on the results of Dance and coworkers represent 90% confidence limits on the random errors associated with each cross section measurement. Apart from these errors the authors estimated the maximum systematic error in the cross section to be $\pm 14\%$ at incident electron energies above 30 eV and a little higher below this energy. The error bars on the Tisone and Branscomb (1966) measurement correspond to 50% confidence limits on the random errors of measurements made relative to the only absolute measurement at 100 eV. Tisone and Branscomb (1966) also quoted a systematic error in their measurements of $+38$, $-58\%$.

Considering, therefore, the quoted errors, the experimental data of the two groups of investigators are in reasonable agreement. It is seen from Figure 7.12 that the variation in the general shape of the electron detachment cross section, $\sigma_{ed}$, with electron energy, $\varepsilon$, is similar to that of ionization. The cross section maximizes at an energy of $\sim 20$ times the threshold energy (0.75 eV), relatively much further above threshold than the ionization of the neutral atom.

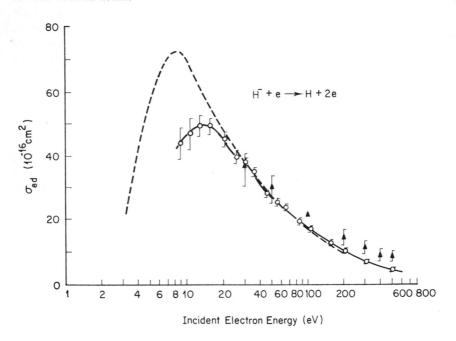

Cross section $\sigma_{ed}$ for electron detachment from $H^-$ by electron impact (Dance, Harrison, and Rundel (1967). O experimental results of Dance, Harrison, and Rundel (1967). ▲ experimental results of Tisone and Branscomb (1966). --- theoretical calculations by McDowell and Williamson (1963) (see text).

**Figure 7.12**

The detachment of the extra electron from $H^-$ by electron impact has been a topic for a number of theoretical calculations, apparently with discordant results (Geltman, 1960; McDowell and Williamson, 1963; Rudge, 1964; Smirnov and Chibisov, 1966; Rogalski, 1966; Bely and Schwartz, 1969). In general, the Bethe–Born approximation, with a semiempirical Coulomb correction (Geltman, 1960) applied to correct for the repulsion between the electron and the negative ion, is in qualitative agreement with experiment. The theoretical results of McDowell and Williamson (1963), consisting of two separate calculations which have been joined together at 40 eV (see Dance and coworkers, 1967), are shown in Figure 7.12. The agreement with the experimental results of Dance and coworkers (1967) at energies greater than 20 eV is seen to be good; below 20 eV the calculation predicts larger cross sections than the measured values. At energies $\gtrsim 10$ eV and with the Coulomb effect removed, the functional form of the cross section measured by Dance and coworkers showed a $\varepsilon^{-1}\ln\varepsilon$ behaviour in accord with the Born approximation prediction, although Tisone and Branscomb (1968) argued that the energy dependence of their $\sigma_{ed}$ for $H^-$ from 100 to 500 eV was not consistent with the slope predicted theoretically by the Born approximation for the high-energy limit.

Tisone and Branscomb (1967) used their crossed-beam apparatus, employed for the study of electron detachment from $H^-$ by electron impact, to the measurement of the absolute cross section for the process

$$O^- + e \rightarrow O + 2e \tag{7.14}$$

in the incident electron energy range 15 to 400 eV, and later (Tisone and Branscomb, 1968) in the range 7.1 to 487.1 eV. The energy of the mass-analysed $O^-$ ion beam was 2.5 keV. At an incident electron energy of 97.1 eV the cross section for process (7.14) was found to be $6.3\pi a_0^2$, a value much smaller than that found for the $H^-$ ion. For $H^-$ $\sigma_{ed}$ was found equal to $23.1\pi a_0^2$ at 98.4 eV (Tisone and Branscomb, 1968). As in the case of $H^-$, the Bethe–Born approximation with the semiclassical Coulomb correction was in qualitative agreement with the experimental results above 20 eV.

### 7.1.4 *Associative and Non-associative Collisional Detachment*

A collision between a negative ion $A^-$ and a neutral atom B at kinetic energies of a few electron volts or less may lead to the ejection of the electron through the process of associative detachment

$$A^- + B \rightarrow AB + e \tag{7.15}$$

Process (7.15) is the inverse of dissociative attachment. If the electron affinity of A is less than the dissocation energy of AB, process (7.15) will be exoergic and thus possible with zero relative kinetic energy. It might, therefore, be important as a mechanism for electron detachment from negative ions by collision at low temperatures (down to a few hundred degrees or less).

For sufficient relative kinetic energy of $A^-$ and B the final state of the nuclei may lie in a continuum, either because the final electronic state is repulsive or because the electron carries away too little energy to leave the nuclei bound, and thus non-associative collisional detachment,

$$A^- + B \rightarrow A + B + e \qquad (7.16)$$

may take place. Process (7.16) proceeds only if the initial relative kinetic energy of $A^-$ and B exceeds the electron affinity of A. Non-associative collisional detachment, therefore, is not likely to occur at very low energies in contrast to the process of associative detachment.

It has been shown in previous chapters that a low-energy electron–molecule collision may result in the formation of an unstable compound negative ion state which subsequently decays to a neutral molecule by emission of an electron. Processes (7.15) and (7.16) can be considered to proceed similarly via unstable compound states.

The theory of low-energy collisions between negative (and positive) ions and neutral a oms is particularly difficult since the approximations which are valid at high energies do not apply when the interaction time, $T$, is longer than or of the order of the time of internal motions, $\tau(=h/\Delta E$, where $\Delta E$ is the net change in potential energy of the system during the collision), associated with the changes in energy of the system. When discussing the dependence of the cross section for inelastic collisions such as excitation, ionization, charge transfer, and detachment, a rule called the 'adiabatic criterion' proved useful (Massey, 1949). A collision is called adiabatic if

$$\frac{T}{\tau} = \frac{l\Delta E}{hv} \gg 1 \qquad (7.17)$$

where $T$ is the interaction (or collision) time, $l$ is the interaction distance and $v$ is the relative particle velocity before collision. In such a collision the colliding particles can continuously change potential energy adiabatically with no net change in energy even when the incident particle has sufficient energy to cause such a change. In the adiabatic criterion, therefore, the probability for a transition is high if the perturbation caused by the collision is rapid compared to the periods of internal motions associated with the change in energy of the system. The adiabatic criterion is an oversimplified expression of a complex process but it proved useful in a number of cases.

Recent theoretical discussions on collisional detachment (see, for example, Demkov, 1964; Smirnov and Firsov, 1965; Lopantseva and Firsov, 1966; Herzenberg, 1967; Bardsley, 1967a, b; Chen, 1967; Chen and Peacher, 1967, 1968; Dalgarno and Browne, 1967) have been restricted to atomic encounters; in most cases $H^-$–H. Certain of these theoretical discussions (for example, Demkov, 1964; Herzenberg, 1967; Bardsley, 1967a, b) involved resonance intermediate states. Herzenberg (1967), in particular, discussed the collisional detachment processes (7.15) and (7.16) assuming that the electronic state of the A⁻–B system is stable at large separations, $R$, of A⁻ and B and that it changes adiabatically as $R$ decreases; at very small $R$ ($\sim 10^{-8}$ cm) the electronic state was assumed to become an unstable compound state, able to emit an electron, but still continuing to change adiabatically. He has derived expressions for the total cross section for electron detachment, and for the cross sections for detachment leaving the nuclei in a single discrete final state. Bardsley (1967a, b) used the perturbed stationary-state approximation to calculate the cross section for electron detachment and charge exchange in H⁻–H collisions, expanding the total wave function for the scattering system in terms of the electronic wave functions for the $^2\Sigma_g^+$ and $^2\Sigma_u^+$ states of H$_2^-$. Bardsley used the resonant widths and potential energy curves for the $^2\Sigma_g^+$ and $^2\Sigma_u^+$ states calculated by Bardsley, Herzenberg and Mandl (1966). His result for H⁻–H is compared in Figure 7.13 with the experimental data of Hummer and coworkers (1960). The result of the Born approximation calculation of McDowell and Peach (1959) is also shown in Figure 7.13. The cross section calculated by Bardsley (1967a, b) decreases monotonically over a whole range of energies in agreement with the experiment. The energy range over which the calculation is more likely to be valid ranges from 50 to 2000 eV. At low energies ($< 1$ eV) the distortion due to the polarization attraction is very important, enhancing the total detachment cross section due to associative detachment. Electron detachment may cause a decrease in the measured cross section for charge transfer between neutral atoms and negative ions and need be

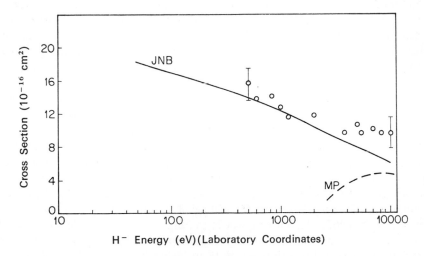

**Figure 7.13** Cross section for electron detachment in H⁻–H collisions (Bardsley, 1967b). (o) experimental results of Hummer and coworkers (1960). (——) calculations by Bardsley (1967b). (– – – –) Born approximation calculation of McDowell and Peach (1959)

considered in measurements of calculations of charge-transfer cross sections (see Figure 7.15 later this section).

At thermal and epithermal energies the total cross section for electron detachment can become very large ($> 10^{-15}$ cm$^2$) due to Langevin spiralling, arising from the long range polarization attraction ($R^{-4}$ potential) between $A^-$ and $B$. The cross section for spiralling or orbiting collisions is (McDaniel, 1964)

$$\sigma_{\text{spir}} = \frac{4\pi}{v}\left(\frac{e^2\alpha}{M_r}\right)^{\frac{1}{2}} \tag{7.18}$$

where $e$ is the electronic charge, $\alpha$ is the polarizability of the atom of molecule, $M_r$ is the reduced mass of the colliding particles, and $v$ is the initial velocity of the particles (the relative velocity outside the region in which the interaction becomes appreciable).

### 7.1.4.1 Experimental results on associative collisional detachment

Thermal-energy associative detachment reactions of negative ions have been reported for a number of binary systems, and their rate constants have been determined (e.g. Fehsenfeld, Ferguson and Schmeltekopf, 1966; Moruzzi and Phelps, 1966; Fehsenfeld and coworkers, 1967; see also Ferguson, 1968). Observation of the decrease in the primary ion intensity as a function of the increasing density of the minor reactant in a binary gas mixture offers a means to measure the rate of associative detachment reactions. Thermal-energy associative detachment reaction rate constants involving $O^-$ ions are listed in Table 7.1. It is to be noted that although the observation of ion destruction in a number of experiments is consistent with (7.15), it does not provide direct evidence concerning the end-product of the reaction. Moruzzi, Ekin and Phelps (1968), using similar techniques to those developed by Pack and Phelps (1966a, b), succeeded in detecting the detached electron currents produced in three associative detachment reactions involving $O^-$. The average energy of the $O^-$ ions in these experiments ranged from thermal up to 0.16 eV. The measured rate coefficients at near-thermal ion energies are included in Table 7.1. From the data listed in Table 7.1 the cross sections for associative detachment at thermal and near thermal energies are seen to be very large. Theoretical calculations for associative detachment reaction rate coefficients for the systems listed in Table 7.1 are lacking. However, theoretical proposals (Chen, 1967; Dalgarno and Browne, 1967; Herzenberg, 1967) putting the rate coefficients close to those for orbiting collisions in a polarization potential (Equation (7.18)) seem to be reasonably valid (see data in Table 7.1). The high collisional detachment cross sections very clearly demonstrate the fragility of negative ions.

Table 7.1 Rate constants for thermal-energy associative detachment reactions involving $O^-$ ions

| Reaction[a] | Rate constant (cm³/sec) at 300°K | Reference[d] |
|---|---|---|
| $O^- + N \to NO + e$ | $2 \times 10^{-10}$ | 1 |
| | $7.2 \times 10^{-10}$ | 3 |
| $O^- + O \to O_2 + e$ | $2 \times 10^{-10}$ | 1 |
| | $6.3 \times 10^{-10}$ | 3 |
| $O^- + H_2 \to H_2O + e$ | $8.6 \times 10^{-10}$ | 1 |
| | $7.5 \times 10^{-10b}$ | 2 |
| | $1.6 \times 10^{-9}$ | 3 |
| $O^- + N_2 \to N_2O + e$ | $< 1 \times 10^{-12}$ | 1 |
| | $9.6 \times 10^{-10}$ | 3 |
| $O^- + O_2 \to O_3 + e$ | $< 1 \times 10^{-12}$ | 1 |
| $O^- + CO \to CO_2 + e$ | $5.6 \times 10^{-10}$ | 1 |
| | $6.5 \times 10^{-10}$ | 2 |
| | $10 \times 10^{-10}$ | 3 |
| $O^- + NO \to NO_2 + e$ | $2.2 \times 10^{-10}$ | 1 |
| | $2.2 \times 10^{-10c}$ | 2 |
| | $9.3 \times 10^{-10}$ | 3 |

[a] For a number of other reactions, see Fehsenfeld and coworkers (1967) and Ferguson (1968).
[b] At $6 \leqslant E/P \leqslant 20$ V/cm torr. No variation with ion energy was observed up to ~0.16 eV.
[c] The rate coefficient was found to decrease with electron energy in the range 0.039 to 0.16 eV.
[d] References:
1. Fehsenfeld, Ferguson and Schmeltekopf (1966). The values listed were taken from Ferguson (1968) and differ somewhat from the original ones reported by Fehsenfeld, Ferguson and Schmeltekopf (1966).
2. Moruzzi, Ekin and Phelps (1968).
3. Polarization value (Equation (7.18)); Reference 1 above.

## 7.1.4.2 Experimental results on non-associative collisional detachment

Electron swarm and electron beam techniques have been employed in the study of this process. The experimental arrangement of Phelps and coworkers described in Section 6.3.1.5 in connection with electron attachment studies has been employed by Phelps and coworkers to study electron detachment from swarms of negative ions in $O_2$ (Chanin, Phelps and Biondi, 1962; Pack and Phelps, 1966a, b). The detachment experiments of Phelps and coworkers furnished the thermal detachment frequency for $O_2^-$ as a function of gas temperature (see Section 6.5.2) and this information was combined with their electron attachment measurements to yield an electron affinity for $O_2$ equal to $0.43 \pm 0.02$ eV. Electron attachment and detachment studies in $O_2$ have been made also by Sukhum, Prasad and Craggs (1967) using the Townsend method. These authors discussed other efforts to study electron detachment from $O_2^-$ by the swarm method.

In studies of electron detachment by beam methods a mass-analysed monoenergetic beam of negative ions usually of sufficiently low energy ($< 5$ keV) that ionization of the target gas is negligible, passes through a target gas and the detached electrons in ion–molecule collisions are collected. Such studies employ techniques similar to those used in charge-transfer experiments. Experimental techniques and literature have been reviewed by a number of authors (see, for example, Bates, 1962; McDaniel, 1964 and Hasted, 1964) and for this reason this topic will not be discussed here. It might, however, be noted that a number of systems have been studied, especially $H^-$, $O^-$ and alkali and halogen ions in rare gases. For some recent work on the detachment of electrons from negative alkaline metal ions in collisions with inert gases, see Bydin (1966). Energy spectra of electrons detached from the negative ions $I^-$, $Br^-$ and $Cl^-$ in collisions with rare gas atoms have been investigated experimentally by Bydin (1967).

### 7.1.5 *Other Modes of Destruction of Negative Ions*

### 7.1.5.1 Detachment by electric fields

Negative ions subjected to a strong electric field while passing through the gap between two electrodes may have their outermost electron(s) detached. When a charge-to-mass analysis is performed on the emerging beam, the intensities of the neutral and negative ion components can be measured. Riviere and Sweetman (1960) studied the effect of an electric field on $He^-$ and $H^-$ ions at energies of 1.0 and 0.33 MeV, respectively. The mass-analysed ion beam from a Van-de-Graaff accelerator was passed through the gap between two electrodes and the fraction of beam ions which had undergone detachment was measured as a function of the electric field strength. Figure 7.14 shows the fraction of the $He^-$ and $H^-$ ions undergoing detachment as a function of the intensity of the electric field in the high-field gap. $He^-$ ($EA(He) = 0.075$ eV; Table 7.2) was completely neutralized for an electric field of $4.5 \times 10^5$ volts/cm. As is seen in Figure 7.14 no detectable amount of detachment from $H^-(EA(H) = 0.75$ eV; Table 7.2) occurred at fields up to $5.4 \times 10^5$ volts/cm. Detachment by electric fields is a very important method in determining the binding energies of electrons in weakly bound atomic negative ions.

### 7.1.5.2 Charge transfer (exchange)

At low energies collisions of stable negative ions with atoms or molecules can lead to the destruction of the primary ion and at the same time to the production of new negative ions by the electron exchange process

$$A^- + B \rightarrow B^- + A \qquad (7.19)$$

or by the negative ion–molecule reaction process

$$A^- + BC \rightarrow B^- + AC \tag{7.20}$$

where A, B and C can be atoms, molecules, or radicals. Reaction (7.19) can proceed when $EA(B) > EA(A)$ and has been repeatedly observed (see, for example, Henglein and Muccini, 1959, 1961; Bailey, 1961; Kraus, Müller-Duysing and Neuert, 1961; Wobschall, Graham and Malone, 1965; Paulson, 1966; Kraus, 1966; Snow, Rundel and Geballe, 1969).

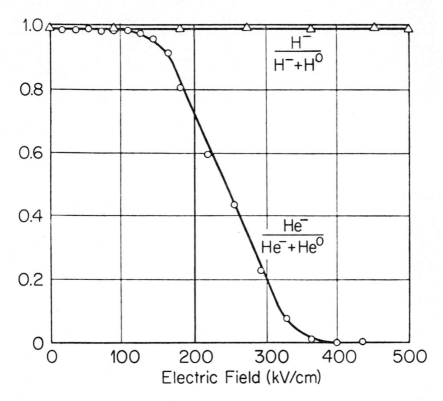

**Figure 7.14** Fraction of $H^-$ and $He^-$ ions undergoing detachment plotted as a function of the intensity of the electric field in the high-field gap (Riviere and Sweetman, 1960)

Reaction (7.20) has been observed also (see, for example, Muschlitz, 1957; Henglein and Muccini, 1959; Trepka, 1963; Jäger and Henglein, 1966; Paulson, 1966; Vogt and Neuert, 1967; Stockdale, Compton and Reinhardt, 1968; Dillard and Franklin, 1968) and energy balance for this process requires that $EA(B) - D(BC) - EA(A) + D(AC) \geqslant 0$.

**Table 7.2** Electron affinities of atoms[a]

| Ion | Electron affinity (eV) | Reference[o] | Method[n] |
|---|---|---|---|
| H⁻($^1S$) | 0.65[b] (0.71)[c] | 1 | a |
| | ~0.733 | 2 | a |
| | [0.7542] | 3, 4 | b |
| | 0.76 | 5 | a |
| | 0.77±0.02[d] | 6 | — |
| | 0.8±0.1 | 7 | c |
| He⁻ | −0.56 | 8 | a |
| | −0.53 | 5 | a |
| | <0 | 2 | a |
| | >0.033 | 9 | b |
| | 0.06±0.005 | 10 | d |
| | ≥0.069 | 11 | b |
| | >0.075[e] | 12 | b |
| | [0.08±0.002] | 13 | e |
| | 0.19 | 14 | a |
| | 0.19[b] | 1 | a |
| Li⁻($^1S$) | 0.34 | 5 | a |
| | ~0.392 | 2 | a |
| | 0.58±0.05 | 15 | b |
| | 0.59 | 8 | a |
| | ~0.60 | 16 | — |
| | 0.62 | 17 | b |
| | 0.64[b] | 1 | a |
| | 0.65 to 1.05[f] | 18 | c |
| | 0.82 | 14 | a |
| Be⁻ | −0.68 | 8 | a |
| | −0.57 | 5 | a |
| | −0.38 | 19 | a |
| | −0.19 | 14 | a |
| | <0 | 2 | a |
| B⁻($^3P$) | ~0 | 2 | a |
| | 0.12 | 5 | a |
| | 0.16 | 8 | a |
| | 0.2 | 20 | a |
| | 0.25 | 21 | b |
| | 0.3±0.05 | 15 | b |
| | 0.33 | 14, 19 | a |
| | 0.39[b] | 1 | a |
| | 0.82 | 22 | a |

Table 7.2 (*continued*)

| Ion | Electron affinity (eV) | Reference[o] | Method[n] |
|---|---|---|---|
| C⁻($^4S$) | >0.83 | 2 | a |
| | 1.12±0.06 | 23 | — |
| | 1.17 | 15 | b |
| | 1.21 | 22 | a |
| | 1.24 | 14 | a |
| | [1.25±0.03] | 24 | f |
| | 1.32[b] | 1 | a |
| | 1.33 | 8 | a |
| | 1.35 | 20 | a |
| | 1.37 | 5 | a |
| | 1.38 | 19 | a |
| | 1.34 to 1.74 | 25 | g |
| N⁻ | −0.32 | 8 | a |
| | −0.31[b] | 1 | a |
| | −0.27±0.11 | 15 | b |
| | −0.15 | 19 | a |
| | <0 | 2 | a |
| | 0.04 | 5 | a |
| | 0.05 | 14 | a |
| | 0.15 | 20 | a |
| | 0.54 | 22 | a |
| O⁻($^2P$) | >0.70 | 2 | a |
| | 1.22±0.14 | 15 | b |
| | 1.32[c] | 1 | a |
| | 1.39 | 8 | a |
| | 1.45 | 26 | h |
| | 1.46±0.02[g] | 27 | i |
| | [1.465±0.005] | 23 | f |
| | 1.465 | 19 | a |
| | 1.47 | 14, 22 | a |
| | 1.478±0.002 | 28 | i |
| | 1.48±0.1 | 29 | f |
| | 1.5 | 30 | g |
| | 2.00 | 20 | a |
| | 3.80 | 5 | a |
| F⁻($^1S$) | ∼3.03 | 2 | a |
| | 3.36[c] | 1 | a |
| | [3.448±0.005] | 31 | i |
| | 3.47 | 19 | a |
| | 3.48 | 32 | j |
| | 3.50 | 14 | a |
| | 3.55±0.09 | 33 | i |
| | 3.56±0.09 | 34 | c |
| | 3.57 | 35 | c |

Table 7.2 (*continued*)

| Ion | Electron affinity (eV) | Reference[o] | Method[n] |
|-----|------------------------|--------------|-----------|
| F⁻($^1S$) | 3.58[h] | 36 | c |
| | 3.62±0.11 | 37 | — |
| | 3.62 | 22 | a |
| | 3.843 | 38 | b |
| | 3.94 | 5 | a |
| | 4.1 | 20 | a |
| Ne⁻ | −0.63[c] | 1 | a |
| | −0.57 | 14 | a |
| | −1.20 | 5 | a |
| | −1.03 | 8 | a |
| | <0 | 2 | a |
| Na⁻($^1S$) | 0.08 | 5 | **a** |
| | 0.14[b] | 1 | a |
| | 0.22 | 8 | a |
| | >0.3 | 2 | a |
| | 0.41 | 39 | k |
| | 0.47 | 14 | a |
| | 0.54 | 17 | b |
| | 0.78 | 40 | b |
| | 0.84 | 41, 42 | l |
| Mg⁻ | −0.87 | 5 | a |
| | −0.69 | 8 | a |
| | −0.61 | 19 | a |
| | −0.56[c] | 1 | a |
| | −0.32 | 14 | a |
| | <0 | 2 | a |
| Al⁻($^3P$) | −0.16 | 5 | a |
| | 0.27 | 8 | a |
| | 0.28 | 19 | a |
| | 0.52 | 40 | b |
| | 0.52 | 14 | a |
| | 0.75 | 20 | a |
| | 1.19 | 22 | a |
| | >1.7 | 2 | a |
| Si⁻($^4S$) | 0.60 | 5 | a |
| | 1.36 | 19 | a |
| | 1.39 | 40 | b |
| | 1.40 | 8 | a |
| | 1.46 | 14 | a |
| | ~1.6 | 2, 20 | a |

**Table 7.2** (*continued*)

| Ion | Electron affinity (eV) | Reference[o] | Method[n] |
|---|---|---|---|
| P⁻($^3P$) | 0.15 | 5 | a |
| | 0.62 | 8 | a |
| | 0.72 | 19 | a |
| | 0.77 | 14 | a |
| | 0.78 | 40 | b |
| | 1.10 | 20 | a |
| | 1.12 | 42 | l |
| | 1.33 | 22 | a |
| | >2.68 | 2 | a |
| S⁻($^2P$) | 1.25 | 42 | l |
| | 2.03 | 8 | a |
| | 2.06 | 5 | a |
| | [2.07±0.07] | 43 | f |
| | 2.08 | 19 | a |
| | 2.09±0.06 | 44 | h |
| | >2.1 | 2 | a |
| | 2.12 | 40 | b |
| | 2.15 | 14 | a |
| | 2.33[h] | 36 | c |
| | 2.37[i] | 45 | c |
| | 2.55 | 20 | a |
| | 2.79 | 22 | a |
| Cl⁻($^1S$) | ~3.1 | 2 | a |
| | 3.10 | 42 | l |
| | 3.56 | 40 | b |
| | [3.613±0.003] | 31 | i |
| | 3.64 | 46 | h |
| | 3.68 | 32 | j |
| | 3.69 | 19 | a |
| | 3.70 | 5 | a |
| | 3.70 | 14 | a |
| | 3.75±0.09 | 34 | c |
| | 3.80 | 35 | c |
| | 3.81[h] | 36 | c |
| | 3.84 | 22 | a |
| | 3.88 | 38 | b |
| | 4.25 | 20 | a |
| Ar⁻ | −1 | 5 | a |
| | <0 | 2 | a |
| K⁻($^1S$) | 0.22 | 39 | k |
| | 0.47 | 17 | b |
| | 0.82 | 41, 42 | l |
| | 0.902 | 47 | b |

**Table 7.2** (*continued*)

| Ion | Electron affinity (eV) | Reference[o] | Method[n] |
|---|---|---|---|
| Sc$^-$($^3F$) | $-0.14$ | 47 | b |
| Ti$^-$($^4F$) | 0.40 | 47 | b |
| V$^-$($^5D$) | 0.94 | 47 | b |
| Cr$^-$($^6S$) | 0.98 | 47 | b |
| Mn$^-$($^5D$) | $-1.07$ | 47 | b |
| Fe$^-$($^4F$) | 0.58 | 47 | b |
| Co$^-$($^3F$) | 0.94 | 47 | b |
| Ni$^-$($^2D$) | 0.25 | 5 | a |
|  | 1.28 | 47 | b |
| Cu$^-$($^1S$) | 1.17 | 5 | a |
|  | $1.5\pm0.5^j$ | 48 | c |
|  | 1.8 | 47 | b |
| Br$^-$($^1S$) | [$3.363\pm0.003$] | 31 | i |
|  | 3.45 | 32 | j |
|  | $3.49\pm0.02$ | 49 | c |
|  | $3.51\pm0.06$ | 34 | c |
|  | $3.53\pm0.12^k$ | 50 | — |
|  | 3.55 | 46 | h |
|  | 3.717 | 38 | b |
|  | $3.82\pm0.15$ | 51 | c |
| Rb$^-$($^1S$) | $0.42^l$ | 17 | b |
|  | 0.16 | 39 | k |
| Ag$^-$($^1S$) | 0.95 | 5 | a |
|  | $2.0\pm0.2^j$ | 48 | c |
| Sb$^-$ | $\geqslant2.0$ (?) | 52 | — |
| I$^-$($^1S$) | [$3.063\pm0.003$] | 31 | i |
|  | [$3.076\pm0.005$] | 53 | f |
|  | $3.13\pm0.12^m$ | 50 | — |
|  | 3.14 | 32 | j |
|  | $3.14\pm0.07$ | 54 | c |
|  | $3.17\pm0.07$ | 34 | c |
|  | 3.23 | 55 | c |
|  | 3.23 | 46 | h |
|  | $3.29^h$ | 36 | c |
|  | 3.33 | 35 | c |
|  | 3.355 | 38 | b |

**Table 7.2** (*continued*)

| Ion | Electron affinity (eV) | Reference[o] | Method[n] |
|-----|------------------------|--------------|-----------|
| Cs⁻($^1S$) | 0.13 | 39 | k |
|  | 0.39[l] | 17 | b |
| Au⁻($^1S$) | 2.1 | 52 | — |
|  | 2.8±0.1[j] | 48 | c |
| Hg⁻ | 1.54 | 52 | — |
|  | 1.79 | 5 | a |

[a] For values prior to 1953, see Pritchard (1953); recommended values are in brackets.
[b] Two-parameter method; see Kaufman (1963).
[c] Three-parameter method; see Kaufman (1963).
[d] Determined theoretically and on the basis of the photodetachment cross section data of Smith and Burch (1959b).
[e] For He⁻$(1s2s2p)\,^4P_{5/2}$.
[f] The authors believe that $EA$(Li) is near the upper limit given in the Table.
[g] Photon impact and measurement of the threshold for the process $O_2 + h\nu \rightarrow O^+ + O^-$.
[h] The electron affinity of F, Cl, I, S and CN were obtained assuming $EA$(Br) = 3.56 eV.
[i] Assuming EA(Br) = 3.6 eV.
[j] Assuming EA(I) = 3.07 eV.
[k] Photoionization of $Br_2$.
[l] Estimated value; see Weiss (1968).
[m] Photoionization of $I_2$.
[n] Methods:
    a) Empirical extrapolation.
    b) Calculation.
    c) Surface ionization.
    d) Decay in an electric field.
    e) Laser photodetachment.
    f) Photodetachment.
    g) Electron impact.
    h) Equilibrium.
    i) Photoabsorption.
    j) Lattice energy.
    k) Charge exchange.
    l) Statistical theory.

[o] References:
    1. Kaufman (1963).
    2. Geltman (1956).
    3. Hylleraas and Midtdal (1956).
    4. Pekeris (1958, 1962).
    5. Glockler (1934).
    6. Weisner and Armstrong (1964).
    7. Khvostenko and Dukel'skii (1960).
    8. Crossley (1964).
    9. Holøien and Geltman (1967).
    10. Smirnov and Chibisov (1966).
    11. A. W. Weiss, quoted in Holøien and Geltman (1967).
    12. Holøien and Midtdal (1955).
    13. Brehm, Gusinow and Hall (1967).
    14. Edlén (1960).

15. Clementi and McLean (1964).
16. Ya'akobi (1966).
17. Weiss (1968).
18. Scheer and Fine (1969).
19. Edie and Rohrlich (1962).
20. Bates and Moiseiwitsch (1955).
21. Schaefer and Harris (1968).
22. Johnson and Rohrlich (1959).
23. Branscomb and coworkers (1958).
24. Seman and Branscomb (1962).
25. Fineman and Petrocelli (1958, 1962).
26. Page (1961a).
27. Elder, Villarejo and Inghram (1965).
28. Berry, quoted by Elder, Villarejo and Inghram (1965).
29. Smith and Branscomb (1955a).
30. Chantry and Schulz (1964).
31. Berry and Reimann (1963).
32. Cubicciotti (1959).
33. Jortner, Stein and Treinin (1959).
34. Bailey (1958),
35. Bakulina and Ionov (1955).
36. Bakulina and Ionov (1959).
37. Stamper and Barrow (1958).
38. Tandon, Bhutra and Tandon (1967).
39. Bydin (1964).
40. Clementi and coworkers (1964).
41. Gáspár and Molnár (1955).
42. Gombás and Ladányi (1960).
43. Branscomb and Smith (1956).
44. Ansdell and Page (1962).
45. Bakulina and Ionov (1957).
46. Page (1960).
47. Clementi (1964).
48. Bakulina and Ionov (1964).
49. Doty and Mayer (1944).
50. Morrison and coworkers (1960).
51. Glockler and Calvin (1936).
52. Buchel'nikova (1958).
53. Steiner, Seman and Branscomb (1962).
54. Sutton and Mayer (1935).
55. Glocker and Calvin (1935).

While most of the electron exchange reactions reported to date are concerned with positive ions, negative ion–molecule reactions are becoming of increasingly recognized importance in radiation research. This development has been accelerated considerably by the recent advances in mass spectrometry. Many atomic and molecular ions are formed more easily by charge transfer from atomic and molecular negative ions than by electron impact. Molecular negative ions of simple species formed by electron collision have short autoionization lifetimes, owing to the fact that (thermal) electron capture leads to high-lying vibrational levels of the negative ion from which electron ejection is fast (see Chapter 6). However, in electron transfer processes low-lying levels of negative ions are reached so that the potential energy is below that of the ground vibrational level of the neutral molecule, and the ions are stable with respect to spontaneous ejection of the attached electron. The investigation of electron transfer reactions between negative ions and neutral molecules is important also in studies of electron affinities of molecules. For example, through process (7.19) one can establish whether $EA(A)$ is greater or smaller than $EA(B)$. Negative-ion charge transfer reactions, many of which occur very quickly and practically without any activation energy, are expected to resemble those of positive ions, both in relation to the adiabatic criterion and to symmetrical resonance behaviour (Hasted, 1964; Chapter 8).

Ion–molecule reactions are recognized by comparison of the appearance potentials and particularly the shapes of the ionization efficiency curves of the parent and daughter ions, and measurement of the pressure dependence of the primary and secondary ion currents. A large number of such reactions have been studied, and the cross section as a function of the primary ion energy has been measured by a number of workers (see, for example, Vogt and Neuert, 1967; Kraus, 1966; Martin and Bailey, 1968). Most beam experiments on collisions between negative ions with like (or unlike) atoms or molecules made no attempt to separate electrons from negative ion collision products. They yielded cross section functions, dominated at low energies by charge transfer, with the collisional detachment contribution increasing with increasing impact energy. The experimental data of Hummer and coworkers (1960) on the cross section for charge transfer in collisions of $H^-$ ions with H atoms are shown in Figure 7.15 as a function of the primary ion energy. The broken line refers to a perturbed stationary-state calculation by Dalgarno and McDowell (1956), and the solid line refers to the perturbed stationary-state calculation of Bardsley (1967b) who allowed for the effect of electron detachment on the charge transfer cross section. As is seen this correction is most important at higher energies. At energies above $\sim 2000$ eV deviations from the prediction of the simple perturbed stationary-state calculation are expected because of the possibility of electronic transitions to $H_2^-$ states other than the ones considered by Bardsley and also because of the increasing effect of momentum transfer to the target.

### 7.1.5.3 Other negative ion destruction processes

Other collision processes by which negative ions can be destroyed include: (i) collisions with positive ions (Chapter 8); (ii) collisions with excited species; and (iii) collisions with surfaces. This last type of collision provides the most effective means of all modes of negative ion destruction when the work function of the surface exceeds the electron affinity of the atom or the molecule involved.

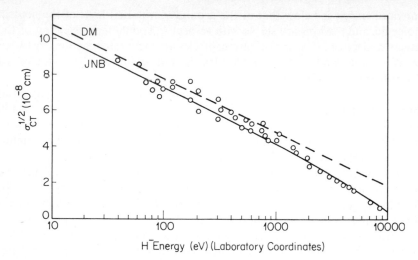

**Figure 7.15** Cross section for charge transfer between $H^-$ and H (Bardsley, 1967b). (o) experimental results of Hummer and coworkers (1960). (– – – –) calculations by Dalgarno and McDowell (1956). (———) results of Dalgarno and McDowell (1956) corrected for the effect of electron detachment by Bardsley (1967b)

## 7.2 Electron Affinities of Atoms and Molecules

The importance of this physical parameter is clearly seen from the many experimental and theoretical efforts that have been devoted to the determination of its numerical value. A number of these methods and the results obtained by their application will now be presented and discussed. Monographs and review articles on electron affinities include those by Massey (1950), Pritchard (1953), Branscomb (1957, 1962), Buchel'nikova (1958), Smirnov (1965), Moiseiwitsch (1965), Vedeneyev and coworkers (1966) and Christophorou and Compton (1967).

Electron affinities are frequently expressed in eV or in kcal/mole. The two units are related to each other and to $cm^{-1}$ (wave number) by

$$1 \text{ eV} = 8066 \text{ cm}^{-1} = 23.069 \text{ kcal/mole}$$

### 7.2.1 *Electron Affinities of Atoms*

The electron affinity $EA(X)$ of an atom X, defined as the difference in potential energy between the ground state of the atom X with a free electron at rest at infinity and the ground state of the negative ion $X^-$, is a measure of the stability of the negative ion $X^-$. For atoms, as stated earlier, this is simply the binding energy of the extra electron on the negative ion, which for a stable negative ion is positive. The binding energy for atomic negative ions ranges from 0 to $\sim 4\,\text{eV}$ (Table 7.2), and the thresholds of photodetachment processes, employed in the determination of $EA(X)$, should be primarily in the infrared or the visible region of the optical spectrum.

In general, the static field of a neutral atom is insufficient by itself to bind an extra electron. As the electron approaches the atom, however, its Coulomb field induces in the atom dipole, quadrupole, and higher multiple moments resulting in an attractive potential. Such a potential has the asymptotic form $e^2 \alpha R^{-4}$ at large electron–atomic nucleus separations $R$ ($\alpha$ is the atomic dipole polarizability) and may be sufficiently strong to bind the extra electron to the atom producing, in this way, a stable negative ion. The polarization field is not sufficiently strong to bind the extra electron in atoms with complete electron shells or subshells (e.g. $\text{He}(1s)$; $\text{Ne}(2p)$; $\text{Ar}(3p)$) where any additional electron must, according to the Pauli exclusion principle, be placed in a shell with a higher principal quantum number. On the other hand, atoms with a single electron vacancy in their outermost shells or subshells (e.g. $\text{H}(1s)$; alkali metals $\{\text{Li}(2s), \text{Na}(3s), K(4s)...\}$; halogens $\{\text{F}(2p)\ \text{Cl}(3p), \text{Br}(4p)...\}$) will (and do) form stable negative ions as a result of the strong polarization field. Further, since the potential for the interaction of an electron with an atom is a short-range one, there can only be a limited number of excited negative ion states. Excited atomic negative ion states are more likely to be metastable and lie very near the continuum (see also discussion in Chapter 6).

The variational calculation of Pekeris (1962) showed that stable excited states of $\text{H}^-$ do not exist, while the analyses of Branscomb (1957, 1962) indicated that the presence of stable excited states is improbable. However, elements of the IV group of the periodic table may constitute an exception forming negative ions in both the $^4S$ and $^2D$ states. Hartree–Fock calculations by Clementi and McLean (1964) indicated that no excited states of the negative ions of the first row of atoms of the periodic table have lower energy than the ground state atom, although they found that the $\text{C}^-(^2D)$ state lies only 0.08 eV above $\text{C}(^3P)$. Clementi and coworkers (1964) asserted that for the second row of atoms along with the ground state negative ions $\text{Al}^-(^3P)$, $\text{Si}^-(^4S)$ and $P^-(^3P)$, the excited states $\text{Al}^-(^1D)$, $\text{Si}^-(^2D)$, $\text{Si}^-(^2P)$ and possibly $P^-(^1D)$ exist.

It might be noted at this point that the electron affinity of an atom is greater in solution than in the gas phase by an amount equal to the difference in energy between the heats of solvation of the atom and the negative ion. Since the heat of solvation of a neutral particle is usually very much smaller compared to that for the ion, the addition of an electron to an atom or a molecule is a much more profitable process in solution than in the gas phase. Thus, negative ions which do not form in the gas phase may do so in the liquid phase. Di-negative atomic and molecular ions have not been observed in the gas phase; they, however, have frequently been detected in the liquid phase (see discussion in Chapter 6).

### 7.2.1.1 Theoretical calculations

With the exception of the lightest atoms, direct theoretical methods on quantitative determinations of electron affinities of atoms are difficult, often too tedious and generally leave much to be desired. Owing to the difficulty of performing exact calculations (and unequivocal experiments), the prediction of electron affinities of atoms have relied heavily on empirical and semiempirical methods, which have only been of limited success. The various theoretical calculations may be divided into two broad groups: extrapolation methods, and quantum-mechanical and other calculations.

A. *Extrapolation methods.* A reasonable semiempirical method which has been quite extensively used is based on extrapolation according to electron series. To a good approximation, the ionization energies of the members of an isoelectronic series of atoms and ions (i.e. a series of atomic species with the same number of electrons) can be represented by the quadratic expression

$$I(Z) = aZ^2 + bZ + c \tag{7.21}$$

where $I(Z)$ is the ionization energy of the ground state of the member of the sequence with atomic number $Z$, and $a$, $b$ and $c$ are constants which can be determined empirically. Glockler (1934) suggested that relation (7.21) could be extended to the negative ion member of the isoelectronic sequence for which $Z = Z_-$, the electron affinity of the parent atom given by (Moiseiwitsch, 1965)

$$EA = I(Z_-) = 3I(Z_0) - 3I(Z_1) + I(Z_2) \tag{7.22}$$

where $Z_0$, $Z_1$ and $Z_2$ are the atomic numbers of the neutral atom and the singly and doubly charged positive ions of the isoelectronic sequence. Glockler's original quadratic extrapolation formula has been employed (with some improvements, see later this section) by a number of workers, but unfortunately uncertainties in the spectroscopic ionization potentials which must be used in this extrapolation introduced large uncertainties in the calculated values of the electron affinity. Glockler's original estimates of the electron affinities for a number of atoms as well as more recent extrapolation values are listed in Table 7.2. These can be compared directly, by refering to Table 7.2, with electron affinity values obtained with other methods. Geltman (1956) argued that the result of the extrapolation depends heavily on the order of extrapolation, the nature of the dependence varying greatly with $Z$. Treating the order of extrapolation, $n$, as a continuous variable he obtained electron affinities, $EA$, corresponding to the value of $n$ for which

$$\frac{dEA(n)}{dn} = 0 \qquad (7.23)$$

His results for a number of atoms are listed in Table 7.2 and show no striking improvement over the use of the quadratic extrapolation.

Several efforts have been made to improve Glockler's extrapolation method by introducing additional terms in Equation (7.21). If we expand in powers of $Z - \sigma$, where $\sigma$ is a screening parameter, we arrive at (Moiseiwitsch, 1965)

$$I(Z) = d(Z - \sigma)^2 + e + \sum_{m=1}^{\infty} \frac{a_m}{(Z - \sigma)^m} \qquad (7.24)$$

where $d$, $e$ and $a_m$ are constants which can be determined either theoretically or by fitting to the experimental data on ionization energies. If we take $d$ to be given by the hydrogenic form $1/n^2$, where $n$ is the principal quantum number of the weakly bound electron of the isoelectronic sequence, and take $a_m = 0$ for $m \geqslant 2$, we have the expression

$$I(Z) = \frac{1}{n^2}(Z - \sigma)^2 + e + \frac{a_1}{Z - \sigma} \qquad (7.25)$$

in units of Rydberg. Edlen (1960) restricted his attention to the above three-term expression and determined from known ionization energies electron affinity values in reasonable agreement with other data (see Table 7.2) except for small $Z$. Johnson and Rohrlich (1959) and Edie and Rohrlich (1962) carried out similar calculations using more terms in expression (7.24) and their results are listed in Table 7.2 also. Other semi-empirical approaches based on Glockler's original idea include those by Kaufman (1963) and Crossley (1964) where estimation of some of the coefficients in Equation (7.25) has been made theoretically.

It should be noted that the extrapolation method according to isoelectron shells is suitable only for atoms with large electron affinities, since only for these cases does the series converge fast enough to permit confinement to the first few terms. The data of the several extrapolation methods listed in Table 7.2, although in poor quantitative agreement with each other and the known experimental data, reveal the proper trend of electron affinity with $Z$.

B. *Quantum-mechanical and other calculations.* A number of variational calculations have been performed especially for the atomic hydrogen negative ion. The most accurate calculation on $H^-$ is that by Pekeris (1962) who obtained a value equal to 0.75415 eV for the electron affinity of H. Pekeris also concluded that no stable excited $H^-$ ions exist. Application of the variational method to several other negative atomic ions had little success because in the majority of such calculations simple trial functions were used. More elaborate calculations for $Li^-$ have been performed using trial functions allowing for configuration interaction (Weiss, 1961). Weiss (1968) also has recently calculated electron affinities for Li, Na and K (Table 7.2) by the method of superposition of configurations. The variational method in the Hartree–Fock approximation was used by Clementi and McLean (1964) and Clementi and coworkers (1964). The results of this latter group of workers are included in Table 7.2 and are seen to be in good agreement with the experimental data for C, F, S and Cl. The value of Clementi and McLean (1964) for the $EA$ of O is somewhat lower than that obtained from photodetachment experiments (Branscomb and coworkers, 1958). In addition, Clementi (1964) computed electron affinities of the iron series atoms (Table 7.2). One should note that the accuracy of the variational methods improves very slowly with increase of the number of variable parameters in the trial function used.

A considerably less accurate method for determining electron affinities of atoms is based on the statistical model of an atom. The electron affinity values for S and Cl obtained by Gombás and Ladányi (1960) using this method are rather small compared with the experimental values given in Table 7.2.

Pritchard (1953) elaborated on the calculations of $EA$ of atoms from lattice energies. The electron affinity $EA(X)$ of an atom X can be calculated if the lattice energy $U$ of a crystal consisting of $A^+$ and $X^-$ ions, the heat of sublimation $S$ per molecule AX, the dissociation energy $D(AX)$ of the molecule AX, and the ionization potential $I(A)$ of atom A are known. The above quantities are related by

$$EA(X) = I(A) + D(AX) + S - U \qquad (7.26)$$

References to calculations of lattice energies for the alkali halides, LiH, some oxides of alkali metals, and some oxides and sulphides of alkaline earth atoms are cited in the review article of Moiseiwitsch (1965).

Recently Tandon, Bhutra and Tandon (1967) evaluated the electron affinity of the halogen atoms using the expression

$$EA(X) = D(AX) + I(A) - \frac{e^2}{R}\left[1 - \left(\frac{KR^3}{e^2} + 3\right)^{-1}\right] \qquad (7.27)$$

which relates the electron affinity of the atom X to the molecular constants of the polar diatomic molecule AX (X = halogen). In expression (7.27) $D(AX)$ is the dissociation energy of AX at the equilibrium internuclear distance $R$, $I(A)$ is the ionization potential of the atom A, $e$ is the electron charge, and $K$ is the force constant for infinitesimal amplitude. Tandon and coworkers (1967) used the molecular constants determined by Tandon and coworkers (1966) for a number of AX molecules and estimated through Equation (7.27) electron affinities for the halogens. Their values (Table 7.2) are seen to be higher than the most reliable experimental data.

### 7.2.1.2 Experimental methods

A variety of experimental methods have been employed to determine the electron affinity of atoms. Pritchard (1953) summarized a number of relatively indirect methods which are unsatisfactory except for the halogens; the halogen atoms, owing to their large electron affinity, have been the subject of many studies. In spite of the many and varying in principle experimental methods, few yielded reliable results, while some of the most reliable techniques (such as the photoabsorption method of Berry and coworkers) are of restricted application. Neither the theoretical nor the experimental results seem to be in satisfactory agreement with possibly the exception of the halogens and a few other atoms such as C, O and S (see Table 7.2). A brief discussion of the most pertinent experimental methods follows.

A. *Electron detachment methods.* The processes of photoabsorption by and photodetachment from a negative ion discussed in Section 7.1 constitute the basis of the most reliable methods to date for obtaining the electron affinity of atoms. The photodetachment thresholds measured by Branscomb and his colleagues and Berry's photoabsorption onsets for the halogens are in excellent agreement (Table 7.2). Unfortunately, such measurements have been restricted to a limited number of atoms.

The photodetachment and photoabsorption results for the electron affinity of C, O, F, S, Cl, Br and I can form the basis for comparisons of other experimental and theoretical data. The extrapolation values, for example, are seen to be in fair agreement with the photodetachment data with the exception of the calculations by Glockler (1934) and Geltman (1956) which in certain cases diverge significantly. In good agreement with the photodetachment data are also the calculations by Edlén (1960), Kaufman (1963) and Crossley (1964), the agreement being a function of $Z$. Further, in reasonable agreement with the photodetachment data on S and the photoabsorption data on Br and Cl are the experimental results obtained from lattice energies and surface ionization, although the latter methods yielded somewhat higher values.

The photodetachment process was discussed in Section 7.1.2 and will not be discussed further here. It might be added, however, that the electron affinity of atoms may be more easily determined from photodetachment experiments where the wavelength of the exciting photon beam is fixed and an energy analysis is made of the detached electrons.

B. *Charge-transfer methods.*  The determination of the electron affinity of atoms from charge-transfer reactions does not seem to be accurate. It is possible, however, with this method to establish whether the electron affinity of an atom is greater or smaller than the electron affinity of another. Estimates of electron affinities of molecules and radicals from charge-transfer reactions and charge-transfer complexes are discussed in Section 7.2.2.

C. *Surface ionization methods.*  The determination of atomic electron affinities by the surface ionization method is an indirect one in which the equilibrium constant for the reaction

$$X + e \rightleftarrows X^-  \qquad\qquad (7.28)$$

is measured near a hot filament. The atoms X are formed by dissociation of gaseous $X_2$ or AX molecules.

The surface ionization method has been refined over a period of years. The original work of Mayer and coworkers (Sutton and Mayer, 1934; Mitchell and Mayer, 1940; McCallum and Mayer, 1943; Vier and Mayer, 1944; Doty and Mayer, 1944) and Glockler and Calvin (1935, 1936) was followed by a number of investigations including those of Bakulina and Ionov (1955, 1957) and Bailey (1958). Zandberg and Ionov (1959) reviewed the surface ionization method in some detail.

Determinations of the differences in the electron affinities of two atoms provide another way of obtaining the electron affinity of an atom X when the electron affinity of the other, Y, is known. This latter surface ionization method may be more accurate since the relative measurement is independent of several experimental parameters such as the pressure of the surrounding gas and the work function of the filament, both of which are difficult to determine accurately. Studies of this nature were originated independently by Bakulina and Ionov (1955, 1957) and Bailey (1958).

The results of Bakulina and Ionov (1955, 1957) on the halogens and sulphur are listed in Table 7.2 and are seen to be in satisfactory agreement with the photodetachment values. In Table 7.2 the results of Bailey (1958) on the halogens are also listed. Bailey's values are in satisfactory agreement with those of Bakulina and Ionov (1955, 1957) and the earlier measurements made by Mayer and coworkers. Other determinations of electron affinities by the surface ionization method are listed in Table 7.2.

A similar device to that of Mayer and coworkers was developed in the 1960's by Page (1960). Page and coworkers applied their method ('magnetron') to the determination of the electron affinities of a large number of atoms, molecules and radicals. Their values for S and O (see

Table 7.2) agree very well with the photodetachment results, while their results for I, Br and Cl (see Table 7.2) are in fair agreement with those of Berry and coworkers. The work of Page and coworkers on radicals and molecules is discussed in Section 7.2.2.

D. *Electron impact methods.* The electron affinity of an atom whose ion is formed by a resonance dissociative electron attachment or by an ion-pair process in electron impact studies of simple molecules can be determined if the corresponding dissociation energy is known and appropriate kinetic energy measurements are made of the charged particles produced. Diatomic and triatomic molecules yielding the same negative ion provide independent ways to determine the electron affinity of a certain (common) atom. Difficulty in determining the appearance potential of the ions formed, uncertainty in the energy scale calibration, energy spread in the electron beam itself, and the effects of gas temperature on the measured kinetic energy thresholds, limit the accuracy of the electron impact method.

Considering the kinetics of dissociative electron attachment to a diatomic molecule AX and neglecting the effect of thermal motion of the target molecules, conservation of energy and momentum requires (Chantry and Schulz, 1964)

$$E_{X^-} = (1 - \beta)[\varepsilon - (D(AX) - EA(X))] \tag{7.29}$$

where $E_{X^-}$ is the kinetic energy of the ion $X^-$, $\beta = M_{X^-}/M_{AX}$, $M_{X^-}$ and $M_{AX}$ are the masses of $X^-$ and AX, respectively, and $\varepsilon$ is the electron energy. The determination of the kinetic energy of the ions $X^-$ produced by electrons of known energy $\varepsilon$ is usually made by measuring the maximum retarding potential, $MRP$, which the ions can penetrate. A linear extrapolation of the plot $MRP$ versus $\varepsilon$ to zero $MRP$ with a slope determined by Equation (7.29) has been interpreted as leading to the proper value of $D(AX) - EA(X)$. Thus, knowledge of $D(AX)$ yields $EA(X)$ and vice versa.

Electron impact studies aiming at the determination of atomic electron affinities have been considerable, especially on atomic oxygen. Measurements have been made of $EA(O)$ from electron impact studies on dissociative electron attachment and ion-pair formation in a number of diatomic ($O_2$ (Schulz, 1962; Hagstrum, 1951, 1955); CO (Hagstrum, 1951, 1955; Fineman and Petrocelli, 1962); NO (Hagstrum, 1951, 1955)) and triatomic ($CO_2$ (Craggs and Tozer, 1960; Schulz, 1962); $N_2O$ (Schulz, 1961); $NO_2$ (Fox, 1960); $SO_2$ (Schulz, 1962)) molecules. The results of the various investigators have been discussed by Craggs and Tozer (1960) and by Schulz (1962) and clearly show the intrinsic uncertainty of this method. Although most of the electron affinity values for atomic oxygen obtained with electron impact methods lie around 1.5 eV (in agreement with the photodetachment value of 1.465 eV; Table 7.2), a number of them lie

$\sim 0.5\,\text{eV}$ higher at $\sim 2.0\,\text{eV}$. These higher values led to the suggestion (Schulz, 1962) that they are the result of the existence of an excited state in atomic oxygen at $\sim 0.5\,\text{eV}$ above the ground state. This, however, would require photodetachment to result from this excited state, a very unlikely possibility.

The effect of gas temperature in the above mentioned electron impact studies has, as a rule, not been considered. However, the gas temperature may greatly affect the determinations of electron affinities by electron impact techniques in two ways: firstly, through the effect of thermal motion of the target molecules on the negative ion energy distribution and, secondly, through the effect of temperature on the internal excitation of vibrational and rotational motion of the molecule which in certain cases (see Chapter 6) can seriously affect the energy threshold. Both effects can lead to a serious overestimate of the electron affinity, as obtained from dissociative electron attachment studies.

In electron impact experiments, it has been customary to neglect the molecular thermal motion and to ascribe any apparent spread in the ion energies above that expected from the spread in electron energies as originating from ions which enter the retarding field at various angles. However, Chantry and Schulz (1964) showed that this can lead to serious errors in determinations of the electron affinity of atoms by electron impact when the ions are produced with energies above thermal. Chantry and Schulz showed that when suitable integration over the Maxwell distribution of thermal velocities of the AX molecules is made, the ion-energy distribution is

$$\frac{\mathrm{d}N}{N} = \left(\frac{1}{4\pi\beta kTE_{X^-}}\right)^{\frac{1}{2}}\left[\exp\,-\left(\frac{1}{\beta kT}(E^{\frac{1}{2}}-E_{X^-}^{\frac{1}{2}})^2\right)\right]\mathrm{d}E \qquad (4.30)$$

where $E_{X^-}$ (assumed $\gtrsim 3kT$) is now interpreted as the most probable ion energy. The width at half maximum of the above distribution is $(11\beta kTE_{X^-})^{\frac{1}{2}}$. For $O^-$ from $O_2$ at $350°\text{K}$ and $E_{X^-} = 2\,\text{eV}$, Chantry and Schulz estimated a width at half-height equal to $0.56\,\text{eV}$, showing that $O^-$ ions are actually produced with energies well above $E_{X^-}$. Thus the assumption normally made that the measured maximum ion kinetic energy corresponds to $E_{X^-}$ as given by Equation (7.29) is incorrect unless the ions are produced with truly zero energy. Under certain assumptions as to the collection geometry, gas temperature, and angular distribution of the products of dissociative electron attachment in $O_2$, Chantry and Schulz re-interpreted Schulz's data on $O^-/O_2$ (Schulz, 1962) taking into account the thermal motion of $O_2$ molecules. They obtained an electron affinity for atomic oxygen equal to $\sim 1.5\,\text{eV}$ (rather than $2\,\text{eV}$ as obtained originally by Schulz) in close agreement with the photodetachment value.

The analysis of Chantry and Schulz could not explain the $O^-/CO_2$ data which were consistent with $EA(O) = 2.0$ eV. In the case of $CO_2$ the $O^-$ ions are formed with zero kinetic energy. Here the effect of internal excitation of the rotational–vibrational motion of the molecule under study must be considered. It has been stated in Chapter 6 that the cross section for dissociative electron attachment from higher vibrational levels of a molecule may be very much higher than for the $v = 0$ level, and that similar effects may arise from rotational excitation. Such an effect is a function of gas temperature and can greatly decrease the measured threshold for the dissociative attachment process, leading to an overestimate of the electron affinity. The onset measurements, therefore, should be extrapolated to $0°K$ for accurate electron affinity determinations (see Chapter 6 and Schulz and Spence (1969)).

E. *Other experimental methods.* Determinations of electron affinities from electron attachment–detachment experiments and electron detachment from negative ion beams passing through electric fields have been discussed in Section 7.1. Radiative attachment continua, photoionization of molecules, ion beams, and sublimation have also been applied for the measurement of atomic electron affinities. A brief discussion of emission continua due to radiative attachment has been presented in Section 7.1.2. Berry and David (1964) made direct spectroscopic observations of the spectra due to radiative capture of electrons to gaseous atoms of Cl, Br and I. They reported sharp onsets of radiation at $3427 \pm 4$, $3327 \pm 4$, $3688 \pm 4$, $3249 \pm 3$ and $4046 \pm 4$Å due to capture of zero energy electrons by $Cl(^2P_{3/2})$, $Cl(^2P_{1/2})$, $Br(^2P_{3/2})$, $Br(^2P_{1/2})$ and $I(^2P_{3/2})$, respectively. The onsets for the radiative attachment continua into the $^3P_{3/2}$ states are in excellent agreement with the photoabsorption thresholds (Berry and Reimann, 1963).

The primary process of photoabsorption $(X_2 + h\nu \rightarrow X_2^*)$ may be followed by molecular dissociation into a positive and a negative ion pair $(X_2^* \rightarrow X^+ + X^-)$. This may be used to determine the electron affinity $EA(X)$ from

$$EA(X) = I(X) + D(X_2) - E(X_2) \tag{7.31}$$

where $D(X_2)$ is the minimum energy required to dissociate the molecule into a pair of atoms, $I(X)$ is the ionization potential of X, and $E(X_2)$ is the threshold energy required to produce an ion pair $(X^+, X^-)$. Morrison and coworkers (1960) determined the electron affinity of the Br and I atoms from a study of the photoionization efficiency curves for $Br_2$ and $I_2$. Their values are listed in Table 7.2.

Finally, in ion beam experiments, $He^-$ ions have been detected (see discussion in Section 7.1) and Fogel', Kozlov and Kalmykov (1959) reported observing $N^-$ ions upon passing a beam of $N^+$ through a target gas.

Honig (1954), applying a sublimation method, reported an electron affinity for the C atom equal to 1.2 eV.

The data listed in Table 7.2 clearly show that the chlorine atom has the largest known electron affinity of all atoms. They also show that the greater the number of electrons in the outermost shell (subshell) of the neutral atom that have the same spin quantum number as the additional electron, the greater is the binding energy for the extra electron. The electron affinity values considered to be the most reliable are included in brackets in Table 7.2.

### 7.2.2 *Electron Affinity of Molecules*

The determination of the electron affinity of molecules is a much more demanding problem than for atoms. There are only few molecules whose electron affinity is known with precision. Reviews and tabulations of the electron affinity of molecules include those by Pritchard (1953), Field and Franklin (1957), and more recently those by Moiseiwitsch (1965), Christophorou and Compton (1967), Gutmann and Lyons (1967) and Page and Goode (1969). Briegleb (1964) has reviewed electron affinity determinations of organic molecules obtained from charge-transfer spectra. Experimental and theoretical data on the electron affinity of molecules and radicals are summarized in Table 7.3. As is seen, in most cases the available information is indeed fragmentary, the uncertainty in the electron affinity values being several tenths of an electron volt or more. Not included in Table 7.3 are some semi-empirical determinations of the electron affinities of various gaseous radicals given by Gaines and Page (1966).

#### 7.2.2.1 Theoretical calculations

There have been a number of recent theoretical calculations on the hydrogen molecule ion $H_2^-$ (Chapters 4 to 6). For the purpose of illustrating the changes in the equilibrium configurations of $H_2$ and $H_2^-$ and the differences between the electron affinity $EA(H_2)$ and the vertical detachment energy $VDE(H_2)$ for $H_2$, the potential energy curves for $H_2$ and $H_2^-$ as calculated by Dalgarno and McDowell (1956) are shown in Figure 7.16. It is seen that the equilibrium configuration of the molecule has changed from $1.4a_0$ to $5.4a_0$, while the dissociation energy has been reduced from 4.5 to 0.15 eV by the addition of an extra electron. The differences between the electron affinity and the vertical detachment energy of $H_2$ are clearly seen. The $EA(H_2)$ is equal to $-3.58$ eV, while the $VDE(H_2)$ is equal to $+0.84$ eV. However, we have seen in Section 7.1.2 that for certain diatomic radicals (OH, SH) possessing positive electron affinities, no such drastic changes occur, and at least for these systems the electron affinity and the vertical detachment energy were found to be very nearly equal.

**Table 7.3** Electron affinities of molecules and radicals[a,b]

| Molecule or radical | Electron affinity (eV) | Reference and method[g,h] | Molecule or radical | Electron affinity (eV) | Reference and method[g,h] |
|---|---|---|---|---|---|
| $H_2$ | $-3.58$ | 1a | AgI | 3.29 | 41g |
| | $-2.85$ | 2a | ICl | 1.43 | 38j |
| | $-0.72$ | 3a | $I_2$ | 1.56 | 38j |
| CH | $\geqslant 1.4$ | 4b | | 1.8 | 42i |
| | $\sim 1.65$ | 3b | | 2.42 | 40i |
| $C_2$ | $\sim 4.0\ (3.1)$ | 5 | $H_2O$ | $\sim 0.9$ | 43i |
| CN | 2.8 | 6c | $HO_2$ | 4.6 | 44i |
| | $3.16 \pm 0.04$ | 7c | $CH_2$ | $-0.95$ | 3b |
| | $3.7 \pm 0.2$ | 8c | $C_3$ | $2.5\ (1.8)$ | 5 |
| | 3.6 | 9 | $CO_2$ | $\sim 3.8$ | 3g |
| | $3.2 \pm 0.17$ | 10b | $NH_2$ | $1.2 \pm 0.43$ | 18c |
| | $3.82 \pm 0.02$ | 11e | | 1.21 | 16 |
| OH | 1.75 | 12e | $N_3$ | 2.34 | 45i, 46i |
| | 1.80 | 13b | | 3.5 | 43g |
| | $[1.83 \pm 0.04]$ | 14e | $NO_2$ | 1.62 | 33g |
| | $1.9 \pm 0.12$ | 15c | | $\leqslant 3.0\ (?)$ | 47e |
| | 2.15 | 16 | | $\geqslant 3.82 \pm 0.06$ | 48b |
| | 2.64 | 17c | | 3.9 | 13b |
| $N_2$ | $<0$ | 18 | | 3.99 | 20c |
| NO | $>0$ | 19f | | $3.10 \pm 0.05$ | 49e |
| | $0.9 \pm 0.1$ | 20c | $O_3$ | 2.89 | 50g |
| $O_2$ | $0.15 \pm 0.05$ | 21e | | $\sim 3.00$ | 24b |
| | $[0.43 \pm 0.02]$ | 22f | $SiCl_2$ | $\geqslant 2.6$ | 51b |
| | $0.46 \pm 0.02$ | 23f | SCN | 2.16 | 6c |
| | $\geqslant 0.58$ | 24b | $SO_2$ | $\sim 1.1$ | 35b |
| | 0.68 | 25, 26 | | 2.8 | 52 |
| | $0.95 \pm 0.07$ | 16g | $ClO_2$ | 2.8 | 36i |
| SiC | $\sim 4$ | 27h | $SeO_2$ | $2.3\ (?)$ | 52 |
| SH | 1.6 | 28b | $BF_3$ | 2.17 | 3b |
| | $\geqslant 1.8$ | 29b | $CH_3$ | $-1.03$ | 53a |
| | 2.3 | 30c, 31c | | 1.08 | 16 |
| | $[2.32 \pm 0.01]$ | 32e | | 1.1 | 54 |
| | 2.6 | 33g, 34g | | 1.1 | 55c |
| SO | $\geqslant 1.1$ | 35b, 29b | | 1.8 | 56a |
| ClO | 4.5 | 36i | $CCl_3$ | 1.44 | 58c |
| | 2.2-3.1 | 37i | | $\geqslant 2.1 \pm 0.35$ | 59b |
| | 2.91 | 3i | | 2.34 | 3i |
| $Cl_2$ | 1.3 | 38j | $NO_3$ | 3.88 | 33g |
| | $\leqslant 1.7$ | 39b | $ClO_3$ | 3.96 | 33g |
| | 2.54 | 40i | $MoO_3$ | $2.73\ (?)$ | 60 |
| $Br_2$ | 1.47 | 38j | $HNO_3$ | 1.99 | 61j |
| | 2.62 | 40i | $CH_3O$ | 0.39 | 62 |
| SeH | 2.2 | 34g | $CH_3S$ | 1.34 | 62 |
| AgCl | 3.79 | 41g | $CHCl_3$ | 1.75 | 58c |
| AgBr | 3.57 | 41g | $CCl_4$ | 2.12 | 58c |

## Table 7.3 (*continued*)

| Molecule or radical | Electron affinity (eV) | Reference and method[g, h] | Molecule or radical | Electron affinity (eV) | Reference and method[g,h] |
|---|---|---|---|---|---|
| $ClO_4$ | 5.82 | 52 | Fluoro-benzo-quinonyl | | |
| $C_2H_4$ | −4.26 | 63a | | | |
| Ethylene | −1.69 | 64a | | | |
| | −1.81 | 53a | $(CH_3CO)_2$ | ≥1.1 | 57f |
| $(CHO)_2$ | ≥1.6 | 65c | Biacetyl | ≥0.39 | 70f |
| $SF_5$ | ~3.39 | 66b | $C_6H_6$ | −1.63 | 77a |
| | 3.66±0.04 | 67c | | −1.62 | 63a |
| $CH_3CO_2$ | 3.3 | 13b | | −1.59 | 58a |
| | 3.36 | 68f | | −1.42 | 78a |
| $C_2H_5$ | 1.4 | 69i | | ≥−1.4 | 79k |
| $C_2Cl_5$ | 1.52 | 58c | | −1.40 | 53a, 64a |
| $C_2Cl_6$ | 1.47 | 58c | | −0.43 | 80i |
| $C_2H_5O$ | 0.61 | 62 | | −0.36 | 81l |
| $C_2H_5S$ | 1.60 | 62 | $C_6H_5O$ | 1.17 | 3i |
| $H_2C=CH$ | 0.14 | 63a | $C_6H_5F$ | ≥−1.2 | 79k |
| —$CH_2$— | 0.24 | 53a | $C_6H_5Cl$ | ≥−0.9 | 79k |
| Allyl | 0.53 | 64a | $C_6H_5Br$ | ≥−0.8 | 79k |
| | 2.1 | 54 | $C_6H_4O_2$ | $1.37\pm0.08^d$ | 76c |
| $SF_6$ | 1.5±0.21 | 67c | p-Benzo-quinone | $1.94^d$ | 71a |
| | ≥1.1 | 57f | $o$-$C_6H_4Cl_2$ | ≥−0.4 | 79k |
| | ≥1.29 | 70f | $C_6H_3FO_2$ Fluoro-benzo-quinone | 2.2 (1.5± 0.26) | 76c |
| $C_4H_4O$ Furan | −0.64 | 71a | | | |
| $C_4H_2O_3$ | 0.57 | 72j^c | $C_6Cl_4O_2$ | 2.5±0.26 | 76c |
| Maleic anhydride | 1.43 | 71a | p-Chloranil | 2.59 | 61j |
| $n$-$C_3H_7$ | 1.0 | 69i | $C_6Br_4O_2$ | 2.6 | 61j |
| $iso$-$C_3H_7$ | ~1.0 | 69i | p-Bromanil | | |
| $trans$- $CH_2=CH$ —$CH=$ $CH_2$ $trans$- Butadiene | −0.34 | 53a | $C_6I_4O_2$ p-Iodanil | 2.55 | 61j |
| | −0.32 | 64a | $C_6H(CN)_2O_2$ 2,3-Dicyano-benzoquinonyl | 1.82±0.09 | 76c |
| $C_4H_5N$ Pyrrole | −1.88 | 71a | $C(NO_2)_4$ | 1.78 | 61j |
| $C_2(CN)_4$ | 2.88±0.06 | 73c | $C_4H_9O$ | 0.65 | 62 |
| | ~6.5 | 74c | $C_6H_5CH_2$— | 0.69 | 53a |
| $iso$-$C_3H_7O$ | 0.69 | 62 | Benzyl | 0.71 | 63a |
| $C_5H_5N$ Pyridine | −0.675 | 71a | | 0.79 | 64a |
| $C_6H_5$ | 2.2 | 75c | | 0.90 | 75c |
| $C_6H_3O_2$— Benzoquinonyl | 2.00±0.04 | 76c | | 1.8 | 54 |
| $C_6H_2FO_2$— | 2.4±0.14 | 76c | $C_6H_5CHO$ Benzaldehyde | 0.16 | 71a |
| | | | | $0.42 (0.45)^e$ | 82c |
| | | | $C_6H_5C=CH$ Phenylacetylene | −1.25 | 64a |

Table 7.3 (*continued*)

| Molecule or radical | Electron affinity (eV) | Reference and method[g,h] | Molecule or radical | Electron affinity (eV) | Reference and method[g,h] |
|---|---|---|---|---|---|
| $C_6H_5NH_2$ | $-1.30$ | 71a | $C_{10}H_7NH_2$ 2-Naphthylamine | $-0.15$ | 71a |
| $C_6H_5NO_2$ | $\geqslant 0.4$ | 57f | $C_{10}H_7CHO$ 1-Naphthaldehyde | $0.745\ (0.669)^e$ | 82c |
|  | $\geqslant 0.51$ | 70f | $C_{10}H_7CHO$ 2-Naphthaldehyde | $0.620\ (0.615)^e$ | 82c |
| $C_6H_5CH_3$ | $\geqslant -1.3$ | 79k | $C_{11}H_9$ | 1.12 | 53a |
| p-$C_6H_4CH_3Cl$ | $\geqslant -1.1$ | 79k | α-Naphthylmethyl | 1.6 | 54 |
| $C_8H_4O_3$ Phthalicanhydride | 0.99 | 71a | $C_{12}H_8$ Biphenylene | $-0.41$ | 77a |
| $C_6H_5CH{=}CH_2$ | $-0.29$ | 53a |  | 0.42 | 78a |
| Styrene | $-0.54$ | 63a | $C_{12}H_{10}$ Biphenyl | $-0.78$ | 77a |
|  | $-0.55$ | 64a |  | $-0.37$ | 53a, 63a |
| $C_6H_4(NO_2)_2$ m-Dinitrobenzene | 1.43 | 61j |  | 0.085 | 88 |
| $C_6H_5COCH_3$ Acetophenone | 0.334 | 82c |  | 0.41 | 86 |
| sym-$C_6H_3(NO_2)_3$ | 0.69 | 83i |  | 0.70 | 81l |
| $C_9H_7N$ Quinoline | 0.36 | 71a | $C_{13}H_9N$ Acridine | 1.01 | 71a |
| $C_6H_5CH{=}$ CHCHO Cinnamaldehyde | $(0.823)^e$ | 82c | $C_{10}H_{12}O_2$ Trimethyl-quinonylmethyl | $0.8\pm0.09$ | 76c |
| $C_{10}H_8$ Naphthalene | $-0.38$ | 77a | $C_{13}H_{11}$ Diphenylmethyl | 1.21 | 53a |
|  | $-0.25$ | 63a | $C_{14}H_{10}$ Anthracene | 0.42 | 89c |
|  | $-0.20$ | 64a |  | 0.43 | 63a |
|  | $-0.14$ | 53a |  | 0.49 | 77a |
|  | $-0.08$ | 78a |  | $0.552\ (0.556)^e$ | 84c, 85c |
|  | $-0.01$ | 72f |  | 0.57 | 90c, 91c, f |
|  | $0.152\ (0.148)^e$ | 84c, 85c |  | 0.58 | 78a |
|  | 0.10 | 71a |  | 0.61 | 64a |
|  | 0.65 | 86 |  | 0.64 | 53a |
|  | 0.67 | 81l |  | 0.74 | 88 |
|  | 0.68 | 80i |  | 0.74 | 71a |
| $C_{10}H_8$ Azulene | $\geqslant 0.46$ | 70f |  | 1.19 | 86 |
|  | $0.587\ (0.656)^e$ | 84c, 85c |  | 1.38 | 81l |
|  | 0.92 | 64a |  | 1.41 | 80a |
| $C_{10}H_6O_2$ 1,4-Naphtho-quinone | $\geqslant 0.6$ | 87f | $C_{14}H_{10}$ Phenanthrene | $-0.20$ | 77a |
|  | 1.60 | 71a |  | $-0.06$ | 53a, 68a |
| $C_{10}H_6O_2$ 1,2-Naphtho-quinone | 1.52 | 71a |  | 0.00 | 63a |
| $C_{10}H_2O_6$ Pyromellitic dianhydride | 1.84 | 71a |  | 0.17 | 78a |
| $C_{10}H_7NH_2$ 1-Naphthylamine | $-0.20$ | 71a |  | 0.20 | 89c |
|  |  |  |  | 0.22 | 71a |
|  |  |  |  | 0.25 | 64a |
|  |  |  |  | $0.308\ (0.307)^e$ | 84c, 85c |
|  |  |  |  | 0.69 | 86, 80i |

Table 7.3 (*continued*)

| Molecule or radical | Electron affinity (eV) | Reference and method[g,h] | Molecule or radical | Electron affinity (eV) | Reference and method[g,h] |
|---|---|---|---|---|---|
| $C_{14}H_8O_2$ Anthraquinone | 1.28 | 71a | | 0.419 (0.397)[e] | 84c, 85c |
| | | | | 0.47 | 78a |
| $C_{14}H_8O_3$ | 1.33 | 71a | | 0.50 | 71a |
| 1-Hydroxy-9,10-anthraqui-none | | | | 0.98 | 80i |
| | | | $C_{18}H_{12}$ | −0.14 | 77a |
| | | | 3,4-Benzo-phenanthrene | 0.33 | 89c |
| $C_{14}H_{12}$ | 1.33 | 81l | | 0.37 | 78a |
| *trans*-Stilbene | | | | 0.542 (0.545)[e] | 84c, 85c |
| $C_{14}H_9CHO$ | 0.655 (0.712)[e] | 82c | $C_{18}H_{14}$ | −0.35 | 77a |
| 9-Anthraldehyde | | | *o*-Diphenyl-benzene | | |
| $C_{16}H_{10}$ | 0.39 | 89c | | | |
| Pyrene | 0.417 | 92a | $C_{18}H_{14}$ | −0 51 | 77a |
| | 0.50 | 90c | *m*-Diphenylben-zene | | |
| | 0.55 | 64a | | | |
| | 0.57 | 78a | $C_{20}H_{12}$ | 0.41 | 92a |
| | 0.579 (0.591)[e] | 84c, 85c | 1,2-Benzopyrene | 0.486 (0.534)[e] | 85c |
| | 0.68 | 77a | | 1.22 | 80a |
| | 0.74 | 71a | $C_{20}H_{12}$ | 0.676 | 92a |
| | 1.16 | 81l | 3,4-Benzopyrene | 0.829 (0.680)[e] | 85c |
| | 1.23 | 80i | | 1.59 | 80i |
| $C_{16}H_{14}$ | 1.47 | 81l | $C_{20}H_{12}$ | 0.727 | 92a |
| 1,4-Diphenyl-butadiene | | | Perylene | 0.80 | 77a |
| | | | | 0.88 | 78a |
| $C_{18}H_{12}$ | 0.82 | 77a | | 1.02 | 71a |
| Tetracene | 0.95 | 78a | | 1.03 | 72d |
| | 0.88 | 90c | | 1.06 | 88 |
| | 1.15 | 88 | | 1.75 | 80i |
| | 1.14 | 71a | $C(C_6H_5)_3$ | 1.44 | 63a |
| | 1.78 | 80i | Triphenylmethyl | 1.65 | 53a |
| $C_{18}H_{12}$ | 0.46 | 89c | | 2.08 | 93a |
| 1,2-Benzan-thracene | 0.61 | 78a | | 2.1 | 16, 54 |
| | 0.62 | 77a | | 2 56 | 94l |
| | 0.66 | 71a | | 2.62 | 81l |
| | 0.696 (0.63)[e] | 84c, 85c | $C_{22}H_{12}$ | 0.50 | 77a |
| | 1.33 | 80i | 1,12-Benzo-perylene | 0.573 | 92a |
| $C_{18}H_{12}$ | −0.28 | 77a | | 0.73 | 78a |
| Triphenylene | −0.033 | 92a | | 1.43 | 80i |
| | 0.11 | 71a | $C_{22}H_{12}$ | 1.78 | 80i |
| | 0.12 | 78a | Anthranthrene | | |
| | 0.14 | 89c | $C_{22}H_{14}$ | 0.446 | 92a |
| | 0.284 (0.285)[e] | 84c, 85c | 1,2,3,4-Diben-zanthracene | 0.58 | 78a |
| | 0.69 | 80i | | 1.17 | 80i |
| $C_{18}H_{12}$ | 0.04 | 77a | $C_{22}H_{14}$ | 0.50 | 92a |
| Chrysene | 0.33 | 89c | 1,2,5,6-Diben-zanthracene | 0.63 | 71a |
| | | | | 0.65 | 77a, 78a |

**Table 7.3** (*continued*)

| Molecule or radical | Electron affinity (eV) | Reference and method[g,h] | Molecule or radical | Electron affinity (eV) | Reference and method[g,h] |
|---|---|---|---|---|---|
| | 0.98 | 80i | 1,2,4,5-Dibenzo-pyrene | | |
| | 0.676 (0.595)[e] | 85c | | | |
| $C_{22}H_{14}$ | 0.463 | 92a | Graphite sheet | 3.87 | 77a |
| 2,7,8-Diben-zanthracene | 0.62 | 78a | | 4.23 | 53a |
| | 0.75 | 77a | Fumaronitrile | 0.79±0.1 | 73c |
| | 0.686 (0.591)[e] | 85c | o-Phthalonitrile | 1.1±0.13 | 73c |
| | 1.10 | 80i | sym-Tetracyano-benzene | 2.2±0.22 | 73c |
| $C_{22}H_{14}$ Pentaphene | 0.63 | 77a | | | |
| | 0.73 | 78a | Hexacyano-benzene | 2.5±0.14 | 73c |
| $C_{22}H_{14}$ Picene | 0.442 | 92a | | | |
| | 0.490 (0.542)[e] | 85c | sym-Tetracyano-pyridine | 2.17±0.07 | 73c |
| | 0.59 | 78a | | | |
| | 0.60 | 71a | 7,7,8,8-Tetra-cyanoquino-dimethane | 2.9±0.19 | 73c |
| | 0.98 | 80i | | | |
| $C_{22}H_{14}$ 4,5,6-Dibenzo-phenanthrene | 0.52 | 78a | | | |
| | | | Hexacyanobuta-diene | 3.3±0.1 | 73c |
| $C_{24}H_{12}$ Coronene | 0.18 | 77a | 2,3,5,6-Tetra-cyanophenyl | 2.4±0.04 | 73c |
| | 0.385 | 92a | | | |
| | 0.54 | 78a | Tetracyano-benzene | 2.21 | 73c |
| | 1.34 | 80i | | | |
| $C_{22}H_{14}$ Pentacene | 1.01 | 77a | 9,10-Dimethylan-thracene | 0.71 | 88 |
| | 1.035 | 92a | | | |
| | 1.19 | 78a | Acenaphthylene | 0.98 | 88 |
| $C_{24}H_{14}$ 2,3,4-Dibenzo-pyrene | 0.683 | 92a | Fluoranthene | 0.63 | 95c |
| $C_{24}H_{14}$ | 1.51 | 80i | | 0.92 | 88 |

[a] A list of electron affinity values determined by the magnetron method can be found in Page and Goode (1969). Also, Jensen (1969) has recently reported $EA(BO_2) = 4.1 \pm 0.2$ eV.

[b] Values in brackets are recommended.

[c] See Briegleb (1964) for electron affinities for a large number of organic compounds obtained from charge-transfer spectra and from polarographic half-wave reduction potentials.

[d] Recent work by Collins and coworkers (1970) indicated that these values are probably too high.

[e] Common intercept (see References 82, 84 and 85).

[f] See Briegleb (1964) for values obtained through method d below.

[g] Methods:
  a) Theoretical calculation.
  b) Electron impact.
  c) Equilibrium; surface ionization.
  d) Semi-empirical calculation ($EA = \alpha + \beta E_1$); $\alpha$ and $\beta$ are constants and $E_1$ is the energy which corresponds to the longest wavelength absorption maximum.
  e) Photodetachment; photoionization.

f) Electron attachment.
g) Lattice energies.
h) Photoelectric work function.
i) Estimated.
j) Charge transfer spectra.
k) Electron scattering.
l) Kinetics of electrode processes.

[h] References:

1. Dalgarno and McDowell (1956).
2. Eyring, Hirschfelder and Taylor (1936).
3. Buchel'nikova (1958).
4. Smith (1937).
5. Honig (1954).
6. Napper and Page (1963).
7. Page (1968).
8. Bakulina and Ionov (1959).
9. See Pritchard (1953).
10. Herron and Dibeler (1960).
11. Berkowitz, Chupka and Walter (1969).
12. Smith and Branscomb (1955b).
13. Tsuda and Hamill (1966).
14. Branscomb (1966).
15. Kay and Page (1966).
16. Pritchard (1953).
17. Page and Sugden (1957).
18. Ion never observed.
19. Bradbury (1934).
20. Farragher, Page and Wheeler (1964).
21. Burch, Smith and Branscomb (1958, 1959).
22. Pack and Phelps (1966a).
23. Phelps and Pack (1961).
24. Curran (1961c).
25. Evans and Uri (1949).
26. Evans, Hush and Uri (1952).
27. Phillip (1958).
28. Rosenbaum and Neuert (1954).
29. Kraus (1961).
30. Ansdell and Page (1962).
31. Page (1958).
32. Steiner (1968).
33. Yatsimirskii (1947).
34. West (1935).
35. Kraus, Müller-Duysing and Neuert (1961).
36. Weiss (1947b).
37. Skinner (1947).
38. Jortner and Sokolov (1961).
39. Baker and Tate (1938).
40. Geiger (1955).
41. Flengas and Rideal (1956).
42. Mulliken (1952a).
43. Gray and Waddington (1956).
44. Weiss (1935).
45. Buckner (1950).
46. Weiss (1947a).
47. Smith and Seman (1962).
48. Curran (1962).
49. Warneck (1969).
50. Nikoeskii and coworkers (1950).

51. Vought (1947).
52. See tables 10 and 11 in Pritchard (1953).
53. Hush and Pople (1955).
54. Hush and Oldham, quoted in Hush and Pople (1955).
55. Page (1962).
56. Sklar (1939).
57. Compton and coworkers (1966).
58. Gaines, Kay and Page (1966).
59. Curran (1961b).
60. See table 11 in Pritchard (1953).
61. Batley and Lyons (1962).
62. Page, quoted in Freeman (1968).
63. Ehrenson (1962).
64. Hoyland and Goodman (1962).
65. Compton and Bouby (1967).
66. Curran (1961a).
67. Kay and Page (1964).
68. Wentworth, Chen and Steelhammer (1968).
69. Matsen, Robertson and Chuoke (1947).
70. Chaney and coworkers (1970); Chaney and Christophorou (1970).
71. Kunii and Kuroda (1968).
72. Briegleb (1964).
73. Farragher and Page (1967).
74. Page and Kay (1963).
75. Gaines and Page (1963).
76. Farragher and Page (1966).
77. Hedges and Matsen (1958).
78. Scott and Becker (1962).
79. Compton, Christophorou and Huebner (1966).
80. Slifkin (1963).
81. Lyons (1950).
82. Wentworth and Chen (1967).
83. Briegleb and Czekalla (1955).
84. Wentworth, Chen and Lovelock (1966).
85. Becker and Chen (1966).
86. Blackredge and Hush, quoted in Hush and Pople (1955).
87. Collins and coworkers (1970).
88. Chaudhuri, Jagur-Grodzinski and Szwarc (1967).
89. Wentworth and Becker (1962).
90. Lyons, Morris and Warren (1968b).
91. Lyons, Morris and Warren (1968a).
92. Ehrenson, quoted in Briegleb (1964).
93. Swift (1938).
94. Bent (1930).
95. Michl (1969).

In spite of the fact that simple molecules have not yet been treated fully theoretically, another group of molecules namely the polycyclic aromatic hydrocarbons have attracted much theoretical attention. The interest in this group of molecules stems from the relative simplicity of their $\pi$-electron structures and the considerable importance of their electron affinity in modern biology (Chapter 9). Most of the calculations on these systems are simple quantum-mechanical treatments where a determination is made of the energy of the lowest unoccupied molecular orbital. The various

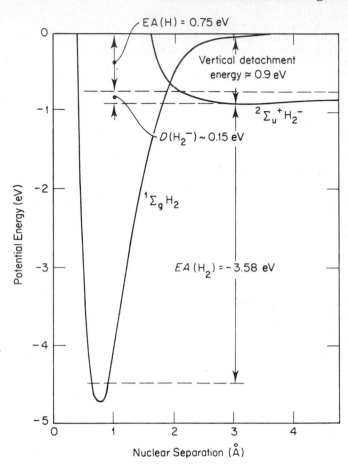

**Figure 7.16** Potential energy curves for $H_2$ and $H_2^-$ (Dalgarno and McDowell, 1956). Note that $EA(H_2) = -3.58$ eV, $EA(H) = 0.75$ eV, and $VDE(H_2) \simeq 0.9$ eV

calculations such as those by Pople (1953, 1957), Hush and Pople (1955), Hedges and Matsen (1958) and Hoyland and Goodman (1962) differ basically in the degree of refinement and the number of assumptions made. The results of the above calculations are included in Table 7.3. Also listed in Table 7.3 are the results of Ehrenson (1962) and Scott and Becker (1962) who used the so-called LCAO–MO $\omega$-technique (see Ehrenson, 1962).

### 7.2.2.2 Experimental methods

The following experimental methods provided information on the electron affinity of molecules.

A. *Photodetachment methods.* Electron affinity values for a number of diatomic and triatomic molecules and radicals obtained by this method have already been referred to in the preceding sections of this chapter. They are included in Table 7.3.

B. *Electron attachment-detachment methods.* The results of Phelps and coworkers and Christophorou and coworkers discussed in this chapter and in Chapter 6 are included in Table 7.3.

C. *The pulse-sampling method.* This is an equilibrium (swarm) method which grew out of attempts by Lovelock (see, for example, Lovelock, 1961a, b; 1963) to develop a detector for gas chromatography. The technique has been subsequently refined and applied to the determination of the electron affinity of complex molecules, especially aromatic hydrocarbons (Wentworth and Becker, 1962, 1964; Becker and Wentworth, 1963; Becker and Chen, 1966; Wentworth, Chen and Lovelock, 1966). Although the original work has been criticized by Stockdale, Hurst and Christophorou (1964a, b), the technique seems to have reached since then a reasonable degree of physical foundation (see Wentworth, Chen and Lovelock, 1966). Electron affinities obtained by this method are listed in Table 7.3. A discussion of the electron affinity values obtained by Wentworth, Chen and Lovelock (1966) for a number of aromatic hydrocarbons has been given by Chaudhuri, Jagur-Grodzinski and Szwarc (1967), who also made a comparison of the electron affinities of aromatic hydrocarbons in the gas phase with those they have obtained in solution (see also Lyons, Morris and Warren, 1968a, b).

D. *The 'magnetron' method.* This is an equilibrium surface ionization method, developed by Page (1960). It is based on the original method of Mayer and coworkers and is credited for a large body of information on the electron affinity of molecules and radicals (Table 7.3). The experimental arrangement consists of a coaxial filament-grid-anode assembly mounted in a solenoid coil. The thermionic electrons from the axial filament are collected at the anode in the absence of a magnetic field. If, however, a magnetic field is applied so as to constrain the electrons into spiral paths between the filament and the collector plate, the electron current that reaches the anode plate is negligible. When an electron acceptor is introduced into the system (usually at a low pressure ($\sim 10^{-3}$ torr)), both negative ions and electrons reach the anode in the absence of the magnetic field, but in the presence of a sufficiently high magnetic field the electrons are captured by the grid while the heavier negative ions are unaffected and pass through the grid to the anode. Hence, from the anode currents with and without an applied magnetic field it is possible to deduce the ratio of electrons to ions which leave the filament under the assumed equilibrium conditions. The equilibrium constant for negative ion formation can be evaluated if the rate at which the electrons and negative ions leave the filament is proportional to their concentration in the neighbourhood of the filament and if the corresponding rate at which the electron attaching molecules reach the filament is calculated from kinetic theory. The electron affinity can then be obtained from the equilibrium constant and the measured filament temperature. For resonance non-dissociative electron attachment, the apparent electron affinity at the mean filament temperature $T_f$ is corrected to 0°K by the relationship

$$EA(T_f) = (EA)_0 + 3/2kT_f \tag{7.32}$$

E. *Electron impact methods.*  Electron impact methods provided estimates for the electron affinity of a number of molecules. Curran (1961c) reported $EA(O_2) \geqslant 0.58$ eV from negative ion studies of $O_3$. This value is higher than that of 0.43 eV obtained by Pack and Phelps (1966a) in high pressure swarm experiments and may suggest that the two electron affinity values involve different ($v = 0$ and $v = 1$) vibrational states of $O_2^-$. Also, Curran (1961b) estimated $EA(CCl_3) \geqslant 2.1 \pm 0.35$ eV from an electron impact study of $CCl_3F$, while Curran (1962) estimated $EA(NO_2) \geqslant 3.82 \pm 0.06$ eV from similar studies on $NO_2$.

Electron scattering experiments on benzene, five benzene derivatives (see Table 7.3), and naphthalene (Compton and coworkers, 1968) provided lower limits to the electron affinity of these systems. It must be pointed out that these experiments yield the 'vertical attachment energy' (Christophorou and Compton, 1967) and not the electron affinity.

F. *Ion-beam methods.*  Khvostenko and Dukel'skii (1958) observed negative ions of mass two in $H_2O$–Sb vapour mixtures bombarded with 80 eV electrons, which they interpreted as $H_2^-$ ions. No $N_2^-$ ions have been observed by Fogel', Kozlov and Kalmykov (1959) in a beam of $N^+$ passing through a target gas.

G. *Determinations based on lattice energies.*  Molecular negative ions have been investigated by calculation of lattice energies. The relation (7.26) has been applied to derive electron affinities of molecules as in the case of atoms. Electron affinities for a number of molecules obtained from lattice energies can be found in Pritchard (1953), Moiseiwitsch (1965) and Table 7.3.

H. *Determinations based on charge-transfer (exchange) reactions.*  Charge-transfer reactions establish whether the electron affinity of a given species is greater or equal to that of another. Numerous such studies have been made. For example, Henglein and Muccini (1959) reported that $EA(SO_2) > EA(SO)$, $EA(SO_2) > EA(C_6H_5NO_2)$ and $EA(NO_2) > EA(O)$, while Kraus, Müller-Duysing and Neuert (1961) reported a number of other such inequalities (e.g. $EA(SO_2) \geqslant EA(CS)$, $EA(SO_2) \geqslant EA(NH_2)$, $EA(CS_2) \geqslant EA(NH_2)$, etc.). It should be noted that simple diatomic and triatomic negative ions which, owing to their short autoionization lifetimes, are not observed in single collision beam experiments may be observed in charge-transfer reactions. This comes about as a result of the electron-transfer process which allows low-lying levels of the ion to be reached so that its potential energy is below that of the ground vibrational level of the neutral molecule. The excess energy (difference in *EA* of acceptor and donor) probably appears as vibrational energy of both the donor and acceptor molecules.

I. *Determinations based on charge-transfer (donor–acceptor) complexes.*
The formation of a charge-transfer complex in solution resulting from an intermolecular interaction whereby an electron is partially or wholly transferred from one component (Donor (D)) of the complex to the other (Acceptor (A)) is usually accompanied by the appearance of a new, characteristic, and quite intense absorption band in the visible or the ultra-violet region of the optical spectrum. The theory of charge-transfer interaction between D and A (Mulliken, 1950, 1951, 1952a, b) relates the energy $(hv)_{CT}$ of the longest wavelength charge-transfer band (energy corresponding to the most probable electronic transition) to the ionization potential $(I)_D$ of D and the electron affinity $(EA)_A$ of A. Hastings and coworkers (1953) gave the expression

$$(hv)_{CT} = (I)_D - (EA)_A - \Delta + \left( \frac{\gamma}{(I)_D - (EA)_A - \Delta} \right) \qquad (7.33)$$

where $\Delta$ is predominantly a Coulomb attraction term (it also includes the sum of all other energy terms such as polarization energy), and $\gamma$ is approximately a constant quantity. The last two terms in Equation (7.33) account for the various interactions that alter the energies of the ground and excited states of the complex and hence the absorption frequency. However, when $\gamma$ is small or when $(I)_D$ is sufficiently large ($> 7.5$ eV), the last term in Equation (7.33) can be neglected; hence,

$$(hv)_{CT} = (I)_D - (EA)_A - \Delta \qquad (7.34)$$

Expression (7.34) has been confirmed in a series of complexes of the same acceptor with different donors (see, for example, McConnel, Ham and Platt (1953); Hastings and coworkers (1953); Briegleb and Czekalla (1959)) and for a series of complexes of the same donor with different acceptors (Förster, 1959) in solution. The experimental finding that $(hv)_{CT}$ varies linearly with $(I)_D$ (or $(EA)_A$) implies that $\Delta$ is a relatively insensitive function of D (or A). However, deviations do exist (McGlynn, 1960).

Electron affinities of molecular acceptors may be determined from spectroscopic data on charge-transfer complexes through Equation (7.34).

**Table 7.4** Simple relations between I, $EA$, $E_{CT}$, $^1L_{a,b}$ and $\chi$ for aromatic hydrocarbons[a]

| Relation | Explanation of symbols and remarks | Reference[b] |
|---|---|---|
| $I = a + bE_1$ (eV) <br> $EA = a + cE_1$ (eV) | These relations result from simple Hückel calculations. $a$, $b$ and $c$ are constants and $E_1$ is the energy of the first excited $\pi$-singlet state. | |
| $I = 4.39 + 0.857\ E_1$ (eV) | $E_1$ is the energy of the longest wavelength absorption maximum. The value of 4.39 corresponds to the ionization potential of graphite (Reference 3). The ionization potential of polycyclic aromatic hydrocarbons decreases with increasing molecular size and the ionization potential of graphite may be considered as the limiting value of the series. The other constant (0.857) was chosen to give the ionization potential of naphthalene (8.2 eV) at $E_1 = 4.53$ eV. | 1, 2 |
| $I = 4.39 + 1.15\ E_1'$ (eV) | $E_1' = \frac{1}{2}(E_{1T} + E_{1S})$ where $E_{1T}$ and $E_{1S}$ are the energies of the first excited $\pi$-triplet ($^3L_a$) and $\pi$-singlet ($^1L_b$) states, respectively. | 4 |
| $I = 5.156 + 0.775\ E_{1,0}$ (eV) | $E_{1,0}$ is the energy of the $0 \rightarrow 0$ transition of the first singlet absorption band. | 5 |
| $I = 5.11 + 0.701\ E_1$ (eV) | For selected hydrocarbons (see Reference 6). | 6 |
| $EA = 2.991 - 0.700\ E_1$ (eV) | $EA$ is the average of the various values for $EA$ reported for each molecule as listed in Reference 6. | 6 |
| $I + EA = $ constant <br> $(I + EA)/2 = \chi$ | Determined from self-consistent molecular orbital calculations with a number of approximations; $\chi$ is the electronegativity. | 7, 8 |
| $(I + EA)/2 = 4.17$ (eV) | From data on eight aromatic hydrocarbons. | 9 |
| $I = 5.13 + 1.72\ E_{CT}$ (eV) | $E_{CT}$ is the peak of the charge-transfer complex spectrum using chloranil as an electron acceptor (see other relationships in References 10 and 6). | 10 |

[a] See Table 7.3 for data on $EA$ and Appendix II for data on $I$. From these tabulations one can clearly see the restricted validity of the relationships listed in this table.
[b] References: (1) Matsen (1956); (2) Hedges and Matsen (1958); (3) Braun and Bush (1947); (4) Birks and Slifkin (1961); (5) Becker and Wentworth (1963); (6) Briegleb (1964); (7) Hush and Pople (1955); (8) Pople (1953, 1957); (9) Becker and Chen (1966); (10) Förster (1959).

Using one electron donor with various acceptors, we have

$$EA(A_i) - EA(A_j) = (h\nu_j)_{CT} - (h\nu_i)_{CT} - [\Delta_{iD} - \Delta_{jD}] \qquad (7.35)$$

where $i$ and $j$ refer to the $i$th and $j$th acceptor for a given donor. It is often assumed that the last term in (7.35) is equal to zero; this can be tested by determining $EA(A_i) - EA(A_j)$ for a series of donors. $(\Delta_{iD} - \Delta_{jD})$ must be constant with respect to changing the donor. However, it must be stressed that the numerous determinations of molecular electron affinities from charge-transfer spectra (see reviews by Briegleb (1961, 1964)) being functions of the particular D,A pair and the environment are often very approximate due to lack of sufficient corrections.

Finally, a number of empirical and semi-empirical simple relations have appeared in the literature relating the ionization potential, electron affinity, the energy of the charge-transfer complex, the energy of the first excited $\pi$-singlet sate, and the electronegativity of aromatic hydrocarbons. Such correlations are often based on limited data and thus are of restricted validity. However, a number of these simple relations have been collected in Table 7.4 since they indicate the interdependence of these basic molecular parameters.

## 7.3 References

D. A. Ansdell and F. M. Page (1962). *Trans. Faraday Soc.*, **58**, 1084.

T. L. Bailey (1958). *J. Chem. Phys.*, **28**, 792.

T. L. Bailey (1961). In *Proc. 2nd International Conference on the Physics of Electronic and Atomic Collisions*. University of Colorado Press, Boulder, Colorado, p. 54.

R. F. Baker and J. T. Tate (1938). *Phys. Rev.*, **53**, 683.

I. N. Bakulina and N. I. Ionov (1955). *Dokl. Akad. Nauk. SSSR*, **105**, 680.

I. N. Bakulina and N. I. Ionov (1957). *Soviet Phys.-Doklady*, **2**, 423.

I. N. Bakulina and N. I. Ionov (1959). *Russian J. Phys. Chem.*, **33**, 286.

I. N. Bakulina and N. I. Ionov (1964). *Soviet Phys.-Doklady*, **9**, 217.

J. N. Bardsley (1967a). In *5th International Conference on the Physics of Electronic and Atomic Collisions, USSR, Leningrad*, 1967, Publishing House 'Nauka', p. 340.

J. N. Bardsley (1967b). *Proc. Phys. Soc. (London)*, **91**, 300.

J. N. Bardsley, A. Herzenberg and F. Mandl (1966). *Proc. Phys. Soc. (London)*, **89**, 305, 321.

D. R. Bates (Ed.) (1962). *Atomic and Molecular Processes*, Academic Press, New York.

D. R. Bates and H. S. W. Massey (1943). *Phil. Trans.*, **A239**, 269.

D. R. Bates and B. L. Moiseiwitsch (1955). *Proc. Phys. Soc. (London)*, **A68**, 540.

M. Batley and L. E. Lyons (1962). *Nature*, **196**, 573.

R. S. Becker and E. Chen (1966). *J. Chem. Phys.*, **45**, 2403.

R. S. Becker and W. E. Wentworth (1963). *J. Am. Chem. Soc.*, **85**, 2210.

O. Bely and S. B. Schwartz (1969). *J. Phys. B (Proc. Phys. Soc., London)*, **2**, 159.

H. E. Bent (1930). *J. Am. Chem. Soc.*, **52**, 1498.

J. Berkowitz, W. A. Chupka and T. A. Walter (1969). *J. Chem. Phys.*, **50**, 1497.

R. S. Berry and C. W. David (1964). In M. R. C. McDowell (Ed.), *Proc. Third International Conference on the Physics of Electronic and Atomic Collisions, London, 1963*, North-Holland Publishing Co., Amsterdam, p. 543.

R. S. Berry and C. W. Reimann (1963). *J. Chem. Phys.*, **38**, 1540.

R. S. Berry, C. W. Reimann and G. N. Spokes (1961). *J. Chem. Phys.*, **35**, 2237.

R. S. Berry, C. W. Reimann and G. N. Spokes (1962). *J. Chem. Phys.*, **37**, 2278.

J. B. Birks and M. A. Slifkin (1961). *Nature*, **191**, 761.

N. E. Bradbury (1934). *J. Chem. Phys.*, **2**, 827.

L. M. Branscomb (1957). In L. Marton (Ed.), *Advances in Electronics and Electron Physics*, Vol. IX, Academic Press, New York, pp. 43–94.

L. M. Branscomb (1962). In D. R. Bates (Ed.), *Atomic and Molecular Processes*, Academic Press, New York, pp. 100–140.

L. M. Branscomb (1966). *Phys. Rev.*, **148**, 11.

L. M. Branscomb, D. S. Burch, S. J. Smith and S. Geltman (1958). *Phys. Rev.*, **111**, 504.

L. M. Branscomb and B. E. F. Pagel (1958). *Monthly Notices Roy. Astron. Soc.*, **118**, 258.

L. M. Branscomb and S. J. Smith (1955a). *Phys. Rev.*, **98**, 1028.

L. M. Branscomb and S. J. Smith (1955b). *Phys. Rev.*, **98**, 1127.

L. M. Branscomb and S. J. Smith (1956). *J. Chem. Phys.*, **25**, 598.

L. M. Branscomb and S. J. Smith (1958); quoted in Branscomb (1966).

A. Braun and G. Bush (1947). *Helv. Phys. Acta.*, **20**, 33.

B. Brehm, M. A. Gusinow and J. L. Hall (1967). *Phys. Rev. Letters*, **19**, 737.

G. Briegleb (1961). *Electronen-Donator-Acceptor Komplexe*, Springer-Verlag, Berlin.

G. Briegleb (1964). *Angew. Chem. Intern. Ed.*, **3**, 617.

G. Briegleb and J. Czekalla (1955). *Z. Elektrochem.*, **59**, 184.

G. Briegleb and J. Czekalla (1959). *Z. Electrochem.*, **63**, 6.

N. S. Buchel' nikova (1958). *Usp. Fiz. Nauk*, **65**, 351.

E. H. Buckner (1950). *Rec. Trav. Chem.*, **69**, 329.

D. S. Burch, S. J. Smith and L. M. Branscomb (1958). *Phys. Rev.*, **112**, 171; see also erratum, *Phys. Rev.*, **114**, 1652 (1959).

Yu. F. Bydin (1964). *Soviet Phys.—JETP*, **19**, 1091.

Yu. F. Bydin (1966). *Soviet Phys.—JETP*, **23**, 23.

Yu. F. Bydin (1967). *JETP Letters*, **6** (No. 9), 297.

S. Chandrasekhar (1958). *Astrophys. J.*, **128**, 114.

S. Chandrasekhar and D. D. Elbert (1958). *Astrophys. J.*, **128**, 633.

E. L. Chaney and L. G. Christophorou (1970). *Oak Ridge National Laboratory Report* ORNL-TM-2613.

E. L. Chaney, L. G. Christophorou, P. M. Collins and J. G. Carter (1970). *J. Chem. Phys.*, **52**, 4413.

L. M. Chanin, A. V. Phelps and M. A. Biondi (1962). *Phys. Rev.*, **128**, 219.

P. J. Chantry and G. J. Schulz (1964). *Phys. Rev. Letters*, **12**, 449.

J. Chaudhuri, J. Jagur-Grodzinski and M. Szwarc (1967). *J. Phys. Chem.*, **71**, 3063.

J. C. Y. Chen (1967). *Phys. Rev.*, **156**, 12.

J. C. Y. Chen and J. L. Peacher (1967). In *Proc. 5th International Conference on the Physics of Electronic and Atomic Collisions, Leningrad, USSR, 1967*, Publishing House 'Nauka', p. 335.

J. C. Y. Chen and J. L. Peacher (1968). *Phys. Rev.*, **168**, 56.

L. G. Christophorou and R. N. Compton (1967). *Health Phys.*, **13**, 1277.

E. Clementi (1964). *Phys. Rev.*, **135**, A980.

E. Clementi and A. D. McLean (1964). *Phys. Rev.*, **133**, A419.

E. Clementi, A. D. McLean, D. L. Raimondi and M. Yoshimine (1964). *Phys. Rev.*, **133**, A1274.

P. M. Collins, L. G. Christophorou, E. L. Chaney and J. G. Carter (1970). *Chem. Phys. Letters*, **4**, 646.

R. N. Compton and L. Bouby (1967). *C. R. Acad. Sci., Paris*, **264**, 1153.

R. N. Compton, L. G. Christophorou and R. H. Huebner (1966). *Phys. Letters*, **23**, 656.

R. N. Compton, L. G. Christophorou, G. S. Hurst and P. W. Reinhardt (1966). *J. Chem. Phys.*, **45**, 4634.

R. N. Compton, R. H. Huebner, P. W. Reinhardt and L. G. Christophorou (1968). *J. Chem. Phys.*, **48**, 901.

J. W. Cooper and S. T. Manson (1969). *Phys. Rev.*, **177**, 157.

J. Cooper and R. N. Zare (1968). *J. Chem. Phys.*, **48**, 942.

J. D. Craggs and B. A. Tozer (1960). *Proc. Roy. Soc. (London)*, **A254**, 229.

R. J. S. Crossley (1964). *Proc. Phys. Soc. (London)*, **83**, 375.

D. Cubicciotti (1959). *J. Chem. Phys.*, **31**, 1646.

R. K. Curran (1961a). *J. Chem. Phys.*, **34**, 1069.

R. K. Curran (1961b). *J. Chem. Phys.*, **34**, 2007.

R. K. Curran (1961c). *J. Chem. Phys.*, **35**, 1849.

R. K. Curran (1962). *Phys. Rev.*, **125**, 910.

A. Dalgarno and J. C. Browne (1967). *Astrophys. J.*, **149**, 231.

A. Dalgarno and M. R. C. McDowell (1956). *Proc. Phys. Soc. (London)*, **A69**, 615.

D. F. Dance, M. F. A. Harrison and R. D. Rundel (1967). *Proc. Roy. Soc. (London)*, **299**, 525.

Yu. N. Demkov (1964). *Soviet Phys.—JETP*, **19**, 762.

J. G. Dillard and J. L. Franklin (1968). *J. Chem. Phys.*, **48**, 2349, 2353.

P. M. Doty and J. E. Mayer (1944). *J. Chem. Phys.*, **12**, 323.

J. W. Edie and F. Rohrlich (1962). *J. Chem. Phys.*, **36**, 623.

B. Edlén (1960). *J. Chem. Phys.*, **33**, 98.

S. Ehrenson (1962). *J. Phys. Chem.*, **66**, 706, 712.

F. A. Elder, D. Villarejo and M. G. Inghram (1965). *J. Chem. Phys.*, **43**, 758.

M. G. Evans and N. Uri (1949). *Trans. Faraday Soc.*, **45**, 224.

M. G. Evans, N. S. Hush and N. Uri (1952). *Quart. Revs.*, **6**, 186.

H. Eyring, J. O. Hirschfelder and H. S. Taylor (1936). *J. Chem. Phys.*, **4**, 479.

A. L. Farragher and F. M. Page (1966). *Trans. Faraday Soc.*, **62**, 3072.

A. L. Farragher and F. M. Page (1967). *Trans. Faraday Soc.*, **63**, 2369.

A. L. Farragher, F. M. Page and R. C. Wheeler (1964). *Discussions Faraday Soc.*, **37**, 203.

F. C. Fehsenfeld, E. E. Ferguson and A. L. Schmeltekopf (1966). *J. Chem. Phys.*, **45**, 1844.

F. C. Fehsenfeld, A. L. Schmeltekopf, H. I. Schiff and E. E. Ferguson (1967). *Planetary Space Sci.*, **15**, 373.

E. E. Ferguson (1968). In L. Marton (Ed.), *Advances in Electronics and Electron Physics*, Vol. 24, Academic Press, NewYork, pp. 1–50.

F. H. Field and J. L. Franklin (1957). *Electron Impact Phenomena*, Academic Press, New York.

M. A. Fineman and A. W. Petrocelli (1958). *Bull. Am. Phys. Soc.*, **3**, 258.

M. A. Fineman and A. W. Petrocelli (1962). *J. Chem. Phys.*, **36**, 25.

S. N. Flengas and E. Rideal (1956). *Proc. Roy. Soc.*, **A233**, 443.

Ya. M. Fogel', V. F. Kozlov and A. A. Kalmykov (1959). *Soviet Phys.—JETP*, **9**, 963.

R. Förster (1959). *Nature*, **183**, 1253.

R. E. Fox (1960). *J. Chem. Phys.*, **32**, 285.

G. Freeman (1968). *Rad. Res. Revs.*, **1**, 1.

A. F. Gaines and F. M. Page (1963). *Trans. Faraday Soc.*, **59**, 1266.

A. F. Gaines and F. M. Page (1966). *Trans. Faraday Soc.*, **62**, 3086.

A. F. Gaines, J. Kay and F. M. Page (1966). *Trans. Faraday Soc.*, **62**, 874.

R. Gáspár and B. Molnár (1955). *Acta Phys. Acad. Sci. Hung.*, **5**, 75.

W. Geiger (1955). *Z. Phys.*, **140**, 608.

S. Geltman (1956). *J. Chem. Phys.*, **25**, 782.

S. Geltman (1958). *Phys. Rev.*, **112**, 176.

S. Geltman (1960). *Proc. Phys. Soc. (London)*, **75**, 67.

S. Geltman and M. Krauss (1960). *Bull. Am. Phys., Soc.*, **5**, 339.

G. Glockler (1934). *Phys. Rev.*, **46**, 111.

G. Glockler and M. Calvin (1935). *J. Chem. Phys.*, **3**, 771.

G. Glockler and M. Calvin (1936). *J. Chem. Phys.*, **4**, 492.

S. Golub and B. Steiner (1968). *J. Chem. Phys.*, **49**, 5191.

P. Gombás and K. Ladànyi (1960). *Z. Physik*, **158**, 261.

P. Gray and T. C. Waddington (1956). *Proc. Roy. Soc.*, **A235**, 481.

F. Gutmann and L. E. Lyons (1967). *Organic Semiconductors*, John Wiley and Sons, New York.

H. D. Hagstrum (1951). *Rev. Mod. Phys.*, **23**, 185.

H. D. Hagstrum (1955). *J. Chem. Phys.*, **23**, 1178.

J. L. Hall and M. W. Siegel (1968). *J. Chem. Phys.*, **48**, 943.

J. L. Hall, E. J. Robinson and L. M. Branscomb (1965). *Phys. Rev. Letters*, **14**, 1013.

J. B. Hasted (1964). *Physics of Atomic Collisions*. Butterworths, London.

S. H. Hastings, J. L. Franklin, J. C. Schiller and F. A. Matsen (1953). *J. Am. Chem. Soc.*, **75**, 2900.

R. M. Hedges and F. A. Matsen (1958). *J. Chem. Phys.*, **28**, 950.

A. Henglein and G. A. Muccini (1959). *J. Chem. Phys.*, **31**, 1426.

A. Henglein and G. A. Muccini (1961). In *Chemical Effects of Nuclear Transformations*, Vol. I, Ed. by International Atomic Energy Agency, Vienna.

H. Henning (1962). *Z. Physik*, **169**, 467.

J. T. Herron and V. H. Dibeler (1960). *J. Am. Chem. Soc.*, **82**, 1555.

A. Herzenberg (1967). *Phys. Rev.*, **160**, 80.

E. Holøien and S. Geltman (1967). *Phys. Rev.*, **153**, 81.

E. Holøien and J. Midtdal (1955). *Proc. Phys. Soc. (London)*, **A 68**, 815.

R. E. Honig (1954). *J. Chem. Phys.*, **22**, 126.

J. R. Hoyland and L. Goodman (1962). *J. Chem. Phys.*, **36**, 12, 21.

D. G. Hummer, R. F. Stebbings, W. L. Fite and L. M. Branscomb (1960). *Phys. Rev.*, **119**, 668.

N. S. Hush and J. Oldham (1955). Quoted in Hush and Pople (1955).

N. S. Hush and J. A. Pople (1955). *Trans. Faraday Soc.*, **51**, 600.

E. A. Hylleraas and J. Midtdal (1956). *Phys. Rev.*, **103**, 829.

K. Jäger and A. Henglein (1966). *Z. Naturforsch.*, **21a**, 1251.

D. E. Jensen (1969). *Trans. Faraday Soc.*, **65**, 2123.

T. L. John (1960). *Astrophys. J.*, **131**, 743.

H. R. Johnson and F. Rohrlich (1959). *J. Chem. Phys.*, **30**, 1608.

J. Jortner and U. Sokolov (1961). *Nature*, **190**, 1003.

J. Jortner, G. Stein and A. Treinin (1959). *J. Chem. Phys.*, **30**, 1110.

M. Kaufman (1963). *Astrophys. J.*, **137**, 1296.

J. Kay and F. M. Page (1964). *Trans. Faraday Soc.*, **60**, 1042.

J. Kay and F. M. Page (1966). *Trans. Faraday Soc.*, **62**, 3081.

V. I. Khvostenko and V. M. Dukel'skii (1958). *Soviet Phys.—JETP*, **7**, 709.

V. I. Khvostenko and V. M. Dukel'skii (1960). *Soviet Phys.—JETP*, **10**, 465.

K. Kraus (1961). *Z. Naturforsch.*, **16a**, 1378.

K. Kraus (1966). *Ann. Physik*, **18**, 288.

K. Kraus, W. Müller-Duysing and H. Neuert (1961). *Z. Naturforsch.*, **16a**, 1385.

T. L. Kunii and H. Kuroda (1968). *Theoret. Chim. Acta (Berlin)*, **11**, 97.

G. B. Lopantseva and O. B. Firsov (1966). *Soviet Phys.—JETP*, **23**, 648.

J. E. Lovelock (1961a). *Nature*, **189**, 729.

J. E. Lovelock (1961b). *Anal. Chem.*, **33**, 162.

J. E. Lovelock (1963). *Anal. Chem.*, **35**, 474.

L. E. Lyons (1950). *Nature*, **166**, 193.

L. E. Lyons, G. C. Morris and L. J. Warren (1968a). *Australian J. Chem.*, **21**, 853.

L. E. Lyons, G. C. Morris and L. J. Warren (1968b). *J. Phys. Chem.*, **72**, 3677.

J. D. Martin and T. L. Bailey (1968). *J. Chem. Phys.*, **49**, 1977.

H. S. W. Massey (1950). *Negative Ions*, Cambridge University Press, London.

H. S. W. Massey (1949). *Rept. Progr. Phys.*, **12**, 248.

F. A. Matsen (1956). *J. Chem. Phys.*, **24**, 602.

F. A. Matsen, W. W. Robertson and R. L. Chuoke (1947). *Chem. Revs.*, **41**, 273.

K. J. McCallum and J. E. Mayer (1943). *J. Chem. Phys.*, **11**, 56.

H. McConnell, J. S. Ham and J. P. Platt (1953). *J. Chem. Phys.*, **21**, 66.

E. W. McDaniel (1964). *Collision Phenomena in Ionized Gases.* John Wiley and Sons, New York.

M. R. C. McDowell and G. Peach (1959). *Proc. Phys. Soc. (London)*, A**74**, 463.

M. R. C. McDowell and J. H. Williamson (1963). *Phys. Letters*, **4**, 159.

S. P. McGlynn (1960). *Rad. Res. Suppl.*, **2**, 300.

J. Michl (1969). *J. Molec. Spectry.*, **30**, 66.

J. J. Mitchell and J. E. Mayer (1940). *J. Chem. Phys.*, **8**, 282.

B. L. Moiseiwitsch (1965). In D. R. Bates and I. Estermann (Eds.), *Advances in Atomic and Molecular Physics*, Vol. I, Academic Press, New York, p. 61.

J. D. Morrison, H. Hurzeler, M. G. Inghram and H. E. Stanton (1960). *J. Chem. Phys.*, **33**, 821.

J. L. Moruzzi, J. W. Ekin, Jr. and A. V. Phelps (1968). *J. Chem. Phys.*, **48**, 3070.

J. L. Moruzzi and A. V. Phelps (1966). *J. Chem. Phys.*, **45**, 4617.

R. S. Mulliken (1950). *J. Am. Chem. Soc.*, **72**, 600.

R. S. Mulliken (1951). *J. Chem. Phys.*, **19**, 514.

R. S. Mulliken (1952a). *J. Am. Chem. Soc.*, **74**, 811.

R. S. Mulliken (1952b). *J. Phys. Chem.* **56**, 801.

E. E. Muschlitz Jr. (1957). *J. Appl. Phys.*, **28**, 1414.

R. Napper and F. M. Page (1963). *Trans. Faraday Soc.*, **59**, 1086.

G. P. Nikoeskii, L. I. Kazarnovskaya, Z. A. Bogdasasyan and I. A. Kazarnovskii (1950). *Dokl. Akad. Nauk. SSSR*, **72**, 713.

J. L. Pack and A. V. Phelps (1966a). *J. Chem. Phys.*, **44**, 1870.

J. L. Pack and A. V. Phelps (1966b). *J. Chem. Phys.*, **45**, 4316.

F. M. Page (1958). *Rev. Inst. Franc. Petrole*, **15**, 112.

F. M. Page (1960). *Trans. Faraday Soc.*, **56**, 1742.

F. M. Page (1961a). *Trans. Faraday Soc.*, **57**, 359.

F. M. Page (1961b). *Trans. Faraday Soc.*, **57**, 1254.

F. M. Page (1962). *Chem. Abstr.*, **57**, 11952a.

Γ. M. Page (1968). *J. Chem. Phys.*, **49**, 2466.

F. M. Page and G. C. Goode (1969). *Negative Ions and the Magnetron*, Wiley-Interscience, London.

F. M. Page and J. Kay (1963). *Nature*, **199**, 483.

F. M. Page and T. M. Sugden (1957). *Trans. Faraday Soc.*, **53**, 1092.

B. E. J. Pagel (1956). *Monthly Notices Roy. Astron. Soc.*, **116**, 608.

B. E. J. Pagel (1959). *Monthly Notices Roy. Astron. Soc.*, **119**, 609.

J. F. Paulson (1966). *Advan. Chem. Ser.*, **58**, 28 (American Chemical Society, Washington, D. C., 1966).

C. L. Pekeris (1958). *Phys. Rev.*, **112**, 1649.

C. L. Pekeris (1962). *Phys. Rev.*, **126**, 1470.

A. V. Phelps and J. L. Pack (1961). *Phys. Rev. Letters*, **6**, 111.

H. R. Phillip (1958). *Phys. Rev.*, **111**, 440.

J. A. Pople (1953). *Trans. Faraday Soc.*, **49**, 1375.

J. A. Pople (1957). *J. Phys. Chem.*, **61**, 6.

H. P. Popp (1965). *Z. Naturforsch.*, **20a**, 642.

H. O. Pritchard (1953). *Chem. Revs.*, **52**, 529.

A. C. Riviere and D. R. Sweetman (1960). *Phys. Rev., Letters*, **5**, 560.

E. J. Robinson and S. Geltman (1967). *Phys. Rev.*, **153**, 4.

M. Rogalski (1966). *Acta Phys. Polon.*, **29**, 15.

F. Rohrlich (1956). *Phys. Rev.*, **101**, 69.

D. E. Rothe (1969). *Phys. Rev.*, **177**, 93.

O. Rosenbaum and H. Neuert (1954). *Z. Naturforsch.*, **9a**, 990.

M. R. H. Rudge (1964). *Proc. Phys. Soc.* (*London*), **83**, 1.

R. D. Rundel, M. F. A. Harrison and D. F. Dance (1967). In *Proc. 5th International Conference on the Physics of Atomic and Electronic Collisions, Leningrad, USSR, 1967*, Publishing House 'Nauka', p. 36.

H. F. Schaefer, III and F. E. Harris (1968). *Phys. Rev.*, **170**, 108.

M. D. Scheer and J. Fine (1969). *J. Chem. Phys.*, **50**, 4343.

G. J. Schulz (1961). *J. Chem. Phys.*, **34**, 1778.

G. J. Schulz (1962). *Phys. Rev.*, **128**, 178.

G. J. Schulz and D. Spence (1969). *Phys. Rev. Letters*, **22**, 47.

D. R. Scott and R. S. Becker (1962). *J. Phys. Chem.*, **66**, 2713.

M. L. Seman and L. M. Branscomb (1962). *Phys. Rev.*, **125**, 1602.

H. A. Skinner (1947). *Nature*, **160**, 716.

A. L. Sklar (1939). *J. Chem. Phys.*, **7**, 984.

M. A. Slifkin (1963). *Nature*, **200**, 877.

B. M. Smirnov (1965). *High Temperature*, **3**, 716.

B. M. Smirnov and M. I. Chibisov (1966). *Soviet Phys.—JETP*, **22**, 585.

B. M. Smirnov and O. B. Firsov (1965). *Soviet Phys.—JETP*, **20**, 156.

L. G. Smith (1937). *Phys. Rev.*, **51**, 263.

S. J. Smith (1960). In N. R. Nilsson (Ed.), *Proc. Fourth International Conference on Ionization Phenomena in Gases, Uppsala*, 1959. Vol. I, North-Holland Publishing Co., Amsterdam, p. 219.

S. J. Smith and L. M. Branscomb (1955a). *J. Res. Nat. Bur. Stand.*, **55**, 165.

S. J. Smith and L. M. Branscomb (1955b). *Phys. Rev.*, **99**, A1657.

S. J. Smith and L. M. Branscomb (1960). *Rev. Sci. Instr.*, **31**, 733.

S. J. Smith and D. S. Burch (1959a). *Phys. Rev. Letters*, **2**, 165.

S. J. Smith and D. S. Burch (1959b). *Phys. Rev.*, **116**, 1125.

Smith and Seman (1962). In H. Maecker (Ed.), quoted in Branscomb in *Proc. Fifth International Conference on Ionization Phenomena in Gases, Munich*, 1961, Vol. I, North-Holland Publishing Co., Amsterdam, p. 1.

W. R. Snow, R. D. Rundel and R. Geballe (1969). *Phys. Rev.*, **178**, 228.

J. G. Stamper and R. F. Barrow (1958). *Trans. Faraday Soc.*, **54**, 1592.

B. Steiner (1968). *Phys. Rev.*, **173**, 136.

B. Steiner (1968). *J. Chem. Phys.*, **49**, 5097.

B. Steiner, M. L. Seman and L. M. Branscomb (1962). *J. Chem. Phys.*, **37**, 1200.

B. Steiner, M. L. Seman and L. M. Branscomb (1964). In M. R. C. McDowell (Ed.), *Atomic Collision Processes*, North-Holland Publishing Co., Amsterdam, p. 537.

J. A. Stockdale, R. N. Compton and P. W. Reinhardt (1968). *Phys. Rev. Letters*, **21**, 664.

J. A. Stockdale, G. S. Hurst and L. G. Christophorou (1964a). *Nature*, **202**, 459.

J. A. Stockdale, G. S. Hurst and L. G. Christophorou (1964b). *Nature*, **203**, 1270.

N. Sukhum, A. N. Prasad and J. D. Craggs (1967). *Brit. J. Appl. Phys.*, **18**, 785.

P. P. Sutton and J. E. Mayer (1934). *J. Chem. Phys.*, **2**, 145.

P. P. Sutton and J. E. Mayer (1935). *J. Chem. Phys.*, **3**, 20.

E. Swift, Jr. (1938). *J. Am. Chem. Soc.*, **60**, 1403.

S. P. Tandon, M. P. Bhutra and K. Tandon (1966). *Indian J. Phys.*, **40**, 49.

S. P. Tandon, M. P. Bhutra and K. Tandon (1967). *Indian J. Phys.*, **41**, 70.

G. C. Tisone and L. M. Branscomb (1966). *Phys. Rev. Letters*, **17**, 236.

G. C. Tisone and L. M. Branscomb (1967). In *Proc. 5th International Conference on the Physics of Atomic and Electronic Collisions, Leningrad, USSR, 1967*, Publishing House 'Nauka', p. 39.

G. C. Tisone and L. M. Branscomb (1968). *Phys. Rev.*, **170**, 169.

L. Von Trepka (1963). *Z. Naturforsch.*, **18a**, 1122.

S. Tsuda and W. H. Hamill (1966). *Advan. Mass Spect.*, **3**, 249, W. L. Mead (Ed.), the Institute of Petroleum, London.

V. I. Vedeneyev, L. V. Gurvich, V. N. Kondrat'yev, V. A. Medvedev and Ye. L. Frankevich (1966). *Bond Energies, Ionization Potentials and Electron Affinities*, St. Martin's Press, New York, English translation.

D. T. Vier and J. E. Mayer (1944). *J. Chem. Phys.*, **12**, 28.

D. Vogt and H. Neuert (1967). *Z. Physik*, **199**, 82.

R. H. Vought (1947). *Phys. Rev.*, **71**, 93.

P. Warneck (1969). *Chem. Phys. Letters*, **3**, 532.

J. D. Weisner and B. H. Armstrong (1964). *Proc. Phys. Soc.* (*London*), **83**, 31.

A. W. Weiss (1961). *Phys. Rev.*, **122**, 1826.

A. W. Weiss (1968). *Phys. Rev.*, **166**, 70.

J. Weiss (1935). *Trans. Faraday Soc.*, **31**, 966.

J. Weiss (1947a). *Trans. Faraday Soc.*, **43**, 119.

J. Weiss (1947b). *Trans. Faraday Soc.*, **43**, 173.

W. E. Wentworth and R. S. Becker (1962). *J. Am. Chem. Soc.*, **84**, 4263.

W. E. Wentworth and R. S. Becker (1964). *Nature*, **203**, 1268.

W. E. Wentworth, E. Chen and J. E. Lovelock (1966). *J. Phys. Chem.*, **70**, 445.

W. E. Wentworth and E. Chen (1967). *J. Phys. Chem.*, **71**, 1929.

W. E. Wentworth, E. Chen and J. C. Steelhammer (1968). *J. Phys. Chem.*, **72**, 2671.

C. D. West (1935). *J. Phys. Chem.*, **39**, 493.

E. P. Wigner (1948). *Phys. Rev.*, **73**, 1002.

R. Wildt (1939). *Astrophys. J.*, **89**, 295.

D. Wobschall, J. R. Graham, Jr. and D. P. Malone (1965). *J. Chem. Phys.*, **42**, 3955.

B. Ya'akobi (1966). *Phys. Letters*, **23**, 655.

K. B. Yatsimirskii (1947). Quoted in Moiseiwitsch (1965).

E. Ya. Zandberg and N. I. Ionov (1959). *Soviet Phys. Uspekhi*, **67** (2), 255 (Usp. Fiz. Nauk, **57**, 581 (1959)).

# 8 Heavy neutral and charged particle interactions

## 8.1 Introduction

Early experimental and theoretical studies on heavy charged particle interactions were principally devoted to the examination of the range of positive ions in matter and the ionization produced as these are slowed down and are absorbed in matter. Certain aspects of such studies have been discussed in the first and second chapters in connection with energy-loss mechanisms, stopping power calculations, and total ionization produced in gases by the complete absorption of high- and low-energy positive ions. In recent years, studies of charged and neutral particle interactions have widened in scope covering areas such as excitation, dissociation, charge-transfer and ion–molecule reactions. Many of these studies came about largely because of the importance of heavy particle collision processes in controlled thermonuclear research and space exploration. Undoubtedly, basic problems in radiation sciences such as radiation damage, radiation detection, and fundamental processes accompanying the interaction of ionizing radiation with matter pressed for advances in the field.

In Chapter 3 certain photophysical processes which involve positive ions and metastable species, such as in the Penning ionization, were discussed. An expression for the excitation and ionization cross section for positive ions in the Bethe–Born approximation was given in Chapter 5. Certain negative ion–molecule reactions were elaborated upon in Chapter 6, whereas in Chapter 7 attention was focused on the processes of associative and collisional detachment from negative ions. In this chapter an outline is given of some other types of physical processes involving interactions of charged and neutral particles. No attempt will be made to review the voluminous literature on heavy-particle collisions. There are excellent books and review articles to which the reader is referred for a detailed discussion of the enormous growth in the field, and for a description of the many and imaginative experimental techniques which have come into use in recent years (e.g. Bates, 1962b; Biondi, 1963; McDaniel, 1964; Hasted, 1964; McDowell, 1966; Stebbings, 1966; Giese, 1966; Dalgarno, 1967; Bates and Estermann, 1968; Ferguson, 1968). It might be mentioned that measurements of the angular, charge, and energy distribution of the products of heavy-particle collisions yielded information on the kinetics of the collision processes, while the identification of the states of the products has been investigated with optical and allied techniques.

Low-energy (thermal to a few eV) ion–neutral reactions are studied principally by flowing afterglow methods, mass spectrometers, drift tubes, and ion cyclotron resonance techniques. At higher energies such reactions are predominantly studied by ion-beam methods. Although no discussion of experimental methods is made in this chapter, the principle of the so-called merging beams technique, applied to the study of low-energy ion–neutral (ion) collisions, will be outlined. In this new technique, developed by Trujillo, Neynaber and Rothe (1966) (see detailed references in the review by Neynaber (1969)), two beams of particles are caused to travel in the same direction along a common axis. The laboratory energies of the two particle beams are in the keV range, but when they are made nearly equal the resulting interaction energy, $W$, between the particles in the two beams in the centre-of-mass system becomes very small. $W$ is given by

$$W = \tfrac{1}{2}M_r(v_2 - v_1)^2$$

or

$$W = M_r\left[\left(\frac{E_2}{m_2}\right)^{\frac{1}{2}} - \left(\frac{E_1}{m_1}\right)^{\frac{1}{2}}\right]^2 \tag{8.1}$$

where $m_2$ and $m_1$ are the respective particle masses, $M_r$ is the reduced mass, and $v_2, v_1$ and $E_2, E_1$ are, respectively, the corresponding laboratory particle velocities and kinetic energies. For identical particles ($m_1 = m_2 = m$),

$$W = \tfrac{1}{2}(E_2^{\frac{1}{2}} - E_1^{\frac{1}{2}})^2 \tag{8.2}$$

and when the energy difference, $\Delta E(= E_2 - E_1)$, between the two beams is small compared to the beam energies themselves,

$$W \simeq \frac{\Delta E^2}{8E} \tag{8.3}$$

where $E \simeq E_2 \simeq E_1$.

Trujillo, Neynaber and Rothe (1966) defined an energy deamplification factor, $D$, as $D \equiv \Delta E / W$. Hence, for small $\Delta E$,

$$D = \frac{\Delta E}{W} \simeq \frac{8E}{\Delta E} \gg 1 \tag{8.4}$$

From Equation (8.3) it is seen that the interaction energy $W$ is small compared to either $E_1$ or $E_2$ or $\Delta E$. Taking $E_1 = 5000$ eV and $E_2 = 5100$ eV, for example, one obtains $W \simeq 0.25$ eV. Also, it can be shown that

$$\frac{\delta W}{W} \simeq \frac{2\delta E}{WD} = \frac{2\delta E}{\Delta E} \tag{8.5}$$

where $\delta E$ is the variation in $\Delta E$, and $\delta W$ is the variation in $W$. Equation (8.5), then, shows that from the variation $\delta E$ in $\Delta E$, one can determine approximately the variation $\delta W$ in $W$; for $\delta E = \pm 1.5$ eV, through Equation (8.5) one finds $\delta W = \pm 0.0075$ eV. Thus, the technique allows investigation of reactions at very low interaction energies whilst the beam energies themselves are kept conveniently high. The technique is suitable for the study of low-energy, two-body reactions with either of the beams being neutral or charged (see experimental details in Trujillo, Neynaber and Rothe (1966) and Neynaber (1969)).

In Table 8.1 a number of interaction processes involving atomic positive ions (with the exception of process (vi) which is for positive molecular ions) are enumerated. These can be conveniently divided into three broad classes: (i) excitation, ionization and stripping reactions, (ii) positive ion–electron and positive ion–negative ion recombination reactions, (iii) charge-transfer (exchange) reactions. Each of the above groups of processes will be elaborated upon briefly. The cross section for a given process is often written as $_{ij}\sigma_{mn}$, where the indices $ij$ and $mn$ refer to the charge states of the particles before and after the collision (Hasted, 1962). It is evident from Table 8.1 that even for the simple case of an atomic positive ion–neutral atom collision, many types of inelastic processes of the general form

$$A^+ + B \rightarrow A^{m+} + B^{n+} + (m+n-1)e \tag{8.6}$$

can ensue. A satisfactory understanding of heavy particle collision processes necessitates quantum calculations.

## 8.2 Excitation, Ionization and Stripping Reactions

Fast heavy positive ions passing through matter dissipate a large fraction of their energy in excitation and ionization. As we have discussed in previous chapters, measurement of the cross sections for these inelastic processes as a function of the particle energy is essential for a quantitative understanding of the fundamental processes through which fast positive ions slow-down and affect matter. Excitation cross section functions for optically allowed

**Table 8.1** Reactions involving positive ions[a,b,c]

| Process | Reaction process | Description |
|---------|------------------|-------------|
| Excitation, ionization and stripping reactions | | |
| (i) | $A^{+*}+B\rightarrow A^{+}+B^{*}$ | Excitation transfer |
| (ii) | $A^{+}+B\rightarrow A^{+}+B^{+}+e$ | Ionization (electron ejection from the target) |
| (iii) | $A^{+}+B\rightarrow A^{2+}+e+B$ | Stripping (electron ejection from the projectile) |
| Positive ion–electron and positive ion–negative ion recombination processes | | |
| (iv) | $A^{+}+e\rightarrow A^{(*)}+h\nu$ | Radiative recombination |
| (v) | $A^{+}+e\rightleftarrows A^{**}\rightarrow A^{(*)}+h\nu$ | Dielectronic recombination |
| (vi) | $AB^{+}+e\rightarrow A^{(*)}+B^{(*)}$ | Dissociative recombination |
| (vii) | $A^{+}+e+B\rightarrow A^{(*)}+B$ | Three-body recombination |
| (viii) | $A^{+}+e+e\rightleftarrows A^{*}+e$ | Electron stabilized recombination |
| (ix) | $A^{+}+B^{-}\rightarrow A^{*}+B^{*}(AB+h\nu)$ | Positive ion-negative ion neutralization |
| (x) | $A^{+}+B^{-}+C\rightarrow A+B+C(AB+C)$ | Three-body ion recombination |
| Charge transfer (exchange) reactions | | |
| (xi) | $A^{+}+B\rightarrow A+B^{+}$ | Single charge transfer (asymmetrical reaction) |
| | $A^{+}+A\rightarrow A+A^{+}$ | Single charge transfer (symmetrical resonance)[d] |
| (xii) | $A^{+}+B\rightarrow A^{-}+B^{2+}$ | Double charge transfer |
| (xiii) | $A^{+}+B\rightarrow A+B^{2+}+e$ | Transfer ionization |
| (xiv) | $A^{+}+B\rightarrow A^{*}+B^{+}(A+B^{+*})$ | Charge transfer to an excited level |

[a] Charge-transfer processes involving negative ions and negative ion–molecule reactions are discussed in Chapters 6 and 7. As usual, the asterisk* indicates that the system is left with excitation energy, and (*) indicates that the system may or may not be left with excitation energy.

[b] Other processes such as rotational and vibrational excitation as well as dissociation may ensue if either of the structures is molecular. Additionally, reactions chemical in nature, known as ion–molecule reactions, may occur. In such reactions a heavy particle may be transferred from one structure to the other or the two structures may combine into a single stable system.

[c] One should also note other processes involving interactions between neutral particles such as $A+B\rightarrow AB+h\nu$ (radiative association), $A+B\rightarrow AB^{+}+e$ (associative ionization), and $A+B\rightarrow A^{-}+B^{+}$ (capture by a neutral atom).

[d] There may be cases where the system that loses an electron and that which gains the electron are different, but the energy balance is exact or close, e.g.,

$$N_{2}(X^{1}\Sigma_{g}^{+}v=0)+O^{+}(^{2}D)\rightarrow N_{2}^{+}(A^{2}\Pi_{u}v=1)+O(^{3}P)+0.0\text{ eV (Bates, 1962b)}.$$

This is called asymmetrical (or accidental) resonance charge transfer.

transitions in heavy-particle collisions are studied in the conventional fashion by observing the emitted radiation which accompanies the radiative de-excitation of the excited target. One should note, however, that like other inelastic processes, excitation by heavy-particle impacts can be caused in many different ways. Further, the projectile itself may be excited in a collision with or without simultaneous excitation of the target and the excitation process may be accompanied by ionization. Measurements of the intensity of specific emission lines or bands as a function of incident energy have been made for a number of targets and atomic projectiles (see references cited in Section 8.1 and discussions in Chapters 1, 3 and 5).

The cross section functions for excitation and ionization by atomic particles behave in the usual manner; they rise from threshold with increasing impact energy to a maximum and decrease thereafter slowly and monotonically with increasing impact energy. In Figures 8.1 and 8.2 typical cross

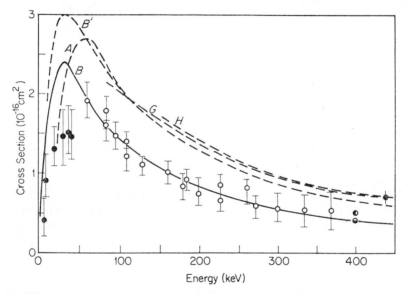

H : (B) Calculated values (Born approximation) by Bates and Griffing (1953).
● Fite and coworkers (1960).  ○ Gilbody and Ireland (1964).
$e$ + H : ◖ Fite and Brackmann (1958).  ◖ Rothe and coworkers (1962).
$e$ + H$_2$ : ✦ Tate and Smith (1932).

**Figure 8.1** Cross sections for ionization of atomic hydrogen and molecular hydrogen by proton impact (Gilbody and Ireland, 1964). H$_2$: (B′) Calculated values by Bates and Griffing (1953), (G) Gilbody and Lee (1963); (A) Afrosimov, Il' in and Fedorenko (1958); (H) Hooper and coworkers (1962). For comparison, electron impact ionization cross sections for the same relative velocity of impact are shown on the figure

section data are given which show the broad features of such cross section functions. A number of features such as the behaviour of the cross section at low energies and the importance of multiple ionization are not yet well understood.

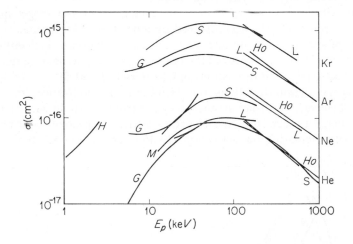

**Figure 8.2** Total ionization cross section functions for protons in rare gases (Hasted, 1964). (S) Solov'ev and coworkers (1962). (Ho) Hooper and coworkers (1962). (L) Gilbody and Lee (1963). (G) Gilbody and Hasted (1957). (M) Calculated values by Mapleton (1958). (H) Ar data of Hasted (1964)

Certain differences between electron–atom (molecule) and heavy particle–atom (molecule) inelastic collisions involving excitation and ionization should be noted. The cross section, $\sigma_i$, for single ionization of atoms and molecules by electron impact rises from zero at the threshold to a peak value at ~100 eV and decreases monotonically as the impact energy is increased, falling to one-half of its maximum value at ~500 eV. Close to threshold, $\sigma_i$ varies more or less linearly with excess energy above threshold for a restricted energy range (Section 5.5). For heavy-particle impacts, although the overall behaviour of the ionization cross section is similar to that for electron impact, certain differences prevail. Firstly, the energy scale is now greatly expanded due to the much larger masses of the colliding particles and hence the lower velocities for the heavy particles compared with electrons of the same kinetic energy. The cross section (Figures 8.1 and 8.2) peaks in the tens-of-keV region and remains large up to impact energies of the order of 1 MeV. Secondly, in heavy-particle collisions diffraction effects are negligible except at very small angles. In the energy range where the cross section is appreciable the de Broglie wavelength for atomic particles, in contrast to that for electrons, is small compared to atomic dimensions. Thirdly, in contrast to the case for electrons, heavy particles making ionizing collisions lose little energy and are scattered through small angles only. The ions produced in heavy-particle impacts are scattered at nearly 90° with respect to the projectile beam and they normally receive very small kinetic energy. Fourthly, at low energies the ionization cross section functions for atomic particles do not show the more or less linear energy rise observed for electron impacts, but they rather follow a power law dependence. At sufficiently high energies the ionization cross section for electrons and singly charged positive ions of the same relative velocity of impact are found to be essentially the same, in agreement with the Bethe–Born approximation (Section 5.5).

Much work has been done on ionizing and stripping collisions between heavy particles at high energies but little information exists on such processes at low energies ($\lesssim 500$ eV). The reader is referred to the sources cited in Section 8.1 for literature and further discussion on this topic as well as on elastic scattering of fast ions and of atoms by atoms and molecules. The latter throws light on the interactions between atomic systems at small nuclear separations.

## 8.3 Positive Ion–Electron and Positive Ion–Negative Ion Recombination Processes

The processes of positive ion–electron and positive ion–negative ion recombination (Table 8.1) determine the mutual neutralization of charged species in ionized gases. As in electron attachment processes (Chapter 6), recombination between charged particles may be either a two- or a three-body process, and, further, in some reactions an unstable intermediate may be formed which dissociates, autoionizes or is stabilized by collision with a third body. The probability of interaction in these processes is increased due to the attractive Coulomb forces between the colliding particles.

### 8.3.1 *Positive Ion–Electron Recombination Processes*

Most of the processes of positive ion–electron recombination differ essentially in the way the excess energy due to recombination is given away. The positive ion–electron recombination process is usually described by a rate coefficient, $\alpha_e$, defined by

$$\frac{dn_e}{dt} = -\alpha_e n_e n_+ \tag{8.7}$$

where $n_e$ and $n_+$ are the number densities of free electrons and positive ions, respectively. When $n_e \simeq n_+$ (as in plasmas)

$$\frac{dn_e}{dt} = -\alpha_e n_e^2 \tag{8.8}$$

At large electron and ion densities the recombination process is more likely to dominate over other processes involving electrons and neutrals, electron attachment for example.

### 8.3.1.1 Radiative recombination ($A^+ + e \to A^{(*)} + h\nu$)

This process is analogous to that of radiative attachment except that the neutral atom is replaced by a positive ion. The atom A may be left in its ground state or in its electronic excited states. If $\varepsilon_e$ is the kinetic energy of the initially free electron and $\varepsilon_i$ and $\varepsilon_x$ are, respectively, the ionization and excitation energies of the atomic states involved in the reaction, then

$$h\nu = \varepsilon_e + (\varepsilon_i - \varepsilon_x) \tag{8.9}$$

Since the electron kinetic energy distribution is comparatively wide, the emission spectrum will take the form of a continuum. Such are the emissions originating in many discharge sources. The excited atomic system A* may subsequently undergo radiative de-excitation. When A is formed in its ground state the process is the reverse of photoionization, and in this case radiative recombination coefficients can be deduced from photoionization data by application of the principle of detailed balancing. In certain cases such estimates have been made for the lowest excited states of certain atomic species. Exact calculations of recombination coefficients have been performed for the atomic hydrogen positive ion $H^+$. Bates, Kingston and McWhirter (1960, 1962) have calculated that for $H^+$, $\alpha_e$ is 10.2, 5.66, 2.36 and $1.11 \times 10^{-13}$ cm³/sec at 250°K for principle quantum numbers $n$ equal to 1, 2, 5 and 10, respectively. The recombination coefficient $\alpha_\Sigma(Z = 1, T)$, for capture of an electron into all $n$ levels of H, was reported by the same authors to be $48.0 \times 10^{-13}$ cm³/sec at 250°K (see Bates and coworkers (1960, 1962) for values on $\alpha_e(Z = 1, T, n)$ for $n = 1, 2, \ldots 10$ and $T$ from 250° to 64,000°K). When the mean thermal energy is small compared with the ionization energy of the $n$th level, $\alpha_e(Z = 1, T, n)$ varies as $n^{-1} T^{-\frac{1}{2}}$. The temperature variation of $\alpha_\Sigma(Z = 1, T)$ is approximately as $T^{-0.7}$ (Bates and Dalgarno, 1962).

Generally, the rates for radiative recombination are slow compared to those for certain other recombination processes such as dissociative recombination (Section 8.3.1.3). Radiative recombination is an important process in the earth's upper atmosphere.

### 8.3.1.2 Dielectronic recombination $(A^+ + e \rightleftarrows A^{**} \rightarrow A^{(*)} + h\nu)$

In this process—indirect form of radiative recombination for ions having more than one ground configuration—the doubly excited atom $A^{**}$ may revert to the initial conditions (in times of the order of $10^{-14}$ sec), a process analogous to autoionization. Alternatively, $A^{**}$ may undergo radiative decay giving a quantum $h\nu$ and an excited or unexcited neutral atom. Since radiative lifetimes are very much longer ($\gtrsim 10^{-9}$ sec) than autoionization lifetimes, the probability of stabilizing $A^{**}$ by radiation is small. For instance, dielectronic recombination rates to normal $N^+$ and $O^+$ ions are much less than those for radiative recombination (Bates, 1962a). However, for certain other ions dielectronic recombination may be faster (Bates and Dalgarno, 1962).

### 8.3.1.3 Dissociative recombination $(AB^+ + e \rightarrow A^{(*)} + B^{(*)})$

For molecular positive ions the excess energy of recombination may be transferred to molecular vibration resulting in dissociation. The fragments may be left with or without electronic excitation energy. The process is analogous to that of dissociative attachment and may be similarly visualized as proceeding via the formation of an unstable intermediate which subsequently dissociates or autoionizes. The dissociative recombination coefficient, $\alpha_{DR}$, is defined as the product of the corresponding cross section, $\sigma_{DR}$, with the electron velocity, $v$, averaged over the Maxwell velocity distribution of the electrons at the electron temperature $T$. In terms of the electron energy, $\varepsilon(=\frac{1}{2}mv^2)$,

$$\alpha_{DR} = \langle \sigma_{DR}(\varepsilon)v \rangle$$

$$= \left(\frac{2^3}{m\pi(kT)^3}\right)^{\frac{1}{2}} \int \sigma_{DR}(\varepsilon) e^{-\varepsilon/kT} \varepsilon \, d\varepsilon \tag{8.10}$$

Two distinct modes of dissociative recombination have been distinguished: direct and indirect. The former (Bates, 1950) involves a single radiationless transition, whereas the latter (Bardsley, 1967, 1968; Chen and Mittleman, 1967) occurs through an intermediate Rydberg state of the molecule. Bardsley (1968), for example, has attempted to apply, with appropriate changes, the resonance formalism for dissociative attachment (Chapter 6) to the process of dissociative recombination.

With $He_2^+$ probably as the only exception (Ferguson, Fehsenfeld and Schmeltekopf, 1965), electrons recombine roughly $10^5$ times as efficiently with molecular ions by dissociative recombination as they do by radiative recombination with atomic ions. For example, the dissociative recombination coefficients $\alpha_{DR}$ for the reactions $O_2^+ + e \rightarrow O + O$, $NO^+ + e \rightarrow N^+ + O$ and $N_2^+ + e \rightarrow N + N$ are, respectively, $2 \times 10^{-7}$ cm$^3$/sec, $4 \times 10^{-7}$ cm$^3$/sec and $1 \times 10^{-7}$ cm$^3$/sec (Ferguson, 1968), whilst radiative recombination coefficients are of the order of $10^{-12}$ cm$^3$/sec (Section 8.3.1.1). As a consequence of the large rates for dissociative recombination, ion–atom interchange reactions ($A^{\pm} + BC \rightarrow AB^{\pm} + C$) which convert atomic into molecular ions, may assume a very important role in controlling electron and ion number densities. For instance, the efficiency of the overall sequential process, $A^+ + BC \rightarrow AB^+ + C$ followed by $AB^+ + e \rightarrow A + B$, may be very much greater than that of the radiative recombination process $A^+ + e \rightarrow A + h\nu$.

### 8.3.1.4 Three-body recombination ($A^+ + e + B \rightarrow A^{(*)} + B$)

The excess energy of recombination between a positive ion and an electron may be removed by a neutral atom or molecule acting as a third body. Little experimental information exists on this process, and the values of $\sim 10^{-11} P$ cm$^3$/sec for He and $\sim 2 \times 10^{-10} P$ cm$^3$/sec for air (Hasted, 1964) may be considered representative of the recombination coefficients via this process; $P$ is the gas pressure in torr.

### 8.3.1.5 Electron stabilized recombination ($A^+ + e + e \rightleftarrows A^* + e$)

Such a process has been proposed in order to explain volume loss of electrons and positive ions from plasmas of moderate and high density. A third-body electron carries away excess kinetic energy. It is essentially the inverse of electron impact ionization of an excited atom.

### 8.3.2 *Positive Ion–Negative Ion Neutralization*

### 8.3.2.1 Two-body ionic recombination

Here the recombining ions $A^+$ and $B^-$ can be either atomic or molecular, and under certain circumstances the recombination process can be very efficient. For negative ions of opposite sign to recombine, the total energy of the system must be decreased as a result of recombination. When $A^+$ and $B^-$ are atomic, the required decrease in energy is $I(A) - EA(B)$. The probability of recombination depends on the ability of the system to dispose of the excess energy of mutual neutralization, either through emission of electromagnetic radiation or through electronic excitation of the neutralized atoms, or through distribution as kinetic energy among the collision products; when either of the reactants is molecular, excess energy may go into internal motion. The requirement, however, that linear and angular momentum be conserved makes the disposal of the recombination energy by an increase in kinetic energy of the recombining ions highly improbable. Accordingly, two-body recombination between positive and negative ions will proceed via the mutual neutralization through charge exchange process

$$A^+ + B^- \rightarrow A^* + B^* \tag{8.11}$$

or via the radiative recombination process

$$A^+ + B^- \rightarrow AB + h\nu \tag{8.12}$$

If both or one of the ions is molecular, dissociative mutual neutralization

$$AB^+ + C^- \rightarrow A + B + C \tag{8.13}$$

may take place.

The neutralization process (8.11) occurs by charge exchange. The extra electron from $B^-$ may be captured into any state of the neutral atom A and the neutral atom B may be left in any state for which energy is conserved. A portion of the neutralization energy may appear as kinetic energy of the neutral atoms.

The radiative recombination process (8.12) requires a radiative transition between two electronic states of AB. Since such radiative transitions have lifetimes of $\sim 10^{-8}$ sec, the reactants must remain close for at least that length of time for the reaction to occur. Ions with thermal energies (say, at room temperature) take about $10^{-13}$ sec to traverse a typical molecular diameter. Hence, within this period of time the probability that a photon will be emitted to de-excite the system is only $\sim 10^{-5} (= 10^{-13} \text{ sec}/10^{-8} \text{ sec})$. This probability corresponds to a recombination coefficient of $\sim 10^{-14}$ cm$^3$/sec. For faster ions the process is even slower. It might then be concluded that process (8.12) is highly inefficient and that except at low pressures this mode of recombination may normally be neglected in comparison with process (8.11) and the process of three-body ionic recombination (Section 8.3.2.2). Early calculations by Bates and Massey (1943) put the recombination coefficient for $O^+ + O^- \rightarrow O^* + O^*$ at room temperature as high as $10^{-8}$ cm$^3$/sec, and possibly higher. A discussion of the theory of ion–ion recombination processes can be found in McDaniel (1964) (Chapter 12).

8.3.2.2 Three-body ionic recombination
$(A^+ + B^- + C \rightarrow A + B + C \text{ (or } AB + C))$

This is the most important ion–ion recombination mechanism at high pressures (above a few torr). The excess energy of recombination between positive and negative ions is removed by a neutral atom or molecule acting as a third body. An alternative process leading to three-body ionic recombination,

$$A^+ + C \rightleftarrows (AC)^+$$

followed by

$$(AC)^+ + B^- \rightarrow AB + C \tag{8.14}$$

has been examined theoretically by Fueno, Eyring and Ree (1960).

## 8.4 Charge Transfer (Exchange) Reactions

Let us now consider the single charge transfer reactions

$$A^+ + A \rightarrow A + A^+ \tag{8.15}$$

and

$$A^+ + B \rightarrow A + B^+ + \Delta E \tag{8.16}$$

where one electron and very little kinetic energy is transferred from the target atom to the projectile ion. Process (8.15) is known as symmetrical resonance charge transfer and process (8.16) as asymmetrical non-resonance charge transfer. In Equation (8.16) $\Delta E$ is called the energy defect and it represents the energy change in the charge-transfer reaction; it is the difference in the ionization energies of atoms A and B.

For the symmetrical resonance charge transfer process (8.15) $\Delta E = 0$, and the cross section $\sigma_{CT}$ is expected to increase monotonically with decreasing energy and attain a large value (of the order of $10^{-15}$ cm$^2$) at thermal energy. With increasing impact velocity $v$ the cross section $\sigma_{CT}$ will decrease as (see Hasted, 1964, Chapter 12)

$$\sigma_{CT}^{\frac{1}{2}} = a - b \ln v \tag{8.17}$$

where $a$ and $b$ are constants which contain the ionization potential $I$. The first quantum-mechanical treatment of this process was given by Massey and Smith (1933). Subsequently, this process has been studied theoretically by, among others, Bates, Massey and Stewart (1953), Fetisov and Firsov (1959), Firsov (1951) and Rapp and Francis (1962). For the intermediate energy range (below $\sim 10^8$ cm/sec and above $\sim 10^5 M_r^{-\frac{1}{2}}$ cm/sec (Rapp and Francis, 1962); $M_r$ is the reduced mass of the collision pair in amu) these theoretical treatments have been quite successful in describing the experimental results.

An extension of the theory on the symmetrical resonance charge transfer to the asymmetrical non-resonance charge transfer process (8.16) has been attempted by Rapp and Francis (1962), who succeeded in accounting qualitatively for the features observed in experiment. However, the theoretical interpretation of charge-transfer reactions is difficult, especially at low energies and for collisions involving molecular encounters. Generally, the energy dependence of the cross section for a particular charge transfer reaction depends greatly on the magnitude of the energy defect $\Delta E$, and it may conveniently be discussed in terms of Massey's adiabatic criterion (Massey, 1949). According to this criterion, the probability of charge transfer is small if the time involved in the electronic transition, $h/\Delta E$, is much shorter than the collision time, $l/v$, where $l$ is the 'adiabatic parameter' representing the range of interaction of $A^+$ and B, and $v$ is the relative velocity of approach. When the two times are comparable, the probability of charge exchange can be large. The adiabatic criterion has been found (Hasted, 1964, 1968) to give reasonable cross section values for typical single charge transfer processes. One then would expect the cross section for the non-resonance charge transfer reaction (8.16) to be small at low impact energies $\left( v \ll \dfrac{l \Delta E}{h} \right)$, rise to a maximum near the energy determined by

$$\frac{h}{\Delta E} = \frac{l}{v} \tag{8.18}$$

and subsequently decrease again. When $\Delta E$ is small or zero (accidental resonance (see Table 8.1) or symmetrical resonance), the maximum cross section will occur at very low energy.

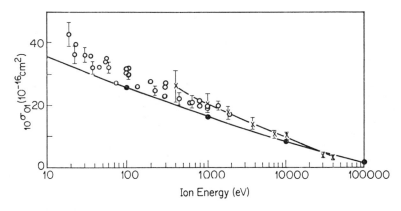

○  Fite, Smith, and Stebbings (1962).

x  Fite and coworkers (1960).

●  Calculated values by Dalgarno and Yadav (1953).

**Figure 8.3** Cross section for charge transfer between protons and atomic hydrogen (Fite, Smith and Stebbings, 1962)

The general features of the cross section functions for processes (8.15) and (8.16) can be seen in Figures 8.3 and 8.4 where the cross section for the symmetrical resonance process $H^+ + H \rightarrow H + H^+$ and the cross sections for charge transfer for H atoms and ions in nitrogen at high energies are shown, respectively.

Exothermic molecular charge transfer reactions have large cross sections over a wide impact energy range. Dissociative charge transfer

$$A^+ + BC \rightarrow A + B^+ + C \qquad (8.19)$$

and ion–molecule (chemical interchange) reactions

$$A^+ + BC \rightarrow AB^+ + C \qquad (8.20)$$

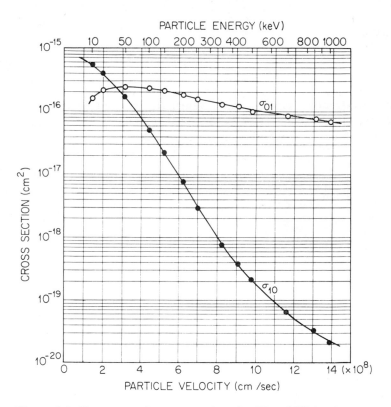

**Figure 8.4** Charge transfer cross sections for H and $H^+$ in nitrogen gas (Barnett and Reynolds, 1958). $\sigma_{01}$ and $\sigma_{10}$ are, respectively, the cross sections for electron loss and electron capture

may take place if either or both of the reactants are molecular. It is possible in such reactions for energy to transfer from the relative motion to internal degrees of freedom of the reactants. In this case a long-lived complex may be formed which survives dissociation for long periods of time. A number of reactions have been reported to proceed through such long-lived intermediates at thermal and epithermal energies (see, for example, Durup and Durup, 1967). It might be pointed out that in contrast to the case of asymmetrical charge-exchange reactions, ion–molecule reactions are thought of not involving an electronic transition. A full quantum mechanical treatment of ion–molecule reactions is lacking. Classically, the cross section, $\sigma_{\text{interchange}}$, for the ion–atom interchange (chemical interchange) process (8.20) is generally expressed as the product

$$\sigma_{\text{interchange}} = p\sigma_0 \tag{8.21}$$

where $\sigma_0$ is the cross section for orbiting collisions given by (Langevin, 1905; Gioumousis and Stevenson, 1958)

$$\sigma_0(v) = \frac{2\pi}{v}\left(e^2\,\frac{\alpha}{M_r}\right)^{\frac{1}{2}} \tag{8.22}$$

and $p$ is the probability that the interchange will occur in such an orbiting collision. $M_r$ is the reduced mass of the ion and the molecule, $\alpha$ is the molecular polarizability, and $v$ is the relative velocity of approach. The probability $p$ is usually taken equal to one. However, if a long-lived intermediate is formed, $p$ may be much less than unity.

## 8.5 References

V. V. Afrosimov, R. N. Il'in and N. V. Fedorenko (1958). *Soviet Phys.—JETP*, **7**, 968.

J. N. Bardsley (1967). In *Proceedings 5th International Conference on the Physics of Electronic and Atomic Collisions*, Publishing House 'Nauka', Leningrad, p. 338.

J. N. Bardsley (1968). *J. Phys. B (Proc. Phys. Soc.)*, **1**, 365.

C. F. Barnett and H. K. Reynolds (1958). *Phys. Rev.*, **109**, 355.

D. R. Bates (1950). *Phys. Rev.*, **78**, 492.

D. R. Bates (1962a). *Planetary Space Sci.*, **9**, 77.

D. R. Bates (Ed.) (1962b). *Atomic and Molecular Processes*, Academic Press, New York.

D. R. Bates and A. Dalgarno (1962). In D. R. Bates (Ed.), *Atomic and Molecular Processes*, Academic Press, New York, pp. 245–271.

D. R. Bates and I. Estermann (Eds.) (1968). *Advances in Atomic and Molecular Physics*, Vols. 4 and 5, Academic Press, New York.

D. R. Bates and G. W. Griffing (1953). *Proc. Phys. Soc. (London)*, **66A**, 961.

D. R. Bates, A. E. Kingston and R. W. P. McWhirter (1960). Quoted in Bates and Dalgarno (1962); *Proc. Roy. Soc. (London)*, **A267**, 297 (1962); **A270**, 155 (1962).

D. R. Bates and H. S. W. Massey (1943). *Phil. Trans. Roy. Soc.*, **A239**, 269.

D. R. Bates, H. S. W. Massey and A. L. Stewart (1953). *Proc. Roy. Soc. (London)*, **A216**, 437.

M. A. Biondi (1963). In L. Marton (Ed.), *Advances in Electronics and Electron Physics*, Vol. 18, Academic Press, New York, pp. 67–165.

J. C. Y. Chen and M. H. Mittleman (1967). In *Proceedings 5th International Conference on the Physics of Electronic and Atomic Collisions*, Publishing House 'Nauka', Leningrad, p. 329.

A. Dalgarno (1967). *Rev. Mod. Phys.*, **39**, 850.

A. Dalgarno and H. N. Yadav (1953). *Proc. Phys. Soc. (London)*, **A66**, 173.

M. Durup and J. Durup (1967). In *Advances in Mass Spectrometry*, Vol. 4, The Institute of Petroleum, p. 677.

E. E. Ferguson (1968). In L. Marton (Ed.), *Advances in Electronics and Electron Physics*, Vol. 24, Academic Press, New York, pp. 1–50.

E. E. Ferguson, F. C. Fehsenfeld and A. L. Schmeltekopf (1965). *Phys. Rev.*, **138**, A381.

I. K. Fetisov and O. B. Firsov (1959). *Soviet Phys.—JETP*, **37**, 95.

O. B. Firsov (1951). *Soviet Phys.—JETP*, **21**, 1001.

W. L. Fite and R. T. Brackmann (1958). *Phys. Rev.*, **112**, 1141, 1151.

W. L. Fite, A. C. H. Smith and R. F. Stebbings (1962). *Proc. Roy. Soc. (London)*, **A268**, 527.

W. L. Fite, R. F. Stebbings, D. G. Hummer and R. T. Brackmann (1960). *Phys. Rev.*, **119**, 663.

T. Fueno, H. Eyring and T. Ree (1960). *Can. J. Chem.*, **38**, 1693.

C. F. Giese (1966). In *Advances in Chemical Physics*, Vol. X, Interscience, New York, pp. 247–273.

H. B. Gilbody and J. B. Hasted (1957). *Proc. Roy. Soc. (London)*, **A240**, 382.

H. B. Gilbody and J. V. Ireland (1964). *Proc. Roy. Soc. (London)*, **A277**, 137.

H. B. Gilbody and A. R. Lee (1963). *Proc. Roy. Soc. (London)*, **A274**, 365.

G. Gioumousis and D. P. Stevenson (1958). *J. Chem. Phys.*, **29**, 294.

J. B. Hasted (1962). In D. R. Bates (Ed.), *Atomic and Molecular Processes*, Academic Press, New York, pp. 696–720.

J. B. Hasted (1964). *Physics of Atomic Collisions*, Butterworths, Washington, Chaps. 7, 10 to 12, 14.

J. B. Hasted (1968). In D. R. Bates and I. Estermann (Eds.), *Advances in Atomic and Molecular Physics*, Vol. 4, Academic Press, New York, pp. 237–266.

J. W. Hooper, D. S. Harmer, D. W. Martin and E. W. McDaniel (1962). *Phys. Rev.*, **125**, 2000.

P. Langevin (1905). *Ann. Chim. Phys.*, **5**, 245.

R. A. Mapleton (1958). *Phys. Rev.*, **109**, 1166.

H. S. W. Massey (1949). *Rept. Progr. Phys.*, **12**, 248.

H. S. W. Massey and R. A. Smith (1933). *Proc. Roy. Soc. (London)*, **A142**, 142.

E. W. McDaniel (1964). *Collision Phenomena in Ionized Gases*, John Wiley and Sons, New York, Chaps. 6, 12.

M. R. C. McDowell (Ed.) (1964). *Atomic Collision Processes*, North-Holland Publishing Co., Amsterdam.

R. H. Neynaber (1969). In D. R. Bates and I. Estermann (Eds.), *Advances in Atomic and Molecular Physics*, Vol. 5, Academic Press, New York, p. 57.

D. Rapp and W. E. Francis (1962). *J. Chem. Phys.*, **37**, 2631.

E. W. Rothe, L. L. Marino, R. H. Neynaber and S. M. Trujillo (1962). *Phys. Rev.*, **125**, 582.

E. S. Solov'ev, R. N. Il'in, V. A. Oparin and N. V. Fedorenko (1962). *Soviet Phys.—JETP*, **15**, 459.

R. F. Stebbings (1966). In *Advances in Chemical Physics*, Vol. X, Interscience, New York, pp. 195–246.

J. T. Tate and P. T. Smith (1932). *Phys. Rev.*, **39**, 270.

S. M. Trujillo, R. H. Neynaber and E. W. Rothe (1966). *Rev. Sci. Instr.*, **37**, 1655.

# 9 Biophotophysics and bioelectronics

## 9.1 Introduction

In this chapter we will ask a simple question: to what extent do the physical (and chemical) properties of molecules help us to understand their biological behaviour? More specifically, can any or a combination of the physical quantities we have elaborated upon in the preceding chapters be used to explain the biological action of molecules? There is no simple answer to these questions. Life is probably too complex to be understood by quantum mechanics and gaseous studies, and yet, unavoidably, we have no chance at all of understanding some of life's secrets without *molecular studies on the physical level*. Molecules remain the unit of life.

The geometrical and electronic structures of molecules determine their physical (and chemical) properties. Molecular parameters directly related to molecular structure, such as the ionization potential, electron affinity, electronic energy levels, metastable states, and electron densities are being considered as the determining factors in molecular interactions. It is thus not surprising that various attempts have been made to relate studies on luminescence, electronic excitation energy transfer, molecular interactions with low-energy electrons, charge-transfer (donor–acceptor) reactions, and molecular geometry to the biological action of such systems—biological action proceeds via charge and/or energy exchange (see, for example, Coulson (1953); Badger (1948, 1954); Clar (1964); Pullman (1964); Pullman and Pullman (1955a,b; 1963a); Lacassagne (1966); Mason (1958a,b; 1959, 1960, 1966); Chalvet and coworkers (1966); Szent-Györgyi (1968a,b); see, also, references later in this chapter).

Free electrons are believed to exist in living systems. Their role in molecular interactions has been controversial but undeniable. Szent-Györgyi (1968a) talks of 'an "invisible fluid" of electrons, which being more mobile than molecules carry energy, charge and information and act as the fuel of life'.

In Chapter 6 we have seen that many molecular species, simple and complex, readily interact with free electrons to form stable negative ions—parent ions and/or fragment ions. The cross sections for such processes at thermal and epithermal energies in the gaseous state are often exceedingly high ($\gtrsim 100\text{Å}^2$). The role of free radicals and ions formed in electron–molecule collisions as well as the subsequent interaction processes that the reaction products undergo is still largely unknown. Suggestions pointing to their importance have been made (see, for example, Blaunstein and Christophorou, 1968).

Many scattered facts as to the role of certain photophysical and electron–molecule interactions in biological systems, especially with respect to chemical carcinogenesis, have at times stirred considerable discussion. In spite of the fact that no single physical property of the molecule is expected to resolve fundamental biological questions, restricted correlations between certain biological actions and a number of molecular physical characteristics have been reported. Their generalization, however, was not notably successful and has been handicapped by lack of knowledge of both the physical and the biological aspects of such problems. Physical molecular properties are not presently accurately known for a large enough number of compounds* to render reasonable weight to studies of this nature. However, should a number of molecular physical properties be accurately determined for a large number of molecular systems, one would hope, in their proper combination and usage, to establish the basic physical framework on which the life scientist would be able to orient himself and build his own line of thought. The interdisciplinary nature of the problem is evident.

To our opinion presently, there are two specific areas where physicists, chemists, and biologists interact considerably. These areas, referred to in the title of this chapter, are the fields of biophotophysics and bioelectronics† which will now be elaborated upon briefly.


### 9.2  Biophotophysics

Even though the physics of polyatomic molecules is still poorly understood, the recent progress in the field has been impressive, demonstrating the need for further intense studies. The processes of absorption and emission of

---

*See, for example, Chapter 7 on electron affinities and Appendix II on ionization potentials. In Chapter 7 a number of simple empirical and semi-empirical relations have been given which relate the ionization potential, electron affinity, the energy of the charge-transfer complex, the energy of the first excited $\pi$-singlet state, and electronegativity of aromatic hydrocarbons. Such correlations are of very restricted validity and may be misleading when they are not used with proper caution .They, however, indicate certain trends in the interdependence of these basic molecular parameters.
†The term bioelectronics has been introduced by Szent-Györgyi (1968a, b).

light and the associated electronic energy levels (at least those of the lowest excited electronic states) of a number of organic structures of wide biological interest are reasonably well known. Luminescence is a fundamental molecular property of organic molecules possessing $\pi$-electron structures, also of proteins and amino acids. The effect of substitution, molecular structure and composition, as well as that of certain environments on molecular absorption and emission spectra, has been investigated in various ways (Chapter 3). The role of excited electronic states, especially that of the lowest states of a given multiplicity (triplet states may be important intermediates in the utilization of light in biological systems), is also well recognized in biophotophysics, and so is the role of certain transitions unique in systems containing 'lone-pair' electrons (organic molecules with N, O, S, etc. atoms; Chapter 3). The quenching of triplet states by molecular oxygen has recently been proposed (Birks, 1969) to lead to the formation of singlet excited oxygen, a system of unusual molecular electronic properties*, considered important in certain types of molecular reactions.

Among the luminescent organic molecules which have been the subject of physical studies, a great number are of high biological significance. For example, the carcinogenic activity of some members of aromatic hydrocarbons has attracted a great deal of interest to the physical properties of this class of molecules. Strong luminescence is often characteristic of strong aromatic carcinogens (e.g. 3,4-benzopyrene, 1,2-benzanthracene), the majority of which have their first excited $\pi$-singlet state located in a very narrow energy range ($\sim 3.0$ to $\sim 3.2$ eV). Hieger in the 1930's pointed out that the fluorescence of carcinogenic tars was different from that of non-carcinogenic tars. Indeed, the strong fluorescence of 3,4-benzopyrene and 1,2-benzanthracene(s) led to their identification in tars (see a discussion on chemical carcinogens in Schoental (1964)).

Attempts have been made to relate the energy of fluorescence, phosphorescence, separation of first excited states, spectral shifts, and quantum yields of fluorescence to certain biological functions of organic molecules. These efforts have, at best, been met with limited success. Jones (1940, 1941, 1943) investigated the ultraviolet absorption spectra of 370 aromatic hydrocarbons, both substituted and unsubstituted, and attempted to relate the observed bathochromic shifts (shifts of the absorption spectrum to longer wavelengths) upon alkyl substitution and the dependence of this shift upon position of the substituent to the carcinogenic indices of methyl substituted 1,2-benzanthracenes. Larger shifts seemed to have been associated with more

---

*The ground state of $O_2$ is the triplet $^3\Sigma_g{}^-$ and the first excited electronic state is the singlet $^1\Delta_g$ which lies only 0.98 eV (Chapter 5) above the ground state. The radiative lifetime of $O_2*$ in $^1\Delta_g$ is extraordinarily long (60 minutes, Jones and Harrison (1958); 45 minutes, Badger, Wright and Whitlock (1965)).

active molecules, but no definite conclusions could be reached. Bruce and Todd (1939) and Bruce (1941) tried to relate the fluorescence intensity to carcinogenicity unsuccessfully. Schoental and Scott (1949) examined the absorption and fluorescence spectra of several polycyclic aromatic hydrocarbons in solution, and Birks and Cameron (1959) studied the fluorescence spectra of 41 organic compounds, including 29 known carcinogens in the crystalline state. No correlation was found in any of these investigations between the position and/or intensity of the fluorescence spectra and the carcinogenic activity. Birks (1959) pointed out that aromatic carcinogens have their first excited $\pi$-singlet state in the range 3.04 to 3.20 eV and proceeded to relate this observation to the carcinogenic activity of these compounds (see later this section). However, a number of non-carcinogenic aromatic hydrocarbons have their first excited $\pi$-singlet state in this energy range, and other studies (Shpol'skii, Il'ika and Basilevich, 1948; Steele, Cusachs and McGlynn, 1967) showed lack of any wide correlation of significance between the 0–0 bands of fluorescence $((h\nu_{0,0})_f)$ and phosphorescence $((h\nu_{0,0})_p)$ of certain groups of aromatic hydrocarbons and their carcinogenic index. Carcinogenic aromatic hydrocarbons have been shown by Christophorou (1963) to form transient photodimers (excimers). This property, however, is of general occurrence common to carcinogenic and non-carcinogenic molecules (Christophorou, 1963).

Steele, Cusachs and McGlynn (1967) reported that the carcinogenic index of methylated derivatives of 1,2-benzanthracene, 3,4-benzophenanthrene, chrysene, and triphenylene showed some correlation with the energy difference $(h\nu_{0,0})_f - (h\nu_{0,0})_p$ or $S_1(^1L_b) - T_1(^3L_a)$ where $S_1(^1L_b)$ and $T_1(^3L_a)$ are, respectively, the energies of the first excited $\pi$-singlet and first excited $\pi$-triplet states of the molecule. Steele and coworkers found that the activity of these compounds reaches its maximum when

$$S_1(^1L_b) - T_1(^3L_a) \simeq 1\,\text{eV} \qquad (9.1)$$

and since, according to Steele and coworkers, all four groups of compounds considered by them undergo an initial common metabolic action (aromatic hydroxylation) which requires an activation and participation of $O_2$, they proceeded to suggest that their finding is consistent with the involvement of the process

$$S_1(^1L_b) + {}^3O_2(^3\Sigma_g^-) \rightarrow T_1(^3L_a) + {}^1O_2(^1\Delta_g) \qquad (9.2)$$

in the interaction, as first suggested by Kautsky (1939). These workers then suggest that the activity of the four groups of hydrocarbons considered by them reaches a maximum when Equation (9.1) is fulfilled and thus (9.2) proceeds to yield singlet $O_2$ (the energy for the transition ${}^3O_2(^3\Sigma_g^-) \rightarrow {}^1O_2(^1\Delta_g)$ is 0.98 eV). The correlation may be accidental and ample exceptions can be found.

In the efforts discussed above the authors carefully pointed out factors which might have affected their comparisons, and also unknown parameters which could not be assessed and realistically accounted for. Understandably, such studies cannot be general enough to be of any practical value, and more sophisticated treatments in connection with other molecular properties are needed in order to evaluate the relative importance of electronic energy levels and the associated energy exchanges which accompany the de-excitation of excited molecules. It may then be of interest to mention a few other aspects of the problem, such as molecular geometry (shape and size) considered to be important in steric effects in molecular complexing. It is known, for example, that special types of molecular coupling, such as the photo-association of two molecules, is affected by molecular geometry and orientation. Of unique interest also are the recently widely studied phenomena of excimer and exciplex formation (Chapter 3). These are molecular interactions of general occurrence in organic and biological systems (Chapter 3; Birks, 1970) and distinctly demonstrate the role of excited states and electronic structure in molecular complexing. (The role of excimer and exciplex formation in biology is believed to be important, but as yet undetermined.) Excimer and exciplex formation is shared by both carcinogenic and non-carcinogenic organic molecules.

Additionally, attention should be drawn to the processes of simultaneous transitions in molecular pairs, intramolecular radiationless transitions, and quenching processes, electronic energy transfer mechanisms, and to theoretical studies on the electronic structure of biologically important substances*. The last two topics are of particular interest†. Electronic energy transfer via the weak coupling mechanism (Förster type) is being widely studied. Other types of energy transfer such as via strong coupling and via excimer or exciplex formation and dissociation are also under intense theoretical and experimental scrutiny.

Owing to the fact that the first excited $\pi$-singlet state of a large number of conjugated aromatic hydrocarbons lies in a very limited energy range (3.0 to 3.2 eV) and that this energy range suitably overlaps the emission spectrum of tryptophan, a specific suggestion has been made (Birks, 1961), namely that energy is transferred via a dipole–dipole mechanism from tryptophan to the hydrocarbon which results in the formation of a complex, responsible for the carcinogenic action of these molecules. The probability

---

*Auger disruptions in systems containing heavy atom centres (such as the porphyrins) may be important in the radiation damage of a biological system.
†Detailed discussions of chemiluminescence, bioluminescence, photosynthesis, and of the phenomenon of vision can be found in the book *Light and Life* by McElroy and Glass (1961), and examples of energy transfer in biological systems can be found in Volume 27 of the Discussions of the Faraday Society which is devoted to energy transfer with special reference to biological systems.

for such a process, as has been discussed in Chapter 3, is proportional to
the overlap integral $J$ given by

$$J = \int f_s(\lambda)\varepsilon_A(\lambda)\lambda^4 \, d\lambda \tag{9.3}$$

where $f_s(\lambda)$ is the quantum intensity of the fluorescence emission of
tryptophan and $\varepsilon_A(\lambda)$ is the molar extinction coefficient of the hydrocarbon
at the wavelength $\lambda$. Thus the carcinogenic action should increase with $J$,
a condition which actually has been found not to be generally satisfied. In
spite of the fact that such a process may be important, it cannot be
generalized in the manner originally proposed by Birks (1961).

   With respect to theoretical studies on the electronic structure of molecules,
one should realize the difficulty in performing such calculations and the very
general assumptions on which the obtained results are based. In spite of this
difficulty, the numerical results of certain simple molecular orbital calculations led to definite propositions (see, for example, Pullman and Pullman,
1955a,b; 1964 and Mason, 1958a,b; 1959; 1960; 1966) as to the mode of
action of aromatic carcinogens. The claim, for example, has been put
forward by the Pullmans that their quantum-mechanical calculations
(LCAO–MO) indicated that certain specific regions of aromatic molecules
are of particular importance for specific types of chemical and biological
reactions. The location of these regions—so-called $K$ and $L$ regions—on the
molecular periphery is illustrated for the case of 1,2-benzanthracene in
Figure 9.1. These regions are essential reactive centres, their reactivity being
determined by the proper magnitude of the *localization energy* given in
units of the resonance integral in the Hückel method. According to the
Pullmans, a polycyclic aromatic hydrocarbon is carcinogenic when it
possesses a strong reactive $K$ region. (Through this strong region the
carcinogen binds to the cell by means of electron transfer.) Should the

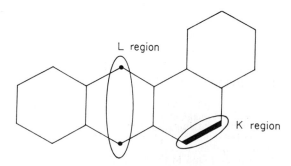

**Figure 9.1** The $K$ and $L$ regions of carcinogenesis
(Pullman, 1964)

molecule possess an $L$ region this must be 'rather unreactive' for the molecule to be carcinogenic. There has been a long and as yet inconclusive controversy concerning these definitive propositions. Certainly, more sophisticated and refined calculations and accurate *physical* experimental work are needed. Some further comments will be presented in the next section where Mason's ideas as well as other notions on the importance of charge-transfer and electron–molecule interactions will be presented.


## 9.3 Bioelectronics

Szent-Györgyi in his recent book on bioelectronics stresses the multiple role of electrons in biology, which ranges from their involvement in binding atoms into molecules to their being probably responsible 'for the charming subtlety of biological reactions'. He points out that intermolecular electron transfer may play a major role in biological regulation, defense, and cancer. There are ample examples which illustrate this.

Following Szent-Györgyi's (1941) original suggestion that proteins may possess semiconductor properties, Evans and Gergely (1949) presented theoretical arguments that the $\pi$-electron orbitals of proteins may indeed overlap in such a manner as to produce a band-type structure and calculated values of 3 to 4 eV for the band width between the conduction and valence bands of proteins. Indirect evidence has been accumulating to indicate that in the living cell electrons may exist in a free state and electron transport may occur along protein chains (see, for example, Eley, 1959). The notion is also generally accepted that in the energy-transfer process of photosynthesis the free electron is a necessary component (e.g. Bradley and Calvin, 1956; Platt, 1959; Arnold and Clayton, 1960). It has been suggested further (Lovelock, 1961) that the electron capturing intermediates and coenzymes in the energy-transfer processes of the living cell, function as an ordered sequence of reversible electron traps. Electrons are thought of being involved, also, in energy transfer in mitochondria (e.g. Green, 1959; Lehninger, 1964) and the Krebs cycle (Lovelock, 1961; Stockdale and Sangster, 1966). Szent-Györgyi (1968a) points out that the cell has a rich source of transferable electrons in its N, S and O atoms.

But perhaps the widest discussion as to the role of electron and charge transfer is on the carcinogenic action of certain groups of organic molecules. Pullman and Pullman (1955b) first examined the possible correlation of electron-donor (acceptor) capacities of aromatic molecules—based on their ionization and reduction potentials—and their carcinogenic activity, and they concluded that limited, but not wide, correlations could be observed. Subsequently, Mason (1958a, b; 1959; 1960) suggested that the initial mechanism in chemical carcinogenesis is that of charge transfer from a

protein to the carcinogenic molecule. Assuming a band structure of
proteins and a calculated value of $3.235 \pm 0.195$ eV for the band width
between the conduction and valence bands in proteins together with energy
levels of the hydrocarbons, Mason suggested a charge-transfer complex
formation between the hydrocarbon and the protein. All that is required,
then, is for an empty molecular orbital to match in energy with the highest
filled band in a protein as shown in Figure 9.2. An electron can then move

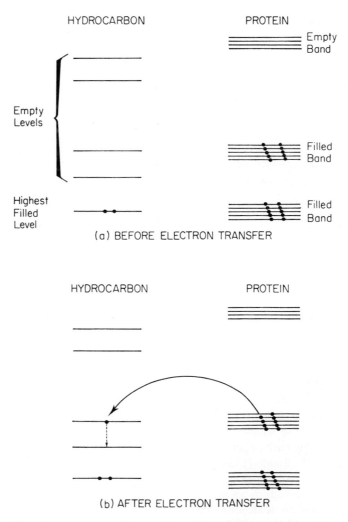

**Figure 9.2** Mason's theory of carcinogenesis. (a) before electron transfer; (b) after electron transfer

from the protein to the carcinogen because of orbital and band overlap, and this electron-transport process results in the formation of a carcinogen–protein complex responsible for the carcinogenic activity. This mechanism has been criticized (see Pullman, 1964) in view of the fact that overlap is

**Table 9.1** Comparison of the thermal electron attachment rate $(\alpha w)_{th}$, the Pullmans' $|k|$ coefficient, the ionization potential $I$ and the electron affinity $EA$ for aromatic hydrocarbons

| Compound | $(\alpha w)_{th}{}^a$ $(\sec^{-1} \text{torr}^{-1})$ | $|k|^b$ | $I^c$ (eV) | $EA^d$ (eV) |
|---|---|---|---|---|
| Benzene | $<3 \times 10^3$ | 1 | $9.24^e, 9.33^f$ | $-1.2$ |
| Naphthalene | $\leqslant 10^4$ | 0.618 | $8.12^e, 8.25^f$ | 0.10 |
| Azulene | $1 \times 10^9$ | — | $7.42^e, 7.74^f$ | 0.66 |
| Anthracene | $1.52 \times 10^8$ $1.45 \times 10^8$ | 0.414 | $7.38^e, 7.56^f$ | 0.63 |
| Phenanthrene | $7 \times 10^5$ | 0.605 | $8.03^f$ | $\sim 0.1$ |
| Triphenylene | $2.74 \times 10^7$ | 0.684 | $8.19^f$ | $\sim 0.06$ |
| Chrysene | $3.25 \times 10^7$ | 0.520 | $8.01^f$ | 0.35 |
| 3,4-Benzophenanthrene | $1.47 \times 10^8$ | 0.566 | $7.76^g$ | 0.41 |
| 1,2-Benzanthracene | $6.64 \times 10^8$ $2.11 \times 10^9$ | 0.452 | $7.35^g$ | 0.61 |
| Pyrene | $1.4 \times 10^7$ $4.4 \times 10^8$ | 0.445 | $7.72^f$ | 0.55 |
| Perylene | $3.4 \times 10^8(?)$ | 0.347 | $6.83^g$ | 0.92 |
| 1,2,3,4-Dibenzanthracene | | 0.499 | $7.43^g$ | 0.52 |
| 1,2,5,6-Dibenzanthracene | | 0.473 | $7.42^g$ | 0.61 |
| 1,2,7,8-Dibenzanthracene | | 0.492 | $7.42^g$ | 0.63 |

[a] From Table 6.8.
[b] From Table IV of A. Pullman (1964).
[c] From Appendix II.
[d] From Table 7.3. The values listed are the arithmetic averages of the corresponding values in the table with the exception of a few values which diverge considerably and which have not been included in the averaging.
[e] Mean of photoionization and ultraviolet absorption values listed in Appendix II.
[f] Mean of electron impact values listed in Appendix II.
[g] Values taken from ultraviolet data listed in Hedges and Matsen (1958).

not a necessary criterion for electron transfer, but rather the ionization potential of the donor and the electron affinity of the acceptor. Pullman (1964) also pointed out that it is incorrect to equate the energy of the lowest filled band of the protein with the highest filled level of the hydrocarbon (Figure 9.2) since this would predict much higher ionization potentials for the hydrocarbon. However, binding energies as well as polarization energies make the latter comment questionable. A more complete discussion of charge-transfer energetics can be found in Mason (1966).

Experimental studies of charge-transfer complexes specific to the above question have been made (e.g. Szent-Györgyi, Isenberg and Baird, 1960; Allison and Nash, 1963a; Epstein and coworkers, 1964; Jones, Bersohn and Niece, 1966; Van Duuren, 1966). There exists, however, a long and somewhat confusing controversy concerning these experimental and theoretical studies (see, for example, Pullman and Pullman, 1963b; Allison and Nash, 1963b; Pullman, 1964). It must be emphasized that up until recently the experimental information on the electron accepting and/or donating capacities of the molecules considered in these discussions has been indirect, derived predominantly from charge-transfer (donor–acceptor) complexes. Although the existence of electron transport remains unquestionable, there is need for accurate experimental determinations of the ionization potentials, electron affinities, and electron capture cross sections for these compounds before its role is understood. Further, the implicit assumption in the majority of these studies that the hydrocarbon molecules act as electron donors may not be valid in view of recent direct physical measurements (Christophorou and Blaunstein, 1969; Table 9.1) which demonstrate that higher aromatic hydrocarbons act as strong electron acceptors.

Electron capture by molecular systems in the gas phase has been discussed in detail in Chapter 6. In assessing the ability of compounds to capture low-energy free electrons, one should carefully distinguish between dissociative and non-dissociative electron capture. If, upon electron capture the molecule breaks up—such a dissociative process can and does proceed often with an exceedingly high cross section even with 'zero' energy electrons if the molecule contains atoms or groups of atoms whose electron affinity exceeds the corresponding bond dissociation energy—the electron affinity of the parent molecule has little, if any, effect on the process and its cross section. If, however, a parent negative ion is formed, the electron affinity of the parent molecule determines the stability of the parent ion (once de-excited) but not necessarily its electron accepting capacity. This distinction is essential and has not been carefully considered in the past. An absolute measurement of the electron accepting capacity of a molecule is its electron capture cross section and not its electron affinity or any theoretical coefficients related to the electron affinity. Such absolute measurements, as has already been stated, have recently been made for a number of organic vapours

(Table 9.1; Chapter 6). These measurements reflect basic molecular properties presently being used as the basis of theories of biological action.

From the data listed in Table 9.1 on the thermal electron attachment rate $(\alpha w)_{th}$, the Pullmans' $|k|$ coefficient, the ionization potential $I$, and the electron affinity $EA$, for several aromatic hydrocarbons, no direct correlation between the molecular electron accepting capacity as manifested by the value of $(\alpha w)_{th}$ and either $|k|$ or $I$ or $EA$ is evident. There is, however, a general trend for $(\alpha w)_{th}$ to increase with $EA$ and to decrease with $|k|$ or $I$. The $|k|$ coefficients are theoretical quantities which, when accurately determined, allow an estimate of the molecular electron affinity but not necessarily of the molecular electron accepting capacity. It must be pointed out that physical quantities such as those listed in Table 9.1 should be used with caution when applied to the reality of a biological system. (In this respect, similar studies in the liquid or the dissolved state would be most valuable.) Additionally, care must be taken to assess the role of dissociative electron attachment to compounds containing halogens or other groups of molecules such as $NO_2$, which possess high electron affinities and are used in studies of charge-transfer complexes. The ability of certain substances such as diiodophenol or dinitrophenol to disturb energy transfer in oxidative metabolism could be due to such a molecular dissociation leading to radical and fragment negative ion formation, and thus acting as an irreversible electron trap.

The preceding discussion merely touches the problem concerned. If, however, it has helped point out the need of further physical studies in this so profoundly important area of research, it has served its purpose well.

## 9.4 References

A. C. Allison and T. Nash (1963a). *Nature*, **197**, 758.
A. C. Allison and T. Nash (1963b). *Nature*, **199**, 467.
W. Arnold and R. K. Clayton (1960). *Proc. Natl. Acad. Sci. U. S.*, **46**, 769.
G. M. Badger (1948). *Brit. J. Cancer*, **II**, 309.
G. M. Badger (1954). *Advan. Cancer Res.*, **II**, 73.
R. M. Badger, A. C. Wright and R. F. Whitlock (1965). *J. Chem. Phys.*, **43**, 4345.
J. B. Birks (1959). *Discussions Faraday Soc.*, **27**, 243.
J. B. Birks (1961). *Nature*, **190**, 232.
J. B. Birks (1969). Private communication.
J. B. Birks (1970). *Photophysics of Aromatic Molecules*, Interscience, London, 1970.
J. B. Birks and A. J. W. Cameron (1959). *Proc. Roy. Soc. (London)*, **A249**, 297.
R. P. Blaunstein and L. G. Christophorou (1968). *J. Chem. Phys.*, **49**, 1526.
D. F. Bradley and M. Calvin (1956). *Proc. Natl. Acad. Sci. U. S.*, **42**, 710.
W. E. Bruce (1941). *J. Am. Chem. Soc.*, **63**, 304.
W. E. Bruce and F. Todd (1939). *J. Am. Chem. Soc.*, **61**, 157.
O. Chalvet, P. Daudel, R. Daudel, C. Moser and C. Prodi (1966). In L. de Broglie (Ed.), *Wave Mechanics and Molecular Biology*, Addison-Wesley Publishing Co., Reading, Mass., p. 106.

L. G. Christophorou (1963). PhD thesis, University of Manchester.
L. G. Christophorou and R. P. Blaunstein (1969). *Rad. Res.*, **37**, 229.
E. Clar (1964). *Polycyclic Hydrocarbons*, Vols. I and II, Academic Press, New York.
C. A. Coulson (1953). *Advan. Cancer Res.*, **I**, 1.
D. D. Eley (1959). *Research*, **12**, 293.
S. S. Epstein, I. Bulon, J. Koplan, M. Small and N. Mantel (1964). *Nature*, **204**, 750.
M. D. Evans and J. Gergely (1949). *Biochim. Biophys. Acta*, **3**, 188.
D. E. Green (1959). *Discussions Faraday Soc.*, **27**, 206.
R. M. Hedges and F. A. Matsen (1958). *J. Chem. Phys.*, **28**, 950.
I. Hieger (1930). *Biochem. J.*, **24**, 505.
A. V. Jones and A. W. Harrison (1958). *J. Atmospheric Terrest. Phys.*, **13**, 45.
J. B. Jones, M. Bersohn and G. C. Niece (1966). *Nature*, **211**, 309.
R. N. Jones (1940). *J. Am. Chem. Soc.*, **62**, 148.
R. N. Jones (1941). *J. Am. Chem. Soc.*, **63**, 151.
R. N. Jones (1943). *Chem. Rev.*, **32**, 1.
H. Kautsky (1939). *Trans. Faraday Soc.*, **35**, 216.
A. Lacassagne (1966). In L. de Broglie (Ed.), *Wave Mechanics and Molecular Biology*, Addison-Wesley Publishing Co., Reading, Mass., p. 101.
A. L. Lehninger (1964). *The Mitochondrion*, W. A. Benjamin, Inc., New York.
J. E. Lovelock (1961). *Nature*, **189**, 729.
R. Mason (1958a). *Nature*, **181**, 820.
R. Mason (1958b). *Brit. J. Cancer*, **12**, 469.
R. Mason (1959). *Discussions Faraday Soc.*, **27**, 129.
R. Mason (1960). *Radiation Res. Suppl.*, **2**, 452.
R. Mason (1966). In L. de Broglie (Ed.), *Wave Mechanics and Molecular Biology*, Addison-Wesley Publishing Co., Reading, Mass., p. 75.
W. D. McElroy and B. Glass (Eds.) (1961). *A Symposium on Light and Life*, The Johns Hopkins Press, Baltimore.
J. R. Platt (1959). *Science*, **129**, 372.
A. Pullman (1964). *Biopolymers Symposia No.* **1**, 47.
A. Pullman and B. Pullman (1955a). *Advan. Cancer Res.*, **3**, 117.
A. Pullman and B. Pullman (1955b). *Cancérisation par les Substances Chimiques et Structure Moléculaire*, Masson, Paris.
B. Pullman and A. Pullman (1963a). *Quantum Biochemistry*, Interscience Publishers, New York.
B. Pullman and A. Pullman (1963b). *Nature*, **199**, 467.
R. Schoental (1964). *Polycyclic Hydrocarbons* (by E. Clar), Vol. I, Academic Press, London, Chap. 18.
R. Schoental and E. J. Y. Scott (1949). *J. Chem. Soc.*, **96**, 1683.
E. V. Shpol'skii, A. A. Il'ina and V. V. Basilevich (1948). *Izvest. Akad. Nauk S. S. S. R.*, **12**, 519.
R. H. Steele, L. C. Cusachs and S. P. McGlynn (1967). *Int. J. Quantum Chem.* **Is**, 179.
J. A. Stockdale and D. F. Sangster (1966). *J. Am. Chem. Soc.*, **88**, 2907.
A. Szent-Györgyi (1941). *Science*, **93**, 609.
A. Szent-Györgyi (1957). *Bioenergetics*, Academic Press, New York.
A. Szent-Györgyi (1968a). *Bioelectronics*, Academic Press, New York.
A. Szent-Györgyi (1968b). *Science*, **161**, 988.
A. Szent-Györgyi, I. Isenberg and S. Baird, Jr., (1960). *Proc. Natl. Acad. Sci. U.S.*, **46**, 1444.
B. L. Van Duuren (1966). *Nature*, **210**, 622.

# Appendix I[a]

[a]Taken from W. Finkelnburg and W. Humbach, *Naturwissenschaften*, **42**, 35 (1955).

| Atomic number | Element | I | $I^+$ | $I^{2+}$ |
|---|---|---|---|---|
| 1 | H | 13.595 | | |
| 2 | He | 24.580 | 54.400 | |
| 3 | Li | 5.390 | 75.62 | 122.42 |
| 4 | Be | 9.320 | 18.21 | 153.85 |
| 5 | B | 8.296 | 25.15 | 37.92 |
| 6 | C | 11.264 | 24.376 | 47.86 |
| 7 | N | 14.54 | 29.60 | 47.426 |
| 8 | O | 13.614 | 35.15 | 54.93 |
| 9 | F | 17.418 | 34.98 | 62.65 |
| 10 | Ne | 21.559 | 41.07 | 63.5 $\pm$0.1 |
| 11 | Na | 5.138 | 47.29 | 71.8 $\pm$0.1 |
| 12 | Mg | 7.644 | 15.03 | 78.2 $\pm$0.1 |
| 13 | Al | 5.984 | 18.82 | 28.44 |
| 14 | Si | 8.149 | 16.34 | 33.46 |
| 15 | P | 10.55 | 19.65 | 30.16 |
| 16 | S | 10.357 | 23.4 | 34.8 |
| 17 | Cl | 13.01 | 23.80 | 39.9 |
| 18 | Ar | 15.755 | 27.6 | 40.90 |
| 19 | K | 4.339 | 31.81 | 45.9 |
| 20 | Ca | 6.111 | 11.87 | 51.21 |
| 21 | Sc | 6.56 | 12.89 | 24.75 |
| 22 | Ti | 6.83 | 13.57 | 28.14 |
| 23 | V | 6.74 | 14.2 | 29.7 |
| 24 | Cr | 6.764 | 16.49 | 31 |
| 25 | Mn | 7.432 | 15.64 | 33.69 |
| 26 | Fe | 7.90 | 16.18 | 30.64 |
| 27 | Co | 7.86 | 17.05 | 33.49 |
| 28 | Ni | 7.633 | 18.15 | 36.16 |
| 29 | Cu | 7.724 | 20.29 | 36.83 |
| 30 | Zn | 9.391 | 17.96 | 39.70 |
| 31 | Ga | 6.00 | 20.51 | 30.70 |
| 32 | Ge | 7.88 | 15.93 | 34.21 |
| 33 | As | 9.81 | 18.7 $\pm$0.1 | 28.3 |
| 34 | Se | 9.75 | 21.5 | 32.0 |
| 35 | Br | 11.84 | 21.6 | 35.9 |
| 36 | Kr | 13.996 | 24.56 | 36.9 |
| 37 | Rb | 4.176 | 27.56 | 40 |
| 38 | Sr | 5.692 | 11.026 | 43.6 |
| 39 | Y | 6.38 | 12.23 | 20.5 |
| 40 | Zr | 6.835 | 12.92 | 24.8 |
| 41 | Nb | 6.88 | 13.90 | 28.1 |
| 42 | Mo | 7.131 | 15.72 | 29.6 |
| 43 | Tc | 7.23 | 14.87 | 31.9 |
| 44 | Ru | 7.36 | 16.60 | 30.3 |
| 45 | Rh | 7.46 | 15.92 | 32.8 |
| 46 | Pd | 8.33 | 19.42 | |

## Appendix I (*continued*)

| Atomic number | Element | I | I$^+$ | I$^{2+}$ |
|---|---|---|---|---|
| 47 | Ag | 7.574 | 21.48 | 36.10 |
| 48 | Cd | 8.991 | 16.904 | 44.5 |
| 49 | In | 5.785 | 18.86 | 28.0 |
| 50 | Sn | 7.332 | 14.6 | 30.7 |
| 51 | Sb | 8.64 | 16.7 | 24.8 |
|  |  |  | $\pm$0.5 |  |
| 52 | Te | 9.01 | 18.8 | 31 |
|  |  |  | $\pm$0.5 |  |
| 53 | I | 10.44 | 19.0 | 33 |
| 54 | Xe | 12.127 | 21.2 | 32.1 |
| 55 | Cs | 3.893 | 25.1 | 34.6 |
|  |  |  |  | $\pm$0.7 |
| 56 | Ba | 5.810 | 10.00 | 37 |
|  |  |  |  | $\pm$1 |
| 57 | La | 5.61 | 11.43 | 19.17 |
| 58 | Ce | (6.91) | 12.3 | 19.5 |
| 59 | Pr | (5.76) |  |  |
| 60 | Nd | (6.31) |  |  |
| 61 | Pm |  |  |  |
| 62 | Sm | 5.6 | (11.2) |  |
| 63 | Eu | 5.67 | 11.24 |  |
| 64 | Gd | 6.16 | (12) |  |
| 65 | Tb | (6.74) |  |  |
| 66 | Dy | (6.82) |  |  |
| 67 | Ho |  |  |  |
| 68 | Er |  |  |  |
| 69 | Tm |  |  |  |
| 70 | Yb | 6.2 | 12.10 |  |
| 71 | Lu | 6.15 | 14.7 |  |
| 72 | Hf | 5.5 | 14.9 |  |
| 73 | Ta | 7.7 | 16.2 |  |
|  |  |  | $\pm$0.5 |  |
| 74 | W | 7.98 | 17.7 |  |
|  |  |  | $\pm$0.5 |  |
| 75 | Re | 7.87 | 16.6 |  |
|  |  |  | $\pm$0.5 |  |
| 76 | Os | 8.7 | 17 |  |
|  |  |  | $\pm$1 |  |
| 77 | Ir | 9.2 | 17.0 |  |
|  |  |  | $\pm$0.3 |  |
| 78 | Pt | 8.96 | 18.54 |  |
| 79 | Au | 9.223 | 20.5 |  |
| 80 | Hg | 10.434 | 18.751 | 34.2 |
| 81 | Tl | 6.106 | 20.42 | 29.8 |
| 82 | Pb | 7.415 | 15.03 | 31.93 |
| 83 | Bi | 7.287 | 19.3 | 25.6 |
| 84 | Po | 8.2 | 19.4 | 27.3 |
|  |  | $\pm$0.4 | $\pm$1.7 | $\pm$0.8 |

Appendix I (*continued*)

| Atomic number | Element | I | I$^+$ | I$^{2+}$ |
|---|---|---|---|---|
| 85 | At | 9.2 | 20.1 | 29.3 |
|  |  | ±0.4 | ⊥1.7 | ±0.9 |
| 86 | Rn | 10.745 | 21.4 | 29.4 |
|  |  |  | ±1.8 | ±1.0 |
| 87 | Fr | 3.98 | 22.5 | 33.5 |
|  |  | ±0.10 | ±1.8 | ±1.5 |
| 88 | Ra | 5.277 | 10.144 |  |
| 89 | Ac | 6.89 | 11.5 |  |
|  |  | ±0.6 | ±0.4 |  |
| 90 | Th |  | 11.5 | 20.0 |
|  |  |  | ±1.0 |  |
| 91 | Pa |  |  |  |
| 92 | U | 4 |  |  |

# *Appendix II*[a]

[a] See also recent tabulations of ionization potentials in V. I. Vedeneyev, L. V. Gurvich, V. N. Kondrat'yev, V. A. Medvedev and Ye. L. Frankevich, *Bond Energies, Ionization Potentials and Electron Affinities*, Edward Arnold Publishers Ltd., 1966; J. L. Franklin, J. G. Dillard, H. M. Rosenstock, J. T. Herron, K. Draxl and F. H. Field, *Ionization Potentials, Appearance Potentials and Heats of Formation of Gaseous Positive Ions*, Nat. Stand. Ref. Data Ser., Nat. Bur. Stand. (U.S.)-NSRDS-NBS 26 (1969).

| Molecule or radical | Ionization potential (eV) | Reference and method[b,c] | Molecule or radical | Ionization potential (eV) | Reference and method[b,c] |
|---|---|---|---|---|---|
| $H_2$ | 15.42 | 1a | $Br_2$ | 10.55 | 4a |
| | 15.37 | 2c | | 10.31 | 26b |
| HF | 15.77 | 3c | | 10.92 | 6c |
| HCl | 12.74 | 4a | $I_2$ | 9.28 | 4a |
| | 12.85 | 4b | | 9.41 | 6c |
| | 12.84 | 5b | | 9.50 | 17c |
| | 12.78 | 6c | $H_2O$ | 12.50 | 1a |
| HBr | 11.62 | 4a | | 12.59 | 4a |
| | 11.93 | 4b | | 12.61 | 27a |
| | 12.04 | 5b | | 12.60 | 28b |
| | 11.69 | 6c | | 12.61 | 4b |
| HI | 10.38 | 4a | | 12.59 | 2c |
| | 10.39 | 4b | | 12.60 | 29c, 3c |
| | 10.71 | 5b | | 12.61 | 30c |
| | 10.37 | 7c | | 12.67 | 7c |
| | 10.44 | 3c | | 12.76 | 6c |
| | 10.48 | 6c | $NH_2$ | 11.4 | 31c |
| OH | 13.18 | 8c | $HO_2$ | 11.53 | 24c |
| | 13.7 | 7c | $H_2S$ | 10.42 | 27a |
| NO | 9.20 | 9a | | 10.46 | 13a |
| | 9.23 | 10a, 11a, 12a | | 10.47 | 32b |
| | | | | 10.47 | 4b |
| | 9.25 | 13a | | 10.5 | 28b |
| | 9.4 | 14a | | 10.4 | 17c |
| | 9.24 | 4b | | 10.45 | 3c |
| | 9.26 | 15b | | 10.5 | 7c |
| | 9.25 | 16c | | 10.62 | 33c |
| | 9.3 | 17c | $N_2O$ | 12.82 | 27a |
| | 9.4 | 18c | | 12.83 | 9a |
| $N_2$ | 15.58 | 19a | | 12.90 | 4a |
| | 15.6 | 1a | | 12.72 | 4b |
| | 15.58 | 7b | | 12.9 | 17c |
| | 15.52 | 20c | $NO_2$ | 9.78 | 34a |
| | 15.57 | 2c | | 9.80 | 35a |
| | 15.6 | 21c, 22c | | 9.91 | 36c |
| | 15.7 | 18c, 23c | | 9.93 | 37c |
| $O_2$ | 12.04 | 10a | $SO_2$ | 12.32 | 27a |
| | 12.08 | 4a | | 12.34 | 4a |
| | 12.10 | 1a, 11a | | 12.42 | 10a |
| | 12.2 | 4b | | 12.05 | 38b |
| | 12.1 | 18c | | 12.11 | 4b |
| | 12.15 | 24c | | 12.44 | 39c |
| | 12.21 | 25c | $NH_3$ | 10.07 | 9a |
| $F_2$ | 15.7 | 19a | | 10.13 | 10a |
| $Cl_2$ | 11.48 | 4a | | 10.15 | 19a, 40a |
| | 11.80 | 6c | | 10.16 | 27a |

## Appendix II (*continued*)

| Molecule or radical | Ionization potential (eV) | Reference and method[b,c] | Molecule or radical | Ionization potential (eV) | Reference and method[b,c] |
|---|---|---|---|---|---|
| | 10.23 | 4a | | 11.41 | 27a, 59a, 7b |
| | 10.25 | 13a | | | |
| | 10.35 | 29a | | 11.35 | 60b |
| | 10.34 | 41c | | 11.39 | 55c |
| | 10.40 | 3c | | 11.40 | 61c |
| | 10.52 | 6c | | 11.42 | 7c |
| CO | 13.98 | 11a | | 11.43 | 22c |
| | 14.01 | 4b, 7b | | 11.46 | 62c |
| | 13.98 | 21c | | 11.50 | 63c |
| | 14.1 | 17c, 18c, 23c | $C_2D_2$ | 11.42 | 59a |
| | | | Acetylene-$d_2$ | | |
| $CH_2$ | 10.4 | 42b | HCHO | 10.87 | 4a |
| | 11.82 | 43c | Formaldehyde | 10.90 | 64a |
| | 11.90 | 44c | | 10.83 | 45b |
| $CO_2$ | 13.68 | 27a | | 10.88 | 65c |
| | 13.79 | 4a | $CH_4$ | 12.8 | 1a |
| | 13.73 | 45b | | 12.98 | 4a |
| | 13.79 | 4b | | 12.99 | 27a |
| | 13.78 | 46c, 47c | | 12.72 | 2c |
| | 13.85 | 22c, 48c | | 12.99 | 55c |
| | 13.88 | 20c | | 13.00 | 66c |
| $CS_2$ | 10.07 | 27a | | 13.04 | 22c |
| | 10.08 | 13a | | 13.07 | 67c |
| | 10.08 | 4b, 38b | | 13.10 | 68c, 54c, 69c |
| | 10.13 | 49c | | | |
| | 10.15 | 48c | | 13.11 | 47c |
| HCN | 13.86 | 7c | | 13.12 | 7c |
| | 13.91 | 6c | | 13.16 | 6c, 70c |
| CNCl | 12.49 | 50c | $CH_2D_2$ | 13.14 | 67c |
| CNBr | 11.95 | 50c | $CD_4$ | 13.25 | 67c |
| CNI | 10.98 | 50c | $CH_3F$ | 12.61 | 72c |
| $CH_3$ | 9.82 | 51a | | 12.72 | 7c |
| | 9.84 | 52b | | 12.85 | 70c |
| | 9.80 | 53c | $CH_3Cl$ | 11.28 | 4a |
| | 9.85 | 54c | | 11.17 | 71b |
| | 9.86 | 43c | | 11.22 | 4b |
| | 9.87 | 55c | | 11.25 | 32b |
| | 9.95 | 56c | | 11.33 | 7b |
| | 9.96 | 7c | | 11.28 | 55c |
| | 10.07 | 57c | | 11.3 | 69c |
| | 10.11 | 58c | | 11.34 | 72c |
| $CD_3$ | 9.83 | 52b | | 11.35 | 7c |
| CH≡CH | 11.25 | 9a | | 11.42 | 70c |
| Acetylene | 11.36 | 11a | | 11.46 | 6c |
| | 11.40 | 4a | | 11.6 | 74c |

**Appendix II** (*continued*)

| Molecule or radical | Ionization potential (eV) | Reference and method[b,c] | Molecule or radical | Ionization potential (eV) | Reference and method[b,c] |
|---|---|---|---|---|---|
| $CH_3Br$ | 10.53 | 4a | | 10.46 | 46c |
| | 10.54 | 73a | | 10.48 | 63c |
| | 10.49 | 71b | | 10.50 | 55c |
| | 10.54 | 7b, 4b | | 10.56 | 7c, 81c |
| | 10.5 | 69c, 72c | | 10.58 | 47c |
| | 10.53 | 70c | | 10.60 | 67c |
| | 10.56 | 55c | | 10.62 | 22c |
| | 10.6 | 7c | | 10.68 | 20c |
| | 10.73 | 6c | $CD_2=CD_2$ | 10.59 | 67c |
| | 10.8 | 74c | $CH{\equiv}CC{\equiv}CH$ | 10.74 | 7b, 83b |
| $CH_3I$ | 9.5 | 10a | Diacetylene | 10.2 | 61c |
| | 9.54 | 13a, 4b | | 10.9 | 62c |
| | 9.49 | 71b, 32b | $CH_3CN$ | 12.22 | 84a |
| | 9.51 | 35c, 70c | Acetonitrile | 11.96 | 85b |
| | 9.55 | 72c | | 12.39 | 6c |
| | 9.59 | 55c | | 12.46 | 7c |
| | 9.6 | 69c | $CH_3OH$ | 10.52 | 10a |
| | 9.67 | 6c | | 10.83 | 27a |
| $CH_2Cl_2$ | 11.35 | 4a | | 10.85 | 13a |
| | 11.4 | 74c | | 10.80 | 32b |
| $CH_2Br_2$ | 10.8 | 74c | | 10.88 | 7c |
| $CHCl_3$ | 11.42 | 4a | | 10.95 | 6c, 48c |
| $CF_3Cl$ | 12.92 | 73a | | 10.97 | 86c |
| | 12.8 | 75c | $HCONH_2$ | 10.16 | 79a |
| | 13.1 | 76c | Formamide | 10.25 | 87a, 19a |
| $CCl_2F_2$ | 11.7 | 77c | | 10.2 | 88b |
| $CCl_4$ | 11.47 | 4a | | 10.25 | 87c |
| | 11.00 | 77c | $CH_2=CHF$ | 10.37 | 73a, 81a |
| $CH_2=C=O$ | 9.60 | 26b | Fluoroethylene | 10.45 | 81c |
| Ketene | 9.40 | 78c | $CH_2=CF_2$ | 10.30 | 73a |
| HCOOH | 11.05 | 4a, 79a | 1,1-Difluoro- | 10.31 | 81a |
| Formic acid | 11.3 | 80b | ethylene | 10.32 | 81c |
| | 11.33 | 4b | $CHF=CF_2$ | 10.14 | 73a |
| | 11.40 | 47c | $CF_2=CF_2$ | 10.12 | 73a |
| | 11.51 | 65c, 6c | Tetrafluoro- | | |
| $CH_2=CH_2$ | 10.46 | 9a | ethylene | | |
| Ethylene | 10.47 | 10a | $CHCl=CHF$ | 9.86 | 81a |
| | 10.48 | 27a | *cis* | 10.14 | 81c |
| | 10.50 | 81a | $CHCl=CHF$ | 9.87 | 81a |
| | 10.52 | 13a | *trans* | 10.30 | 81c |
| | 10.41 | 60b | $CHF=CClF$ | 9.86 | 81a |
| | 10.45 | 82b | *cis* | 10.17 | 81c |
| | 10.50 | 32b | $CHF=CClF$ | 9.83 | 81a |
| | 10.51 | 22b, 4b, 7b | *trans* | 9.96 | 81c |
| | 10.4 | 28c | $CH_2=CHCl$ | 10.00 | 73a, 81a, |

## Appendix II (*continued*)

| Molecule or radical | Ionization potential (eV) | Reference and method[b,c] | Molecule or radical | Ionization potential (eV) | Reference and method[b,c] |
|---|---|---|---|---|---|
| Chloroethylene | 10.00 | 82b | | 9.41 | 41c, 6c, 33c |
| | 9.95 | 89b | | 9.65 | 105c |
| | 10.10 | 81c | $CH_3CHO$ | 10.20 | 64a |
| $CD_2=CDCl$ | 10.02 | 81a | Acetaldehyde | 10.21 | 4a |
| | 10.10 | 81c | | 10.25 | 14a |
| $CHCl=CHCl$ | 9.65 | 73a | | 10.18 | 91b |
| cis-1,2-Di- | 9.66 | 90a | | 10.23 | 26b |
| chloroethylene | 9.61 | 89b | | 10.25 | 92c |
| | 9.67 | 90c | | 10.26 | 65c |
| $CHCl=CHCl$ | 9.63 | 73a | | 10.28 | 6c |
| trans-1,2-Di- | 9.64 | 90a | $CH_2OCH_2$ | 10.49 | 27a |
| chloroethylene | 9.91 | 89b | Ethylene oxide | 10.57 | 4a |
| | 10.00 | 90c | | 10.65 | 93c |
| $CH_2=CCl_2$ | 9.79 | 73a | $CH_3NO_2$ | 11.08 | 84a |
| 1,1-Dichloro- | | | Nitromethane | 11.34 | 36c |
| ethylene | | | $C_2H_6$ | 11.49 | 27a |
| $CCl_2=CHCl$ | 9.45 | 73a | Ethane | 11.65 | 4a, 87a |
| Trichloroethylene | 9.47 | 4a | | 11.5 | 66c |
| | 9.94 | 6c | | 11.59 | 94c |
| $CCl_2=CCl_2$ | 9.32 | 73a | | 11.60 | 46c |
| Tetrachloro- | | | | 11.65 | 7c |
| ethylene | | | | 11.66 | 55c |
| $CH_2=CHBr$ | 9.80 | 73a | | 11.76 | 22c |
| Bromoethylene | 9.82 | 81a | | 11.78 | 20c |
| | 9.97 | 81c | $CH_2=CHCHO$ | 10.10 | 4a |
| $CHBr=CHBr$ | 9.45 | 73a, 90a | Acrolein | 10.06 | 89b |
| cis-1,2-Di- | 9.69 | 90c | | 10.11 | 7b |
| bromoethylene | | | | 10.14 | 92c |
| | | | | 10.25 | 33c |
| $CHBr=CHBr$ | 9.46 | 90a | | 10.34 | 6c |
| trans-1,2-Di- | 9.47 | 73a | $CH_3COOH$ | 10.37 | 84a |
| bromoethylene | 9.54 | 90c | Acetic acid | 10.38 | 64a |
| | | | | 10.33 | 26b |
| $CHBr=CBr_2$ | 9.27 | 73a | | 10.66 | 65c |
| Tribromoethylene | | | | 10.70 | 6c |
| $C_2H_5-$ | 8.78 | 26b | | 10.72 | 95c |
| Ethyl | 8.25 | 53c | $C_2H_5Cl$ | 10.97 | 4a |
| | 8.34 | 55c | Ethyl chloride | 10.98 | 87a |
| | 8.60 | 57c | | 10.90 | 4b |
| | 8.72 | 7c | | 11.18 | 6c |
| $CH_3C\equiv CH$ | 10.36 | 4a, 116b | | 11.2 | 96c |
| Propyne | 10.39 | 7c | $C_2H_5Br$ | 10.24 | 10a, 71b |
| | 10.48 | 62c | Ethyl bromide | 10.29 | 4a, b |
| $CH_3NH_2$ | 8.97 | 4a | | 10.21 | 26b |
| Methylamine | 9.18 | 27a | | 10.49 | 6c |

## Appendix II (*continued*)

| Molecule or radical | Ionization potential (eV) | Reference and method[b,c] | Molecule or radical | Ionization potential (eV) | Reference and method[b,c] |
|---|---|---|---|---|---|
| | 10.7 | 96c | | 9.18 | 104c |
| $C_2H_5I$ | 9.33 | 4a | | 9.24 | 6c |
| Ethyl iodide | 9.30 | 71b | $CH_2=C=$ | 9.57 | 104c |
| | 9.35 | 4b, 26b | $CHCH_3$ | | |
| | 9.47 | 6c | 1,2-Butadiene | | |
| | 9.60 | 96c | $n\text{-}C_3H_7-$ | 8.13 | 55c |
| $-(CH_2)_3-$ | 10.06 | 84a | | 8.15 | 53c |
| Cyclopropane | 10.09 | 4a | $iso\text{-}C_3H_7-$ | 7.52 | 53c |
| | 10.23 | 20c | | 7.57 | 55c |
| | 10.53 | 97c | $(CH_3)_2NH$ | 8.24 | 19a |
| $CH_2=CHCH_3$ | 9.73 | 4a, 98a | Dimethylamine | 8.36 | 27a |
| Propene | 9.60 | 82b | | 8.4 | 26b |
| | 9.65 | 22b | | 8.93 | 41c |
| | 9.70 | 7b | | 9.21 | 33c |
| | 9.78 | 21c | | 9.55 | 7c |
| | 9.80 | 7c | | 9.5 | 105c |
| | 9.84 | 22c, 20c | $C_2H_5NH_2$ | 8.86 | 40a |
| $C_2H_5CN$ | 11.84 | 19a | Ethylamine | 9.19 | 27a |
| Propionitrile | 11.85 | 6c | | 9.50 | 14a |
| $C_2H_5OH$ | 10.48 | 84a | | 9.19 | 41c |
| Ethyl alcohol | 10.50 | 13a | | 9.32 | 6c |
| | 10.63 | 27a | | 9.60 | 105c |
| | 10.65 | 6c, 86c | $C_4H_4NH$ | 8.20 | 19a |
| $-CH=CHOCH=$ | 8.89 | 4a | Pyrrole | 8.90 | 26b |
| $CH-$ | 9.01 | 99b | | 8.97 | 100c |
| Furan | 9.00 | 100c | | 9.20 | 106c |
| $(CH_3)_2O$ | 10.00 | 4a | $CH_3COCH_3$ | 9.67 | 27a |
| Dimethyl ether | 10.06 | 47c | Acetone | 9.69 | 13a |
| $CH_3CONH_2$ | 9.65 | 79a | | 9.71 | 64a |
| Acetamide | 9.77 | 87a | | 9.73 | 4a |
| | 10.36 | 65c | | 9.75 | 14a |
| | 10.39 | 33c | | 9.71 | 4b |
| $-CH=CHSCH=$ | 8.86 | 19a | | 9.89 | 33c |
| $CH-$ | 8.91 | 99b, 26b | | 9.92 | 6c |
| Thiophene | 9.10 | 100c | $C_2H_5CHO$ | 9.98 | 84a |
| | 9.2 | 101c | Propionaldehyde | 10.06 | 6c |
| $(CH_3)_2S$ | 8.69 | 19a | | 10.14 | 92c |
| Dimethyl sulphide | 8.73 | 26b | $C_3H_8$ | 11.07 | 27a |
| $CH{\equiv}CCH_2CH_3$ | 10.34 | 62c | Propane | 11.08 | 4a |
| 1-Butyne | | | | 11.09 | 55c |
| $CH_3C{\equiv}CCH_3$ | 9.85 | 26b, 7c, 62c | | 11.20 | 66c |
| 2-Butyne | | | | 11.21 | 22c, 7c |
| $CH_2=CHCH=$ | 9.07 | 13a, 102a | $-CH=$ | 8.55 | 102a |
| $CH_2$ | 9.08 | 27a | $CHCH_2CH=$ | 8.58 | 7b, 99b |
| 1,3-Butadiene | 8.71 | 103b | $CH-$ | 8.7 | 7c |

Appendix II (*continued*)

| Molecule or radical | Ionization potential (eV) | Reference and method[b,c] | Molecule or radical | Ionization potential (eV) | Reference and method[b,c] |
|---|---|---|---|---|---|
| Cyclopentadiene | 8.9 | 106c | alcohol | | |
| CH₃CH= | 9.73 | 4a, 87a | CH₃CH= | 8.59 | 102a |
| CHCHO | 9.81 | 87c | CHCH=CH₂ | 8.68 | 109c |
| Crotonaldehyde | | | *cis*-1,3-Pentadiene | | |
| C₂H₅COOH | 10.09 | 4a | CH₂= | 9.08 | 6c |
| Propionic acid | 10.47 | 6c | CHC(CH₃)= | | |
| CH₃CO₂CH₃ | 10.27 | 19a | CH₂ | | |
| Methyl acetate | 10.51 | 65c | Isoprene | | |
| n-C₃H₇Cl | 10.82 | 19a | –CH= | 9.01 | 87a |
| | 10.96 | 6c | CH(CH₂)₃– | 9.27 | 109c |
| iso-C₃H₇Cl | 10.78 | 19a | Cyclopentene | | |
| | 10.65 | 70c | C₃H₇NH₂ | 8.78 | 40a |
| n-C₃H₇Br | 10.18 | 19a | n-Propylamine | 9.40 | 14a |
| | 10.24 | 55c | | 9.17 | 6c |
| | 10.29 | 6c | | 9.6 | 105c |
| iso-C₃H₇Br | 10.08 | 19a | iso-C₃H₇NH₂ | 8.72 | 19a |
| | 10.11 | 6c | iso-propylamine | 8.86 | 27a |
| n-C₃H₇I | 9.26 | 19a | | 9.5 | 105c |
| | 9.41 | 6c | (CH₃)₃N | 8.32 | 41c |
| iso-C₃H₇I | 9.17 | 19a | Trimethylamine | 9.02 | 33c |
| CH₂=CHC₂H₅ | 9.58 | 4a | | 9.2 | 105c |
| 1-Butene | 9.61 | 98a | | 9.3 | 7c |
| | 9.65 | 107c, 108c | C₃H₇CHO | 9.86 | 19a |
| | 9.76 | 109c, 22c | n-Butyraldehyde | 10.01 | 6c |
| CH₃CH=CHCH₃ | 9.13 | 110a | iso-C₃H₇CHO | 9.74 | 19a |
| *cis*-2-Butene | 9.24 | 87b | CH₃COC₂H₅ | 9.50 | 14a |
| | 9.29 | 22c | Methyl ethyl | 9.53 | 79a |
| | 9.34 | 109c | ketone | 9.54 | 13a |
| | 9.41 | 108c | | 9.74 | 65c |
| CH₃CH=CHCH₃ | 9.13 | 110a | | 9.76 | 6c |
| *trans*-2-Butene | 9.2 | 82b | –CH₂CH₂OCH₂– | 9.49 | 111a |
| | 9.13 | 108c | CH₂– | | |
| | 9.27 | 109c, 22c | Tetramethylene | | |
| CH₂=C(CH₃)₂ | 9.23 | 73a | oxide | | |
| 2-Methyl propene | 9.26 | 109c | C₃H₇COOH | 10.16 | 19a |
| | 9.35 | 22c | Butyric acid | 10.22 | 6c |
| C₃H₇CN | 11.67 | 19a | C₄H₁₀ | 10.50 | 27a |
| n-Butyronitrile | | | n-Butane | 10.63 | 4a |
| C₃H₇OH | 10.20 | 87a | | 10.3 | 66c |
| n-Propyl alcohol | 10.50 | 14a | | 10.34 | 94c |
| | 10.15 | 26b | | 10.80 | 22c |
| | 10.42 | 87c, 86c | iso-C₄H₁₀ | 10.55 | 110a |
| | 10.46 | 6c | | 10.57 | 87a |
| iso-C₃H₇OH | 10.15 | 4a | | 10.78 | 27a |
| iso-Propyl | 10.27 | 86c | | 10.34 | 94c |

## Appendix II (*continued*)

| Molecule or radical | Ionization potential (eV) | Reference and method[b,c] | Molecule or radical | Ionization potential (eV) | Reference and method[b,c] |
|---|---|---|---|---|---|
| | 10.79 | 109c | Diethyl sulphide | 8.48 | 26b |
| $CH_2CHCH=$ | 8.23 | 112b | $–CH=CH(CH_2)_4–$ | 8.72 | 27a |
| $CHCH=CH_2$ | | | Cyclohexene | 8.95 | 4a, 26b |
| Hexatriene | | | | 9.18 | 109c |
| $NHCH_2CH_2–$ | 8.41 | 27a | | 9.24 | 6c |
| $CH_2CH_2$ | | | | 9.7 | 106c |
| Pyrrolidine | | | $C_4H_9NH_2$ | 8.71 | 40a |
| $–CH_2OCH_2CH_2$ | 9.13 | 19a | n-Butylamine | 8.79 | 27a |
| $OCH_2–$ | 9.52 | 6c | | 9.19 | 6c |
| p-Dioxane | | | $C_4H_9NH_2$ | 8.64 | 19a |
| $–(CH_2)_4–$ | 10.58 | 97c | t-Butylamine | 8.83 | 27a |
| Cyclobutane | | | $(C_2H_5)_2NH$ | 8.01 | 40a |
| $C_4H_9Br$ | 10.13 | 4a, 19a | Diethylamine | 8.51 | 27a |
| n-Butyl bromide | 10.12 | 6c | | 8.44 | 41c |
| iso-$C_4H_9Br$ | 9.98 | 19a | $CH_3COC_3H_7$ | 9.39 | 19a |
| iso-Butyl bromide | 10.24 | 6c | 2-Pentanone | 9.47 | 64a |
| $C_4H_9I$ | 9.21 | 19a | | 9.59 | 6c |
| n-Butyl iodide | 9.32 | 6c | $C_5H_{12}$ | 10.33 | 110a |
| $–(CH_2)_5–$ | 10.51 | 110a | n-Pentane | 10.35 | 87a |
| Cyclopentane | 10.53 | 84a | | 10.5 | 66c |
| | 10.92 | 97c | | 10.55 | 22c |
| $CH_2=CHC_3H_7$ | 9.50 | 110a, 98a | iso-$C_5H_{12}$ | 10.30 | 110a |
| 1-Pentene | 9.66 | 22c | | 10.32 | 87a |
| | 9.67 | 109c | | 10.1 | 66c |
| $C_2H_5C(CH_3)=$ | 9.12 | 87a | | 10.60 | 109c |
| $CH_2$ | 9.20 | 109c | $CH_3C(CH_3)_3$ | 10.35 | 87a |
| Ethyl methyl | | | Neopentane | 10.37 | 110a |
| ethylene | | | | 10.29 | 7c |
| $(CH_3)_2C=$ | 8.67 | 87a | $CH_2=CHC_4H_9$ | 9.45 | 98a |
| $CHCH_3$ | 8.68 | 73a | 1-Hexene | 9.46 | 110a |
| 2-Methyl 2-butene | 8.75 | 82b | | 9.59 | 22c |
| | 8.80 | 32b | $(CH_3)_2C=$ | 9.08 | 87a |
| | 8.89 | 109c | $CHCOCH_3$ | 8.89 | 87c |
| $C_4H_9OH$ | 10.04 | 19a | Mesityl oxide | | |
| n-Butyl alcohol | 10.10 | 26b | $(CH_3)_2C=$ | 8.30 | 110a, 82b |
| | 10.30 | 86c | $C(CH_3)_2$ | 8.53 | 109c |
| iso-$C_4H_9OH$ | 10.17 | 86c | Tetramethyl | | |
| t-$C_4H_9OH$ | 9.7 | 26b | ethylene | | |
| | 9.92 | 86c | $–(CH_2)_6–$ | 9.79 | 27a |
| $(C_2H_5)_2O$ | 9.53 | 4a | Cyclohexane | 9.88 | 4a |
| Diethyl ether | 9.55 | 111a | | 10.3 | 7c |
| | 9.61 | 27a | | 10.50 | 97c |
| | 9.58 | 55c | $CH_3COC_4H_9$ | 9.44 | 79a |
| | 9.72 | 6c | 2-Hexanone | 9.56 | 64a |
| $(C_2H_5)_2S$ | 8.43 | 19a | | 9.58 | 6c |

## Appendix II (*continued*)

| Molecule or radical | Ionization potential (eV) | Reference and method[b,c] | Molecule or radical | Ionization potential (eV) | Reference and method[b,c] |
|---|---|---|---|---|---|
| $C_6H_{14}$ | 10.17 | 110a | $C_4H_4N_2$ | 9.35 | 113a |
| n-Hexane | 10.18 | 87a | 1,3-Pyrimidine | 9.47 | 114a |
| | 10.10 | 66c | | 9.91 | 115c |
| | 10.43 | 22c | $C_4H_4N_2$ | 9.27 | 114a |
| | 10.54 | 6c | 1,4-Pyrazine | 9.29 | 113a, 116b |
| iso-$C_6H_{14}$ | 10.12 | 87c | | 10.01 | 115c |
| | 10.34 | 109c | $C_5H_5N$ | 9.23 | 4a |
| $(CH_3)_2CHCH-$ | 10.00 | 110a | Pyridine | 9.28 | 27a |
| $(CH_3)_2$ | 10.02 | 87a | | 9.40 | 79a |
| 2,3-Dimethyl | 10.24 | 109c | | 9.27 | 117b |
| butane | | | | 9.7 | 118b |
| $(C_2H_5)_2CHCH_3$ | 10.05 | 110a | | 9.76 | 100c |
| 3-Methyl pentane | 10.08 | 87a | | 9.80 | 7c |
| | 10.30 | 109c | $C_6H_6$ | 9.24 | 64a |
| $-CH(CH_3)(CH_2)_5-$ | 9.85 | 84a | Benzene | 9.25 | 13a, 27a, |
| Methyl cyclo- | 10.19 | 97c | | | 119a |
| hexane | | | | 9.19 | 120b |
| $CH_2=CHC_5H_{11}$ | 9.54 | 22c | | 9.24 | 121b |
| 1-Heptene | | | | 9.25 | 122b, 117b |
| $(C_3H_7)_2O$ | 9.28 | 111a | | 9.20 | 53c |
| n-Propyl ether | | | | 9.21 | 21c, 123c, |
| $(C_2H_5)_3N$ | 7.50 | 40a | | | 36c |
| Triethylamine | 7.84 | 27a | | 9.24 | 124c |
| | 7.68 | 53c | | 9.38 | 125c |
| | 7.85 | 41c | | 9.43 | 22c |
| $C_7H_{16}$ | 10.06 | 110a | | 9.50 | 126c |
| n-Heptane | 10.08 | 87a | | 9.52 | 127c, 100c, |
| | 10.20 | 27a | | | 6c |
| | 10.00 | 66c | $C_6H_5F$ | 9.19 | 4a |
| | 10.35 | 22c | Fluorobenzene | 9.20 | 73a |
| $CH_2=CHC_6H_{13}$ | 9.52 | 22c | | 9.21 | 119a |
| 1-Octene | | | | 9.20 | 4b, 121b |
| $C_8H_{18}$ | 10.24 | 22c | | 9.30 | 128c |
| n-Octane | | | | 9.67 | 6c |
| $(C_4H_9)_2O$ | 9.18 | 111a | $C_6H_5Cl$ | 9.07 | 4a, 73a |
| n-Butyl ether | 9.28 | 70c | Chlorobenzene | 9.42 | 6c |
| $C_9H_{20}$ | 10.21 | 22c | $o$-$C_6H_4Cl_2$ | 9.06 | 73a |
| n-Nonane | | | $o$-Dichlorobenzene | 9.07 | 87a |
| $CH_2=CHC_8H_{17}$ | 9.51 | 22c | $m$-$C_6H_4Cl_2$ | 9.12 | 87a |
| 1-Decene | | | $m$-Dichlorobenzene | | |
| $C_{10}H_{22}$ | 10.19 | 22c | $p$-$C_6H_4Cl_2$ | 8.94 | 87a |
| n-Decane | | | $p$-Dichlorobenzene | 8.95 | 73a |
| $C_4H_4N_2$ | 8.71 | 113a | $C_6H_5Br$ | 8.98 | 4a |
| 1,2-Pyridazine | 8.91 | 114a | Bromobenzene | 9.41 | 6c |
| | 9.86 | 115c | | 10.05 | 128c |

## Appendix II (*continued*)

| Molecule or radical | Ionization potential (eV) | Reference and method[b,c] | Molecule or radical | Ionization potential (eV) | Reference and method[b,c] |
|---|---|---|---|---|---|
| $C_6H_5I$ | 8.73 | 4a | *o*-Bromotoluene | | |
| Iodobenzene | 9.10 | 6c | *m*-$C_6H_4BrCH_3$ | 8.81 | 84a |
| $C_6H_5OH$ | 8.50 | 4a | *p*-$C_6H_4BrCH_3$ | 8.67 | 4a |
| Phenol | 8.52 | 64a | *o*-$C_6H_4ICH_3$ | 8.62 | 84a |
| | 9.01 | 100c | *m*-$C_6H_4ICH_3$ | 8.61 | 84a |
| | 9.03 | 6c | *p*-$C_6H_4ICH_3$ | 8.50 | 84a |
| $C_6H_5SH$ | 8.33 | 87a | $C_6H_5CH=CH_2$ | 8.47 | 19a |
| Phenyl mercaptan | 8.56 | 100c | Styrene | 8.35 | 26b |
| $C_6H_5C \equiv CH$ | 8.82 | 19a | | 8.86 | 6c |
| Phenyl acetylene | 9.15 | 6c | $C_6H_5OCH_3$ | 8.20 | 4a |
| $C_6H_5NH_2$ | 7.69 | 64a | Anisole | 8.56 | 100c |
| Aniline | 7.70 | 4a | $C_6H_5NHCH_3$ | 7.34 | 64a |
| | 7.84 | 123c | *N*-Methylaniline | 7.65 | 123c |
| | 8.0 | 126c | *o*-$C_6H_4(CH_3)NH_2$ | 7.5 | 126c |
| | 8.23 | 100c | *o*-Toluidine | 7.68 | 123c |
| $\alpha$-$C_5H_4NCH_3$ | 9.02 | 87a | *m*-$C_6H_4(CH_3)NH_2$ | 7.57 | 123c |
| $\alpha$-Picoline | 9.66 | 100c | *p*-$C_6H_4(CH_3)NH_2$ | 7.60 | 123c |
| $\beta$-$C_5H_4NCH_3$ | 9.04 | 87a | $C_6H_5CH_2NH_2$ | 8.64 | 64a |
| $\beta$-Picoline | 9.71 | 100c | Benzylamine | 9.04 | 123c |
| $\gamma$-$C_5H_4NCH_3$ | 9.04 | 87a | $C_9H_7N$ | 8.30 | 79a |
| $\gamma$-Picoline | 9.01 | 26b | Quinoline | 8.62 | 113a |
| | 9.56 | 100c | iso-$C_9H_7N$ | 8.55 | 113a |
| $C_6H_5CHO$ | 9.51 | 4a | $C_6H_5COCH_3$ | 9.65 | 79a |
| Benzaldehyde | 9.60 | 64a | Acetophenone | 9.77 | 6c |
| | 9.63 | 92c | *o*-$C_6H_4(CH_3)_2$ | 8.56 | 4a, 73a, 64a |
| | 9.82 | 6c | *o*-Xylene | 8.58 | 121b, 4b |
| $C_6H_5CH_3$ | 8.81 | 64a | | 8.8 | 126c |
| Toluene | 8.82 | 13a, 73a | | 8.96 | 100c |
| | 8.77 | 118b | | 8.97 | 127c |
| | 8.82 | 121b | *m*-$C_6H_4(CH_3)_2$ | 8.56 | 4a, 73a |
| | 8.78 | 124c | | 8.59 | 64a |
| | 8.80 | 53c | | 8.58 | 4b, 121b |
| | 9.01 | 126c | | 8.8 | 126c |
| | 9.20 | 100c, 87c | | 9.01 | 100c |
| | 9.23 | 6c | | 9.02 | 127c |
| *o*-$C_6H_4FCH_3$ | 8.92 | 84a | *p*-$C_6H_4(CH_3)_2$ | 8.44 | 64a |
| *o*-Fluorotoluene | | | | 8.45 | 13a, 73a |
| *m*-$C_6H_4FCH_3$ | 8.92 | 84a | | 8.48 | 4b, 121b |
| *p*-$C_6H_4FCH_3$ | 8.79 | 84a | | 8.4 | 126c |
| *o*-$C_6H_4ClCH_3$ | 8.83 | 87a | | 8.86 | 100c |
| *o*-Chlorotoluene | | | | 8.88 | 127c |
| *m*-$C_6H_4ClCH_3$ | 8.83 | 87a | $C_6H_5C_2H_5$ | 8.76 | 13a |
| *p*-$C_6H_4ClCH_3$ | 8.69 | 4a | Ethyl benzene | 8.75 | 32b |
| | 8.70 | 87a | | 8.77 | 121b |
| *o*-$C_6H_4BrCH_3$ | 8.78 | 4a | | 9.12 | 6c |

**Appendix II** (*continued*)

| Molecule or radical | Ionization potential (eV) | Reference and method[b,c] | Molecule or radical | Ionization potential (eV) | Reference and method[b,c] |
|---|---|---|---|---|---|
| $C_{10}H_8$ | 8.12 | 4a | $C_{10}H_7CH_3$ | 7.96 | 4a, 87a |
| Naphthalene | 8.14 | 64a | 1-Methyl | 8.0 | 32b |
| | 8.15 | 113a | naphthalene | | |
| | 8.10 | 4b, 121b | $C_{10}H_7CH_3$ | 7.96 | 87a |
| | 8.24 | 123c | 2-Methyl | | |
| | 8.26 | 125c | naphthalene | | |
| $C_{10}H_8$ | 7.41 | 129a | $C_6H_5C_6H_5$ | 8.27 | 87a |
| Azulene | 7.43 | 130b | Biphenyl | 8.22 | 123c |
| | 7.72 | 131c | $C_{13}H_9N$ | 7.78 | 79a |
| | 7.76 | 123c | Acridine | | |
| $C_5H_3N(CH_3)_2$ | 8.85 | 87a | $C_6H_5C_4H_9$ | 8.69 | 4a, 110a |
| 2,3-Lutidine | | | n-Butyl benzene | 8.5 | 4b |
| $C_5H_3N(CH_3)_2$ | 8.85 | 87a | | 9.14 | 6c |
| (2,4–) | | | $C_6H_5C_4H_9$ | 8.69 | 110a |
| $C_5H_3N(CH_3)_2$ | 8.85 | 87a | iso-Butyl benzene | | |
| (2,6–) | | | $C_6H_5C_4H_9$ | 8.68 | 110a |
| $C_6H_5N(CH_3)_2$ | 7.14 | 64a | t-Butyl benzene | 8.5 | 32b |
| N,N-Dimethyl | 7.30 | 123c | | 9.35 | 6c |
| aniline | | | $C_6H_4(C_2H_5)_2$ | 8.91 | 127c |
| $\alpha$-$C_{10}H_7NH_2$ | 7.30 | 79a | 1,2-Diethyl benzene | | |
| $\alpha$-Naphthylamine | | | $C_6H_4(C_2H_5)_2$ | 8.99 | 127c |
| $\beta$-$C_{10}H_7NH_2$ | 7.25 | 79a | (1,3–) | | |
| $C_6H_5C_3H_7$ | 8.72 | 4a | $C_6H_4(C_2H_5)_2$ | 8.93 | 127c |
| n-Propyl benzene | 9.14 | 6c | (1,4–) | | |
| $C_6H_5C_3H_7$ | 8.69 | 4a | $C_{14}H_{10}$ | 7.38 | 79a |
| iso-Propyl benzene | 8.6 | 32b | Anthracene | 7.55 | 125c |
| | 8.76 | 121b | | 7.56 | 87c |
| | 9.13 | 6c | $C_{14}H_{10}$ | 8.03 | 125c |
| $C_6H_3(CH_3)_3$ | 8.48 | 73a | Phenanthrene | | |
| 1,2,3-Trimethyl | 8.75 | 127c | $C_{16}H_{10}$ | 7.72 | 132c |
| benzene | | | Pyrene | | |
| $C_6H_3(CH_3)_3$ | 8.39 | 73a | $C_{18}H_{12}$ | 6.88 | 79a |
| 1,3,5-Trimethyl | 8.40 | 4a | Tetracene | 6.95 | 132c |
| benzene | 8.41 | 64a | $C_{18}H_{12}$ | 8.01 | 132c |
| | 8.5 | 126c | Chrysene | | |
| | 8.76 | 100c | $C_{18}H_{12}$ | 8.19 | 132c |
| | 8.79 | 127c | Triphenylene | | |

[b] Methods:
a) Photoionization     b) Ultraviolet absorption     c) Electron impact

## <sup>c</sup> References

(1) N. Wainfan, W. C. Walker and G. L. Weissler, *Phys. Rev.*, **99**, 542 (1955); (2) L. G. Smith and W. Bleakney, *Phys. Rev.*, **49**, 883 (1936); (3) D. C. Frost and C. A. McDowell, *Can. J. Chem.*, **36**, 39 (1958); (4) K. Watanabe, *J. Chem. Phys.*, **26**,542 (1957);(5)W. C. Price, *Proc. Roy. Soc. (London)*, **A167**, 216 (1938); (6) J. D. Morrison and A. J.C. Nicholson, *J. Chem. Phys.*, **20**, 1021 (1952); (7) F. H. Field and J. L. Franklin, *Electron Impact Phenomena and the Properties of Gaseous Ions*, Academic Press, New York, 1957; (8) S. N. Foner and R. L. Hudson, *J. Chem. Phys.*, **25**, 602 (1956); (9) W. C. Walker and G. L. Weissler, *J. Chem. Phys.*, **23**, 1540, 1547 and 1962 (1955); (10) E. C. Y. Inn, *Phys. Rev.*, **91**, 1194 (1953); (11) M. I. Al-Joboury, D. P. May and D. W. Turner, *J. Chem. Soc. (London)*, Part I, 616 (1965); (12) K. Watanabe, F. F. Marmo and E. C. Y. Inn, *Phys. Rev.*, **91**, 1155 (1953); (13) K. Watanabe, *J. Chem. Phys.*, **22**, 1564 (1954); (14) H. Hurzeler, M. G. Inghram and J. D. Morrison, *J. Chem. Phys.*, **28**, 76 (1958); (15)E.Miescher, *J. Quant. Spectry. Radiative Transfer*, **2**, 421 (1962); (16) G. G. Cloutier and H. I. Schiff, *J. Chem. Phys.*, **31**, 793 (1959); (17) H. D. Smyth, *Rev. Mod. Phys.*, **3**, 347 (1931); (18) H.D. Hagstrum, *Rev. Mod. Phys.*, **23**, 185 (1951); (19) K. Watanabe, T. Nakayama and J. R. Mottl, *J. Quant. Spectry. and Radiative Transfer*, **2**, 369 (1962); (20) F. H. Field, *J. Chem. Phys.*, **20**, 1734 (1952); (21) R. E. Fox and W. M. Hickam, *J. Chem. Phys.*, **22**, 2059 (1954); (22) R. E. Honig, *J. Chem. Phys.*, **16**, 105 (1948); (23) J. T. Tate and P. T. Smith, *Phys. Rev.*, **39**, 270 (1932); (24) S. N. Foner and R. L. Hudson, *J. Chem. Phys.*, **23**, 1364 (1955); (25) D. C. Frost and C. A. McDowell, *J. Am. Chem. Soc.*, **80**, 6183 (1958); (26) L. D. Isaacs, W. C. Price and R. G. Ridley, 'On the Threshold of Space', in *Proc. of the Conf. on Chemical Aeronomy*, M. Zelicoff (Ed.), Cambridge, Mass., p. 143 (1956); (27) M. I. Al-Joboury and D. W. Turner, *J. Chem. Soc. (London)*, Part IV, 4434 (1964); (28) T. M. Sudgen and W. C. Price, *Trans. Faraday Soc.*, **44**, 108, 116 (1948); (29) D. C. Frost, C. A. McDowell and D. A. Vroom, *Can. J. Chem.*, **45**, 1343 (1967); (30) M. Cottin, *J. Chim. Phys.*, **56**, 1024 (1959); (31) S. N. Foner and R. L. Hudson, *J. Chem. Phys.*, **29**, 442 (1958); (32) W. C. Price, *Chem. Rev.*, **41**, 257 (1947); (33) I. Omura, K. Higasi and H. Baba, *Bull. Chem. Soc. Japan*, **29**, 504 (1956); (34) T. Nakayama, M. Y. Kitamura and K. Watanabe, *J. Chem. Phys.*, **30**, 1180 (1959); (35) D. C. Frost, D. Mak and C. A. Mc-Dowell, *Can. J. Chem.*, **40**, 1064 (1962); (36) R. J. Kandel, *J. Chem. Phys.*, **23**, 84 (1955); (37) J. E. Collin, *Nature*, **196**, 373 (1962); (38) W. C. Price and D. M. Simpson, *Proc. Roy. Soc. (London)*, **A165**, 272 (1938); (39) R. M. Reese, V. H. Dibeler and J. L. Franklin, *J. Chem. Phys.*, **29**, 880 (1958); (40) K. Watanabe and J. R. Mottl, *J. Chem. Phys.*, **26**, 1773 (1957); (41) J. E. Collin, *Can. J. Chem.*, **37**, 1053 (1959); (42) G. Herzberg, *Can. J. Phys.*, **39**, 1511 (1961); (43) E. W. C. Clarke and C. A. McDowell, *Proc. Chem. Soc. (London)*, February, **69**, (1960); (44) A. Langer and J. A. Hipple, *Phys. Rev.*, **69**, 691 (1946); (45) A. D. Walsh, *Trans. Faraday Soc.*, **43**, 60 (1947); (46) J. L. Franklin and H. E. Lumpkin, *J. Am. Chem. Soc.*, **74**, 1023 (1952); (47) R. R. Bernecker and F. A. Long, *J. Phys. Chem.*, **65**, 1565 (1961); (48) J. E. Collin, *Acad. Roy. Sci. Belg. Bull.*, **45**, 734 (1959); (49) J. D. Morrison, *J. Chem. Phys.*, **19**, 1305 (1951); (50) J. T. Herron and V. H. Dibeler, *J. Am.Chem. Soc.*, **82**, 1555 (1960); (51) F. A. Elder, C. Giese, B.Steiner and M. G. Ingram, *J. Chem. Phys.*, **36**, 3292 (1962); (52) G. Herzberg and J. Shoosmith, *Can. J. Phys.*, **34**, 523 (1956); (53) C. E. Melton and W. H. Hamill, *J. Chem. Phys.*, **41**, 3464 (1964); (54) A. Langer, J. A. Hipple and D. P. Stevenson, *J. Chem. Phys.*, **22**, 1836 (1954); (55) J. M. Williams and W. H. Hamill, *J. Chem. Phys.*, **49**, 4467 (1968); (56) F. P. Lossing, K. U. Ingold and I. H. S. Henderson, *J. Chem. Phys.*, **22**, 621 (1954); (57) J. A. Hipple and D. P. Stevenson, *Phys. Rev.*, **63**, 121 (1943); (58) J. D. Waldron, *Trans. Faraday Soc.*, **50**, 102 (1954); (59) V. H. Dibeler and R. M. Reese, *J. Chem. Phys.*, **40**, 2034 (1964); (60) W. C. Price, *Phys. Rev.*, **47**, 444 (1935); (61) F. H. Coats and R. C. Anderson, *J. Am. Chem. Soc.*, **79**, 1340 (1957); (62) J. L. Franklin and F. H. Field, *J. Am. Chem. Soc.*, **76**, 1994 (1954); (63) J. E. Collin, *Bull. Soc. Chim. Belges*, **71**, 15 (1962); (64) F. I. Vilesov and A. N. Terenin, *Dokl. Akad. Nauk SSSR*, **115**, 744 (1957); (65) K. Higasi, I. Omura and H. Baba, *Nature*, **178**, 652 (1956); (66) M. B. Koffel and R. A. Lad, *J. Chem. Phys.*, **16**, 420 (1948); (67) F. P. Lossing, A. W. Tickner and W. A. Bryce, *J. Chem. Phys.*, **19**, 1254 (1951); (68) L. G. Smith, *Phys. Rev.*, **51**, 263 (1937); (69) H. Branson and C. Smith,

*J. Am. Chem. Soc.*, **75**, 4133 (1953); (70) D. C. Frost and C. A. McDowell, *Proc. Roy. Soc. (London)*, **A241**, 194 (1957); (71) W. C. Price, *J. Chem. Phys.*, **4**, 539, 547 (1936); (72) V. H. Dibeler and R. M. Reese, *J. Res. Natl. Bur. Stand.*, **54**, 127 (1955); (73) R. Bralsford, P. V. Harris and W. C. Price, *Proc. Roy. Soc. (London)*, **A258**, 459 (1960); (74) H. Gutbier, *Z. Naturforsch.*, **9a**, 348 (1954); (75) J. W. Warren and J. D. Craggs, *Mass Spectrometry* (Institute of Petroleum, London, 1952), p. 36; (76) V. H. Dibeler, R. M. Reese and F. L. Mohler, *J. Res. Natl. Bur. Stand.*, **57**, 113 (1956); (77) R. F. Baker and J. T. Tate, *Phys. Rev.*, **53**, 683 (1938); (78) F. A. Long and L. Friedman, *J. Am. Chem. Soc.*, **75**, 2837 (1953); (79) F. I. Vilesov, *Dokl. Akad. Nauk, (Dan) SSSR*, **132**, 632, 1332 (1960); (80) W. C. Price and W. M. Evans, *Proc. Roy. Soc. (London)*, **A162**, 110 (1937); (81) J. Momigny, *Nature*, **199**, 1179 (1963); (82) W. C. Price and W. T. Tutte, *Proc. Roy. Soc. (London)*, **A174**, 207 (1940); (83) W. C. Price and A. D. Walsh, *Trans. Faraday Soc.*, **41**, 381 (1945); (84) K. Watanabe, T. Nakayama and J.R. Mottl, University Hawaii, Final Report on the Ionization Potential of Molecules by a Photo-Ionization Method, Contract No. DA-04-200-ORD480 (1959); (85) J. A. Cutler, *J. Chem. Phys.*, **16**, 136 (1948); (86) I. Omura, H. Baba and K. Higasi, *Bull. Chem. Soc. Japan*, **28**, 147 (1955); (87) R. I. Reed, *Ion Production by Electron Impact*, Academic Press, London, 1962; (88) H. D. Hunt and W. T. Simpson, *J. Am. Chem. Soc.*, **75**, 4540 (1953); (89) A. D. Walsh, *Trans. Faraday Soc.*, **41**, 35, 498 (1945); (90) J. Momigny, *Acad. Roy. Sci. Belg. Bull.*, **46**, 686 (1960); (91) A. D. Walsh, *Proc. Roy. Soc. (London)*, **A185**, 176 (1946); (92) R. I. Reed and M. B. Thornley, *Trans. Faraday Soc. (London)*, **54**, 949 (1958); (93) E. J. Gallegos and R. W. Kiser, *J. Am. Chem. Soc.*, **83**, 773 (1961); (94) D. P. Stevenson and J. A. Hipple, Jr., *J. Am. Chem. Soc.*, **64**, 1588 (1942); (95) K. Hirota, K. Nagoshi and M. Hatada, *Bull. Chem. Soc. Japan*, **34**, 226 (1961); (96) A. Irsa, *J. Chem. Phys.*, **26**, 18 (1957); (97) R. F. Pottie, A. G. Harrison and F. P. Lossing, *J. Am. Chem. Soc.*, **83**, 3204 (1961); (98) B. Steiner, C. F. Giese and M. G. Inghram, *J. Chem. Phys.*, **34**, 189 (1961); (99) W. C. Price and A. D. Walsh, *Proc. Roy. Soc. (London)*, **A179**, 201 (1941); (100) H. Baba, I. Omura and K. Higasi, *Bull. Chem. Soc. Japan*, **29**, 521 (1956); (101) V. I. Khvostenko, *Russ. J. Phys. Chem.*, **36**, 197 (1962); (102) M. J. S. Dewar and S. D. Worley, *J. Chem. Phys.*, **49**, 2454 (1968); (103) T. M. Sudgen and A. D. Walsh, *Trans. Faraday Soc.*, **41**, 76 (1945); (104) J. E. Collin and F. P. Lossing, *J. Am. Chem. Soc.*, **79**, 5848 (1957); (105) J. E. Collin, *Bull. Soc. Chim. Belges*, **62**, 411 (1953); (106) J. Hissel, *Bull. Soc. Roy. Sci. Liège*, **21**, 457 (1952); (107) D. P. Stevenson, *J. Am. Chem. Soc.*, **65**, 209 (1943); (108) V. H. Dibeler, *J.Res. Natl. Bur. Stand.*, **38**, 329 (1947); (109) J. E. Collin and F. P. Lossing, *J. Am. Chem. Soc.*, **81**, 2064 (1959); (110) W. C. Price, R. Bralsford, P. V. Harris and R. G. Ridley, *Spectrochim. Acta*, **14**, 45 (1959); (111) F. I. Vilesov and B. L. Kurbatov, *Dokl. Akad. Nauk, SSSR*, **140**, 1364 (1961); (112) W. C. Price and A. D. Walsh, *Proc. Roy. Soc. (London)*, **A185**, 182 (1946); (113) A. J. Yencha and M. A. El-Sayed, *J. Chem. Phys.*, **48**, 3469 (1968); (114) D. W. Turner, *Adv. Phys. Org. Chem.*, **4**, 31 (1966); (115) I. Omura, H. Baba, K. Higasi and Y. Kanaoka, *Bull. Chem. Soc. Japan*, **30**, 633 (1957); (116) J. E. Parkin and K. K. Innes, *J. Mol. Spectry.*, **15**, 407 (1965); (117) M. A. El-Sayed, M. Kasha and V. Tanaka, *J. Chem. Phys.*, **34**, 334 (1961); (118) W. C. Price and A. D. Walsh, *Proc. Roy. Soc. (London)*, **A191**, 22 (1947); (119) I. D. Clark and D. C. Frost, *J. Am. Chem. Soc.*, **89**, 244 (1967); (120) W. C. Price and R. W. Wood, *J. Chem. Phys.*, **3**, 439 (1935); (121) V. J. Hammond, W. C. Price, J. P. Teegan and A. D. Walsh, *Discussions Faraday Soc.*, **9**, 53 (1950); (122) P. G. Wilkinson, *Can. J. Phys.*, **34**, 596 (1956); (123) J. H. D. Eland, P. J. Shepherd and C. J. Danby, *Z. Naturforsch.*, **21a**, 1580 (1966); (124) V. H. Dibeler, R. M. Reese and M. Krauss, *Advances in Mass Spectroscopy*, Pergamon Press, London 1966, Vol. III, p. 471; (125) M. E. Wacks and V. H. Dibeler, *J. Chem. Phys.*, **31**, 1557 (1959); (126) H. Hartman and M. B. Svendsen, *Z. Phys. Chemie.*, NF **11**, 16 (1957); (127) F. H. Field and J. L. Franklin, *J. Chem. Phys.*, **22**, 1895 (1954); (128) J. R. Majer and C. R. Patrick, *Trans. Faraday Soc.*, **58**, 17 (1962); (129) T. Kitagawa, H. Inokuchi and K. Kodera, *J. Mol. Spectry.*, **21**, 267 (1966); (130) L. B. Clark, *J. Chem. Phys.*, **43**, 2566 (1965); (131) R. J. Van Brunt and M. E. Wacks, *J. Chem. Phys.*, **41**, 3195 (1964); (132) M. E. Wacks, *J. Chem. Phys.*, **41**, 1661 (1964).

# Appendix III

1. Speed of light
   $$c = 2.997930 \times 10^{10} \text{ cm/sec}$$

2. Electronic charge
   $$e = 1.60206 \times 10^{-19} \text{ Coul.} = 4.80286 \times 10^{-10} \text{esu} = 1.60206 \times 10^{-20} \text{emu}$$

3. Electron rest mass
   $$m = 9.1083 \times 10^{-28} \text{ g}$$

4. Proton rest mass
   $$M_p = 1.67239 \times 10^{-24} \text{ g}$$

5. Neutron rest mass
   $$M_n = 1.67470 \times 10^{-24} \text{ g}$$

6. Ratio of proton to electron mass
   $$M_p/m = 1836.12$$

7. Charge-to-mass ratio of electron
   $$e/m = 5.27305 \times 10^{17} \text{ esu/g} = 1.75890 \times 10^7 \text{ emu/g}$$

8. Planck's constant
   $$h = 6.62517 \times 10^{-34} \text{ joule sec} = 6.62517 \times 10^{-27} \text{ erg sec}$$
   $$\hbar = h/2\pi = 1.05443 \times 10^{-34} \text{ joule sec} = 1.05443 \times 10^{-27} \text{ erg sec}$$

9. Boltzmann's constant
   $$k = 1.38044 \times 10^{-23} \text{ joule/}^\circ\text{K} = 1.38044 \times 10^{-16} \text{ erg/}^\circ\text{K}$$

10. Gas constant
    $$R = 8.31662 \times 10^3 \text{ joule/kg-mole }^\circ\text{K} = 8.31662 \times 10^7 \text{ erg/g-mole }^\circ\text{K}$$

11. Avogadro's number (physical scale)
    $$N_A = 6.02486 \times 10^{23} \text{ (g-mole)}^{-1}$$

12. Loschmidt's number (physical scale)
    $$N_L = 2.68719 \times 10^{19} \text{ cm}^{-3}$$

13. First Bohr radius
    $$a_0 = \hbar^2/me^2 = 5.29172 \times 10^{-9} \text{ cm}$$
    (Velocity of electron in first Bohr orbit of hydrogen atom is
    $v_0 = 2.1877 \times 10^8$ cm/sec)

14. The Rydberg
    $$R = 109737.309 \text{ cm}^{-1}$$

15. Mass–energy conversion factors
    1 g $= 5.61000 \times 10^{26}$ MeV
    1 electron mass $= 0.510976$ MeV
    1 atomic mass unit $= 931.141$ MeV
    1 proton mass $= 938.211$ MeV
    1 neutron mass $= 939.505$ MeV

16. Quantum energy conversion factors

$1 \text{ eV} = 1.60206 \times 10^{-19} \text{ joule} = 1.60206 \times 10^{-12} \text{ erg} = 23.069 \text{ kcal/mole}$
$= 8065.75 \text{ cm}^{-1}$

$1 \text{ joule} = 10^7 \text{ erg}$

Wavelength $\lambda$ of photon with energy $E$ (in eV)

$$\lambda(\text{Å}) = \lambda(nm \times 10) = \frac{12397.67}{E(\text{eV})}.$$

*Appendix   IV*

# PERIODIC TABLE

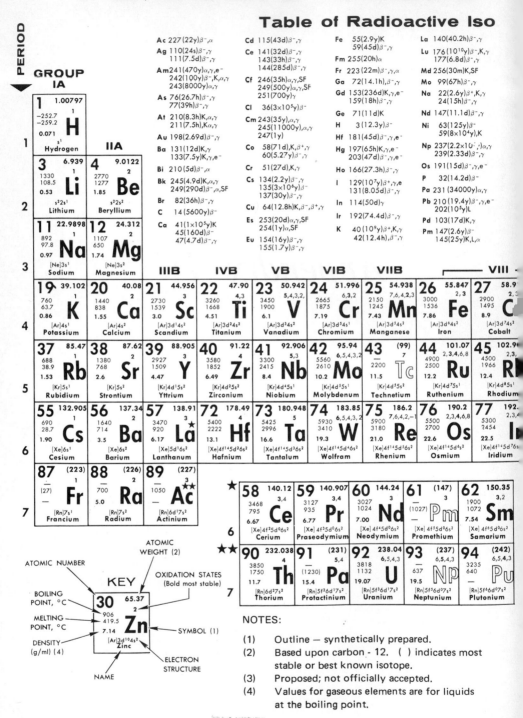

## Table of Radioactive Iso

NOTES:

(1) Outline — synthetically prepared.
(2) Based upon carbon - 12. ( ) indicates most stable or best known isotope.
(3) Proposed; not officially accepted.
(4) Values for gaseous elements are for liquids at the boiling point.

# OF THE ELEMENTS

pes

```
210(138.4d)α,γ        Sr   90(28y)β-
209(100y)α,K,γ             89(51d)β-,γ
143(13.8d)β-               85(64d)K,γ
197(18h)β-,γ          Ta  182(115d)β-,γ
242(3.8 × 10⁵y)α,SF   Tb  160(73d)β-γ
241(13y)β-,α,γ        Tc   99(2×10⁵y)β-
239 (24300y)α,γ,SF        97(10⁵y)K
226(1620y)α,γ         Te  127(9.3h)β-
86(18.6h)β-,γ         Th  232(1.4x10¹⁰y)α,γ,SF
188(16.7h)β-,γ             228(1.91y)β-
186(3.7d)β-,γ         Tl  204(3.56y)β-,K
222(3.82d)α           Tm  170(127d)β-,γ,e-
103(40d)β-,γ          U   238(4.5×10⁹y)α,γ,SF
97(2.9d)K,γ,e-             234(2.5×10⁵y)α,γ,SF
35(87d)β-                  235(7.1x10⁸y)α,γ,SF
122(2.8d)β-,K,β⁺,γ         233(1.6×10⁵y)α,γ
124(60d)β-,e-         W   185(73d)β-,γ
46(84d)β-,γ           Y    90(64h)β-,e-
75(121d)K,γ           Yb  175(4.2d)β-,γ
153(47h)β-,γ               169(31d)K,γ,e-
145(340d)K,γ          Zn   65(245d)K,β⁺,γ
113(119d)K,L,γ,e-     Zr   95(65d)β-,γ,e-
                           93(9×10⁵y)β-,γ
```

Half lives are in parentheses where s, m, h, d and y stand for seconds, minutes, hours, days and years respectively. The symbols describing the mode of decay and resulting radiation are defined as follows:

| | | | |
|---|---|---|---|
| α | alpha particle | L | L-electron capture |
| β- | beta particle | SF | spontaneous fission |
| β⁺ | positron | γ | gamma ray |
| K | K-electron capture | e- | internal electron conversion |

## INERT GASES

| 2 | 4.0026 |
|---|---|
| | 0 |
| −268.9 | |
| −269.7 | He |
| 0.126 | |
| s² | |
| | Helium |

|  IIIA  |  IVA  |  VA  |  VIA  |  VIIA  |

| 5 10.811 | 6 12.0111 | 7 14.0067 | 8 15.9994 | 9 18.9984 | 10 20.183 |
|---|---|---|---|---|---|
| 3 | ±4,2 | ±3,5,4,2 | −2 | −1 | 0 |
| − | 4830 | −195.8 | −183 | −188.2 | −246 |
| (2030) | 3727g | −210 | −218.8 | −219.6 | −248.6 |
| 2.34 B | 2.26 C | 0.81 N | 1.14 O | 1.11 F | 1.20 Ne |
| s²2s²2p¹ | s²2s²2p² | s²2s²2p³ | s²2s²2p⁴ | s²2s²2p⁵ | s²2s²2p⁶ |
| Boron | Carbon | Nitrogen | Oxygen | Fluorine | Neon |

| 13 26.9815 | 14 28.086 | 15 30.9738 | 16 32.064 | 17 35.453 | 18 39.948 |
|---|---|---|---|---|---|
| 3 | 4 | ±3,5,4 | 6,3,4,−2 | ±1,4,5,6,7 | 0 |
| 2450 | 2680 | 280w | 444.6 | −34.7 | −185.8 |
| 660 | 1410 | 44.2w | 119.0 | −101.0 | −189.4 |
| 2.70 Al | 2.33 Si | 1.82w P | 2.07 S | 1.56 Cl | 1.40 Ar |
| [Ne]3s²3p¹ | [Ne]3s²3p² | [Ne]3s²3p³ | [Ne]3s²3p⁴ | [Ne]3s²3p⁵ | [Ne]3s²3p⁶ |
| Aluminum | Silicon | Phosphorus | Sulfur | Chlorine | Argon |

|  IB  |  IIB  |

| 28 58.71 | 29 63.54 | 30 65.37 | 31 69.72 | 32 72.59 | 33 74.922 | 34 78.96 | 35 79.909 | 36 83.80 |
|---|---|---|---|---|---|---|---|---|
| 2,3 | 2,1 | 2 | 3 | 4 | ±3,5 | 6,4,−2 | ±1,4,5 | 0 |
| 2730 | 2595 | 906 | 2237 | 2830 | 613* | 685 | 58 | −152 |
| 1453 | 1083 | 419.5 | 29.8 | 937.4 | 817 | 217 | −7.2 | −157.3 |
| 8.9 Ni | 8.96 Cu | 7.14 Zn | 5.91 Ga | 5.32 Ge | 5.72 As | 4.79 Se | 3.12 Br | 2.6 Kr |
| [Ar]3d⁸4s² | [Ar]3d¹⁰4s¹ | [Ar]3d¹⁰4s² | [Ar]3d¹⁰4s²4p¹ | [Ar]3d¹⁰4s²4p² | [Ar]3d¹⁰4s²4p³ | [Ar]3d¹⁰4s²4p⁴ | [Ar]3d¹⁰4s²4p⁵ | [Ar]3d¹⁰4s²4p⁶ |
| Nickel | Copper | Zinc | Gallium | Germanium | Arsenic | Selenium | Bromine | Krypton |

| 46 106.4 | 47 107.870 | 48 112.40 | 49 114.82 | 50 118.69 | 51 121.75 | 52 127.60 | 53 126.904 | 54 131.30 |
|---|---|---|---|---|---|---|---|---|
| 2,4 | 1 | 2 | 3 | 4,2 | ±3,5 | 6,4,−2 | ±1,4,5,7 | 0 |
| 3980 | 2210 | 765 | 2000 | 2270 | 1380 | 989.8 | 183 | −108.0 |
| 1552 | 960.8 | 320.9 | 156.2 | 231.9 | 630.5 | 449.5 | 113.7 | −111.9 |
| 12.0 Pd | 10.5 Ag | 8.65 Cd | 7.31 In | 7.30 Sn | 6.62 Sb | 6.24 Te | 4.94 I | 3.06 Xe |
| [Kr]4d¹⁰5s⁰ | [Kr]4d¹⁰5s¹ | [Kr]4d¹⁰5s² | [Kr]4d¹⁰5s²5p¹ | [Kr]4d¹⁰5s²5p² | [Kr]4d¹⁰5s²5p³ | [Kr]4d¹⁰5s²5p⁴ | [Kr]4d¹⁰5s²5p⁵ | [Kr]4d¹⁰5s²5p⁶ |
| Palladium | Silver | Cadmium | Indium | Tin | Antimony | Tellurium | Iodine | Xenon |

| 78 195.09 | 79 196.967 | 80 200.59 | 81 204.37 | 82 207.19 | 83 208.980 | 84 (210) | 85 (210) | 86 (222) |
|---|---|---|---|---|---|---|---|---|
| 2,4 | 3,1 | 2,1 | 3,1 | 4,2 | 3,5 | 4,2 | | |
| 4530 | 2970 | 357 | 1457 | 1725 | 1560 | − | − | − |
| 1769 | 1063 | −38.4 | 303 | 327.4 | 271.3 | 254 | (302) | (−71) |
| 21.4 Pt | 19.3 Au | 13.6 Hg | 11.85 Tl | 11.4 Pb | 9.8 Bi | (9.2) Po | − At | − Rn |
| [Xe]4f¹⁴5d¹⁰6s⁰ | [Xe]4f¹⁴5d¹⁰6s¹ | [Xe]4f¹⁴5d¹⁰6s² | [Xe]4f¹⁴5d¹⁰6s²6p¹ | [Xe]4f¹⁴5d¹⁰6s²6p² | [Xe]4f¹⁴5d¹⁰6s²6p³ | [Xe]4f¹⁴5d¹⁰6s²6p⁴ | [Xe]4f¹⁴5d¹⁰6s²6p⁵ | [Xe]4f¹⁴5d¹⁰6s²6p⁶ |
| Platinum | Gold | Mercury | Thallium | Lead | Bismuth | Polonium | Astatine | Radon |

| 63 151.96 | 64 157.25 | 65 158.924 | 66 162.50 | 67 164.930 | 68 167.26 | 69 168.934 | 70 173.04 | 71 174.97 |
|---|---|---|---|---|---|---|---|---|
| 3,2 | 3 | 3,4 | 3 | 3 | 3 | 3,2 | 3,2 | 3 |
| 1439 | 3000 | 2800 | 2600 | 2600 | 2900 | 1727 | 1427 | 3327 |
| 826 | 1312 | 1356 | 1407 | 1461 | 1497 | 1545 | 824 | 1652 |
| 5.26 Eu | 7.89 Gd | 8.27 Tb | 8.80 Dy | 9.05 Ho | 9.33 Er | 9.33 Tm | 6.98 Yb | 9.84 Lu |
| [Xe]4f⁷6d⁰6s² | [Xe]4f⁷5d¹6s² | [Xe]4f⁹5d⁰6s² | [Xe]4f¹⁰5d⁰6s² | [Xe]4f¹¹5d⁰6s² | [Xe]4f¹²5d⁰6s² | [Xe]4f¹³5d⁰6s² | [Xe]4f¹⁴5d⁰6s² | [Xe]4f¹⁴5d¹6s² |
| Europium | Gadolinium | Terbium | Dysprosium | Holmium | Erbium | Thulium | Ytterbium | Lutetium |

| 95 (243) | 96 (247) | 97 (247) | 98 (251) | 99 (254) | 100 (253) | 101 (256) | 102 (254) | 103 (257) |
|---|---|---|---|---|---|---|---|---|
| 6,5,4,3 | 3 | 4,3 | 3 | | | | | |
| 11.7 Am | − Cm | − Bk | − Cf | − Es | − Fm | − Md | − No | − (Lw) |
| [Rn]5f⁷6d⁰7s² | [Rn]5f⁷6d¹7s² | [Rn]5f⁸6d¹7s² | [Rn]5f⁹6d¹7s² | Einsteinium | Fermium | Mendelevium | Nobelium | See note 3 |
| Americium | Curium | Berkelium | Californium | | | | | (Lawrencium) |

*Reprinted by permission of E. H. Sargent and Company.*

# Author Index

Italic numbers indicate the page on which the reference is actually given.

# Subject Index

(B 957)     Printed in Belgium by Ceuterick s.a.
               Brusselse straat 153   3000-Louvain
               Managing director L. Pitsi   Bertemse baan 25   3008-Veltem-Beisem